Lecture Notes in Civil Engineering

Volume 7

Series editors

Marco di Prisco, Politecnico di Milano, Milano, Italy
Sheng-Hong Chen, Wuhan University, Wuhan, China
Giovanni Solari, University of Genoa, Genova, Italy
Ioannis Vayas, National Technical University of Athens, Athens, Greece

Lecture Notes in Civil Engineering (LNCE) publishes the latest developments in Civil Engineering—quickly, informally and in top quality. Though original research reported in proceedings and post-proceedings represents the core of LNCE, edited volumes of exceptionally high quality and interest may also be considered for publication. Volumes published in LNCE embrace all aspects and subfields of, as well as new challenges in, Civil Engineering. Topics in the series include:

– Construction and Structural Mechanics
– Building Materials
– Concrete, Steel and Timber Structures
– Geotechnical Engineering
– Earthquake Engineering
– Coastal Engineering
– Hydraulics, Hydrology and Water Resources Engineering
– Environmental Engineering and Sustainability
– Structural Health and Monitoring
– Surveying and Geographical Information Systems
– Heating, Ventilation and Air Conditioning (HVAC)
– Transportation and Traffic
– Risk Analysis
– Safety and Security

More information about this series at http://www.springer.com/series/15087

Seyhan Fırat · John Kinuthia
Abid Abu-Tair
Editors

Proceedings of 3rd International Sustainable Buildings Symposium (ISBS 2017)

Volume 2

 Springer

Editors
Seyhan Fırat
Gazi University
Ankara
Turkey

Abid Abu-Tair
The British University in Dubai
Dubai
United Arab Emirates

John Kinuthia
University of South Wales
Pontypridd
UK

ISSN 2366-2557   ISSN 2366-2565   (electronic)
Lecture Notes in Civil Engineering
ISBN 978-3-030-09715-8   ISBN 978-3-319-64349-6   (eBook)
https://doi.org/10.1007/978-3-319-64349-6

Printed on acid-free paper

This Springer imprint is published by the registered company Springer International Publishing AG
part of Springer Nature
The registered company address is: Gewerbestrasse 11, 6330 Cham, Switzerland

Foreword

With the UEA having announced its vision 2030 to become an environmentally, socially and economically sustainable community and with Dubai's Sustainable City aiming to be the star tourist attraction for Expo 2020, it was an exciting time to be involved with the 3rd International Sustainable Buildings Symposium—ISBS 2017 held on 15–17th March 2017. This conference with its broad range of research topics and many excellent papers will be a sound basis for establishing sustainable cites, especially those in hot climates. I learned a lot at this conference, and others will find the papers a useful reference.

Geoff Levermore
Emeritus Professor
MACE
University of Manchester, UK

Preface

The first "International Sustainable Buildings Symposium" conferences (ISBS 2010) took place on 26–28th May 2010 at Gazi University, Ankara, Turkey. The plan was to hold the ISBS conference series at different cities every two years. The sequel to ISBS 2010 (ISBS 2015) however also took place at Gazi University on 28–30th May 2015 but at a much improved scale. The decision to hold it here was partly contributed by the development of an efficient conference organizing team within the Department of Civil Engineering in the Technology Faculty of Gazi University. The other reason is that Ankara has plenty to offer, and this also contributed to hosting the ISBS conference two years in a row. Ankara is the capital of Turkey since the establishment of the Republic of Turkey in 1923 and the country's second largest city, İstanbul being the largest. The city of Ankara city a population of 5.150.072 as of 2014. Centrally located in Anatolia, the city is an important commercial and industrial centre of attraction.

Against the background explained, the "3rd International Sustainable Buildings Symposium—ISBS 2017" was held on 15–17th March 2017 far away from Ankara, in Dubai, in the United Arab Emirates (UAE). Dubai is the most populous city in the UAE. It is located on the southeast coast of the Arabian Gulf and is one of the seven emirates that make up the country. Dubai has emerged as a global city and business hub of the Middle East. It is also a major transport hub for passengers and cargo. Dubai is a beautiful city and a major tourist destination. It is famous for sightseeing attractions such as the Burj Khalifa (the world's tallest building) and shopping malls that come complete with mammoth aquariums and indoor ski slopes. Moreover, the city has many cultural highlights as well as all the glamorous modern add-ons. It is in this city that ISBS 2017 was successfully held.

Organizing ISBS 2017 aimed at bringing together researchers and experts from both UAE, Turkey and the rest of the world in dialogue with researchers and authorities of implementing and consulting firms and institutions. Researchers with significant reputation on an international scale were invited to the symposium as keynote speakers. In this context, the symposium facilitated researchers, academic institutions, municipalities, government bodies, non-governmental organizations, other official and private establishments that are active in construction sector and environmental technologies, to discuss the current issues in construction and environmental technology areas. In addition, both national and international companies found the opportunity to introduce their products and services. The effective interaction between local and foreign experts, authorities from private and government agencies, as well as between key players in academic, research and administrative circles in higher education institutions, is a pre-requisite to the holistic achievement of the aspirations of the ISBS series of conference on sustainable buildings.

The conference proceedings are split into two separate volumes: Volume 1 (ISBN 978-3-319-63708-2) contains the sections "Sustainable Buildings and Smart Cities" and "Sustainable Planning—Infrastructure and Resilience", and Volume 2 (the current book) contains the sections "Energy and Environment—Emerging Climate Change and Impact on the Built Environment", "Environmental Policies and Practices" and "Strengthening and Rehabilitation of Structures".

Ankara, Turkey   Prof. Seyhan Fırat
Pontypridd, UK   Prof. John Kinuthia
Dubai, United Arab Emirates   Prof. Abid Abu-Tair

Conference Organization

Chairs

Prof. Dr. Seyhan Fırat (Gazi University, Turkey)
Prof. Dr. Abid Abu-Tair (British University in Dubai, UAE)
Prof. Dr. Hüseyin Yılmaz Aruntaş (Gazi University, Turkey)
Assoc. Prof. Dr. Arzuhan Burcu Gültekin (Ankara University, Turkey)
Assist. Prof. Dr. Hanan Taleb (British University in Dubai, UAE)

Secretariat

Res. Assist. Anıl Özdemir
Res. Assist. Murat Pinarlik
Res. Assist. Pınar Sezin Öztürk Kardoğan
Res. Assist. Rüya Kiliç Demircan

Committee Members

International Organizing Committee
Prof. Dr. Abid Abu-Tair (British University in Dubai, UAE)
Prof. Dr. Seyhan Fırat (Gazi University, Turkey)
Prof. Dr. Halim Boussabaine (British University in Dubai, UAE)
Prof. Dr. Hüseyin Yılmaz Aruntaş (Gazi University, Turkey)
Prof. Dr. Gülgün Yilmaz (Anadolu University, Turkey)
Prof. Dr. Jamal Khatib (University of Wolverhampton, England)
Prof. Dr. John Kinuthia (University of South Wales, Wales, United Kingdom)
Prof. Dr. Hasan Arman (UAE University, UAE)

Assoc. Prof. Dr. Arzuhan Burcu Gültekin (Ankara University, Turkey)
Assoc. Prof. Dr. Alaa Ameer (British University in Dubai, UAE)
Assoc. Prof. Dr. Mukaddes Darwish (Texas Tech University, USA)
Assoc. Prof. Dr. Mürsel Erdal (Gazi University, Turkey)
Assoc. Prof. Dr. Nihat Sinan Işik (Gazi University, Turkey)
Assoc. Prof. Dr. Peiman Kianmehr (American University in Dubai, UAE)
Assoc. Prof. Dr. Rami Haweeleh (American University of Sharjah, UAE)
Assist. Prof. Dr. Burçhan Aydin (Texas A&M University, USA)
Assist. Prof. Dr. Duygu Erten (ÇEDBİK, Turkey)
Assist. Prof. Dr. Emad S.N. Mushtaha (University of Sharjah, UAE)
Assist. Prof. Dr. Hanan Taleb (British University in Dubai, UAE)
Dr. Mohammed Alhaj Hussein (Green Quality Consulting, Saudi Arabia)
Dr. Yasser M. Al-Saleh (INSEAD, UAE)
Res. Assist. Anıl Özdemir (Gazi University, Turkey)
Res. Assist. Murat Pinarlik (Gazi University, Turkey)
Res. Assist. Pınar Sezin Öztürk Kardoğan (Gazi University, Turkey)
Res. Assist. Rüya Kiliç Demircan (Gazi University, Turkey)
Lecturer Emre Aytuğ Özsoy (Anadolu University, Turkey)
Michael DAX (DGNB—German Green Building Council, Germany)

Local Organising Committee

Prof. Dr. Abid Abu-Tair (British University in Dubai, UAE)
Prof. Dr. Halim Boussabaine (British University in Dubai, UAE)
Prof. Dr. Hasan Arman (UAE University, UAE)
Assoc. Prof. Dr. Alaa Ameer (British University in Dubai, UAE)
Assoc. Prof. Dr. Peiman Kianmehr (American University in Dubai, UAE)
Assoc. Prof. Dr. Rami Haweeleh (American University of Sharjah, UAE)
Assist. Prof. Dr. Emad S.N. Mushtaha (University of Sharjah, UAE)
Assist. Prof. Dr. Hanan Taleb (British University in Dubai, UAE)
Dr. Mohammed Alhaj Hussein (Green Quality Consulting, Saudi Arabia)
Dr. Yasser M. Al-Saleh (INSEAD, UAE)

International Scientific Committee

Prof. Dr. Abid Abu-Tair (British University in Dubai, UAE)
Prof. Dr. Alemdar Bayraktar (Karadeniz Technical University, Turkey)
Prof. Dr. Ali İhsan Ünay (Gazi University, Turkey)
Prof. Dr. Alpin Köknel Yener (İstanbul Technical University, Turkey)
Prof. Dr. Birol Kilkiş (Başkent University, Turkey)

Prof. Dr. Bjarne Olesen (Technical University of Denmark, Denmark)
Prof. Dr. Cengiz Duran Atiş (Erciyes University, Turkey)
Prof. Dr. Claude-Alain Roulet (LESO-EPFL, Switzerland)
Prof. Dr. Dejan Mumovic (University College London, England)
Prof. Dr. Demet Irkli Eryildiz (Okan University, Turkey)
Prof. Dr. Derek Clements-Croome (University of Reading, England)
Prof. Dr. Erkan Çelebi (Sakarya University, Turkey)
Prof. Dr. Figen Beyhan (Gazi University, Turkey)
Prof. Dr. Füsun Demirel (Gazi University, Turkey)
Prof. Dr. Gülgün Yilmaz (Anadolu University, Turkey)
Prof. Dr. Gülser Çelebi (Çankaya University, Turkey)
Prof. Dr. H. Yılmaz Aruntaş (Gazi University, Turkey)
Prof. Dr. Hacı Mehmet Şahin (Gazi University, Turkey)
Prof. Dr. Hakim S. Abdelgader (Tripoli University, Libya)
Prof. Dr. Hasan Arman (UAE University, UAE)
Prof. Dr. Hulusi Özkul (İstanbul Technical University, Turkey)
Prof. Dr. Hülagü Kaplan (Gazi University, Turkey)
Prof. Dr. I. Özgür Yaman (Middle East Technical University, Turkey)
Prof. Dr. İlker Bekir Topçu (Eskişehir Osmangazi University, Turkey)
Prof. Dr. Jamal M. Khatib (University of Wolverhampton and Beirut Arab University)
Prof. Dr. John Kinuthia (University of South Wales, Wales, United Kingdom)
Prof. Dr. Kambiz Ramyar (Ege University, Turkey)
Prof. Dr. Khalid El Harrouni (Ecole Nationale d'Architecture, Morocco)
Prof. Dr. Lale Balas (Gazi University, Turkey)
Prof. Dr. Leyla Tokman (Anadolu University, Turkey)
Prof. Dr. Marko Serafimov (St. C. and Methodius University, Macedonia)
Prof. Dr. Mesut Başibüyük (Çukurova University, Turkey)
Prof. Dr. Metin Arslan (Ankara University, Turkey)
Prof. Dr. Muhammad Saleh (Chemical Sciences University, Malaysia)
Prof. Dr. Negim El-Sayed (University of Wolverhampton, England)
Prof. Dr. Nesrin Çobanoğlu (Gazi University, Turkey)
Prof. Dr. Nihal Arioğlu (İstanbul Technical University, Turkey)
Prof. Dr. Nilay Coşgun (Gebze Technical University, Turkey)
Prof. Dr. Ömer Faruk Bay (Gazi University, Turkey)
Prof. Dr. Pieter Jacobus Cornelis Wilde (University of Plymouth, UK)
Prof. Dr. Reşat Ulusay (Hacettepe University, Turkey)
Prof. Dr. S. Feyza Çinicioğlu (Istanbul University, Turkey)
Prof. Dr. Said Kenai (Bilida University, Algeria)
Prof. Dr. Salih Yazicioğlu (Gazi University, Turkey)
Prof. Dr. Semih Eryildiz (Doğuş University, Turkey)
Prof. Dr. Seyhan Fırat (Gazi University, Turkey)
Prof. Dr. Soofia Tahira Elias-Özkan (METU, Turkey)
Prof. Dr. Tülay Tikansak Karadayi (GTU, Turkey)
Prof. Dr. Ulrich Knaack (Technical University Delft, the Netherlands)

Prof. Dr. Zerrin Yilmaz (İstanbul Technical University, Turkey)

Prof. Gregory Keeffe (Manchester Metropolitan University, England)

Assoc. Prof. Dr. A. Güliz Bilgin Altinöz (METU, Turkey)

Assoc. Prof. Dr. Arzuhan Burcu Gültekin (Ankara University, Turkey)

Assoc. Prof. Dr. Alper Büyükkaragöz (Gazi University, Turkey)

Assoc. Prof. Dr. Ayşe Tavukçuoğlu (METU, Turkey)

Assoc. Prof. Dr. Ayşin Sev (Mimar Sinan Fine Art University, Turkey)

Assoc. Prof. Dr. Bahar Demirel (Fırat University, Turkey)

Assoc. Prof. Dr. Bige Tuncer (Singapore University of Technology)

Assoc. Prof. Dr. Cevdet Söğütlü (Gazi University, Turkey)

Assoc. Prof. Dr. F. Nihan Özdemir Sönmez (Gazi University, Turkey)

Assoc. Prof. Dr. Hamza Çinar (Gazi University, Turkey)

Assoc. Prof. Dr. Hasbi Yaprak (Kastamonu University, Turkey)

Assoc. Prof. Dr. İdil Ayçam (Gazi University, Turkey)

Assoc. Prof. Dr. Kutlu Kayihan (Gebze Technical University, Turkey)

Assoc. Prof. Dr. Metin Ipek (Sakarya University, Turkey)

Assoc. Prof. Dr. Mustafa Sahmaran (Gazi University, Turkey)

Assoc. Prof. Dr. Müge Mukaddes Darwish (Texas Tech University, USA)

Assoc. Prof. Dr. Mürsel Erdal (Gazi University, Turkey)

Assoc. Prof. Dr. Nihat Sinan Işik (Gazi University, Turkey)

Assoc. Prof. Dr. Recep Birgül (Muğla Sıtkı Koçman University, Turkey)

Assoc. Prof. Dr. Rudi Stouffs (TU Delft, the Netherlands)

Assoc. Prof. Dr. Şule Tudeş (Gazi University, Turkey)

Assoc. Prof. Dr. Ülkü Duman Yüksel (Gazi University, Turkey)

Assist. Prof. Dr. Ahmet Beycioğlu (Düzce University, Turkey)

Assist. Prof. Dr. Burçhan Aydin (Texas A&M University, USA)

Assist. Prof. Dr. Çiğdem Belgin Dikmen (Bozok University, Turkey)

Assist. Prof. Dr. Gökhan Durmuş (Gazi University, Turkey)

Assist. Prof. Dr. Gülsu Harputlugil (Çankaya University, Turkey)

Assist. Prof. Dr. Hanifi Tokgöz (Gazi University, Turkey)

Assist. Prof. Dr. İlker Tekin (Bayburt University, Turkey)

Assist. Prof. Dr. Kürşat Yildiz (Gazi University, Turkey)

Assist. Prof. Dr. Murat Çavuş (Gaziosmanpaşa University, Turkey)

Assist. Prof. Dr. Yağmur Kopraman (Gazi University, Turkey)

Assist. Prof. Dr. Yeşim Aliefendioğlu (Ankara University, Turkey)

Dr. Ayşe Nur Albayrak (Gebze Technical University, Turkey)

Dr. Ayşem Berrin Çakmakli (Middle East Technical University, Turkey)

Dr. David Tann (University of East London, UK)

Dr. Jeremy Gibberd (CSIR, South Africa)

Dr. Karel Mulder (Delft University of Technology, the Netherlands)

Acknowledgements

The organization of a conference/symposium is a massive task; it cannot be achieved without the hard work and diligence of so many people. First, we would like to thank the symposium co-chair Prof. Seyhan for his excellent leadership and his fantastic team at Gazi University, and they have excelled in organizing the paper submission, reviews and communication with the authors and delegates of ISBS 2017. I would also like to thank the team at the British University in Dubai, who dedicated much time and effort to organize and receive the delegate and look after them over the 3 days of the event. Special thanks go to the Vice Chancellor of the British University in Dubai, Prof. Abdullah Alshamsi, for his support, encouragement and dedication to the event and for participating in all the symposium activities.

Great gratitude also goes to our sponsors, the Emirates National Oil Company our main sponsor and Al Maktoum Foundation our gold sponsor, Gazi University in Ankara, the British University in Dubai and the Turkish Civil Engineering Council.

The symposium was enriched by the participation of the keynote speakers from the UK, the USA and the UAE, and we thank them all wholeheartedly.

The success of 3rd International Symposium of Sustainable Buildings is owed to the authors and delegates who wrote and presented the papers, and we are especially indebted to them all. I also want to thank proceedings co-editors Prof. John Kinuthia and Prof. Seyhan Fırat.

Dubai, United Arab Emirates   Prof. Abid Abu-Tair
June 2017

Contents

Part I   Energy and Environment—Emerging Climate Change and Impact on the Built Environment

Trend Detection in Annual Temperature and Precipitation Using Mann–Kendall Test—A Case Study to Assess Climate Change in Abu Dhabi, United Arab Emirates   3
Aydin Basarir, Hasan Arman, Saber Hussein, Ahmed Murad, Ala Aldahan, and M. Abdulla Al-Abri

The Assessment of Energy of Residential Building of District Korkuteli/ANTALYA   13
Gökhan Durmuş and Sadık Önal

A Dynamic Thermal Simulation in New Residential Housing of Lakhiayta City in Morocco   26
Karima El Azhary, Najma Laaroussi, Mohammed Garoum, Khalid El Harrouni, and Majid Mansour

A Literature Survey on Integration of Wind Energy and Formal Structure of Buildings at Urban Scale   36
Serpil Paltun, Arzuhan Burcu Gültekin, and Gülser Çelebi

The Role and Importance of Energy Efficiency for Sustainable Development of the Countries   53
Serap Pelin Türkoğlu and Pınar Sezin Öztürk Kardoğan

A Study on Some Factors Affecting on CO_2 Curing of Expanded Perlite Based Thermal Insulation Panel   61
Gökhan Durmuş, Onuralp Uluer, Mustafa Aktaş, İbrahim Karaağaç, Ataollah Khanlari, Ümit Ağbulut, and Damla Nur Çelik

Turkey's Renewable Energy Sources and Govermental Incentives   71
Evren Kaydul, Zafer Demir, and Nesrin Çolak

Green Wall Systems: A Literature Review   82
Yasemin Baran and Arzuhan Burcu Gültekin

**Achieving Sustainable Buildings via Energy Efficiency Retrofit: Case
Study of a Hotel Building**   97
Muhsin Kılıç and Ayşe Fidan Altun

**Application of Solid Wastes for the Production of Sustainable
Concrete**   108
Vasudha D. Katare and Mangesh V. Madurwar

**Sustainable Architecture Under the Timeline Frame: Case Study
of Fujairah in UAE**   122
Tahani Yousuf and Hanan Taleb

**Determination of Optimum Insulation Thickness on Different Wall
Orientations in a Hot Climate**   145
Erhan Arslan and Irfan Karagoz

An Examination of the Energy-Efficient High-Rise Building Design   158
H. Handan Yücel Yıldırım, Arzuhan Burcu Gültekin,
and Harun Tanrıvermiş

**On the Feasibility of Using Photovoltaic Panels to Produce the
Electricity Required at a Chemical Treatment Plant on Industrial
Company**   176
Akın Yalçın, Zafer Demir, and Nesrin Çolak

**Behavior of Mortar Samples with Waste Brick and Ceramic Under
Freeze-Thaw Effect**   189
Selçuk Memiş, İ.G. Mütevelli Özkan, M.U. Yılmazoğlu, Gökhan Kaplan,
and Hasbi Yaprak

Part II   Environmental Policies and Practices

Sustainable Design Approaches on Packaging Design   205
Ceyda Özgen

**The Aim of Sustainable Urban Development and Climate Change
Policies**   220
Ayse Nur Albayrak and Seda H. Bostanci

Acoustics Treatment for University Halls   231
Ammar Abdulkareem, Mahmood Abu Ali, and Emad Mushtaha

**Kuznets's Inverted U Hypothesis: The Relationship Between
Economic Development and Ecological Footprint**   242
Ceyda Erden Özsoy

An Examination of Sustainability and United Nations Sustainable Development Goals: Turkey Case 252
Arzuhan Burcu Gültekin, Çiğdem Belgin Dikmen, Ahmet Hilmi Erciyes, and Demet Örgü

Disseminating the Use of Sustainable Design Principles Through Architectural Competitions. 272
Gizem Kuçak Toprak

The Healthcare Industry and Medical Devices in Environmental Sustainability: Medical Device Industry is Going Green 280
Meriç Yavuz Çolak, Levent Çolak, and Elif Gürel

Assesment of Historical Kars Stone Bridge in Terms of Sustainability of Historical Heritage 291
Elif Tektas, Zeynep Yesim İlerisoy, and Mehmer Emin Tuna

Refugee Problems in Turkey and Its Evaluation in the Context of Sustainability 306
Sefik Tas, Mursel Erdal, and Huseyin Yilmaz Aruntas

Understanding Sustainability Through Mindfulness: A Systematic Review 321
Ecem Tezel and Heyecan Giritli

Estidama Pearl Building Rating System of Abu Dhabi and Al Sa'fat of Dubai: Comparison and Analysis 328
Omair Awadh

An Investigation into University Students' Perceptions of Sustainability 338
Begüm Sertyeşilışık, Heyecan Giritli, Ecem Tezel, and Egemen Sertyeşilışık

Properties of Mortars Produced with PKF Press Filter Waste 347
İlker Bekir Topçu and Taylan Sofuoğlu

Evaluation of Occupational Safety Culture in Construction Sector in the Context of Sustainability 361
Mursel Erdal, Nihat Sinan Isik, and Seyhan Fırat

Part III Strengthening and Rehabilitation of Structures

The Investigation of Alkali Silica Reaction of Mortars Containing Borogypsum. 369
Kursat Yildiz

Structural Stability Analysis of Large-Scale Masonry Historic City Walls 384
Rüya Kılıç Demircan, and Ali İhsan Ünay

The Reliability of the ACI 209R-92 Method in Predicting Column
Shortening in High-Rise Concrete Buildings   396
Alaa Habrah and Abid I. Abu-Tair

A Review on the Effect of Environmental Conditions on Concrete
Evaporation and Bleeding   413
İlker Bekir Topçu and Burak Işıkdağ

Triplet Shear Tests on Masonry Units with and Without Seismic
Textile Reinforcement   427
Berna Istegun and Erkan Celebi

Investigation of Buckling Behavior of FRP-Concrete
Hybrid Columns   436
Ferhat Aydın, Tahir Akgül, and Emine Aydın

An Investigation of Concrete Strength of Hybrid Construction
Materials Under the Effect of Heat   449
Ferhat Aydın, Metin İpek, and Kutalmış Akça

Electrical Curing Application on Cement-based Mortar with Different
Stress Intensity   462
Tayfun Uygunoğlu, İsmail Hocaoğlu, and İlker Bekir Topçu

Effect of Particle Size Distribution on the Sintering Behaviour of Fly
Ash, Red Mud and Fly Ash-Red Mud Mixture   469
Sedef Dikmen, Zafer Dikmen, Gülgün Yılmaz, and Seyhan Fırat

Key Success Factors Impacting the Success of Innovation in UAE
Construction Projects   482
Mansour Faried, Malak Saad, and Khalid Almarri

Analysis of Strengthened Composite Beams Under Flexural Stress   506
Emre Ercan, Bengi Arısoy, Ali Demir, and Anıl Özdemir

Wave Propagation and Vibration Isolation in Soils   519
Erkan Çelebi

Concrete Strength Variation Effect on Numerical Thermal
Deformations of FRP Bars-Reinforced Concrete Beams in Hot
Regions   526
Ali Zaidi, Aissa Boussouar, Kaddour Mouattah, and Radhouane Masmoudi

A Study on the Use of Advanced Nondestructive Testing Methods on
Historic Structures   536
Rukiye Tuğla, Rüya Kılıç Demircan, and Gökhan Kaplan

Studying the Historical Structure Damage Due to Soil Hazards and
Examination of Applied Repairment-Strengthening Techniques   550
Rüya Kılıç Demircan, and Pınar Sezin Öztürk Kardoğan

Self-Sensing Behavior Under Monotonic and Cyclic Loadings of ECC Containing Electrically Conductive Carbon-Based Materials 566
Mustafa Şahmaran, Ali Al-Dahawi, Vadood Farzaneh, Oğuzhan Öcal, and Gürkan Yıldırım

Determination of Favorable Time Window for Infrared Inspection by Numerical Simulation of Heat Propagation in Concrete 577
Ersan Güray and Recep Birgül

Determination of Self-Healing Performance of Cementitious Composites Under Elevated CO_2 Concentration by Resonant Frequency and Crack Opening Measurements 592
Süleyman Bahadır Keskin, Kasap Keskin Özlem, Gürkan Yıldırım, Mustafa Şahmaran, and Özgür Anıl

Part I
Energy and Environment—Emerging Climate Change and Impact on the Built Environment

Trend Detection in Annual Temperature and Precipitation Using Mann–Kendall Test—A Case Study to Assess Climate Change in Abu Dhabi, United Arab Emirates

Aydin Basarir[1(✉)], Hasan Arman[2], Saber Hussein[2], Ahmed Murad[2], Ala Aldahan[2], and M. Abdulla Al-Abri[3]

[1] Department of Agribusiness, College of Food and Agriculture, United Arab
Emirates University, Al Ain, UAE
abasarir@uaeu.ac.ae
[2] Department of Geology, College of Science, United Arab Emirates University,
Al Ain, UAE
{harman,S_Hussein,ahmed.murad,aaldahan}@uaeu.ac.ae
[3] Department of Meteorology, National Center of Meteorology and Seismology,
Al Ain, UAE
malabri@ncms.ae

Abstract. Annual average temperature and precipitation can be considered as two important indicators to judge the possibility of future climate change. The main objective of this study is to analyze the possibility of trend in annual average temperature and precipitation for Abu Dhabi and Al Ain cities in the United Arab Emirates (UAE). Mann–Kendall non-parametric tests were run at 5% significance level on annual time series data obtained from two stations located in Abu Dhabi and Al Ain cities for the period of 1972–2014. Significant increasing trend in temperature and no significant trend in precipitation of both cities were detected. Thus, there is a possibility to have higher temperature, but not significant change in precipitation of Abu Dhabi and Al Ain cities in the long run.

Keywords: Trend analysis · Temperature · Precipitation · Climate change
Mann–Kendall non-parametric test

1 Introduction

Climate change has been detected globally in different forms such as downpours, storms, rising temperature and sea level, retreating glaciers etc. [1]. The impacts of climate change can be assessed by analyzing trends in temperature and precipitation [2].

The United Arab Emirates (UAE) is located in the Middle East part of Asia and covers an area of 82,880 km². The country has borders with Saudi Arabia and Oman (Fig. 1) and is member of the Gulf Co-operation Council (GCC). There are seven

© Springer International Publishing AG, part of Springer Nature 2018
S. Fırat et al. (eds.), *Proceedings of 3rd International Sustainable
Buildings Symposium (ISBS 2017)*, Lecture Notes in Civil Engineering 7,
https://doi.org/10.1007/978-3-319-64349-6_1

emirates in the UAE, Abu Dhabi is the largest of them and covers an area of 67,340 km², which is 87% of the total area of the UAE [3].

Fig. 1. Location Map (*blue borders* indicate the study area)

The UAE has arid climate with the average annual summer temperature ranges between 35 and 40 °C with the highest value of 47 °C. The winter is relatively shorter and runs from December to February with temperature drop, especially in the inland areas. The UAE has low rainfall, and it is usually accompanied by thunderstorms in every December and January. During summer, humidity in the UAE ranges between 60 and 100% particularly on the inhabited coasts and gets lower in the inland areas. As for the wind, the monsoon winds blowing across the UAE is stronger in the spring and late summer months, and are of two types: Northern dry wind which helps mitigate the temperature as far as it is not laden with dust, and mostly very humid eastern wind [4–6].

Climate change is expected to occur and effect the UAE via large spatial and temporal variations. The changes in temperature and precipitation are the most important indicators of climate variability impacts. Therefore, this study examined the historical variations of such indicators in Abu Dhabi and Al Ain cities located in Abu Dhabi Emirate.

The main objective of this study is to analyze the possible trends for climate indicators such as annual average temperature and precipitation in Abu Dhabi and Al Ain cities of Abu Dhabi Emirate for the period of 1972–2014.

2 Literature Review

Climate change is considered to be directly linked to temperature and precipitation. Numerous studies have been conducted on global and regional levels to assess temperature and precipitation trends and to quantify the impacts of temperature and precipitation. Most of these studies have found positive trends in different temperature and precipitation indices. Chinchorkar et al. [7], Karmeshu [1], Partal and Kahya [8], Murumkar [2] focused on detecting trend in either temperature or precipitation by using long run annual data. They mostly detected significant trends in the both indicators of the study area they analyzed. The detections can be taken as simple indicator of climate change for the research areas.

In addition to analysis of the annual data some researchers such as Nalley et al. [9], Mapurisa and Chikodzi [10], Hirsch et al. [11], Lanzante et al. [12], Rebetez and Reinhard [13] have studied on monthly or seasonal indicators to detect a possible trend in temperature and precipitation. The significant trends were detected generally in the longer run.

3 Methodology

The annual average temperature and precipitation data for the period of 1972–2014 was obtained from UAE National Center of Meteorology and Seismology (NCMS) [14] for Abu Dhabi and Al Ain cities airports.

Mann–Kendall (M–K) nonparametric method is one of the mostly utilized technique by researchers [1, 2, 7, 8, 10] in analysis of climatic time series data. The test is based on Mann [15] and Kendall [16]. "It is a rank correlation test for two sets of observations between the rank order of the recorded values and their ordered values in time" [9]. The main advantage of the test is that it is a non-parametric test and the normal distribution of data is not required. In case of inhomogeneous time series data analysis, having low sensitivity to sudden breaks is the second most important advantage of the test [17]. According to the null hypothesis (H_o) of the test, there is no trend and the data are randomly and independently ordered. The judgement of null hypothesis is tested by alternative hypothesis (H_1), which assumes the existence of trend [18].

The M–K trend test statistic can be calculated as shown in Eq. (1) [10].

$$S = \sum_{k=1}^{n-1} \sum_{j=k+1} sign\left(x_j - x_k\right) \tag{1}$$

x_j denotes the ordered data values; n is the length of observations and S is the M–K statistics. The sign of the test statistic is [19].

$$sign(x_j - x_k) = \begin{cases} 1 & if \ x_j - x_k > 0 \\ 0 & if \ x_j - x_k = 0 \\ -1 & if \ x_j - x_k < 0 \end{cases} \qquad (2)$$

For n $\geq$ 10 then the S statistic is approximately normally distributed with mean zero ($E(S) = 0$) and variance as in Eq. 3.

$$V(S) = \frac{n(n-1)(2n+5) - \sum_{k=1}^{nk} t_k(k)(k-1)(2k+5)}{18} \qquad (3)$$

t_k denotes the number of ties or duplicates to extent k. Equation 3 is used in case of tied values in time series. nk is the total number of ties in dataset. In case of having $n \geq 10$ the standardized test statistic for M–K can be calculated using Eq. 4.

$$Z_S = \begin{cases} \frac{S-1}{\sqrt{V(S)}}, & (if \ S > 0) \\ 0, & (if \ S > 0) \\ \frac{S+1}{\sqrt{V(S)}}, & (if \ S > 0) \end{cases} \qquad (4)$$

The test statistics Z_S is used to measure the significance of trend. Upward and/or downward trend direction depends on positive and negative values of the test, respectively. The calculated test value is compared to standard normal variate at some statistical significance level (α) to check the validity of the null hypothesis. If calculated $|Z_s|$ value is greater than $Z_{\alpha/2}$ then the null hypothesis is invalid and trend is significant [20, 21].

When running M–K test in XLSTAT 2016 program another statistic called Kendall's tau is obtained. Kendall's tau is a measure of correlation and calculates the strength of the relationship between two variables. The requirement is that the two variables (X and Y) need to be paired and at least ordinal observations structure; then the calculation of the correlation between them will be possible. Just like Spearman's rank correlation, the test is calculated on the ranks of the data [22]. The values are put in order and numbered, 1 for the lowest value, 2 for the next lowest and so on for each variable separately. The statistic takes values between -1 and $+1$, with a positive correlation indicating that the ranks of both variables increase together whilst a negative correlation indicates that as the rank of one variable increases, the other decreases [23].

The Spearman rank correlation test is easier than Kendall's tau and more widely used to measure rank correlation. However, the main advantages of using Kendall's tau are that the distribution of this statistic has slightly better statistical properties and there is a direct interpretation of Kendall's tau in terms of probabilities of observing concordant and discordant pairs [24]. In almost all situations the values of Spearman's rank correlation and Kendall's tau are very close and would invariably lead to the same conclusions [22].

Autocorrelation is defined as the correlation of a variable with itself over successive time intervals. It is essential to consider autocorrelation prior to testing for trends in time series analysis. In the presence of autocorrelation, the chances of detecting significant trends increases even if they are absent and vice versa [1]. Hamed and Rao [21] suggested a modified M-K test In order to eliminate the effect of autocorrelation. The autocorrelation between the ranks of the data is calculated by the test after removing the apparent trend. Following Karmeshu [1] the adjusted variance is given by:

$$Var[S] = \frac{1}{18}[N(N-1)(2N+5)]\frac{N}{NS^*} \tag{5}$$

where,

$$\frac{N}{NS^*} = 1 + \frac{2}{N(N-1)(N-2)}\sum_{l=1}^{p}(N-l)(N-i-1)(N-i-2)p_s(i) \tag{6}$$

N is the number of observations in the sample, NS^* is the effective number of observations to account for autocorrelation in the data, $p_s(i)$ is the autocorrelation between ranks of the observations for lag i, and p is the maximum time lag under consideration [8].

In order to analyze the relations between variables the correlation coefficient test can be used. As explained by Hill et al. [25] "the correlation coefficient between two variables measures the degree of linear association between them." The coefficient values lies between -1 and $+1$. Following Hill et al. [25], the coefficient can be calculated as follows:

$$\rho = \frac{cov(X,Y)}{\sqrt{var(X)var(Y)}} \tag{7}$$

where ρ is correlation coefficient, X and Y are two random variables.

Addinsoft's XLSTAT 2016 software was used to perform the statistical M–K test. The confidence level for null hypothesis was 95% for both temperature and precipitation data of Abu Dhabi and Al Ain cities. In addition to the results obtained from the M–K test, linear trend lines are plotted for each city using Microsoft Excel 2016. Deterministic statistics and correlation coefficient test between the variables were calculated as well. The results of analysis are presented in tables and graphs of the following section.

4 Results and Discussions

The deterministic statistics of time series data for temperature and precipitation obtained from Abu Dhabi and Al Ain stations are given in Table 1. The temperature for Abu Dhabi (TEMPAD) ranged between 26.5 and 28.75 °C with average value of 27.75 °C. Unlike to Abu Dhabi, Al Ain city seems to have lower minimum and higher maximum temperature (TEMPAIN) values. Even though the mean temperature values for both cities are close to each other, standard deviation value for Al Ain seems to be higher than the value for Abu Dhabi. This indicates more changes in temperature of Al Ain.

Table 1. Deterministic statistics of temperature and precipitation (1972–2014)

Variables	Min	Max	Mean	SD
TEMPAD (°C)	26.5	28.75	27.75	0.52
TEMPAIN (°C)	24.7	29.41	27.77	1.16
PERCAD (mm)	1.6	183.10	56.11	47.24
PERCAIN (mm)	0.89	302.50	90.73	75.12

The range between minimum and maximum values of precipitation for both Abu Dhabi (PERCAD) and Al Ain (PERCAIN) cities seems to be huge. There is no stability in precipitation for both cities. Thus, the standard deviations of the precipitation for both cities are extremely high. More changes seem to be occur in Al Ain.

M–K test was run on temperature of Abu Dhabi and Al Ain cities and the results are given in Table 2. The significance value for α was taken as 0.05. The p-values for both cities are less than the critical 0.05 value. Null hypothesis of no trend can be easily rejected and alternative hypothesis of trend in the temperature can easily be accepted.

Table 2. M–K test for temperature

City	M–K Statistics (S)	Kendall's Tau	Var (S)	p-value
Abu Dhabi	375	0.419	9113	0.0001
Al Ain	535	0.596	9117	0.0001

Figures 2 and 3 are graphs for annual average temperature observations and related trend line for Abu Dhabi and Al Ain cities for the period of 1972–2014. The figures show a clear increasing trend in the temperature of both Abu Dhabi and Al Ain cities.

The M–K test was also run on precipitation data of Abu Dhabi and Al Ain cities. The results are shown in Table 3. The p-values for both cities are larger than the critical significance 0.05 value. Thus, null hypothesis of no trend in time series data of both cities can be accepted.

The annual average precipitation data and related linear trend lines are shown on Figs. 4 and 5 for Abu Dhabi and Al Ain, respectively. Even though linear trends seem to be in declining shape, the M–K test statistics are not significant enough to reject the null hypothesis of no trend (Table 3).

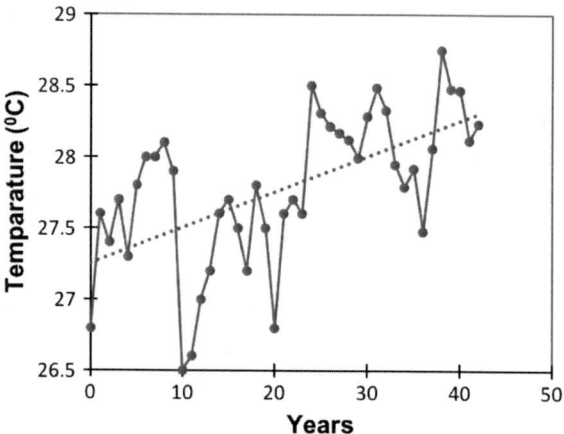

Fig. 2. Linear trend line for temperature data of Abu Dhabi

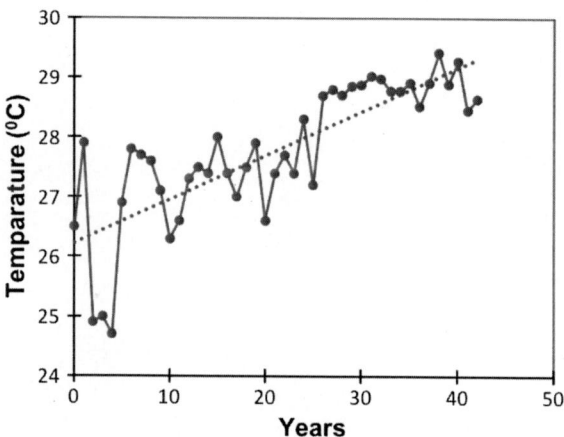

Fig. 3. Linear trend line for temperature data of Al Ain

Table 3. M–K test for precipitation

City	M–K statistics (S)	Kendall's Tau	Var (S)	p-value
Abu Dhabi	−59	−0.065	9118	0.55
Al Ain	−139	−0.154	9118	0.15

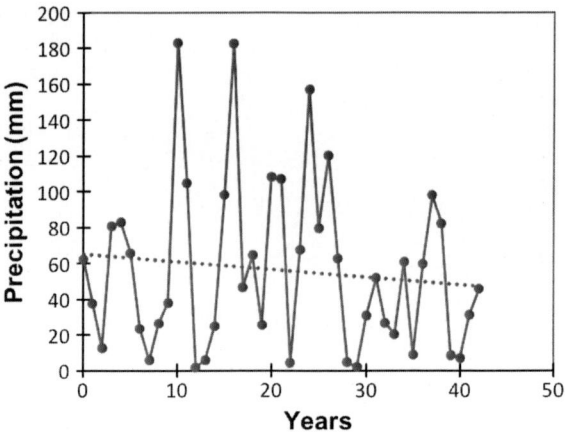

Fig. 4. Linear trend line for precipitation data of Abu Dhabi

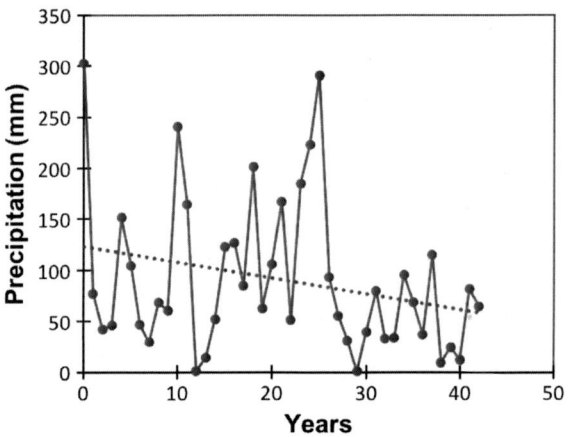

Fig. 5. Linear trend line for precipitation data of Al Ain

The Pearson correlation coefficient test was conducted to determine the relationships between temperature and precipitation trends in Abu Dhabi and Al Ain cities. As shown in the Table 4, statistically significant coefficients indicate strong relationships between the climate indicators.

Table 4. Correlation coefficient between variables

Variables	TEMPAD	TEMPAIN	PERCAD
TEMPAIN	0.673 (<0.0001)*		
PERCAD	−0.296 (0.054)*	−0.241 (0.120)*	
PERCAIN	−0.393 (0.009)*	−0.367 (0.016)*	0.648 (<0.0001)*

*p-values

5 Conclusions

The annual average values of both temperature and precipitation obtained from two stations located in Abu Dhabi and Al Ain cities were analyzed using the M–K trend test. The analysis included for the period of 1972–2014. It seems that there is a conformity in the results obtained from the M–K test and linear trend lines for Abu Dhabi and Al Ain cities. The M–K test indicate an increasing trend in the temperature of both cities and the trend lines on the Figures support such argument. On the other hand, according to M–K test, there is no existence of trend for time series of precipitation for both Abu Dhabi and Al Ain cities.

The results of this study reveal that both Abu Dhabi and Al Ain are experiencing warming trends in temperature. The precipitation for both regions seems to be stable and not experiencing significant trends.

The relationships between the temperature and precipitation trends and large-scale climate circulations are important for the UAE climate. The relationships were analyzed in this study by using Pearson correlation coefficient test. The relationship between temperature and/or precipitation with the other climate change indicators such as humidity and wind speed can also be analyzed in future studies by using the same test.

The findings of this study can be considered as a baseline information to be incorporated in the future when analyzing linkages between temperature and precipitation trends in the UAE and different climatic phenomena.

Acknowledgements. This study was funded by United Arab Emirates University (UAEU) Center Based Research Grant (UCBR) program, National Water Center.

References

1. Karmeshu N (2012) Trend detection in annual temperature and precipitation using the Mann–Kendall test: a case study to assess climate change on select states in the northeastern United States, MS Thesis, University of Pennsilvania
2. Murumkar AR, Arya DS (2014) Trend and periodicity analysis in rainfall pattern of Nira basin, Central India. Am J Clim Change 3:60–70. doi:10.4236/ajcc.2014.31006
3. UAENSB, UAE, National Statistical Bureau (2016) http://www.uaestatistics.gov.ae
4. UAE, M.o.E.a.W., State of Environment of the United Arab Emirates Report. 2015, Ministry of Environment and Water: Abu Dhabi (2015) http://www.moccae.gov.ae/en/home.aspx
5. Stocker TF, Qin D, Plattner GK, Tignor M, Allen SK, Boschung J, Midgley BM, ..., IPCC (2013) Climate change 2013: the physical science basis. In: Contribution of working group I to the fifth assessment report of the intergovernmental panel on climate change (2013)
6. Arman H (2015) Uncertainties, risks and challenges relating to CO_2 emissions and its possible impact on climate change in the United Arab Emirates. Int J Glob Warm 8:1–17. doi:10.1504/IJGW.2015.071575
7. Chinchorkar SS, Sayyad FG, Vaidya VB, Pandye V (2015) Trend detection in annual maximum temperature and precipitation using the Mann–Kendall test: a case study to assess climate change on Anand of central Gujarat. MAUSAM 66:1–6

8. Partal T, Kahya E (2006) Trend analysis in Turkish precipitation data. Hydrol Process 20:2011–2026. doi:10.1002/hyp.5993
9. Nalley D, Adamowski J, Khalil B, Ozga-Zielinski B (2013) Trend detection in surface air temperature in Ontario and Quebec, Canada during 1967–2006 using the discrete wavelet transform. Atmos Res 132:375–398. doi:10.1016/j.atmosres.2013.06.011
10. Mapurisa B, Chikodzi D (2014) An assessment of trends of monthly contributions to seasonal rainfall in south-eastern Zimbabwe. Am J Clim Change 3:50–59. doi:10.4236/ajcc.2014.31005
11. Hirsch RM, Slack JR, Smith RA (1982) Techniques of trend analysis for monthly water quality data. Water Resour Res 18:107–121. doi:10.1029/WR018i001p00107
12. Lanzante JR, Klein SA, Seidel DJ (2003) Temporal homogenization of monthly radiosonde temperature data. Part II: trends, sensitivities, and MSU comparison. J Clim 16:241–262. doi:10.1175/1520-0442(2003)016%3C0241:THOMRT%3E2.0.CO;2
13. Rebetez M, Reinhard M (2008) Monthly air temperature trends in Switzerland 1901–2000 and 1975–2004. Theor Appl Climatol 91:27–34. doi:10.1007/s00704-007-0296-2
14. Anonymus (2015) In climate data for temperature, rainfall, humidity, and wind speed. N.C.o. M.a. Seismology. (NCMS), and D.o.M.a.C.D. UAE, Editors. 2015: Abu Dhabi
15. Mann HB (1945) Nonparametric tests against trend. Econom: J Econom Soc 245–259. doi:10.2307/1907187
16. Kendall MG (1948) Rank correlation methods. Published by C. Griffin, London. doi:10.2307/2333282
17. Tabari H, Marofi S, Aeini A, Talaee PH, Mohammadi K (2011) Trend analysis of reference evapotranspiration in the western half of Iran. Agric Forest Meteorol 151:128–136. doi:10.1016/j.agrformet.2010.09.009
18. Önöz B, Bayazit M (2003) The power of statistical tests for trend detection. Turk J Eng Environ Sci 27:247–251
19. Yue S, Pilon P, Phinney B, Cavadias G (2002) The influence of autocorrelation on the ability to detect trend in hydrological series. Hydrol Process 16:1807–1829. doi:10.1002/hyp.1095
20. Motiee H, McBean E (2009) An assessment of long-term trends in hydrologic components and implications for water levels in Lake Superior. Hydrol Res 40:564–579. doi:10.2166/nh.2009.061
21. Hamed KH, Rao AR (1998) A modified Mann–Kendall trend test for autocorrelated data. J Hydrol 204:182–196. doi:10.1016/S0022-1694(97)00125-X
22. Crichton NJ (1999) Information point: Spearman's rank correlation. J Clin Nurs 8:763–766
23. Vlaminck H, Maes B, Jacobs A, Reyntjens S, Evers G (2001) The dialysis diet and fluid non-adherence questionnaire: validity testing of a self-report instrument for clinical practice information point: Kendall's Tau. J Clin Nurs 10:707–715. doi:10.1046/j.1365-2702.2001.00537.x
24. Conover WJ (1980) Practical nonparametric statistics. Wiley, New York, NY
25. Hill RC, Griffiths WE, Judge GG, Reiman MA (2001) Undergraduate econometrics, vol 4. Wiley, New York

The Assessment of Energy of Residential Building of District Korkuteli/ANTALYA

Gökhan Durmuş[1(✉)] and Sadık Önal[2]

[1] Civil Engineering Department, Faculty of Techonology, Gazi University,
Ankara, Turkey
gdurmus@gazi.edu.tr
[2] Director of Urban Planning and Urban Planning, Korkuteli/Antalya, Turkey
onalsadik07@gmail.com

Abstract. Annual primary energy consumption for sustainable construction operation and maintenance costs is estimated to be approximately 30–40%. The issue of increasing energy efficiency in buildings has become increasingly important. In addition, residential buildings are being surveyed in terms of energy efficiency compared to other buildings (offices, factories, shopping malls, etc.,). In this study, in order to determine the energy performance of buildings, the values of energy performance of 480 dwellings with different typology and area constructed in different districts of Antalya/Korkuteli district were analyzed by Building Energy Performance Program (BEP-TR) program prepared by Ministry of Environment and Urban Planning. The BEP-TR program has been evaluated in terms of energy performance evaluation of heating, cooling, hot water, lighting and greenhouse gas emissions. Each letter code of the energy performance of housing has been investigated statistically average assessed. As a result, while the heating values varied from A to G from energy performance, only class C appeared in the greenhouse gas distribution. In addition, when the averages are examined, the energy class of all houses is "Class C".

Keywords: Energy identity certificate · Energy efficiency
Energy performance of buildings program (BEP-TR)
Energy performance of buildings

1 Introduction

Environmental quality has become increasingly affected by the built environment—as ultimately, buildings are responsible for the bulk of energy consumption and resultant atmospheric emissions in many countries [13]. Operation and maintenance of structures in the world consumption of 30–40% of the primary energy [4, 16]. These structures are globally 40% in the greenhouse gas emission [2]. Looking at the emissions of greenhouse gasses in Turkey, for the year 2012, the energy sector ranks first with 308.6 Mton CO_2 equivalents (70.2%). The CO_2 emission was 357.5 Mtons, which is 81.26% of total greenhouse gas emissions, compared to the total greenhouse gas emissions this year. The CO_2 emission value from electricity generation in the energy sector in Turkey is 116,76 Mton [23]. It is known that the values of CO_2 emissions from fossil fuels (natural gas, coal and petroleum sources) over the last 30 years in Turkey have increased rapidly over the years. It is known that greenhouse gas emission value at the beginning of the harmful effects arising as a result of increasing use of energy increases the atmospheric value. In addition, different investigations have been made for the amounts of energy that structures have consumed in the context of the life cycle.

Gültekin A.B, Çelebi G., in their studies, The open-end model for assessment of the environmental impacts of construction products during their life cycles within the context of life cycle assessment (LCA) methodology was proposed in this study. The model was concretized by a case study of an assessment of the environmental impacts of wallpapers related to maintenance in the usage stage [11].

T. Ramesh, etc., in their studies; a critical review of the life cycle energy analyses of buildings resulting from 73 cases across 13 countries is presented. Results show that operating (80–90%) and embodied (10–20%) phases of energy use are significant contributors to building's life cycle energy demand. Life cycle energy (primary) requirement of conventional residential buildings falls in the range of 150–400 kWh/m^2 per year and that of Office buildings in the range of 250–550 k Wh/m^2 per year. Building's life cycle energy demand can be reduced by reducing its operating energy significantly through the use of passive and active technologies even if it leads to a slight increase in embodied energy [22]. Shukla etc., in their studies; A mudbrick structure in India has been assessed for energy performance [25]. Debnath etc., in their Works; energy analyzes were carried out on single and multi-storey reinforced concrete buildings [7].

Huberman, N., and Pearlmutter D., in their studies; to identify building materials which may optimize a building's energy requirements over its entire life cycle, by analyzing both embodied and operational energy consumption in a climatically responsive building in the Negev desert region of southern Israel—comparing its actual material composition with a number of possible alternatives. It was found that the embodied energy of the building accounts for some 60% of the overall life-cycle energy consumption, which could be reduced significantly by using "alternative" wall infill materials [13].

The use of building's energy consumption in construction and operation stages, assessing energy separately. However, this distinction does not mean that there is a distinction between them in determining the energy performance of the building. For example, building materials (brick, cement, aggregate, glass, insulation materials, etc.,) used in building construction are effective in the amount of energy consumed by the building during use. All of these materials are produced using natural resources with different production processes [22, 25]. Approximately 10–20% of the energy consumption for the buildings is spent in line with the parameters such as building materials used in the construction phase, electricity energy demand, building typology. In recent years energy analysis values have been estimated in terms of different building typologies [22, 25].

The safe and sustainable use of energy, the efficient use of greenhouse gasses, the protection of the environment, the increasing tendency and indecisiveness of oil prices, the shift from fossil sources to new and renewable energy sources, and so on [5, 17]. Depending on the building's function, energy quantities measured in new buildings and measured in existing buildings to meet the standard requirements such as heating, cooling, ventilation, lighting are examined under energy certification [26, 38]. The Energy Performance Regulations of Building, which was published in Turkey in the Official Gazette no. 27078 in 2008, entered into force. According to this regulation; To assess all kinds of energy use; Categorize according to the types of energy efficiency and primary energy use; The buildings to be determined for CO_2 emission; And to evaluate the use of renewable energy sources in new and existing buildings [20]. The regulation envisages the construction of a new building with a capacity of more than 1.000 sqm for the heating, cooling, ventilation, sanitary hot water, electricity and illumination energy needs of the buildings in whole or in part to meet the needs of hydraulics, the wind, the sun, geothermal, biomass, biogas and non-fossil energy-based system solutions such as the tide have become mandatory by regulation by designers.

Turkey, is a big country, having a population of about Urban population (% of total) in Turkey was 73.40 as of 2015. Its highest value over the past 55 years was 73.40 in 2015, while its lowest value was 31.52 in 1960. In Turkey, although the urban population has increased with respect to the city, the situation with the cities is very different. The largest city population is in Marmara with 79% and Central Anatolia with 70%. The least urban population is 49% in the Black Sea region and 53% in the Eastern Anatolia. This situation reveals that the population growth in the cities, the migration from the rural settlements to the cities in our country, continues to grow in the rural population, and even the rural population begins to decline [14]. This indicates that energy consumption differs according to the typologies of the buildings [3].

Citherlet and Defaux have reduced insulation by about 50% in energy consumption by changing the insulation material thickness in residential buildings [10]. The Mitrarate and Vale studies have compared the insulated double wall constructed in New Zealand and the single-walled structures without insulation. They found that energy saving was 40% for the insulated double-walled structure [18]. In the study of Utama and Gheewala, the building constructed with clay and the clay used as the binder material in Indonesia was compared and the result was that the structure constructed with clay was less energy consumption than the cement. In Eskin N. study, the effect of different environmental conditions and building characteristics on the annual energy amount of the buildings was investigated. As a result, the effects of the building features, the building location, the climate zone in which the building is located, and the annual heating and cooling loads energy expenditure are presented [10].

The purpose of this study was to compare the energy performances of buildings with different locations and projects to determine energy performance changes. In addition, the energy identities of these buildings were assessed. It has been seen that the location and building typology of the buildings are being influenced by their energy performance. It has been found that the energy performance of houses built in different locations of a region with the same climatic data shows close affinity.

2 Methodology

The energy performances of the houses designed for 80-storey three-storey 80-story (total 1920 independent sections) duplex living units, each of which was built in a different neighborhood of Antalya province Korkuteli district, were investigated. The mean values of these constructions are given in Table 1 and the statistical values of the buildings are given in Table 2.

Table 1. The average energy performances

Energy performance conditions	Mean (kWh/year)
Primary heating	22.8
Primary cooling	165.2
Primary hot water	31.7
Primary lighting	28.5
Greenhouse gas emulsion	0.5953

Table 2. Housing typology that is widely constructed in Antalya–Korkuteli district

Building construction area (m^2)	Number of floors	Number of housing	Number of independent sections
435–707	3	80	4

Korkuteli district has a great development in the field of construction and settlement especially in recent years because it is located on the plateau of Antalya. Three-storey residential typologies examined in the study Three-storey residential areas consisting of four independent sections of approximately 235 m^2 are the most commonly used building typologies in the district.

2.1   Technical Specifications of Measuring Energy Efficiency Structures

The building construction characteristics examined in the study are as follows:

- Depending on the terrain of the terrain, terrain safety tension is determined as 15 t/m^2.
- Building design and planning operations were carried out as an III-A group within the scope of the Architectural and Engineering Services Specification which entered into force with the decision of the Council of Ministers dated 06.07.1985 and numbered 85/9707.
- The constructed buildings consist of the basement, floor, and normal floor. The average height of the building was 3 m.
- The bricks constituting the exterior walls of the buildings examined in the study were designated as W class vertical holes with dimensions of 29 × 19 × 25 cm^3 (7 N/mm^2 pressure strength) according to TS EN 771-1 [33].
- Vertical hole bricks of 29 × 19 × 13.5 cm^3 (10 N/mm^2 pressure resistance) according to TS EN 771-1 have been used for the interior spaces of the buildings.
- C25 was used from the concrete classes specified in TS 500 [27] and TS EN 206-1 [35] used in building construction. STIII steel grade is specified as concrete steel. Structures are made of reinforced concrete carcass.

- The energy performance is given in Figs. 2, 3, 4, 5, 6 and 7 along with the thickness of the floor and wall materials used in the measured buildings.

2.2   Geographical Structure of the Antalya–Korkuteli District

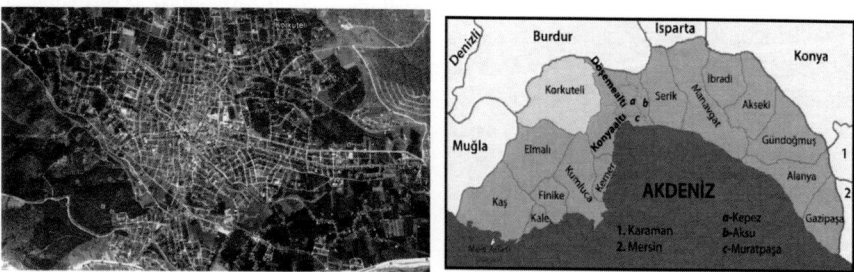

Fig. 1. The maps of the Korkuteli district

The geographical structure of Korkuteli/Antalya Mediterranean, in Antalya Korkuteli town in Muğla province, east Gölhisar and Çavdır towns in Burdur province, in the South Kumluca and Elmalı towns and in the North Bucak and Tefenni towns in Burdur province surrounded by these cities. The map and space image of Korkuteli is shown in Fig. 1.

The area is 2471 km^2. The height above sea level is 1020 m. It's climate is ¼ mediterranean climate ¾ of Lokes region. The cold weather comes from lakes of the region; the hot weather comes from Mediterranean region. The average temperature in winter is −5 °C and in summer 25 °C. A land where plains and hills admired in the beginning of Bey Mountains. There are 53 neighborhoods in Korkuteli and its population is 52,000.

2.3   Building Energy Simulation Program

Prepared December 5, 2008, by the Ministry of Public Works and Settlement (BEP-TR) is a web-based program. The building enables that annual energy consumption per square meter, CO_2 emissions, the datum of the reference values and comparison with a reference building A–G placed an inter-energy class operations. This program is also using the method of calculation of the energy performance of buildings (BEP-HY), all the parameters that affect energy consumption, the impact of energy efficiency and energy performance of buildings is used to determine the class [12, 39].

BEP-TR program enable that residences, office buildings, educational buildings, health buildings, hotels, shopping and commercial centers as well as assessment of the energy performance of new buildings and existing building typologies.

2.4 Energy Inputs

To calculate net energy requirements of buildings ventilation, lightening, heating. and cooling is mainly ın addition to these data; climate data, building geometry the ventilation and thermal properties of the building, building materials and definition of building components are important inputs.

This information, which is necessary for the energy analysis of the building to be revealed, is prepared by analyzing all the information about the building project and the regional location by the user who enters the data system [9]. This information is the result of the project engineer's architectural, static, mechanical and electrical files examined by the relevant engineer [15].

Table 3. The standards for methods used in Europe for building energy performance

Standard no	Standard name
TSE EN ISO 13790 [31]	Energy performance of buildings—calculation of energy use for space heating and cooling
TS EN ISO 13789 [32]	Thermal performance of buildings—transmission heat loss coefficient —calculation method
TS EN 15251 [37]	Indoor environmental input parameters for design and assessment of energy performance of buildings addressing indoor air quality, thermal environment, lighting and acoustics
TS 825 [34]	Thermal insulation requirements for buildings
TS EN ISO 14683 [28]	Thermal bridges in building construction—Linear thermal transmittance—simplified Methods and default values
TS EN ISO 10456 [29]	Building materials and products—Hygrothermal properties—Tabulated design values and procedures for determining declared and design thermal values
TS EN 12524 [36]	Building materials and products. Hygrothermal properties. Tabulated design values
BR 443 [1]	Conventions for U-value calculations
DIN V18599-2 [8]	Energy efficiency of buildings—calculation of the energy needs, delivered energy and primary energy for heating, cooling, ventilation, domestic hot water, and lighting—Part 2: energy needs for heating and cooling of building zones
TS EN ISO 13370 [30]	Thermal performance of buildings—Heat transfer via the ground— calculation methods

2.5 Building Energy Performance Calculation Methods

Standards for methods used in Europe for building energy performance calculation methods are given in Table 3 [6, 19, 21, 24].

3 Findings and Discussion

80 buildings which built in different areas of Korkuteli total 480 flats; have been researched for building energy performance. The program includes; external climate conditions for energy. indoor requirements evaluating all energy use of a building by taking into account bath local conditions and cost effectiveness, classifying in terms of primary energy and carbon dioxide emission, evaluating the feasibility of renewable energy (CO_2) saves, the control of heating and cooling systems and the restriction of greenhouse gas emissions Greenhouse gas emission valves have been calculated after the necessary data is entered to BEP-TR programmed in addition to heating, cooling, ventilation, and lighting.

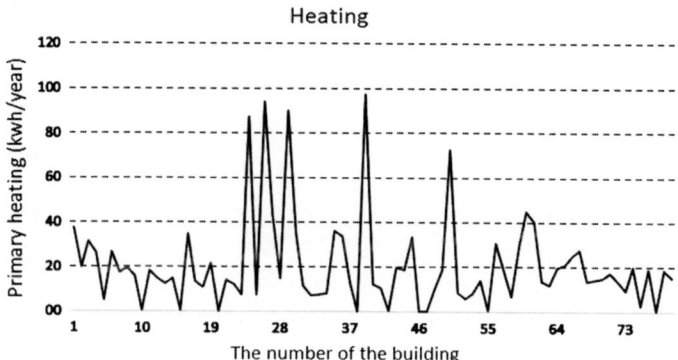

Fig. 2. Building energy performance heating results

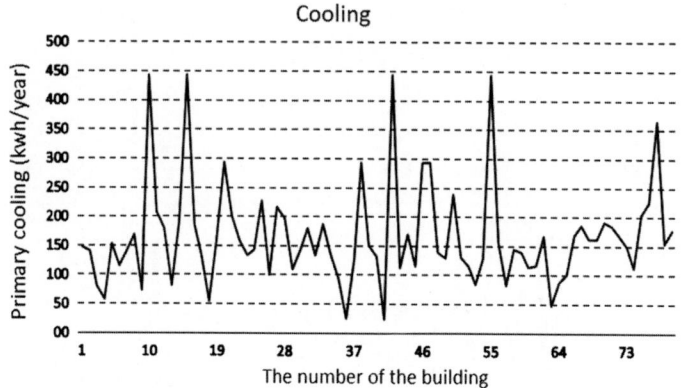

Fig. 3. Building energy performance cooling results

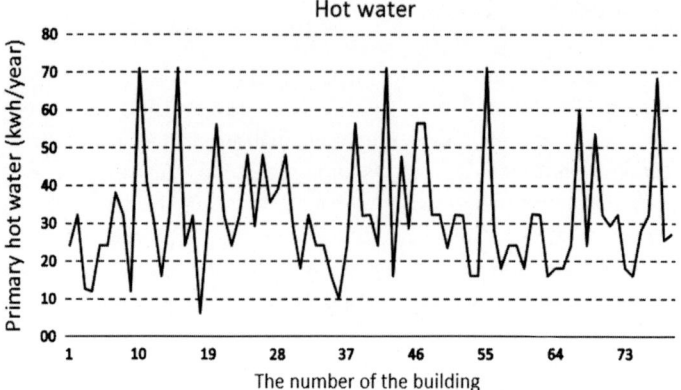

Fig. 4. Building energy performance hot water results

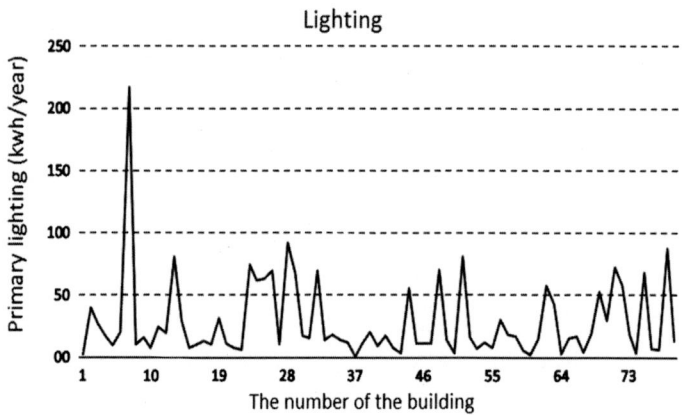

Fig. 5. Building energy performance lighting results

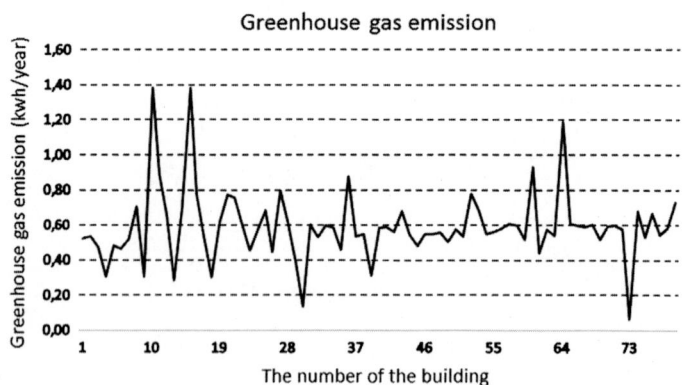

Fig. 6. Building energy performance greenhouse gas emission results

3.1 Findings

The heating, cooling, hot water, lighting and greenhouse gas emission valves of buildings which built in six neighborhood of province are evaluated in terms of the location where they move built. Annual heating data of investigated buildings in Fig. 2, cooling data in Fig. 3, hot water in Fig. 4, lighting in Fig. 5 and greenhouse gas emission in Fig. 6 are given in kWh.

According to Fig. 2, primary heating values of 480 dwellings are distributed according to usage area in m^2 basis. This distribution varies between 0.23 and 179.15 kWh/year. When examined in terms of the primary heating value used in all buildings, it is seen that 10% A, 42.5% B, 22.5% C, 13.8% D, 5% E, 3.8% found out.

According to Fig. 3, primary cooling values of 480 dwellings are distributed according to the usage area in m^2 basis. This distribution varied between 24.2 and 443.6 kWh/year. When examined from the perspective of the primary cooling value used in all buildings, it was found to be 2.5% A, 6.3% B, 75% C and 16.3% D.

According to Fig. 4, primary hot water values of 480 dwellings are distributed according to the usage area in m^2 basis. This distribution ranged from 6.2 to 71 kWh/year. When examined in terms of the primary hot water value used in all the buildings, 2.5% A and 97.5% B was found.

According to Fig. 5, primary heating values of 480 dwellings are distributed according to the usage area in m^2 basis. This distribution ranged from 0.6 to 217 kWh/year. When examined in terms of the primary lighting value used in all the buildings, 37.5% A and 62.5% B were found.

According to Fig. 6, greenhouse gas emulsion values of 480 dwellings are distributed according to the usage area in m_2 basis. This varies from 0.627 to 1.3841 kWh/ year, depending on the distribution. When the greenhouse gas used in all the buildings is analyzed in terms of emulsion value were found to be C class.

It has been determined that the buildings that have been evaluated have appeared for buildings having a construction area ranging from 70 to 700 m^2. Figures 6 and 7 shows that the difference between the values of the building appears to depend on the location of the building for the others, depending on the building m^2 for some buildings. In the multivariate analysis of variance over the values, it is found that the variables are independent of the usage field and the regression coefficient value is very low. The highest cooling value was found to be effective when comparing greenhouse gas emulsion, heating, cooling, hot water and lighting values. This is shown in Fig. 7.

Greenhouse gas emission values generally show close results. In some cases, it has been observed that differences arise from the calculation.

The energy performance values obtained are those calculated and officially used by energy ID certification experts. It is mandatory to obtain an Energy Identity Certificate to obtain a Building Permission Certificate.

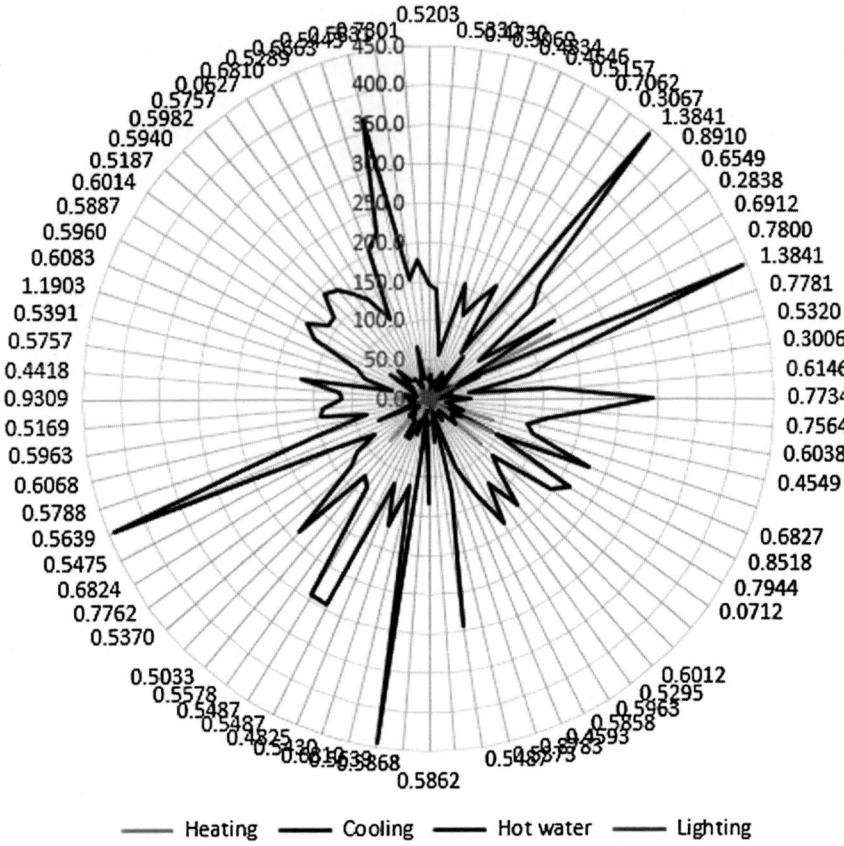

Fig. 7. The effect of greenhouse gas in all buildings

3.2 Discussion

The proposals that have been made in the context of the data obtained for the BEP-TR program in the province of Korkuteli in the province of Antalya and the greenhouse gas emulsion due to building energy performance;

- It has been observed that the energy performance values of a building with the same building form appear in the working direction. As a result of these differences, it is evaluated that the building characteristics are caused by introducing the values of energy identity experts to the BEP-TR program according to the climate conditions of Antalya. Especially in heating and cooling results, it is seen that when the terrestrial climate is evaluated as 3/4 of the terrain of the terrestrial terrain and the results of building energy performance used in practice are not considered to be suitable results. However, in order to calculate buildings energy performance values more appropriately, it is necessary to enter the buildings as data into the BEP-TR program in accordance with the climatic conditions. Especially in the heating and cooling values, it is more appropriate to enter the data and power of the system including near.

The BEP-Tr program accepts averages of energy performance scores in primary heating, cooling, hot water, lighting and greenhouse gas distribution. If at least three c-grades of these performance scores were found to be more effective, they would be more effective.

4 Results

In the study, 480 houses built in Antalya/Korkuteli province were found to have different energy certification results. Particularly in the energy performance evaluations, the calculations made by the experts of different energy identity documents have been interpreted that the heating, cooling and hot water systems of Antalya province are used for Korkuteli district.

- In the province of Korkuteli in Antalya province, building energy performance evaluations were made for a total of 480 apartments in the form of 80 apartments in total, and it was determined that the energy class of the buildings was "C Class" in total. This conclusion proves that buildings are an acceptable building in class C, energy classification, in the evaluation of Energy Identity Certificate.
- However, it has been observed that there are differences in the energy performance data of buildings that are close to building typology. It is envisaged that these result changes originated, in particular, in building m², the building location, and the methods that users follow during the calculation.
- It has been observed that climate and building location are important when energy identity certificate is determined. However, for the BEP-Tr program, it has been concluded that the location of the building cannot be introduced to the system as a whole in the project and that the result obtained by comparing it with the geometric shapes registered in the system may have a positive or negative contribution to the resulting values. Moreover, during the analysis evaluation, it is seen that the system can be registered to the system by comparing it with geometric shapes like rectangular, square, U or H rather geometrically according to the exact structure of the building. This situation is both time consuming for the user and the change in the geometry of the building because the results of the issue of reliability come to mind.
- Some building materials used in the construction of the BEP program are not included as an individual, so it is seen that nearby materials are used to get results. It is also assessed that this may have been the result.

References

1. Anderson B (2006) Conventions for U-value calculations. BRE Scotland, 44
2. Aykal FD, Gümüs B, Akça YÖ (2009) Sürdürülebilirlik kapsaminda yenilenebilir ve etkin enerji kullaniminin yapilarda uygulanmasi. V Yenilenebilir Enerji Kaynaklari Sempozyumu YEKSEM 9:19–22

3. Bansal D (2010) Embodied energy in residential cost effective units-up to 50 m². In: International conference on sustainable built environment (ICSBE-2010), Kandy, Sri Lanka, pp 13–14

4. Bansal D, Singh R, Sawhney R (2014) Effect of construction materials on embodied energy and cost of buildings: a case study of residential houses in India up to 60 m² of plinth area. Energy Build 69:260–266

5. Citherlet S, Defaux T (2007) Energy and environmental comparison of three variants of a family house during its whole life span. Build Environ 42:591–598

6. De Wilde P (2014) The gap between predicted and measured energy performance of buildings: a framework for investigation. Autom Constr 41:40–49

7. Debnath A, Singh S, Singh Y (1995) Comparative assessment of energy requirements for different types of residential buildings in India. Energy Build 23:141–146

8. DIN (2007) Energy efficiency of buildings—calculation of the energy needs, delivered energy and primary energy for heating, cooling, ventilation, domestic hot water and lighting —Part 2: Energy needs for heating and cooling of building zones German Institute for Standardization, 110

9. Durmuş G, Önal S (2014) Uluslararası Standartlarında İnşa Edilen Yapının Enerji Kimliğinin Belirlenmesi: Gaziantep Örneği. EJOSAT: European Journal of Science and Technology. Avrupa Bilim ve Teknoloji Dergisi 1:43–51

10. Eskin N (2011) Konut Dışı Binaların Yıllık Enerji İhtiyaçlarının İncelenmesi. İTÜ, İstanbul, 1–6

11. Gültekin AB, Gülser Ç (2016) Proposal of a model for assessment of the environmental impacts of construction products within the context of life cycle assessment methodology. Düzce Üniversity Bilim ve Teknoloji Dergisi, 4

12. Harputlugil GU (2014) The tools assessment of the building energy performance-energy simulation. Install Eng 144:9

13. Huberman N, Pearlmutter D (2008) A life-cycle energy analysis of building materials in the Negev desert. Energy Build 40:837–848

14. Internet (Date of acces: 02.01.2017) Türkiye'de Kırsal Nüfus ve Kentsel Nüfus

15. Internet (Date of acces: 02016) Energy performance building regulations (BEP-tr). ministry of environment and urban planning

16. Keskin T (2010) Assessment of current status of building sector report

17. Koroneos C, Kottas G (2007) Energy consumption modeling analysis and environmental impact assessment of model house in Thessaloniki—Greece. Build Environ 42:122–138

18. Mithraratne N, Vale B (2004) Life cycle analysis model for New Zealand houses. Build Environ 39:483–492

19. Olesen BW (2007) The philosophy behind EN15251: indoor environmental criteria for design and calculation of energy performance of buildings. Energy Build 39:740–749

20. Özyurt G, Karabalık K (2009) Enerji Verimliliği, Binaların Enerji Performansı ve Türkiye'deki Durum. TMMOB, İnşaat Mühendisleri Odası, Türkiye Mühendislik Haberleri 457:32–34

21. Petersdorff C, Boermans T, Harnisch J (2006) Mitigation of CO_2 emissions from the EU-15 building stock. Beyond the EU directive on the energy performance of buildings (9 pp). Environ Sci Pollut Res 13:350–358

22. Ramesh T, Prakash R, Shukla KK (2010) Life cycle energy analysis of buildings: An overview. Energy Build 42:1592–1600

23. Report NGGI (Date of Acces: 2015) Annual report submission under the frame convention on climate change

24. Rey F, Velasco E, Varela F (2007) Building energy analysis (BEA): a methodology to assess building energy labelling. Energy Build 39:709–716

25. Shukla A, Tiwari G, Sodha M (2009) Embodied energy analysis of adobe house. Renew Energy 34:755–761
26. Talakonukula R, Ravi P, Karunesh KS (2013) Life cycle energy analysis of a multifamily residential house: a case study in Indian context. Open J Energy Effic
27. TSI (06.11.2013) TS500, Requirements for design and construction of reinforced concrete structures. Turkish Standards Institute, 79
28. TSI (09.04.2009) EN ISO 14683, Thermal bridges in building construction- Linear thermal transmittance- Simplified method and default values. Turkish Standards Institute, 35
29. TSI (09.04.2009) EN ISO 10456, Building materials and products—Hygrothermal properties—Tabulated design values and procedures for determining declared and design thermal values (ISO 10456:2007). Turkish Standards Institute, 41
30. TSI (09.04.2009) EN ISO 13370, Thermal performance of buildings—Heat transfer via the ground—Calculation methods Turkish Standards Institute,:61
31. TSI (13.03.2013) EN ISO 13790, Energy performance of buildings—calculation of energy use for space heating and cooling. Turkish Standards Institute, 141
32. TSI (13.03.2013) EN ISO 13789, Thermal performance of buildings—transmission heat loss coefficient- Calculation method. Turkish Standards Institute, 30
33. TSI (13.12.2011) EN 771-1, Specification for masonry units—Part 1: Clay masonry units. Turkish Standards Institute, 49
34. TSI (18.12.2013) TS 825, Thermal insulation requirements for buildings. Turkish Standards Institute, 83
35. TSI (19.04.2002) EN 206-1, Concrete- Part 1: Specification, performance, production and conformity. Turkish Standards Institute, 68
36. TSI (21.11.2000) EN 12524, Building materials and products- Hydrothermal properties- Tabulated design values. Turkish Standards Institute, 12
37. TSI (31.01.2008) EN 15251, Indoor environmental input parameters for design and assessment of energy performance of buildings addressing indoor air quality, thermal environment, lighting and acoustics. Turkish Standards Institute, 44
38. Wagner H, Mathur J, Bansal N (2007) Energy security climate change and sustainable development. Anamaya Publishers, New Delhi
39. Yaka İF, Öna LS, Koçer A, Güngör A (2016) Comparison of building energy performance of different provinces. J Glob Eng Stud 3:8

A Dynamic Thermal Simulation in New Residential Housing of Lakhiayta City in Morocco

Karima El Azhary[1], Najma Laaroussi[1(✉)], Mohammed Garoum[1],
Khalid El Harrouni[2], and Majid Mansour[2]

[1] Materials Energy and Acoustics Team (MEAT), Mohammed V University in
Rabat, EST de Salé, Salé, Morocco
najma.laaroussi@um5.ac.ma

[2] Ecole Nationale d'Architecture, Medinat al Irfane, Rabat, Morocco

Abstract. The aim of this study is to get the most optimal energy efficient heating and cooling material combinations that would result to an energy saving buildings. The regulatory requirement limits of thermal characteristics of residential building's envelope were stated by the TRCM. It is mandatory to respect the thermal transmittance, also known as U-value of external walls that should not exceed 1.20 W/m^2 K and the annual savings in heating and cooling range is 40 kWh/m^2/year in the climate area represented by Casablanca. The simulation results were analyzed to find the combination of parameters yielding the lowest energy consumption and to define potential energy savings for this residential building located in this climate zone.

Keywords: Building simulation · Heating and cooling · Comfort Thermal regulation

1 Introduction

The residential energy consumption remains high and it's increasing overtime in Morocco due to the growth of population and the urbanization [1]. The Moroccan government establishes an ambitious program to reduce residential energy consumption and carbon emissions by building new smart houses. The program aims to achieve 1.2 Mt/year of energy consumption reduction in the building sector by 2020 and decreasing greenhouse gas (GHG) emissions by about 4.5 Mt/year. Therefore, the Moroccan government has introduced a thermal regulation for building envelopes. The implementation of the thermal regulation of construction in Morocco (TRCM) focuses on improving thermal performance and reducing the requirement of buildings in heating and cooling [2].

The need for reducing the energy consumption used for heating or cooling buildings has motivated many studies with the aim at improving the thermal performance of construction. The field of construction and building materials and their application in new works provides an international dissemination of research and development. The description and resulting graphics for a Moroccan public administration building and the complete description of all buildings and simulation results can be found in [3].

© Springer International Publishing AG, part of Springer Nature 2018
S. Fırat et al. (eds.), *Proceedings of 3rd International Sustainable Buildings Symposium (ISBS 2017)*, Lecture Notes in Civil Engineering 7, https://doi.org/10.1007/978-3-319-64349-6_3

The effect of the insulation of the roof and the facades on the cooling energy of a typical house villa situated in hot and arid climate zone like Marrakech was studied in [4]. The results showed that the insulation of the roof is essential for the Marrakech climate as it resulted in reducing the needs for cooling of almost 40%.

Sick et al. [5] used dynamic building simulations performing sensitivity analysis for the influence of various parameters on the heating and cooling demands of different building types in Morocco. In [6], the effect of using three different external wall constructions on the energy consumption, energy cost and the comfort in Egypt. The simulations results showed different performance for each specification across the climatic zones. Gueddouch et al. [7, 8] have developed a dynamic model of the building based on equivalent circuit model with the aim of evaluating the energy consumption of a typical building in Morocco.

Benhamou and Bennouna [9] carried out an experimental study of the thermal performance of an underground heat exchanger for the air conditioning of a well-insulated house located outside of the city of Marrakech. The coupling of a direct thermal calculation with an optimization algorithm to achieve the identification of the thermal characteristics of a building structure was presented in [10], the proposed optimization model was verified and validated against experimental results obtained from a wooden structure with a heated wall.

The renovation of social housing which provides an interesting opportunity for reducing energy consumption and increasing the comfort of residents has been studied by Gagliano et al. [11]. The improvement of thermal insulation of the building envelope and the use of renewable energy sources have been proposed and have led to energy savings and increased interior comfort. For various cities in Morocco, Guechchati et al. [12] studied the impact of passive design parameters on energy consumption for annual cooling and heating in villas according to regions.

Thus, the energy consumption in Morocco's residential sector is growing. The most efficient way to satisfy this demand on energy is to introduce energy-efficiency measures in new buildings.

In this work, a series of energy simulations were performed for the residential housing located in new city (Lakhiayta) in the suburbs of Casablanca. The residential building was simulated as a case study, by using DesignBuilder [13] by applying different insulations and wall compositions in comparison with the ordinary walls using conventional materials in this climatic zone.

2 Methodology and Parameterization

The studied building is being investigated and modelled using the simulation software DesignBuilder [13], which is the transient simulation of systems, including multi-zone buildings to develop comfortable and energy-efficient building designs from concept to completion. The simulations concern the building and its equipment, including control strategies, occupant behavior and energy systems, thus allowing the correct calculation of heat transfer coupled with thermal storage within the building structure like walls and ceilings.

The building type is a simple standard multi-storey residential building and as shown in Fig. 1. The model of apartments is schematized in the Fig. 2.

Fig. 1. The view of residential housing (R + 6)

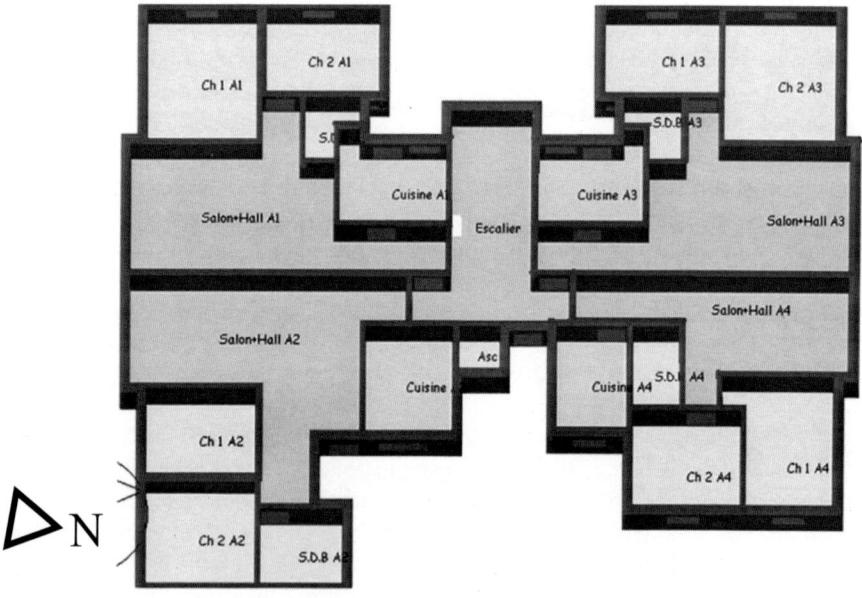

Fig. 2. The typical plan for a flat in the building

The simulation model is made up of 14 thermal zones, two zones for each one of the upper floors representing the four apartments with 60 m^2 of surface, one zone modelling the ground floor is reserved for commercial stores and parking lots and one zone for the staircase.

We estimated that about 5 people live in an apartment. The construction design is documented in Table 1 according to building practice in Morocco. Table 2 gives the U-value corresponding to variation of thickness of polystyrene. The characteristic areas and the volume of building are summarized in Table 3.

Table 1. The design characteristics of construction

Building element	Without insulation U (W/m^{2}K)	With insulation U (W/m^{2}K)	U_{TRCM} (W/m^{2}K)
External wall	1.96	1.33–0.97 (Table 2)	$U \leq 1.20$
Roof	2.04	0.54	$U \leq 0.75$
Floor plate	1.60	1.60	No requirement
First floor intermediate	2.13	0.4	No requirement
Window-to-wall ratio (WWR)	15%		
Windows	Simple glazing, g = 0.86 U = 5.89 (W/m^{2}K)	Double glazing, g = 0.7 U = 3.09 (W/m^{2}K)	$U \leq 5.8$
Orientation of building	Sud-Ouest		

Table 2. The U-value corresponding to variation of thickness of polystyrene

Insulation thickness (cm)	2	3	4	5	6
External wall U (W/m^{2}K)	1.33	1.23	1.12	1.05	0.97

Table 3. The characteristic sizes of construction

Orientation	South	West	North	East	Sum
Gross area (m^2)	300	362.1	259.5	362.1	1283.7
Window area (m^2)	48.7	43.5	48.7	47.7	188.7
WWR facade (%)	16.23	12	18.76	13.17	≈15
Floor area (m^2) without staircase	1426.4	Volume (m^3) without staircase		3993.92	

The Table 4 describes the simulated interior conditions of the building. The dwellings are used by five persons from 17:00 to 07:30 and by 3 persons otherwise as shown in Table 4. Ventilation rate including infiltration is 36 m^3/h per person, corresponding to an air change rate between 1 and 0.4 h^{-1}. All internal gains sum up to 4533 kWh per dwelling per year, evenly distributed. External shading (50%) is used during summer.

Table 4. The interior conditions as simulated for the building

Interior conditions	Week		Week-end	
	7:30–17:00	17:00–7:30	7:30–17:00	17:00–7:30
Occupancy (persons)	3	5	5	5
Ventilation rate including infiltration	Distributed according occupancy			
External shading during summer	50%		50%	
Interior gains sum up	Distributed according occupancy			
Set point temperature heating (°C)	20			
Set point temperature cooling (°C)	24			

3   Results

The parameter variations of the building simulations are the insulation and the type of windows. The construction method used in the simulations was modified by adding layers from 2 to 6 cm of insulation external walls, without changing the construction technique. The increasing of the thickness of the insulating layer of polystyrene decreases the U-value of external walls in the building as shown in Table 2.

3.1   Prescriptive Approach

The minimum technical specifications for thermal performance of buildings in Morocco were set by the TRCM according to the climatic conditions [2]. Then, the prescriptive approach of TRCM was setting technical specifications to be respected by climate zone (roofs, exterior walls, floors and windows).

The recommended setting for zone 1 with a 15% WWR is given in Table 1 and the maximum values to be respected are presented. When introducing the thermal parameters of the envelope referred in Table 1, it shows non-compliance for the not insulated building in the reference case and for the insulated external walls with 2 and 3 cm of thickness (Table 2).

3.2   Performance Approach

The performance approach was setting maximum limits for heating and cooling (kWh/m^2/year) in relation to internal reference temperature (20 °C for heating and 24–26 °C for cooling) [2].

The climate data for the studied Moroccan location were used in this project. The data are available in the software DesignBuilder. The results are displayed as the annual delivered energy for heating, cooling, lighting, cooking and specific electricity (refrigerator, TV, computer).

The following results are generated for the studied building:

– The specific heating and cooling energy demands for heating to 20 °C and cooling to 24 °C room temperatures, respectively.

– The solar heat gains for building in kWh/m² for year.
– The number of hours with room temperature exceeding 24 °C during summer without cooling and lower than 18 °C during winter without heating.

In order to evaluate the heating and cooling loads of the building the set points are fixed to 20 and 24 °C, respectively.

The total energy consumption obtained for the reference case is 53.94 kWh/m²/year, exceeding the requirement of the TRCM which is 40 kWh/m²/year (for Zone 1). This consumption includes heating and cooling energy.

The most dominating consumption is cooling energy, about 45.06 kWh/m²/year. However, the heating energy is very low (8.08 kWh/m²/year). It's clear that the cooling load is a significant part of total consumption as shown in Table 5.

Table 5. The specific heating and cooling energy of the building

Insulation thickness (cm)	Annual solar heat gain (kWh/m²)	Annual cooling energy demand (kWh/m²)	Annual heating energy demand (kWh/m²)	Total annual consumption energy (kWh/m²)
Without insulation (reference case)	Simple glazing 33.68 (46.22)	45.06	8.88	53.94
2	Double glazing 33.68 (27%)	38.05 (16%)	4.03 (55%)	42.08 (22%)
3		37.06 (18%)	3.55% (60%)	40.61 (25%)
4		36.50 (19%)	3.09 (65%)	39.59 (27%)
5		36.10 (20%)	2.81 (68%)	38.28 (29%)
6		35.21 (22%)	2.49 (72%)	37.70 (30%)

There is significant difference in cooling energy demand between the reference case and the case using insulation in building. Although, the observed difference was between 55 and 72% for heating energy consumption and not exceeding 22% for cooling energy demand.

The results in Fig. 3 show that the use of the double glazing reduced the solar heat gains by 25%.

The monthly indoor air temperature in the conditioned building is ranging between 19 and 24 °C with the external temperature disorders reaching the peak on July and the lowest level on January as shown in Fig. 4.

The unconditioned building is ventilated during summer when the ambient air temperature is lower than the room air temperature. The difference between the indoor temperature and the external dry bulb temperature is varying and is at the maximum difference on July.

Applying the thermal insulation for building envelope, the results in Fig. 5 show that a significant decrease in heating and cooling demand has been achieved compared to the reference case. The insulation is of major importance in heating energy demands exceeding 55%. The effect on cooling energy demands is small and is reduced by approximately 22% when adding an insulation layer of 6 cm thickness.

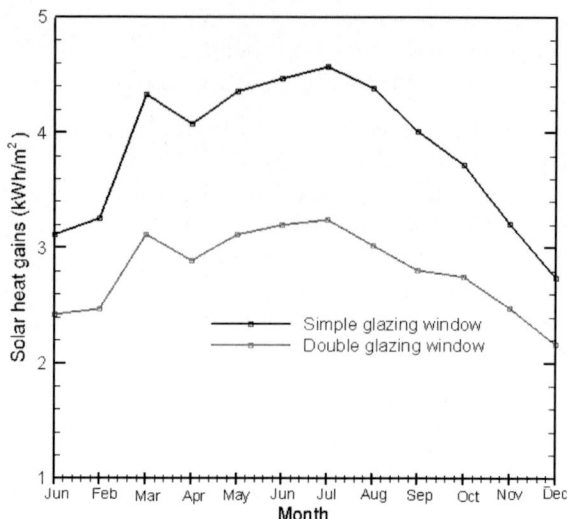

Fig. 3. Comparison of solar heat gains in the building for used of simple and double glazing

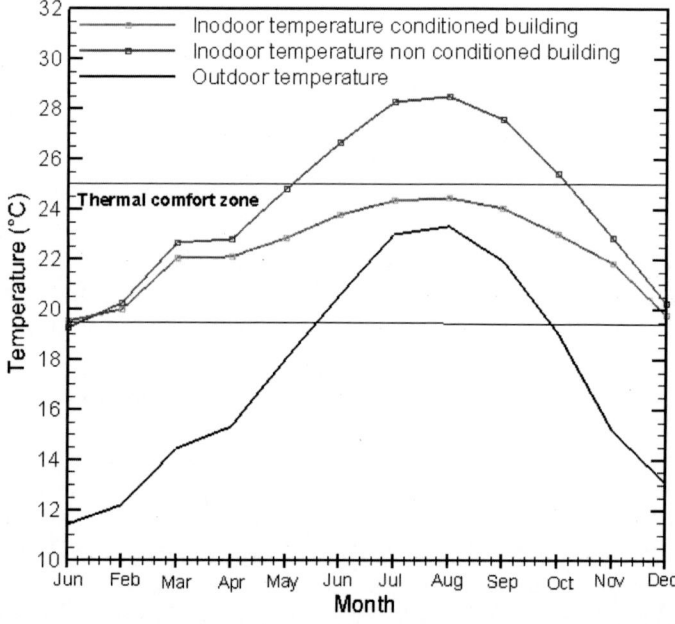

Fig. 4. Mean monthly air temperatures for interior and exterior of the building

By turning off heating and cooling in the simulation, we can count the hours beyond certain room temperatures considered to limit the comfort region.

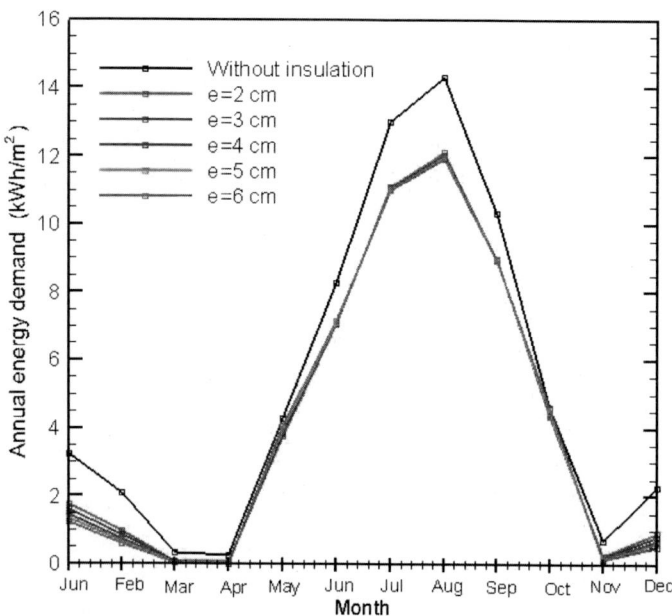

Fig. 5. Annual sum of the cooling and heating energy consumption in building for different thickness of insulation of the external walls

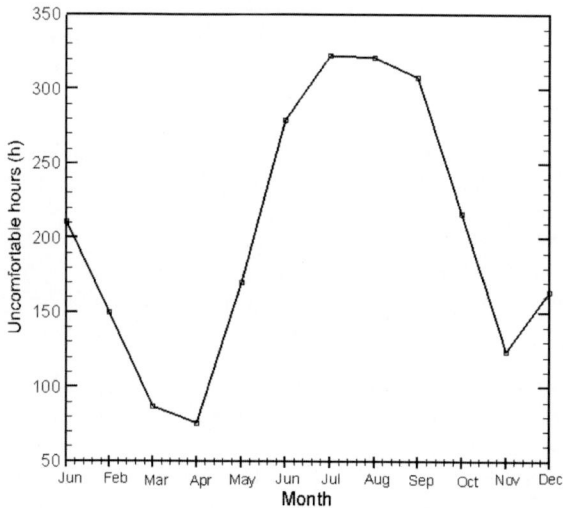

Fig. 6. Monthly number of uncomfortable hours without heating and cooling

The results are shown in Fig. 6. It is observed that Casablanca regions may show high cooling and little heating demands. The observation of the unconditioned building in Fig. 6, reveals the numbers of hours with uncomfortably low or high temperature, respectively in good accordance with heating and cooling demands in the conditioned building (Fig. 5).

4 Conclusion

In this paper, dynamic simulations were carried out with DesignBuilder software on a typical building located in the Casablanca region. The simulations resulted in a significant reduction which reaches 30% of the final total energy consumption for heating and cooling energy compared to the reference case. Results indicate that thermal regulation of buildings in Morocco has a considerable effect on heating and cooling load.

The insulation parameters have the greatest influence in the heating energy demand which is reduced until 72% for 6 cm of insulation of the external walls.

Due to Moroccan regulation, the energy consumption of buildings has to be reduced of the current and this objective could be reached if we build based on these principles and adapting existing buildings.

The regulation aims at allowing improvement to the thermal performance of the building envelope with only little extra costs.

References

1. Ministère de l'Habitat, de l'Urbanisme et de l'Aménagement de l'Espace: Immobilier 2020: Eléments de prospective. N° du dépôt légal: 2011 MO 2633, ISBN:978-9954-30-538-6, p 178
2. Thermal Regulation of Construction in Morocco (TRCM), National Agency for the Development of Renewable Energy and Energy Efficiency (ADEREE) (2014)
3. Sick F, Schade S (2010) Report on TRNSYS Simulations for establishing a Morocco Building Energy Code, Report prepared by order of Deutsche Gesellschaft für Internationale Zusammenarbeit (GIZ) GmbH, D-65760 Eschborn, Germany
4. Sobhy I, Brakez A, Benhamou B (2014) Modélisation dynamique d'une maison typique à Marrakech et propositions pour améliorer ses performances énergétiques, 3 ème AMT, Agadir, Morocco
5. Sick F, Shade S, Mourtada A, Uh D, Grausam M (2014) Dynamic building simulations for the establishment of a Moroccan thermal regulation for buildings. J Green Build 9:145–165
6. Fahmy M, Mahdy MM, Nikolopoulou M (2014) Prediction of future energy consumption reduction using GRC envelope optimization for residential buildings in Egypt. Energy Build 70:186–193
7. Gueddouch T, Saad A, Hmidat A (2016) Analysis of the influence of Moroccan thermal regulation on energy consumption of a typical apartment located in Morocco. Int J Appl Eng Res 11(5):3458–3461 (ISSN: 0973-4562)
8. Gueddouch T, Saad A (2014) Modeling of a building with the aim of the evaluation of its energy consumption: application to a typical building in morocco. Int J Comput Sci Eng 2 (2):1–5
9. Benhamou B, Bennouna A (2013) Energy performances of a passive building in marrakech: parametric study. Energy Procedia 42:624–632
10. Bouache T, Ginestet S, Limam K, Lindner G, Bosschaerts W (2013) Identification of thermal characteristics of a building. Energy Procedia 42:280–288

11. Gagliano A, Nocera F, Patania F, Capizzi G (2013) A case study of energy efficiency retrofit in social housing units. Energy Procedia 42:289–298
12. Guechchati R, Moussaoui MA, Mezrhab A (2014) Evaluation des besoins d'une bâtiment dans différentes ville du Maroc, 3ème Congres de l'Association Marocaine de Thermique, AMT2014, Agadir, Maroc, 21–22 Avril
13. DesingBuilder software, version 3

A Literature Survey on Integration of Wind Energy and Formal Structure of Buildings at Urban Scale

Serpil Paltun[1] , Arzuhan Burcu Gültekin[2](✉) , and Gülser Çelebi[3]

[1] Department of Architecture, Faculty of Architecture, Gazi University, Ankara, Turkey
serpil.paltun@csb.gov.tr
[2] Department of Real Estate Development and Management, Faculty of Applied Sciences, Ankara University, Ankara, Turkey
arzuhanburcu@yahoo.com
[3] Department of Interior Architecture, Faculty of Architecture, Çankaya University, Ankara, Turkey
gulsercelebi@gmail.com

Abstract. Providing the needed and indispensable steady, quality and safe energy is one of the most important issues today. Wind energy is one of the most important renewable energy source. Wind energy has found uses much more in water pumping and obtaining electricity in rural areas until last years. Today, wind energy has taken its place in the energy sector as an alternative source of energy production. High-density building arrays within a city, the combination of indoor and outdoor spaces with different purposes effect wind flow and acceptable wind comfort. In urban areas, wind energy and wind comfort are important requirements. Not only in new urbanisation areas but also for existing urban areas and city centers acceptable wind comfort plays an important role among and around buildings. When viewed from this angle, the aim of this study is enlightening the building aerodynamics, wind effect and wind energy in urban environment, and also giving information about how to analyze the wind comfort and design criteria in dense urban areas.

Keywords: Wind energy · Wind comfort · Building's formal structure

1 Introduction

Displacement of air mass due to differences in temperature and pressure in the atmosphere is defined as the wind. Wind energy that is transformation of solar energy is the conversion of kinetic energy of air mass to mechanical energy [1].

With the urbanization process, the more number of long and narrow streets surrounded by buildings is being to be seen in the construction area of largest cities. The development of these street canyons has brought out the problems related to pollution in recent years. These areas at the pedestrian level called space generally acts as external living space. Therefore, in the plan and design process, to ensure the

© Springer International Publishing AG, part of Springer Nature 2018
S. Fırat et al. (eds.), *Proceedings of 3rd International Sustainable Buildings Symposium (ISBS 2017)*, Lecture Notes in Civil Engineering 7,
https://doi.org/10.1007/978-3-319-64349-6_4

acceptable wind environment at the pedestrian level areas, it is very important to evaluate the ventilation potential and performance [2].

Traditionally, analysis and assessment of wind environment firstly based on wind tunnel tests. Today, numerical modelling (Computational Fluid Dynamics-CFD) that offers the advantages of computer capability is widely accepted. If optimization design will be carried out and if the effect of environmental impact on urban development will be assessed, owing to the fact that these test are very expensive and they require long time, numerical modeling is preferred to wind tunnel testing [3].

Over the past few decades, wind tunnels has emerged as an empirical tool to simulate the natural wind and to identify the wind power acting on the structure in the simulated air flow using models of small scale. The boundary layer wind tunnel model test has become the most reliable method to identify the wind-induced loads and structural response to these loads of high-rise buildings. With the development of advanced data collection, correct aerodynamic model test is possible with widely accepted high frequency force balance technique and instantaneous wind pressure integration. Both of these aerodynamic testing techniques has the advantage of the usage of lightweight but relatively rigid building model [4].

In recent years, the use of fluid dynamics (CFD) in the wind engineering has progressed rapidly. Currently this approach, that is known as the computational wind engineering is recognized as one of the most important research areas. The rapid growth of computational wind engineering applications is expanding the scope of wind engineering as shown in Fig. 1 and covering a variety of phenomenon from micro-climate around the human body to medium-scale climate in urban areas [5].

In computational wind engineering, a wide range of issues has studied with dif-ferent perspectives. In the study of Murakami et al. [5], applications that are related to human space around buildings and environmental problems including the following conditions is mentioned.

- Wind environment in the building complex.
- Thermal and dynamic effects of the wind on the human body.
- Pollutant distribution around a building.
- Thermal comfort in outdoor [5].

In many studies, wind load analysis are made for combination of high-rise, low-rise or high-rise and low-rise buildings and different results obtained by using experiment and computational fluid dynamics (CFD) environment. Roberson and Crowe [6] studied the pressure distribution in a building for turbulent flow conditions experi-mentally. Ahmad and Kumar [7] studied the wind load on the low-rise buildings. And they examined the effect of geometry on pressure for low-rise buildings. Aygün and Başkaya [8] examined the surface pressure formed by wind flow around a high-rise building. Mendis et al. [9] studied the wind load on high-rise building experimentally and numerical according to the Australian conditions. Holmes et al. [10] searched the wind loads of high, medium and low-rise buildings for 15 different region in Asia-Pacific region.

Şafak [11] mentioned the main approaches and assumptions of static and dynamic load calculations for wind load. Huang et al. [12] made the numerical analyses of wind loads on high-rise buildings with steel construction by Computational Fluid Dynamics

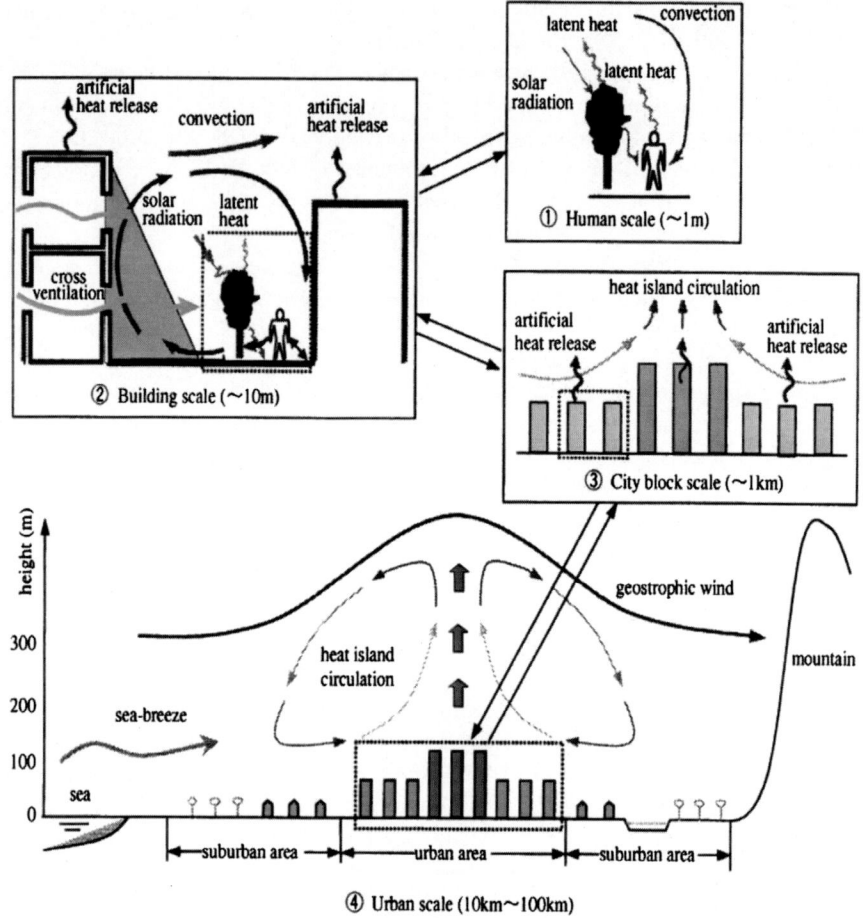

Fig. 1. Various scales related to wind climate [5]

(CFD). Liang et al. [13] studied the wind loads which effect the dynamic torsional loads on high and rectangular buildings experimentally. Huang and Chen [14] studied the wind and static loads on high-rise buildings on the basis of co-frequency pressure measurements. Tominaga et al. [15] studied the wind effects around buildings in a particular region by CFD method. Huang et al. [16] examined the RANS method analyses and kinematic simulation of wind load for high-rise buildings. Cheung and Liu [17] made the CFD analyses of ventilation process of high-rise buildings in their studies. Blocken et al. [18] examined the buildings of Eindhoven University of Technology by modelling them in CFD environment, and studied the wind loads by CFD analyses [19].

Recently, the prediction of wind environment around a high-rise building, using CFD technique, was carried out at the practical design stage. In view of this point, a working group for CFD prediction of wind environment around buildings, which consisted of researchers from several universities and private companies, was organized by the Architectural Institute of Japan (AIJ). During initial stage of that project, the working group carried out cross comparisons of CFD results for flow around a single high-rise building, building block placed within the surface boundary layer and flow within a building complex in urban area. Fig. 2 illustrates the six (a–f) test cases for these cross comparisons. They were carried out to clarify the major factors which affect the prediction accuracy [20].

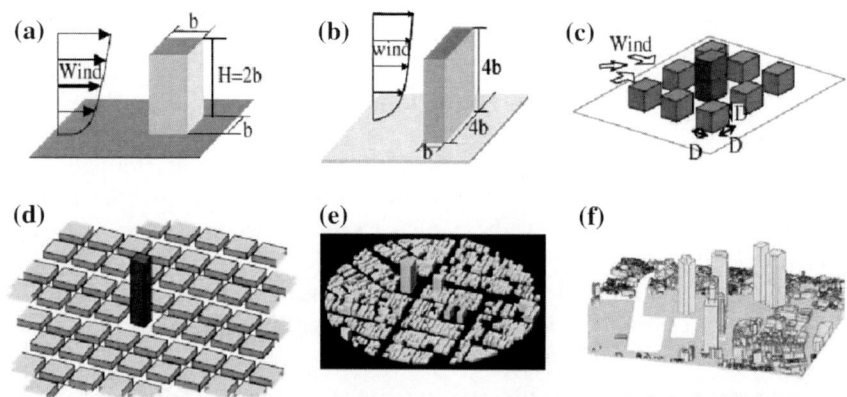

Fig. 2. Six test cases for cross-comparison. **a** The test case A (2:1:1 square prism); **b** the test case B (4:4:1 square prism); **c** the test case C (Simple city block); **d** the test case D (A high-rise building in the city); **e** the test case E (building complexes in actual urban areas (Niigata)); **f** the test case F (building complexes in actual urban areas (Shinjuku)) [20]

Because of the reasons like, reduction of surface evaporation and green spaces in cities, increasing the concrete, asphalt surfaces and structural field, meteorological parameters are changed and this cause climate change at local and regional scale. Therefore, big cities are becoming venues for their unique climate.

Increase of high-rise buildings, reduction of green space in urban areas affects the city's aerodynamics in our country, it is an important requirement that establishing a set of design criteria and standards for building wind safety and comfort and living comfort of people in urban environment.

2 Methodology

The aim of this study is to reveal the literature studies about the wind, wind energy and effects of the different form and settlement of building and external environment elements on the building aerodynamics, and also it is believed that this study will enlighten the procedures how to analyze the wind effect on the buildings in urban scale. Accordingly; first of all, introduction about the study will be given, wind energy and integration of wind energy and building aerodynamics will be explained. After that, results of the study will be explained.

3 Literature of Wind and Wind Energy

Starting from worldwide, when it is considered the causes of the formation of air movement or wind, it is seen that, they are their thermal or dynamic based pressure system, coriolis force formed by the earth's rotation, friction force formed by the roughness of earth and viscosity connected to the air flow [21]. As it is seen from Fig. 3 the records of anemograph, wind's direction and intensity is always variant. This variation feature seen in the wind's direction and intensity is called as turbulence.

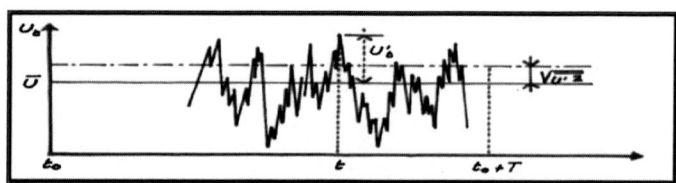

Fig. 3. An example of a anemograph record that measures the wind speed [22]

The buildings that create the urban open space make the same influence of the topography in open rural areas. There occurs a difference between flow types in open rural areas and flow types in urban areas, also wind or turbulent air movements ceased to properly pass the turbulent situation. Thus, type of air movement in urban open space scale is formed again in connection with geometric features as seen at Fig. 4 [22].

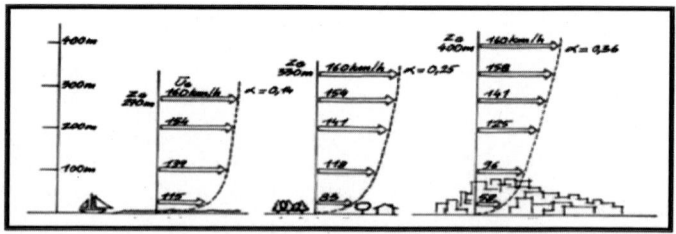

Fig. 4. Earth in the boundary layer wind speed gradients change depending on the configuration of built environment characteristics [22]

Depending on developments in the economic and social fields, based on the fact that supplying the growing energy needs of approximately 4–5% per year worldwide from fossil and nuclear sources that have harmful effects to the environment, it is need to benefit from wind energy that is clean, environmentally friendly and a local source [23].

As a result of the research it is said that the first use of wind energy in the historical process is conducted in the civilizations of China, Tibet, India and Afghanistan. The first written information about the use of wind turbines belong to the simple made horizontal-axis wind turbines in B.C. 200–300 years by Great Alexander. There are information about the use of vertical-axis wind turbines in B.C. 700s in the Persian. From Asia to Europe in the 10th century, the use of wind power developed by western countries. The first wind power, is produced by the Danish Professor Paul La Cour in 1891 [24].

Concerns about the effects of wind on the building dates back to the early human settlements. Design principles that is laid down in the design of many ancient cities as it can be understood from the coming to the fore again like principles of Feng-Shui, used for the development of ancient Chinese dynasties (wind and water), it appears to remain valid today [22].

From the summarize of the book of Aynsley, Melbourne and Vickery named Architectural Aerodynamics that constitutes an important resource about building aerodynamics; Aristotle BC 4th century in "Meteorologik" first time to discuss the mysterious winds, his student Theophratus' weather forecast methods, the ancient Greek who give the names of wind gods to the wind direction in the wind rose transfer us the importance of this issue. The principles that laid out in the book of "Ten Books on Architecture" by The Roman architect and engineer Vitruvius, in the B.C. 1st century spread to other European countries in 15th century. As well as the one who went to Europe in 1573 developed the city planning laws for Spanish cities in the South and Central America, also in world's various climatic zones, Japan, Canada, India, and other similar buildings and culture of the city is seen as designed according to local wind [22].

After the Industrial Revolution, in the city planning, in 1890 Europe, Nuremburg, taking into account the natural lighting in every room, in 1874 Sweden laws say that light and air protect is necessary to protect the health, in order to move the fumes from factories out of the city of Vienna in the 1900s, the prevailing wind is taken into account in city and regional planning. In contemporary practice of urban planning principles similar experimental, parallel to the development of digital design techniques, there are many examples [25].

4 Integration of Wind Energy and Building Form (Building Aerodynamics)

One of the main topics of building aerodynamics is revealing the characteristics of air flow that occurs in or around the building within the urban texture. As it is seen from Fig. 5, the air molecules hitting the wind above surface of the building stop when they hit the surface, change the direction by licking the surface and finally form the trace region behind the building following side surfaces by being separated from the surface of the breaking point [22].

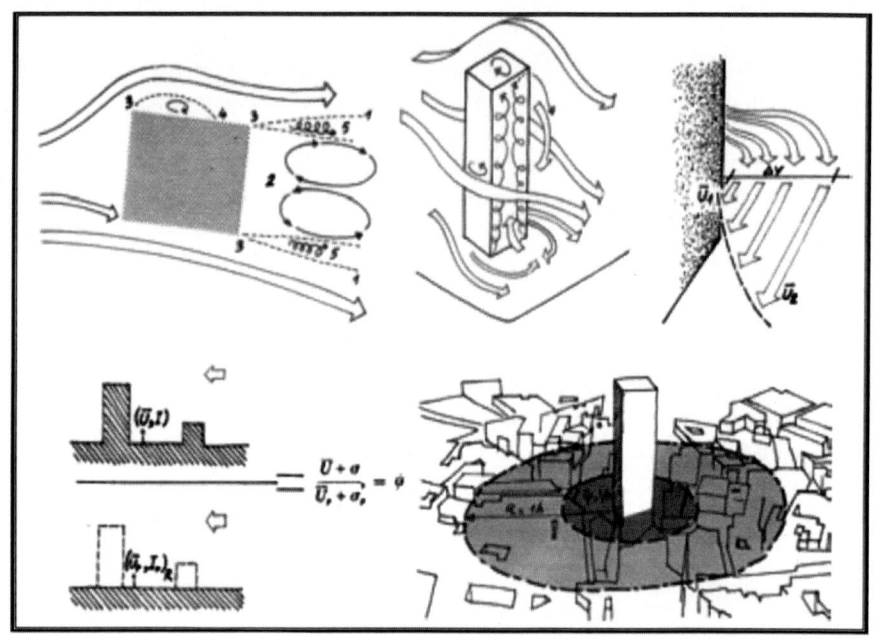

Fig. 5. The air current schemes formed around building [22]

Air molecules pushed by a unit and having different speeds form vortices. So, around the building, uncomfortable areas formed whose speed and direction is variable. This changing formation depending on the building geometry and dimensions can be solved by the work done in the design phase [22]. In this context, to reveal the relationship between building form (geometry) and wind energy, analysis is performed seen in Fig. 6.

While the airflow formed by the wind passing around the side faces of building, vortices occur as seen in Fig. 7 schematically.

Since vortex occur as variable (first one side, then other side), dynamic loads created by vortex also variable and influence on the direction perpendicular to the flow direction of the wind. Since they affect very specific and narrow frequency, vortex can be defined as sinusoidal load [26]. Wind pressure seen on the building shown in Table 1, varies depending on the topography, height, internal pressure, aerodynamic pressure and the building form [27].

High structures, are affected by vibrations caused by wind action. Therefore, wind has an important role in the structural and architectural design. There are different design strategies that enhance the functional performance of the buildings and reduce the negative effects of wind. Between these strategies, the important and effective design approach is the aerodynamic modifications in architecture [28].

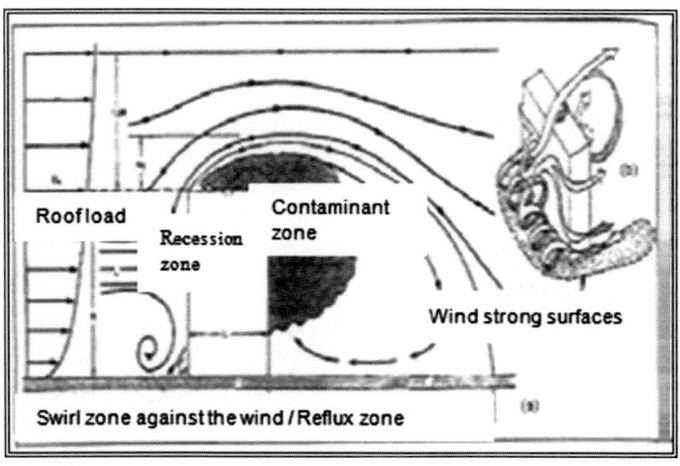

Fig. 6. Air flow around the rectangular building [22]

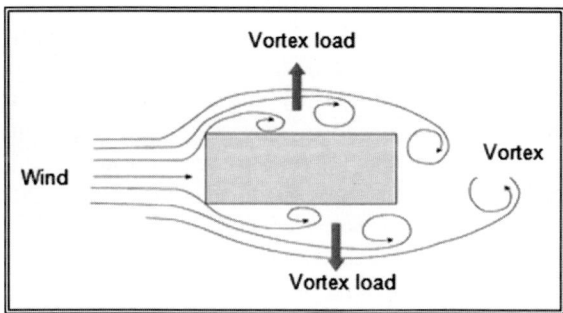

Fig. 7. The formation of vortex load [26]

Table 1. Factors affecting the building and wind interaction [27]

Factors	Impacts
Topography	Sudden changes in topography such as hills, ridges cause an increase in wind speed. Therefore, a building located near the top of a hill, will receive more wind loads than relatively the one in a flat area
Building height	If the height increase from the ground, wind speed also increase. Therefore, if building height increase, wind load effecting on the building increase
Internal pressure	The wind hitting the building causes the pressure increase (positive pressure) or decrease (negative pressure). Internal pressure change occurs by the size and frequency of the openings in the building faces
Aerodynamic pressure	Due to the aerodynamic effects (structure and interaction of wind) the highest loads occur at the roof edge. Winds affecting the structure of the front is usually lower than the wind load acting on the roof
Building form	The building form affect the wind pressure coefficient and thus wind load acting on the building

High-rise buildings, due to the increase in wind power with building height and building form, they show a more sensitive behavior against wind effect. Various forms of high-rise buildings and aerodynamic properties of these forms has been studied by many researchers. The dynamic behavior of a structure in which the face of high winds must be contained within the acceptable limits for structural design and user comfort. The building form aerodynamic is a design criteria that must be decided from the beginning of the design phase of a high-rise building [29].

According to Ilgın and Günel, the effects of wind on the structures can be controlled in two ways:

1. Significant architectural changes: Modifications that affect the architectural concept like buildings became slender with height, to gain a sculptural appearance of the structure of the top, changing the building form, creating opening in the building
2. Small architectural changes: Modifications that do not affect the architectural concept like corner modifications and settlement of the building according to the strong wind direction [28].

These two substances were examined on the following examples. Gaining sculptural view of the top of the structure as shown in the examples reduces the negative impacts of wind on buildings [30].

As it is seen in the example, Petronas Towers in Malaysia shown in Fig. 8 [31], reducing the plan section of the building by 'changing the form' at the high level of the building reduces the effect of wind force by varying the effect of the wind energy.

It is a known fact that, building form has a significant impact on protection of lateral resistance of the building form. If the building form is only limited by the rectangular prism, this form is exposed to effects of lateral wind. To cylinder, ellipsoid, triangular or other structure forms act less lateral force than rectangular prism-shaped structure [31].

Wind pressure acting on the building in the form of cylinder (circular) or the ellipse-shaped is reducing by 20–40% compared to the rectangular prism-shaped building [32]. Because of this, in many famous building, building forms that have favorable effects are preferred.

The Marina City Towers' (Chicago 1964) cylinder form seen in Fig. 9a, Millennium Towers' (Tokyo 2009) conical circular plan seen in Fig. 9b, Toronto City Halls' (Toronto 1965) crescent-shaped form seen in Fig. 9c and U.S. Steel Building's (Pittsburgh 1970) triangular plan seen in Fig. 9d are among the buildings that aerodynamic building forms are prepared [29].

Also, the openings created especially close to the roof section of the building facade are the aerodynamic answers of the building that reduce the negative effects of the wind affecting on the building. The Shanghai World Financial Center (Shanghai 2008) seen in Fig. 10 is one of the best examples about this subject [33].

In this section, it has been mentioned that wind has influence on the building, in order to reveal these effects, various wind tunnel experiments have been carried out. In these studies, it has been demonstrated that wind power increases with building height and form, because of this, wind energy acting on the buildings can be controlled with same aerodynamic modifications during the design stage.

Fig. 8. Petronas Towers in Malaysia [31]

(a)   (b)   (c)   (d)

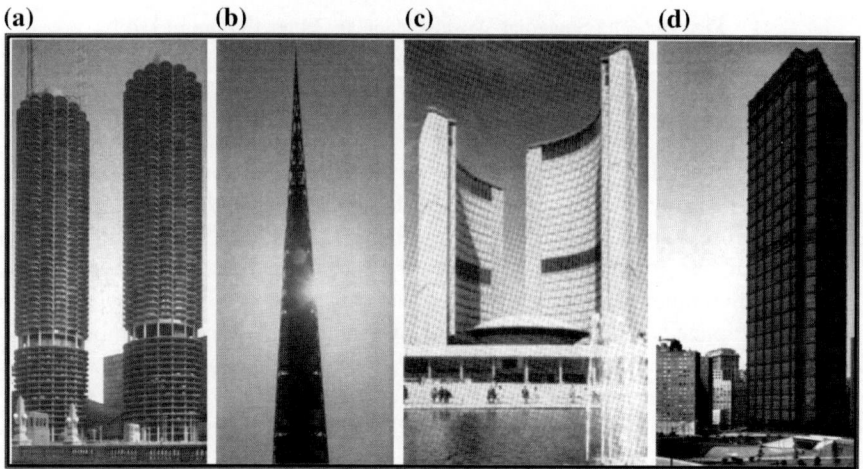

Fig. 9. a The Marina City Towers (Chicago 1964), **b** Millennium Tower (Tokyo 2009), **c** Toronto City Hall (Toronto 1965), **d** The U.S. Steel Building (Pittsburgh 1970) [29]

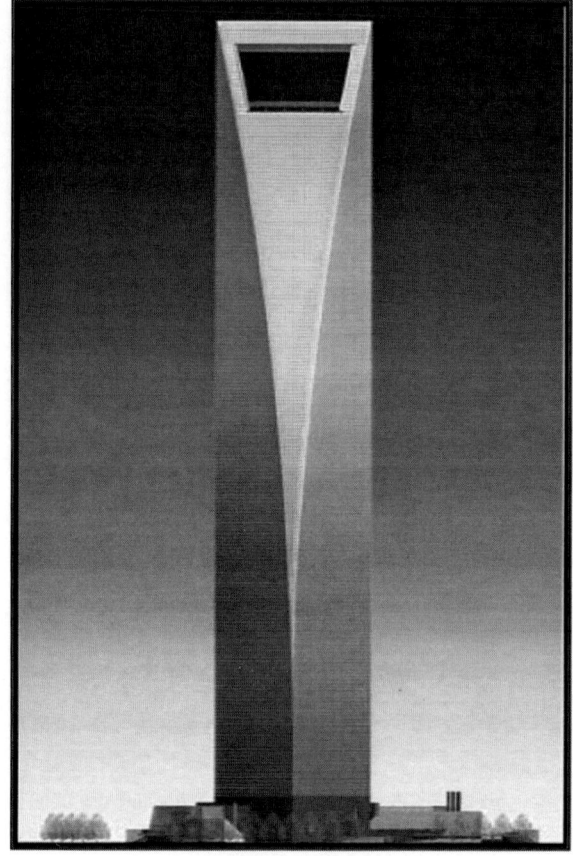

Fig. 10. The Shanghai World Financial Center Building [33]

5   Results

In a wide range, building shape, height and distance between buildings affects the direction and intensity of the wind flow. Airflow between buildings has been investigated by many researchers using different statistical analyzes. Usually, street canyons with typical architectural forms, semi-enclosed areas, the relationship between form and courtyard complex of low-rise buildings in a relatively open field wind environment were examined. Collected and stored measured data comprises information on wind direction and wind speed at different locations. Then statistical methods are used to determine the key factors defining the effects of building form on the air flow. In urban environment, information about how it is the wind flows around buildings, to decide on the best positioning and wind turbines, air pollution dispersion and pedestrian comfort has a great importance in terms of design [34].

In Fig. 11, from the schematic representation of the obstacles to the turbulence, in Fig. 12, it is included in the schematic representation of complex the flow path of the air around the building form.

System integration level and forms of architecture are discussed in different ways by different researchers. In this context, according to Çakmak [35], five relationships can be mentioned about structure-facade integration. These relationships are as seen in Table 2; remote, touching, connected, meshed and unified.

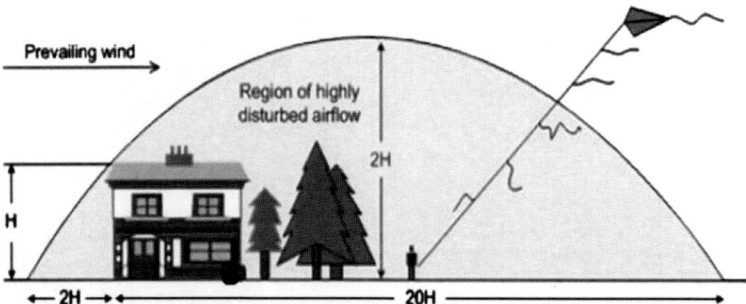

Fig. 11. Turbulence from obstacles results with high turbulent area near these obstacles [34]

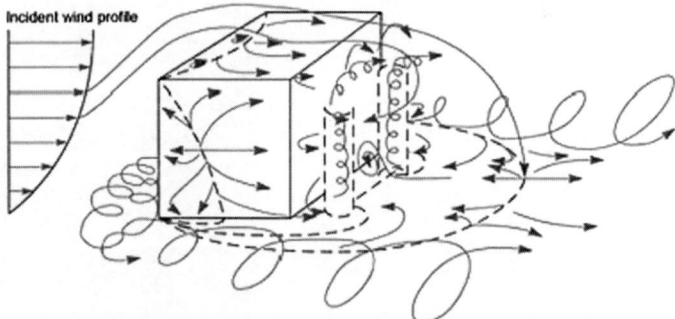

Fig. 12. Wind flow around building form creates mixed air flow path sand waves [34]

According to another aspect, integration level and shape of the building systems and ecosystems can be taken into account like ecological design and physical integration, system integration and continuous integration.

Table 2. Table showing the level of integration of building systems [35]

Integration levels	
Remote Systems are physically separated from each other and still been coordinated form	
Touching One of the system remains above the other and in the position it is touched with this system. It is only influenced by its own weight where it is	
Connected The systems are connected in various ways to each other permanently	
Meshed Systems occupy the same space	
Unified	

Ecological design is the design approach that design system is integrated with our natural environment. They deal with the building not only as a single structure but also with urban and land. The characteristics of the area where the building is discussed in the context of ecology and these properties are integrated physically, systematic and continuous.

In Bachman's study, the integration potential between architectural systems can be classified as physical integration, visual integration, performance integration and integrated integration. According to this view, architectural systems must share the same space, the coming together should be resolved aesthetically and must collaborate or complete each other in a point.

Physical integration Building components must be in harmony with each other. These components share a common volume in a building and they are interacting in different ways in this volume. Physical integration includes the basic components of the system volume and shape interactions in common. In standard application, for example, many buildings on the ground and roof sections are usually divided into separate regions: recessed lighting in the bottom section, later space for groove, in the top section, depth section supporting the above ground at the top [36].

Visual integration All components of the building are coming together to complete the visuality. This expression is available for as well as structure to the overall visual idea also, each room features and even individual components. Color, size, shape and placement properties can be used to create the desired effect [36].

Performance integration Where the physical integration is called as shared space and visual integration is shared image, performance integration is related with common functions. For example, a bearing wall serves as a shell as well as being a bearing wall [36].

Integrated integration All these integration levels seen between systems are emerging often integrated. The level in which physical, visual and performance integration are seen together is called integrated integration [36].

6 Conclusions and Suggestions

Many research studies have been conducted about wind energy and building relationships (building aerodynamics). Owing to the increase of wind speed with height and the increase of wind energy proportional to the height, it is found that wind turbines has been used in more high-rise buildings. As a result of these studies, it is put forward that wind energy has serious effects on high-rise buildings and solutions that convert negative aerodynamic effects to positive is very important in design stages.

As a result, selection of appropriate building form results in reduction of aerodynamic forces around the buildings. This method can lighten the wind response compared with the original building form. Surface roughness at the urban scale, position and height of buildings, vegetation and open space fiction, and thus the effect of the wind affects the city's pedestrian wind comfort with pedestrian comfort conditions.

The world's energy problem extending since the 1970s, development, diversification of architectural systems in recent years and existing energy resources come to the level that they cannot meet the needs of building stock make it necessary to take measures to reduce the environmental damage.

In later years with questioning the relationship between architecture and environment, it is seen that energy is a factor that is driving the design in architecture, because of this architectural systems' diversification and integration of wind energy and building is becoming important from the beginning of the design stage. In this context, buildings must be designed within the framework of a performance-based approach.

To provide environmentally conscious design and thus integration of building and wind, there are some parameters that should be taken into account from the beginning of the design process. These are;

- Climate
- The building location and geographical location.
- The usage of building.
- Intended use of the building.
- Building form.

Today, as a result of the tendency to renewable energy sources, getting the benefit of these sources has accelerated and it has been started to be used effectively by many countries. Wind energy that is leading source of renewable energy has turned into a giant energy sector especially in Denmark, Germany and Spain, Europe and the United States. The wind energy that is first limited with the application in wind farms, nowadays with technological developments is emerging in the building examples that display integration of energy generation and building. Especially in European countries and the USA, in planned high-rise buildings, the designs that building systems and wind energy examples are increasing.

References

1. Akkaya AV, Akkaya EK, Dağdaş A (2002) Environmental assessment of renewable energy sources. IV. National clean energy symposium, İstanbul, pp 37–43
2. Bu Z, Kato S, Ishida Y, Huang (2009) New criteria for assessing local wind environment at pedestrian level based on exceedance probability analysis. Build Environ 44:1501–1508. doi:10.1016/j.buildenv.2008.08.002
3. Zhang A, Gao C, Zhang L (2005) Numerical simulation of the wind field around different building arrangements. J Wind Eng Ind Aerodyn 93:891–904. doi:10.1016/j.jweia.2005.09.001
4. Chan CM, Chui JKL, Huang MF (2009) Integrated aerodynamic load determination and stiffness. Design optimization of tall buildings. Struct Des Tall Spec Build 18:59–80. doi:10.1002/tal.397
5. Murakami S, Ooka R, Yoshida S, Kim S (1999) CFD analysis of wind climate from human scale to urban scale. J Wind Eng Ind Aerodyn 81:57–81. doi:10.1016/S0167-6105(99)00009-4
6. Roberson JA, Crowe CT (1978) Pressure distribution on model buildings at small angles of attack in turbulent flow. In: proc. 3rd US Natl. Conf. on Wind Engineering Research, University of Florida, USA
7. Ahmad s, Kumar K (2001) Interference effect of wind loads on low-rise hip roof buildings. Eng Struct 23(12):1577–1589
8. Aygün C, Başkaya Ş (2003) Experimental investigation of surface pressures caused by wind flow around a multi-story building. Gazi University Engineering architecture faculty magazine 18(4):15–31
9. Mendis P, Ngo T, Haritos N, Hira A, Samali B, Cheung J (2007) Wind loads on tall buildings. EJSE (Special Issues: Loading on Structures):41–54
10. Holmes JD, Tamura Y, Krishna P (2008) Wind loads on low, medium and high rise buildings by Asia-Pasific codes. The 4th International Conference on Advances in Wind and Structures, Jeju, Korea
11. Şafak E (2011) Account of wind loads on high rise buildings. The chamber of civil engineers 4th national steel constructions symposium, Istanbul

12. Huang S, Li QS, Xu S (2006) Numerical evolution of wind effects on a tall steel building by CFD. J Constr Steel Res 63(5): 612–627
13. Liang S, Li QS, Liu S, Zhang L, Gu M (2004) Torsional dynamic wind loads on rectangular tall buildings. Eng Struct 26(1):129–137
14. Huang G, Chen X (2007) Wind load effects and equivalent static wind loads of tall buildings based on synchronous pressure measurements. Eng Struct 29(10):2641–2653
15. Tominaga Y, Mochida A, Yoshie R, Kataoka H, Nozu T, Yoshikawa M, Shirasawa T (2008) AIJ guidelines for practical applications of CFD to pedestrian wind environment around buildings. J Wind Eng Ind Aerod 96(10–11):1749–1761
16. Huang MF, Lau IWH, Chan CM, Kwok, KCS, Li G (2011) A hybrid RANS and kinematic simulation of wind load effects on full scale tall buildings. J Wind Eng Ind Aerod 99 (11):1126–1138
17. Cheung JOP, Liu CH (2011) CFD simulations of natural ventilation behaviour in high-rise buildings in regular and staggered arrangements at various spacings. Energ Buil 43(5):1149–1158
18. Blocken B, Janssen WD, van Hoof T (2012) CFD simulation for pedestrian wind comfort and wind safety in urban areas: general decision framework and case study for the Eindhoven University Campus. Environ Model Softw 30:15–34
19. Kanbur BB, Pınarbaşı A, Koca A (2013) Investigation of the natural ventilation resulting from the effects of wind loads on high-rise buildings in HAD environment. Eng Mach 54 (637):44–53
20. Mochida A, Lun IYF (2008) Prediction of wind environment and thermal comfort at pedestrian level in urban area. J Wind Eng Ind Aerodyn 96:1498–1527
21. Lawson TV (1980) Wind effects on buildings design applications. Applied Science Publishers, London, p 45. doi:10.1016/0141-0296(82)90037-2
22. Istanbul Branch of the Chamber of Mechanical Engineers. http://www.mmoistanbul.org/yayin/tesisat/103/4/
23. Çağlar M, Canbaz M (2002) Turkey wind energy potential. IV. National clean energy symposium, İstanbul, pp 347–355
24. Bozdoğan B (2003) Architectural design and ecology. Master thesis, YTÜ Institute of Science and Technology, İstanbul, pp 2–124
25. Aynsley RM, Melbourne W, Vickery BJ (1977) Architectural aerodynamics, vol 3(1). Applied Science Publication, London, p 32
26. Istanbul High Buildings Wind Directive (2009) Metropolitan Mayor of Istanbul, Istanbul
27. Whole Building Design Guide. http://www.wbdg.org/resources/env_wind.php
28. Ilgın HE, Günel MH (2007) The role of aerodynamic modifications in the form of tall buildings against wind excitation. Metu JFA Ankara 2:17–25
29. You K, Kim Y (2009) The wind-induced response characteristics of a typical tall buildings. Struct Des Tall Spec Build Korea 18:217–233. doi:10.1002/tal.419
30. Kareem A, Kıjewskı T, Tamura Y (1999) Mitigation of motion of tall buildings with recent applications. Wind Struct 2(3):201–251
31. Ali M, Armstrong P (1995) Architecture of tall buildings. McGraw-Hill Book Company, New York, p 52
32. Schueller W (1977) High-rise building structures. Wiley, New York, pp 26–33
33. Dutton R, Isyumov N (1990) Reduction of tall building motion by aerodynamic treatments. J Wind Eng Ind Aerodyn. doi:10.1016/0167-6105(90)90416-A
34. Ishugah TF, Li Y, Wang RZ, Kiplagat JK (2014) Anvances in wind energy resource exploitation in urban environment: a review. Renew Sustain Energy Rev 37:613–626. doi:10.1016/j.rser.2014.05.053

35. Çakmak SP (2006) 20th century change in architectural design strategies: reasons and direction of change. Master thesis, Gazi University Institute of Science and Technology, Ankara, pp 43–50
36. Bachman LR (2003) Integrated buildings. Wiley, New York, pp 3–48

The Role and Importance of Energy Efficiency for Sustainable Development of the Countries

Serap Pelin Türkoğlu[1]([⊠]) and Pınar Sezin Öztürk Kardoğan[2]

[1] Department of Business Administration, Faculty of Economics and
Administrative Sciences, Giresun University, Giresun, Turkey
serappelinozturk@gmail.com
[2] Department of Civil Engineering, Faculty of Technology, Gazi University,
Ankara, Turkey
sezinozturk@gazi.edu.tr

Abstract. Energy, which is needed for every aspect of life, plays a key role for the development of the countries. Countries need to use energy efficiently to be advantageous in the global competition and ensure the sustainable development. Countries using the energy efficiently succeed economically and have leading the field in the competition. The purpose of this study is to put forward to the role and importance of energy efficiency for the sustainable development of the countries. In this study, energy efficiency has been examined conceptually considering the studies in the literature and the role and importance of energy efficiency has been emphasized for the sustainable development of the countries.

Keywords: Energy · Efficiency · Sustainable development

1 Introduction

Energy has been an important element for human development and economic growth. Providing sufficient and affordable energy is required to increase human welfare and living standards. Energy should be considered as an important factor in terms of economic development; since the energy is used as an input in most of the production processes. Energy consumption increases in parallel with economic growth and development. In this way, energy need should be met sufficiently and economically.

In the last 50 years, there have been important developments. Living standards have increased and people have been living longer and healthier. Science and technology increased welfare of society considerably. Energy resources of Middle East, which are abundant and cheap, have especially contributed to these developments. Sufficient global energy resources have importance both for the world and the countries individually in terms of sustainable development, running of the economy and welfare of the society. Thus, consistent use of energy is going to be possible and the energy will be secured [6].

Industrialised countries have been using the energy more efficiently since 1973. Following the oil shock of the 1970s, these countries have made policies to increase energy efficiency in all the sectors of their economies. These policies have contributed to the decrease in the energy intensity. In addition to that, most of the industrialised

© Springer International Publishing AG, part of Springer Nature 2018
S. Fırat et al. (eds.), *Proceedings of 3rd International Sustainable
Buildings Symposium (ISBS 2017)*, Lecture Notes in Civil Engineering 7,
https://doi.org/10.1007/978-3-319-64349-6_5

countries have intensified their energy efficiency works as a part of their own strategies in order to decrease their greenhouse gas emissions [5].

In the second part of this study, energy efficiency has been discussed conceptually. In the third part, the role and importance of energy efficiency in the sustainable development have been explained. In the conclusion part, the study has been overviewed.

2 Conceptual Framework of Energy Efficiency

Energy is one of the primary elements which are needed for social and economic development. Energy is a means to achieve the goals such as health, high level of living standards, sustainable economy and a clean environment [7].

Energy resources of the countries are one of the main factors indicating their development and leadership position in the rivalry. Therefore, efficient use of energy becomes more of an issue for the countries. Energy efficiency is identified as the efficiency scaling the relation between energy inputs and outputs by means of comparison [2].

Patterson [15] explained that energy efficiency means obtaining the same product and service using less energy. Galvin [3] identified energy efficiency with the help of monetary and physical indicators. According to Galvin [3], the ratio between energy input and energy output as monetary value is considered as monetary indicators. Energy input per gross domestic product is an example of this. The ratio between energy input and physical output states physical indicators. Amount of aluminium that is produced or how many kilometres you drive your car in the highway is an example of physical indicators.

Ganda and Ngwakwe [4] states that energy efficiency means the policies, technologies and strategies to solve the problems of residential, commercial and national energy use reducing the financial cost and minimizing greenhouse gas emission which causes global warming.

During the oil shock of 1973, measuring the energy efficiency has been an important part of the energy strategy, especially for those lacking energy. With sharp increase in the world oil prices, many countries have been aware of the need for the understanding of the methods for efficient energy consumption and increasing the energy efficiency. As from late 1980s, because of the increasing worries about the fact that fossil fuels cause global warming, improving the energy efficiency in order to reduce greenhouse gas emission has been an important issue for the countries [1].

Countries need to follow various policies on the basis of sector in order to improve their energy efficiencies. Policies of some sector are as follow [12]:

- New buildings can be 70% more efficient using insulated windows, modern gas furnaces and more efficient air conditioners. It is possible to save energy with the help of localized heating, heat pumps and solar energy. With advanced lighting system, it is possible to save cost at a rate between 30 and 60%.
- When it comes to residences, there have been improvements in fridges, water heaters and dish washers. It is possible to make energy savings with the technological improvements in this field.

- Energy demand in the industry and CO_2 emission can be halted improving the efficiencies of engines, pumps and boilers and heating systems, energy recovery in the process of production, recycling the material that is used and using the materials more efficiently.
- In the transportation sector, efficiency of the gasoline-powered and diesel-powered vehicles can be improved with turbine turbocharger, fuel injection, improved electronic methods, more compact engines, hybrid cars and advanced diesel engines.

3   Ensuring Sustainable Development for the Countries via Using Energy Efficiently

Sustainable development, according to the definition of United Nations World Commission on Environment and Development in 1987, has been stated as meeting the needs of today's society not jeopardizing the needs of the future generations [17]. Sustainable development is a general term which identifies the development of countries considering economic, social and environmental factors. Efficient energy use has some economic, social and environmental aspects. Therefore, energy efficiency has an important role for ensuring sustainable development.

It is a need for the sustainable developments of the countries to obtain the energy resources easily and at a reasonable cost in the long term. To ensure sustainable development, countries need to use energy with maximum benefit and minimum environmental damage [16].

Energy need has been increased worldwide with high population growth rate, technological developments and increasing living standards. Limited reserves and increasing prices of the fossil resources have leaded the countries to take measures in order to protect themselves from a possible energy crisis. Energy sector provides input for the other sectors of the economy. For this reason, stable and consistent energy policy is a must for sustainable development [9].

Because of the worries about high energy prices, global warming and sustainable development, energy efficiency has been an important of the energy strategy in many countries. Increasing the energy efficiency is the cheapest, fastest and the most environmental-friendly way of meeting an important part of energy need. Improved energy efficiency reduces the need for investment in the energy demand [12].

Increasing the energy sufficiency is important from various policy perspectives. Saving from the energy obtained from the fossil fuels is an important target for the countries to prevent the running out of fossil fuels in the near future. Improvement of the energy efficiency is also going to increase energy security of the countries. Especially, reduction of the energy use is a must to prevent the worsening of the environmental quality. Minimizing the costs is another aim to ensure energy efficiency. In terms of cost efficiency, it is important to reduce energy use in the period of high energy prices and use other inputs instead of energy at the same time [10].

Energy is crucial for the countries to ensure their sustainable development. Some of the policy measures to support sustainable development are as follows [6, 8, 11]:

- Increasing the energy efficiency in the industry, construction and transportation sectors,
- Reducing the use less efficient coal-fired power plants and forbidding their constructions,
- Increasing the investment to the renewable energy,
- Phasing out the subventions in the consumption of the fossil fuels,
- Reducing the methane emission in the production of oil and gas,
- Distributing sufficient and reasonable priced energy resources to the regions where there is no energy service,
- Encouraging energy efficiency,
- Providing widely use of advance energy technologies,
- Placing importance on activities of research and development on improving new and advanced energy technologies,
- Encouraging the use environment friendly energy resources such as renewable energy resources,
- Empowering the regulations on energy,
- Reflecting the environmental cost of energy use and consumption to the energy, prices as much as it is possible and necessary,
- Contribution of the free and open trade to the energy market and security,
- Ability of the energy systems to response to the urgent situations in quick and flexible way,
- Improving international cooperation and connections.

If today's demographic, economic, social and technological trends aren't strong and balanced by new government policies, there are significant difficulties in the long-term sustainability of the global energy system. Developments in energy saving in OECD countries accelerated with increases in energy prices in the 1970s [11]. OECD countries have taken an important step in energy conservation and environmental issues with the policies they apply in the energy field. Energy policies for OECD countries may differ [13].

Energy efficiency policies in Turkey increase energy security to benefit from growing economy and reduce greenhouse gas emissions. Turkey has established a comprehensive, strategic and legal framework to promote energy efficiency. The energy efficiency law in Turkey has a wide coverage area, including related regulations. These areas include: increasing and supporting energy efficiency, setting up energy efficiency consulting companies, establishing energy management systems, promoting energy efficiency investments, increasing energy efficiency in transport and buildings, preventing the sale of inefficient devices and raising awareness.

Luxembourg implemented energy performance regulations in 2008 for residential buildings and in 2011 for non-residential buildings. These regulations establish a methodology to calculate energy performance of buildings, to determine minimum energy requirements for new buildings, extensions and renewed elements of existing buildings. Luxembourg is promoting the energy efficiency of buildings with the introduction of energy performance certificates.

New Zealand's "Smart Warming Program" aims to increase the number of hot, dry and energy-efficient homes to protect patient health and prevent loss of productivity.

The first "New Zealand Energy Strategy" and the second 5-year "New Zealand Energy Efficiency and Conservation Strategy" were published in October 2007. However, after the National Party government was elected in 2008, both of these documents were audited. The purpose of doing this is to present a clearer link between energy policies and economic growth.

The program of energy saving targets for energy suppliers is a central component of Ireland's energy efficiency policy. The first "National Energy Efficiency Action Plan" identified potential energy saving programs targeting energy suppliers in 2009.

The "Green Deal" initiated in the United Kingdom in January 2013 is the program of the British government to help increase energy efficiency in homes and businesses. This program saves on energy bills and makes improvements on the cost side possible. "Electricity Market Reform" is the most important and radical change for the UK electricity market. This reform includes regulations that make low carbon production investments attractive and provide safe, affordable electricity supply for the UK.

Norway supports the increase of oil production in order to achieve long-term goals in energy field. It also uses oil and gas in a profitable way, the management of energy resources is based on comprehensive knowledge and facts, and the energy management framework is designed to be sensitive to health, safety and environmental issues.

USA is consistently publishing the "Climate Action Plan" to reduce national and international greenhouse gas emissions. There are three main pillars of this plan: to reduce carbon pollution in USA, to prepare USA for the effects of climate change, and to pioneer the international struggle to combat global climate change and prepare for the effects of climate change. The plan is designed to reduce emissions by 17% below the 2005 level by 2020. In addition, USA regularly updates national and international reports.

Sweden is working to build a vehicle fleet that does not use fossil fuels until 2030, and the energy is taxed at a high rate. The Swedish tax system supports the purchase of environmentally friendly vehicles through tax exemption during the first 5 years. This incentive is reinforced by an extra subsidy for "super environmentally friendly" cars that emit less than 50 g of CO_2 per kilometer of hybrid and electric vehicles. To promote alternative fuels in Sweden, high-mix renewable fuel mixture into gasoline and motor is subject to full tax exemption.

Canada is defeating procedures to regulate projects for the development of natural resources. In 1997, Canada created a building law that sets new buildings and sets a minimum level for energy performance in their design. "The Canadian National Energy Law for Buildings", which assessed the minimum energy performance level of buildings over three floors in 2011, entered into force. This law covers heating, ventilation, air conditioning, water heating, lighting, electric power motor systems and equipments for buildings.

Japan is working to improve energy efficiency through its energy efficient economy. The Japanese government has laid down the law on the rational use of energy in May 2013. The first step in this regulation is to improve the thermal insulation performance of homes and buildings. In this context, the use of energy-saving isolators and windows is helping to reduce the energy consumption associated with air conditioning and water heating. In the second step of the scheme, the introduction of technologies such as smart meters, energy management systems and accumulators are encouraged to reduce energy demand.

Italy is eyeing the natural gas market mechanisms and infrastructure to keep rising gas prices under control. In Italy, research projects based on the system approach of interest to the national electricity system and applied research are being developed. These activities aim to improve the performance of the system in terms of economy, security and environment and to renew.

In South Korea, in 2008, the government enacted the "Basic Energy Law" in the formulation and implementation of a national energy plan for every 5 years for a period of 20 years. The goals of each plan are to: to promote the direction of future energy policies, to identify medium and long-term strategies for safe energy sources, to expand infrastructure to provide domestic energy, and to rationalize the use of energy for the development of the national economy.

France intends to bring innovative energy projects to the market with its research programs. Since 2010, the French Agency for Environment and Energy Management has been responsible for investment programs for low carbon vehicles, intelligent network projects, supporting the testing of renewable energies and green chemistry in real conditions and demonstration premises. These programs are new tools designed to support innovative green projects.

In agreement with the government in line with a framework established by Denmark, energy organizations are helping to provide cost-effective energy savings throughout the economy. In 2020, 50% of electricity consumption from wind power is one of Denmark's energy policy objectives. Renewable energy conversion by 2050 is the long-term goal of Denmark's full conversion.

The use of hydropower as a renewable energy source is crucial for Austria's electricity generation. The use of hydroelectric power in Austria contributes to the provision of energy security and to the independence of electricity supply. With about 70% of the electricity generated from renewable energy sources, Austria is the country with the highest share in the European Union.

Approximately 80% of Australia's total energy consumption is generated by the industry. The Government's "Energy Efficiency Opportunities Program" aims to make the industry more energy efficient. The participating enterprises in this program receive financial assistance of approximately 800 million Australian dollars per year.

The Czech Republic is updating its energy strategy, especially considering the declining production of lignite. The Czech Republic's energy policy aims to facilitate the transformation of the Czech energy sector in order to produce sufficient electricity and cost effectively meet the high supply standards.

In 2008, the government approved an "Energy Efficiency Action Plan" aimed at reducing the use of fossil fuels by 20% by 2020 and growing electricity demand by 5% between 2010 and 2020 in Switzerland. In addition, the Swiss government finances R&D at energy efficiency and promotes professional energy efficiency training as well as consultancy.

In Portugal, the "Energy Efficiency Program in Public Administration" is intended to encourage efficiency in the public sector's energy use, which is mandatory for the central government and is optional for municipalities. Portugal improves the security of natural gas supply with major investments (including the construction of a new interconnection pipeline with Spain, increased underground storage capacity at Carriço, and expansion of capacity at the Sines LNG Terminal).

Energy security is one of the main features of the Polish energy policy. An important aspect of Poland's energy security policy is the diversification of fuel and technology. The government supports the development of clean technologies and the production of coal-based liquid and gaseous fuels. Poland is striving to reduce its energy dependence on Russia and diversify its energy sources. Poland is focused on maximizing the use of existing domestic energy resources.

Spain supplies a large amount of electricity from renewable sources. Until 2020, about 40% of the electricity consumed in Spain will be from renewable sources. In the first half of 2013, about 48% of the electricity demand was supplied from renewable sources.

Germany is increasing its knowledge and awareness in this area by issuing annual reports on energy politics. Germany's energy targets include reducing greenhouse gas emissions by 40% by 2020, at least 80% by 2050, and reducing primary energy consumption by 20% by 2020 and 50% by 2050. In Germany, the share of renewable energy sources in energy consumption will increase by 30% by 2030 and by 60% by 2050.

According to the OECD [12], if governments fail to implement their policy in the framework between the present and 2030, the following situations will be the case:

- Energy consumption will increase by 53%,
- CO_2 emissions related to energy will increase by 55%,
- The poor peoples of the world will continue to lack access to electricity, modern cooking and heating services.

4 Conclusions

Considering that energy is a limited resource, it is important for countries to use the minimum level and to obtain maximum economic output from the energy used. When this process takes place, minimizing the harm to the people and the environment is also considered as a matter to be paid attention to. Increasing energy efficiency for countries leads to reduced environmental pollution and increased competition.

Today, the local, regional and global negative effects of the traditional energy production and consumption technologies on the people, environment and natural resources have come to a serious state. In order to ensure the sustainable development and protect natural balance, it is vital to provide energy from consistent, safe, cheap, clean, good quality and domestic energy resources and use it efficiently [14].

It is important to monitor eco-friendly energy efficiency policies for the economic development and progress of countries. Sustainable economic development based on energy efficiency should form the basis of energy policies of countries. If countries increase their economic activities by improving their environmental performance, they can reach a good level of energy efficiency. As long as the balance between economic growth and energy consumption is balanced, the energy efficiency of countries will increase.

Energy efficiency has become important for the countries to provide sustainable development. Especially for reducing CO_2 emission and effects of the climate change, energy efficiency has a key role for the energy efficiency.

Countries should know that they will not be able to succeed in rivalry and sustainable development without being aware of the comprehending the efficient use of energy. This study casts light upon the future studies in the field of energy efficiency for the countries to ensure sustainable development explaining the role of energy efficiency.

References

1. Ang BW (2006) Monitoring changes in economy-wide energy efficiency: from energy-GDP ratio to composite efficiency index. Energy Policy 34:574–582
2. Cui Q et al (2014) The changing trend and influencing factors of energy efficiency: the case of nine countries. Energy 64:1026–1034
3. Galvin R (2014) Estimating broad-brush rebound effects for household energy consumption in the EU 28 countries and Norway: some policy implications of Odyssee data. Energy Policy 73:323–332
4. Ganda F, Ngwakwe CC (2014) Role of energy efficiency on sustainable development. Environ Econ 5(1):86–99
5. Geller H et al (2006) Polices for increasing energy efficiency: thirty years of experience in OECD countries. Energy Policy 34:556–573
6. Goldemberg J (2000) World energy assessment: energy and the challenge of sustainability. United Nations Publications, New York
7. IAEA. Energy indicators for sustainable development: guidelines and methodologies. Vienna, International Atomic Energy Agency (2005)
8. IEA (2015) Energy and climate change, world energy outlook special report. IEA Publications, Paris
9. Koçaslan G (2010) Sürdürülebilir kalkınma hedefi çerçevesinde Türkiye'nin rüzgar enerjisi potansiyelinin yeri ve önemi (The place and importance of wind power potential of Turkey in the framework of the target of sustainable development). J Soc Stud 4:53–61
10. Mukherjee K (2008) Energy use efficiency in U.S. manufacturing: a nonparametric analysis. Energy Econ 30(1):76–96
11. OECD (1997) Sustainable development: OECD policy approaches for the 21st century. OECD Publications, Paris
12. OECD (2007) OECD contribution to the United Nations commission on sustainable development 15 energy for sustainable development. OECD Publications, Paris
13. OECD/IEA (2013) Energy policy highlights. IEA Publications, Paris
14. Oskay C (2014) Sürdürülebilir kalkınma çerçevesinde rüzgar enerjisinin önemi ve Türkiye'de rüzgar enerjisi yatırımlarına yönelik teşvikler (Importance of wind power in the framework of sustainable development and incentives for the investments of wind power in Turkey). J Niğde Univ Fac Econ Adm Sci 7(1):76–94
15. Patterson MG (1996) What is energy efficiency? Concepts, indicators and methodological issues. Energy Policy 24(5):377–390
16. Rosen MA (1995) The role of energy efficiency in sustainable development, in foundations and applications of general science theory. In: Knowledge tools for a sustainable civilization, interdisciplinary conference, Canadian conference on IEEE, pp 140–148
17. Soubbotina TP (2004) Beyond economic growth: an introduction to sustainable development. World Bank Publications, Washington

A Study on Some Factors Affecting on CO_2 Curing of Expanded Perlite Based Thermal Insulation Panel

Gökhan Durmuş[1]([✉]), Onuralp Uluer[2], Mustafa Aktaş[3],
İbrahim Karaağaç[2], Ataollah Khanlari[4], Ümit Ağbulut[5],
and Damla Nur Çelik[6]

[1] Department of Civil Engineering, Faculty of Technology,
Gazi University, Ankara, Turkey
gdurmus@gazi.edu.tr
[2] Department of Manufacturing Engineering, Faculty of Technology,
Gazi University, Ankara, Turkey
{uluer, ibrahimkaraagac}@gazi.edu.tr
[3] Department of Energy System Engineering, Faculty of Technology,
Gazi University, Ankara, Turkey
mustafaaktas@gazi.edu.tr
[4] Natural and Applied Science Institute, Faculty of Technology,
Gazi University, Ankara, Turkey
[5] Department of Manufacturing Engineering, Faculty of Technology,
Düzce University, Düzce, Turkey
[6] Department of Civil Engineering, Faculty of Engineering,
Antalya Bilim University, Antalya, Turkey

Abstract. Energy, which is needed for every aspect of life, plays a key role for the development of the countries. Countries need to use energy efficiently to be advantageous in the global competition and ensure the sustainable development. Countries using the energy efficiently succeed economically and have leading the field in the competition. The purpose of this study is to put forward to the role and importance of energy efficiency for the sustainable development of the countries. In this study, energy efficiency has been examined conceptually considering the studies in the literature and the role and importance of energy efficiency has been emphasized for the sustainable development of the countries.

Keywords: Energy · Efficiency · Sustainable development

1 Introduction

There are many thermal insulation materials used in thermal insulation technology. Especially in the construction industry, traditional heat insulation applications usually include expanded polystyrene foam (EPS), extruded polystyrene foam (XPS), polyisocyanurate (PIR), rigid polyurethane (PUR), and PUR/PIR. However, recent literature studies have shown that expanded perlite material can be used as the main material in thermal insulation materials [1].

© Springer International Publishing AG, part of Springer Nature 2018
S. Fırat et al. (eds.), *Proceedings of 3rd International Sustainable Buildings Symposium (ISBS 2017)*, Lecture Notes in Civil Engineering 7,
https://doi.org/10.1007/978-3-319-64349-6_6

Perlite is one of the natural insulation materials. Perlite is obtained from pumice, which is a glassy form of rhyolitic or dacitic magma. It contains 2–5% water [2]. When perlite is suddenly heated (at temperatures of 649–816 °C), the created steam forms bubbles within the softened rock to produce a frothy-like structure. The formation of these bubbles allows perlite to expand up to 15–20 times of its original volume [3]. Expanded perlite (EP) has superior properties such a light apparent density, low thermal conductivity, good chemical stability, wide use temperature scope, non-toxic, tasteless, fireproofing, sound absorption [4]. EP thermal properties make it possible to use it as thermal insulation materials [5]. In addition EP can be used as an additive in various mixtures or as a main component for composite e.g. portland cement/perlite composites for blocks, perlite/sodium silicate boards, roof insulation panels made of perlite/fibres/bituminous material, fiber reinforced perlite/cement composites, gypsum/perlite composites and light weight concrete etc. [6]. However, their applications as the main constituent of composites have been limited due to relatively poor mechanical properties low strength [7].

The expanded perlite is used as main material in thermal insulation materials and as aggregate in the production of Portland cement/perlite composites. There are some important considerations related using expanded perlite based thermal insulation materials. These considerations are the type of binder used to bond the expanded perlite particles, additive materials (fibers, waterproofing agents, fire retardant, etc.) to improve thermal, mechanical and physical properties of the thermal insulation panel, pressing pressure and curing conditions. Curing conditions are one of the important parameters to achieve the desired properties. Conventional and microwave techniques are available for curing such a thermal insulation panels [8–11].

Researchers focused on different curing methods for EPHIB. Taherishargh et al. studied mechanical properties of heat-treated expanded perlite–aluminum syntactic foam. Based on the conditions under which the optimum mechanical properties were achieved, the specimens were solution treated at 540 °C for 16 h. Then, the specimens were treated at 160 °C for 10 h to cool down. As a result, when the heat treated sample's density increase, the strength of the heat treated sample has significantly increased [12].

Arifuzzaman and Kim studied expanded perlite/sodium silicate composites. Sodium silicate was used as a binder. The diluted samples were cured in an electric fan forced air oven (Lebec Oven BTC-9090) at 65 °C for 10 h [7]. Demirboğa and Gül, investigated that the thermal conductivity and compressive strength of EP aggregate concrete with mineral admixtures. They prepared a concrete mixture using laboratory counter-current mixer for 5 min. For each mixture, all specimens were cured in lime saturated water at 20 ± 3 °C until 6th, 14th and 27th day. According to their results, the lowest value of thermal conductivity of sample was 155.3 mW/m K [13].

Shastri and Kim studied that a new consolidation process for expanded perlite particles. They cured samples after molding for characterization of formability in an oven at 80 °C for 6 h. It has been found that the compressive strength at a density of 0.3 g/cm^3 of perlite foam is similar to that of gypsum foam having densities between of 0.7–0.9 g/cm^3 [14]. Tian et al. studied the effect of pressing pressure and curing temperature on mechanical properties of EPHIB. They indicate that 0.38 MPa of compression pressure and 105 °C temperature are suitable for EPHIB [4].

Skubic et al. investigated the effect of microwave method expanded perlite based heat insulation boards. They indicated that quick drying can be achieved by microwave method. But, an appropriate temperature control should be applied to avoid hot spot inside the panel [8]. In another study by Skubic et al. sintering behavior of EPHIB theoretically and experimentally. They used microwave method to sinter the panels. Then the panels treated at high temperature. As a result, the high-temperature treatment caused shrinkage in some parts of the insulation plates. Also, the thermal conductivity of panel was obtained as 0.045 W/m K [10].

Erdoğan, studied the usage of perlite to produce geopolymers. He produced geopolymers mixing alkaline activator and expanded perlite material. He stated that the most suitable condition is dry curing for mechanical properties of geopolymers. When the curing temperature was around 1000 °C, the highest pressure, and bending strength was achieved. According to the results, perlite was found to be an effective natural pozzolan and at the same time, it could be used to produce geopolymeric binders [15].

Milano et al. studied the low-pressure thermophysical properties of expanded perlite board at the temperature under atmospheric conditions. Samples were filled with dry air until atmospheric pressure reached to 5 μbar. Temperatures ranged from 17 to 117 °C for samples. The results showed that expanded perlite based board's thermal conductivity was measured between the 0.01 and the 0.02 W/m K for vacuum condition. Also, thermal conductivity of insulation board was measured 0.04–0.05 W/m K for atmospheric pressure condition. The obtained results showed that thermal conductivity under the atmospheric conditions is approximately four times higher than thermal conductivity under vacuum conditions [16].

There are different curing methods in literature studies about expanded perlite material. When the literature studies were reviewed, it was seen that expanded perlite was used as the main material or additive material in heat insulation materials. It is clear that different curing conditions are applied preparing EPHIB. There are some different researchers about curing condition using CO_2 gas in the literature. Shi and Wu studied on some factors affecting CO_2 curing of lightweight concrete products. Samples were put into a curing chamber for CO_2 curing. There was vacuumed to a pressure of around 600 mmHg and maintained for 2 min before CO_2 was injected. The obtained results showed that the accelerated reactions between CO_2 and cement minerals happen mainly during the first 15 min. CO_2 curing degree and strength also increased as the CO_2 pressure and curing time increased [17]. In another study, Baojian et al. investigated CO_2 curing to improve the properties of concrete blocks containing recycled aggregates. Blocks were prepared according to the procedure. Then, blocks were placed in a pressurized 100% CO_2 curing chamber for 6, 12 and 24 h. According to strength, shrinkage of the CO_2 and moist results, The CO_2 cured blocks achieved higher compressive strength and lower drying shrinkage than the corresponding moist cured blocks [18].

He and Liu researched weathering properties of CO_2-cured concrete blocks. They compared steam curing and the CO_2 curing according to energy extensive process and production cost and they defined that CO_2 curing of concrete blocks consumes CO_2 and reduces energy consumption. This situation is so important to prevent of greenhouse gas. Some freshly molded lightweight concrete blocks were taken from a block manufacture plant. Then samples were cured in a CO_2 curing and steam-curing. For

CO_2 curing, samples were left on a rack between 3 and 6 h. Then they were precon-ditioned in a relative dry windy environment for about 4 h. After that, they were placed into a tank for CO_2 curing. It was observed that compressive strength of blocks cured with CO_2 was similar to steam-cured blocks. However, the dimensional stability of the CO_2-cured concrete blocks was better than steam-cured blocks [19]. When literature studies were reviewed, it is clear that curing expanded perlite material with CO_2 gas is a new technology for construction sector. In literature, various methods used to cure expanded perlite-based heat insulation panels. But, no studies have been found in the literature on this curing method of expanded perlite with carbon dioxide gas (CO_2). However, there are some different researches about curing condition using CO_2 gas in the literature but these researches are not related to curing of expanded perlite.

In this study, an alternative to the traditional curing methods expanded perlite-based heat insulation panels cured with carbon dioxide (CO_2) gas. Quick curing is the main aim of this method. Also, the thermal conductivity of manufactured panels measured for 30 days.

2   Materials

2.1   Expanded Perlite

The expanded perlite was provided in Çankırı, Turkey (Fig. 1). Also, the chemical and physical properties of expanded perlite, which was in this study, were given in Tables 1 and 2.

(a)   **(b)**

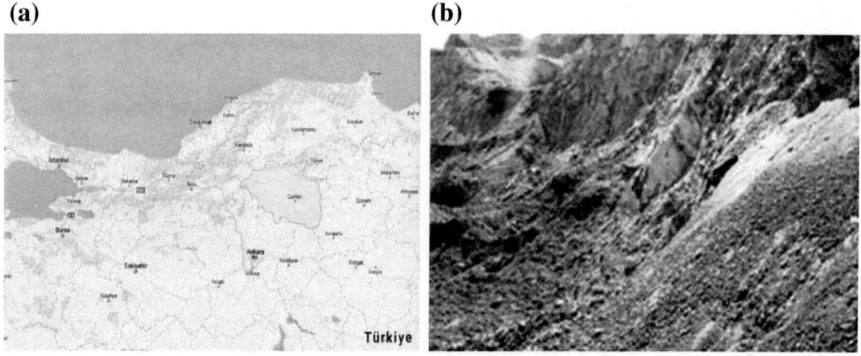

Fig. 1. a Perlite mine location. **b** The view of the perlite mine

2.2   Sodium Silicate

Sodium silicate is one of several compounds containing sodium oxide (Na_2O) and silica (SiO_2). The pure compounds are colorless or white. In generally, sodium silicates are referred to as orthosilicate, metasilicate, disilicate and tetrasilicate, depending on the acid from which they are produced. Sodium silicates are used as a raw material to produce silica gels. Sodium silicate is mainly used in paper, soap, detergent,

Table 1. Chemical properties of expanded perlite

Content	wt%
SiO_2	74
Al_2O_3	14.33
K_2O	4.95
MgO	0.28
CaO	0.50
Fe_2O_3	0.97

Table 2. Physical properties of expanded perlite

Properties	Values
Density, kg/m^3	45–50
Grain size, mm	0–0.5
Thermal conductivity (−40 °C), W/m K	0.035–0.039
Thermal conductivity (−125 °C), W/m K	0.025–0.029
pH	4–9

construction materials, precision casting, anti-corrosive materials, textiles and minerals industries. Sodium silicate (Na_2O_3Si) was used as a binder.

2.3 Mold

The mold made of aluminum. As it can be seen in Fig. 2 the dimensions of mold are $500 \times 300 \times 100$ mm^3. Teflon cover was used to prevent sticking the material to the mold.

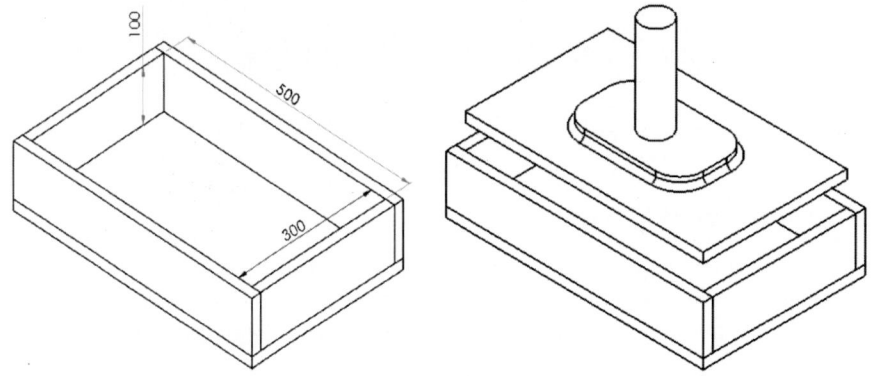

Fig. 2. Aluminum mold

3   Methods

3.1   Compression of Expanded Perlite

First of all, expanded perlite and binder were mixed together by the aid of mechanical mixer throughout 10 min and the mortar was obtained. Then the mortar had been poured into the mold. The mold size was $500 \times 300 \times 100$ mm. The prepared mortar were compressed in thickness direction by 50% and were obtained in dimensions of $500 \times 300 \times 50$ mm. it was compressed by hydraulic press (Fig. 3). The sample was held for 3 min under constant press force (16.67 kPa). Then test sample was obtained in $300 \times 300 \times 50$ mm^3 dimensions by cutting the panel with a cutting machine and this size is used to measure thermal conductivity coefficient (Fig. 4).

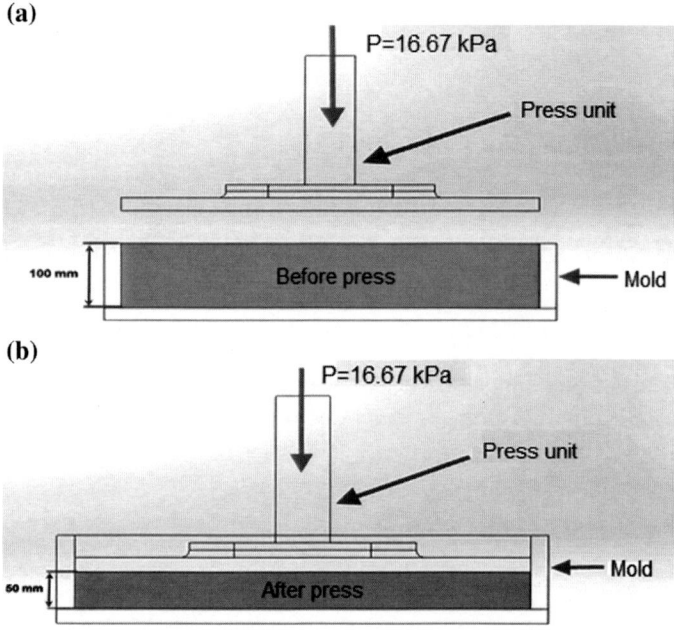

Fig. 3.   Press and molding unit **a** before press. **b** After press

3.2   Curing with CO$_2$ Gas

The molded sample was placed in a vacuum bag. Two nozzles are placed on the bag used for vacuuming. One of the nozzles was vacuumed (5 mbar) and the other was given CO$_2$ gas after vacuuming. The gas was injected into the vacuum bag (Fig. 5) with 10 l/min for 30 s. The CO$_2$ gas has a moisture content of less than 10 ppm in 99.9% purity. The samples were kept for 15 min in a closed vacuum bag. Then the curing procedure was completed.

There is an important issue related to CO$_2$ curing of thermal insulation panel. This issue can be explained as follows. When the CO$_2$ gas reacts sodium silicate, sodium

Fig. 4. Expanded perlite-based heat insulation panel

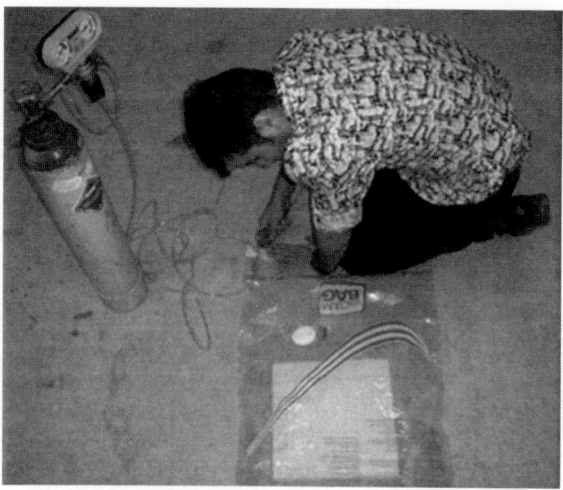

Fig. 5. Carbon dioxide curing

silicate (Na_2O_3Si) hardened by separating sodium carbonate (Na_2CO_3) and silicon dioxide (SiO_2). This equation is formed as;

$$Na_2O_3Si + CO_2 \rightarrow Na_2CO_3 + SiO_2$$

3.3   Measurement of Thermal Conductivity

The thermal conductivity of sample was measured by HFM 300 device. The properties of the thermal conductivity testing device were given in Table 3.

Table 3. The properties of thermal conductivity testing device

Specification	Values
Temperature range (Plates), °C	0–40
Temperature control (Plate)	Peltier
Thermal resistance measuring range, m_2K/W	0.1–8.0
Thermal conductivity measuring range, W/m K	0.001–2.5
Accuracy, %	±1– 3
Variable contact pressure, kPa	0.25

4   Results

Thermal conductivity of cured panel measured periodically for 30 days. The obtained results can be seen in Fig. 6. As it can be seen, the thermal conductivity of cured panels decreased as function of time. In first 15 days, the thermal conductivity of panel has been decreased. Finally, at the end of 30 days the thermal conductivity coefficient of the panel was measured as 77.83 W/m K. At the same time, mechanical strength was also measured according to 40 × 40 × 160 mm size. Thermal insulation panel has 1.27 MPa compressive strength. Quick curing is an important advantage of carbon dioxide curing method. The expanded perlite-based heat insulation panels can be cured just in 15 min. Also, the manufactured panels have better mechanical properties. Thus, a panel has been developed which can be used more practically in heat insulation applications.

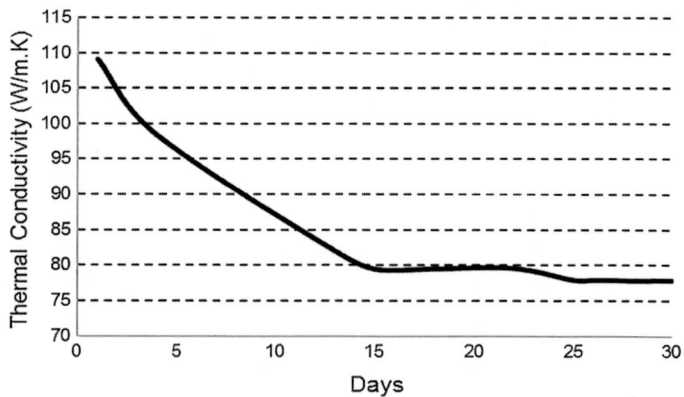

Fig. 6. Thermal conductivity variations of heat insulation panel

5 Conclusion

Expanded perlite-containing composites have been limited due to relatively poor mechanical properties low strength. This curing method, a more practical panel was produced in terms of mechanical properties. Compressive strength of manufactured expanded perlite-based heat insulation board was measured as 1.27 MPa. Also, the energy spent in curing and the curing time were decreased. These are important in terms of energy efficiency and production technique of heat insulation panel. The developed method is also important in terms of environmental factors for the production process. With the work to be done, the properties of the panel can be improved.

Acknowledgements. This study was supported by the Scientific and Technological Research Council of Turkey (Project no.: 115M041). We are indebted to TÜBİTAK for its financial support.

References

1. Arifuzzaman M, Kim HS (2014) Development of new perlite/sodium silicate composites| NOVA. The University of Newcastle's Digital Repository
2. Arifuzzaman M, Kim HS (2015) Novel mechanical behaviour of perlite/sodium silicate composites. Constr Build Mater 93:230–240
3. Demirboğa R, Gül R (2003) Thermal conductivity and compressive strength of expanded perlite aggregate concrete with mineral admixtures. Energy Build 35:1155–1159
4. Dube W, Sparks L, Slifka A (1991) Thermal conductivity of evacuated perlite at low temperatures as a function of load and load history. Cryogenics 31:3–6
5. Erdoğan S (2011) Use of perlite to produce geopolymers. In: Proceedings of 31st cement and concrete science conference novel developments and innovation in cementitious materials
6. Gunning D (1994) Perlite market study report. Gunning and Mc Neal Associates Ltd. Crown Publications Inc, Victoria
7. Milano G, Scarpa F, Timmermans G (1996) Low-pressure thermophysical properties of EPB-expanded perlite board. Therm Conduct 23:362–372
8. Mladenovič A, Šuput J, Ducman V, Škapin AS (2004) Alkali–silica reactivity of some frequently used lightweight aggregates. Cem Concr Res 34:1809–1816
9. Shastri D, Kim HS (2014) A new consolidation process for expanded perlite particles. Constr Build Mater 60:1–7
10. Shi C, Wu Y (2008) Studies on some factors affecting CO_2 curing of lightweight concrete products. Resour Conserv Recycl 52:1087–1092
11. Shi C, Wang D, He F, Liu M (2012) Weathering properties of CO_2-cured concrete blocks. Resour Conserv Recycl 65:11–17
12. Skubic B, Lakner M, Plazl I (2012) Microwave drying of expanded perlite insulation board. Ind Eng Chem Res 51:3314–3321
13. Skubic B, Lakner M, Plazl I (2013) Sintering behavior of expanded perlite thermal insulation board: modeling and experiments. Ind Eng Chem Res 52:10244–10249
14. Sun D, Wang L (2015) Utilization of paraffin/expanded perlite materials to improve mechanical and thermal properties of cement mortar. Constr Build Mater 101:791–796

15. Taherishargh M, Belova I, Murch G, Fiedler T (2014) On the mechanical properties of heat-treated expanded perlite–aluminium syntactic foam. Mater Des 63:375–383
16. Tian YL, Guo XL, Wu DL, Sun SB (2013) A study of effect factors on sodium silicate based expanded perlite insulation board strength. Appl Mech Mater 405:2771–2777
17. Uluer O, Aktaş M, Karaağaç İ, Durmuş G, Khanlari A, Ağbulut Ü, Çelik (2017) Manufacturing of expanded perlite based heat insulation material using theoretical thermal conductivity prediction model results. In: International conference on progress in applied science, pp 156–161
18. Vaou V, Panias D (2010) Thermal insulating foamy geopolymers from perlite. Miner Eng 23:1146–1151
19. Zhan B, Poon C, Shi C (2013) CO_2 curing for improving the properties of concrete blocks containing recycled aggregates. Cement Concr Compos 42:1–8

Turkey's Renewable Energy Sources and Govermental Incentives

Evren Kaydul[1]([⊠]), Zafer Demir[2], and Nesrin Çolak[3]

[1] Anadolu University, Fen Bil. Ens., Eskisehir, Turkey
evrenkaydul@icloud.com
[2] Anadolu University, Porsuk MYO, Eskisehir, Turkey
zaferdemir@anadolu.edu.tr
[3] Anadolu University, Ulaştırma MYO, Eskisehir, Turkey
nesrincolak@anadolu.edu.tr

Abstract. One of most crucial problems the world is facing today is the decrease in available fossil fuels. This decrease is, however, also increasing the value of renewable energy resources. According to International Renewable Energy Agency (IRENA) data, power generation from renewable energy sources increased 100% between the years spanning 2000–2013 and 14% between 2013 and 2015. In the years ahead, this rise is expected to increase exponentially. In light of these changes Turkey has set a target to increase its renewable energy ratio in the energy sources to 81% during the years 2015–2019. In this study, Turkey's renewable energy sources—wind, solar, hydraulic, biomass, and geothermal—along with its potentials for wave energy and this potential's effect on Turkey's installed power will be examined in detail. Governmental incentives that are provided for renewable energy sources and the results these EPDK study will also present suggestions related to power generation from renewable energy sources and incentive process.

Keywords: Turkey renewable energy sources · Incentives
Power generation

1 Introduction

Modern renewable energy can put the world back on track to limiting global warming to 2 °C by the end of this century, the target agreed by all countries at the COP21. Finally, the RE map analysis shows how we can operationalise the United Nations' Sustainable Development Goal call on Energy of "ensuring access to affordable, reliable, sustainable and modern energy for all" [1].

1.1 Renewable Energy Resources

Renewable energy is energy that is derived from natural processes (e.g. sunlight and wind) that are replenished at a higher rate than they are consumed. Solar, wind, geothermal, hydropower, bioenergy and ocean power are sources of renewable energy. The role of renewables continues to increase in the electricity, heating and cooling and transport sectors [2].

© Springer International Publishing AG, part of Springer Nature 2018
S. Fırat et al. (eds.), *Proceedings of 3rd International Sustainable Buildings Symposium (ISBS 2017)*, Lecture Notes in Civil Engineering 7,
https://doi.org/10.1007/978-3-319-64349-6_7

This work uses 5-year segments to investigate the share of renewable energy resources among total resources. It is noted that there has been a world-wide increase in the use of renewable energy sources [3] (Table 1).

Table 1. Share of renewable energy in electricity production

Region	Share of renewable energy in electricity production (%) in 2005	Share of renewable energy in electricity production (%) in 2010	Share of renewable energy in electricity production (%) in 2015
Africa	16.9	17.4	18.9
Asia	13.9	16.1	20.3
CIS	18	16.7	16.1
Europe	20.1	25.7	34.2
Latin America	59.3	57.7	52.4
Middle East	4.3	2.0	2.2
North America	24	25.8	27.7
Pacific	17.9	18.6	25.0

Source World energy resources (2016, p. 9)

2 Renewable Energy Sources in Turkey

2.1 Renewable Energy Sources

On 31 November 2016, renewable energy sources claimed a 43.2% share in Turkey's total installed power capacity of 78,591.8 MW. The utilizations of the sub-categories of renewable energy sources include: wind (5387 with 6 MW at 6.86%); geothermal (775 with 1 MW at 0.99%); biomass (464, with 8 MW at 0.59%); hydroelectricity (26,515 at 9 MW with 33.74%); and solar (792 at 8 MW with 1.01%) [4] (Table 2).

Table 2. Annual changes to renewable energy sources in installed power capacities of renewable energy resources

Source	2013	2014	2015	2016
Wind	2759.6	3629.7	4503.2	5387.6
Geo	310.8	404.9	623.9	775.1
Biomass	236.9	288.1	344.7	464.8
Hydro	22289.1	23643.2	25867.8	26515.9
Solar	0	40.2	248.8	792.8
Other	38447.6	41513.6	41558.3	44655.6
Summary	64044.0	69519.7	73146.7	78591.8

Sources TEİAŞ installed power capacity

An investigation of Turkey's potentials for renewable energy sources demonstrates that Turkey has a wind energy potential of 48,000 MW. This potential equals 1.30% of Turkey's total surface land mass [5].

Turkey ranks fifth in the world and first in Europe with its approximate 31,500 MTt of geothermal energy potential. A 6 TWh portion of this energy can be used for the generation of electricity, while a 19 TWh portion can be used as a direct energy resource [6].

It is predicted that Turkey's animal waste potential is equivalent to a volume of biogas that is roughly equal to 1.5–2 million tons of petroleum (MTEP) [7]. This in turn is equivalent to an energy potential of approximately 17–23 TWh.

Turkey has an annual hydroelectric energy potential of 433 TWh, which that will provide technological utilization of 216 Twh per year, and an annual economic based hydroelectric energy potential of 144 TWh [8].

It has been demonstrated that Turkey's solar energy potential is based on an average annual total sunlight period of 2640 h (or 7.2 h per day) and an average solar radiation intensity value of 1311 kWh/m^2-per year (daily total of 3.6 kWh/m^2) [9].

3 Laws Governing Renewable Energy Sources Renewable Energy Resources

The legal structure that began in 2001 with the establishment of the Energy Marketing Regulatory Authority has resulted in a number of modifications to current laws and the passing and publication of new legislation, guidelines, and by-laws aimed at regulating Turkey's electricity markets. This Authority has been significantly instrumental in meeting the needs of the energy market and in the provision of legislation that has especially streamlined and updated the electricity markets. These legislative efforts have also played an important role in increasing the share of renewable energy resources in Turkey's overall energy consumption markets.

Listed below are the names and descriptions of these legislations and a brief summary of the innovations they have brought to the sector.

3.1   2001—Law no. 4628: Law Governing the Electricity Market

Law 4628 Governing the Electricity Market, which was passed in 2001 and aimed at encouraging the use of renewable energy and domestic energy resources includes the provision of oversight to the Energy Marketing Regulatory Authority (EPDK). This law, which references renewable energy resources in Turkey, is comprised of two legal texts. The law also limits the non-licensed generation of electricity to 500 kW Renewable Energy Resources [10].

3.2   2002—Guidelines Governing the Licensed Electricity Market

Law Number 4628 Governing the Electricity Market lays out the legal procedures and principles of licensing those legal entities that are determined to have the qualifications of competing in a financially sound, stable, and transparent electricity market. As a means of predicting or eliminating problems that could arise in the post-licensing period, amendments made to the law in 2013 established a pre-licensing, probational period in which those license applicant companies can be monitored prior to approval or denial of license [11].

3.3   2003—Guidelines Governing the Legal Procedures and Principles of Water Utilization Agreements for Operations on the Electricity Market

These guidelines have been prepared exclusively for private sector entities investing in energy producing facilities using state-owned hydraulic resources [12].

3.4   2005—Law Number 5346 Governing the Utilizations of Renewable Energy Resources in the Generation of Electric Power Renewable Energy Resources Renewable Energy Resources

Proposals call for the establishment of a fixed price for purchasing of electricity generated at YEK facilities. This Turkish Lira price should not be less than the equivalent of 5 Euro cents/kWh and should not be more than the Turkish Lira equivalence of 5.5 Euro cents/kWh. There is no distinction in prices for parts and equipment that are either domestically manufactured or imported [13].

An amendment made in 2010 determines a different unit price for each kind of renewable energy resource utilized and also authorizes additional incentives for the use of domestic parts and equipment. The amendment also specifies that these incentives will be valid for a period of 10 years and will include all energy generating facilities, including the YEKDEM plants, [14].

3.5   2013—Law Number 6446 Governing the Electricity Market

This law modifies the exemption on obtaining a license for the generation of electricity from renewable energy resources by raising the generated energy derived from renewable energy resources to 1 MW. The law also proposes the organization of a competition for the licensing of wind and solar generating stations [15].

3.6   2013—Guidelines for the Certification and Support Regarding Renewable Energy Resources

These guidelines comprise the principles and methods and monitoring to be utilized in the determination of source of renewable energy resources in the purchasing of the electricity generated by the license holder of the YEK Certificate (Renewable Energy Resource Certificate) issued by the EPDK [16].

3.7 Guidelines Legislating the Competition Organized for Pre-licensing Applications for the Establishing of a Wind and/or Solar Energy Generating Plant

These guidelines establish the competitive method to be employed by Turkish Electricity Transmission Council (TEIAS) if there are more than one applications from any one region. The guidelines have also been designed to determine the details of the Wind and Solar Energy Based Electricity Generation Station Contribution Margins [17].

3.8 2013—Guidelines Relative to the Non-licensed Electricity Market

These guidelines refer to the principles and methods to be employed in the generation of electricity by those legal individuals and entities that are exempted from the requirement to obtain a license [18].

The following guidelines are to be used for the transmission of extra electricity to the system by those legal individuals and entities that are generating less than <1 MW of electricity.

An amendment made to the law in October of 2016 eliminates Article 21 of the law relative to the support of domestically manufactured parts and equipment [19].

3.9 Bulletin Related to the Implementation of the Guidelines Legislating the Non-licensed Generation of Electricity in the Electricity Market

This bulletin has been prepared for the purpose of explaining and implementing the regulations outlined in the Guidelines Relative to the Non-Licensed Generation of Electricity [20].

3.10 2014—Guidelines to the Electricity Marketing Grid

These guidelines delineate the standards to be utilized in ensuring the kind of effective planning, operating, decision making implementations that will lead to a safe and low-cost electricity transmission system. The guidelines also outline the conditions to be established so as to provide consumers with high quality and adequate supplies of electricity [21].

3.11 2015—Guidelines to the Monitoring of Wind Power Capacities of Wind Energy Generating Stations and the Connecting of These Stations to the Forecasting Center

These guidelines explain the legalities involved in attaching wind power generating stations to the Wind Power Monitoring and Forecasting Center (RITM). By law, all stations licensed to generate 10 MW or more of wind energy must be connected to this center [22].

3.12   2016—Guidelines Relative to Renewable Energy Resources Fields/Farms

These guidelines have been published to establish extensive renewable energy generation areas (YEKA). A tender system is used to provide investors with advanced technology areas to be used in the domestic generation of electricity or for the procurement or transfer of know-how within the country [23].

The guidelines call for the generation of 1.7 billion kWh of energy at the station to be established following the tender.

4   Incentive Mechanisms

4.1   State Sponsored Investment Support

Turkey's policies relative to energy investments include the provisions of incentives as per Law Number 3305 published in 2012: Decision Relative to Governmental Investment İncentives. This law delineates the implementation of investment provisions by dividing the country into six regions based on the socio-economic development levels of the provinces included in these regions.

The incentive system includes the provision of various support mechanisms based on the categories of: general, regional, large scale, and strategic investments. Those generating stations benefiting from these support systems are exempted from various customs taxations, the paying of VAT, other taxation deductions, and the employer's share of social security benefit payments. The investor also receives interest payment supports, VAT redemptions, and support for social security [26].

A later Bulletin has removed the purchasing of imported solar panels from the list of incentive supports [27]. This amendment serves to benefit the transfer of currency out of the country and to support domestic manufacture of such panels. The purpose of the latter is to modify Turkey's position from technology consumption to technology production.

From January through August of 2016 a total of 23 foreign investment entities and 1270 domestic capital companies were granted incentive certification for energy sector investments [28].

For the first time in the history of the Turkish Republic, the Mid-Term Plan for the years 2017–2019 prepared through the coordination of the Ministry of Development has included the terms "wind", "solar", and "hydroelectric".

Predictions call for energy imports to total 32 billion dollars in 2017. By the end of 2017 it is expected that energy importation will maintain its nominal increase and will reach a total of 41.3 billion dollars. In order to achieve security in the field of energy, efforts will be expended towards domestic and renewable energy resources, while investments in nuclear technologies will also continue. Efforts directed towards the acquisition of machinery and equipment used in the establishment and operations of such alternative energy resources as wind, solar, and hydroelectric power will also continue [29].

4.2 Renewable Energy Resources Support Mechanism

An amendment to Law 5346 published in 2010 has determined the provision of 7.3 (US dollar) cents/kWh for the generation of electricity through solar and wind power. Other incentives are directed towards the use of domestically manufactured parts and equipment in the establishment of such generating plants. A purchasing price of 11 (USD) cents/kWh has been determined for wind generated energy.

The kWh incentive figures are given in Table 3 [14].

Table 3. Incentives for support mechanism

Production type based on renewable energy source	Prices to be applied (U.S.D. cent/kWh)	Maximum prices can be applied with the addition of local content (U.S.D. cent/kWh)
Hydroelectric energy based production power plant	7.3	7.3 + 2.3 = 9.6
Wind energy based production power plant	7.3	7.3 + 3.7 = 11.0
Geothermal energy based production power plant	10.5	10.5 = 2.7 = 13.2
Biomass energy based production power plant	13.3	13.3 + 5.6 = 18.9
Solar energy based production power plant (PV)	13.3	13.3 + 6.7 = 20.0
Solar energy based production power plant (concentrator)	13.3	13.3 + 9.2 = 22.5

Sources Law Number 5346, p. 9, 10

Table 4 (which shows both the Yekdem Feed-in Tariff Price per MWh and the Market Clearing Price used as a reference), demonstrates that those years with high Market Clearing Price have low Yekdem tariff prices, while, correspondingly, when the Market Clearing Price is high, the Yekdem tariffs drop [30].

Table 4. YEKDEM feed-in tariff price—Market Clearing Price

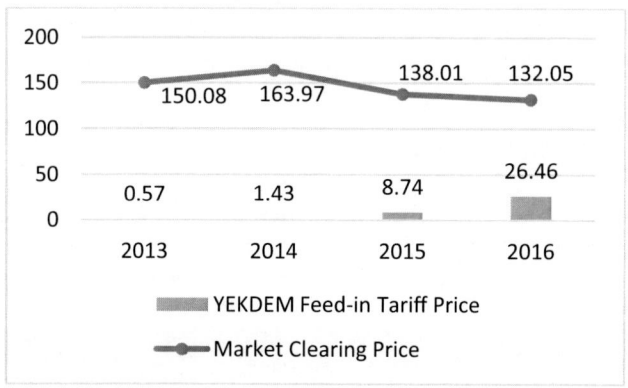

Sources EPİAŞ transparency

The influence of the Renewable Energy Resources Council has strengthened with the provision of incentives, resulting in a relative rise in the retail prices of electricity on the Electricity Market. Commercial electricity is being calculated according to the unit sales price Market Clearing Price (MCP) + Yekdem Feed-In Tariff Price.

The Guidelines to the Certification and Support of Renewable Energy Resources was amended on 29 April 2016 [31] so as to assume the responsibility for the state of unbalance in costs between renewable energy resources and generating plants.

These new guidelines, which provide YEKDEM generating plants with a 2% disparity right, have catalyzed major debate due to the fact that this rate varies according to the kind of resource being utilized, especially so the disparity related wind turbine fields, which is higher than that of other energy resources.

Turkey's Union of Wind Energy Generators have reported that this new tolerance rate has resulted in a 15% decrease in income [32].

During the period spanning January to October 2016 total incentives totaling 381.982.841,17 TL have been paid as LÜYTOB (YEKDEM Non-Licensed Production Feed-In Cost) and 9019, 617,679. 94 of YEKBED (YEKDEM Licensed Production Feed-In Cost) as compensation for the generation of 39,132,779.35 MWh of energy, for a total of 9,401,600,521.11 TL YEKTOB (YEKDEM Total Incentives) [33].

In 2017 there are also other renewable energy resource-based electricity generation plants that are not being supported by incentives and that are producing an annual 17,399.94 MW of electricity. Forecasts call for these plants to generate a total of 47,448.19 GWh of energy [34].

Table 5 shows the installed power of YEKDEM plants by years and by resources utilized.

Table 5. Changes by years to the established power of YEKDEM participating energy generating plants by resources utilized

Sources	2014	2015	2016	2017
Wind	826.4	2732.1	4319.8	5238.7
Solar	0	0	0	12.9
Geo	227.8	389.9	599.2	752.1
Biomass	136.3	185.2	203.7	300.0
Hydro	608.4	2116.3	9960.0	11096.3
Sum	1798.9	5423.5	15082.7	17399.9

Sources EPDK YEKDEM reports

5   Conclusions and Proposals

Table 6 provides Turkey's installed power during the years spanning 2013–2016. While it shows that the potentials of wind power have not transformed into investments up to 2016, it also shows that there has been a 95% increase in Licensed Wind Established Power.

Table 6. Turkey's installed power capacity (MW)

Sources	2013	2014	2015	2016
Fuel-oil + naphta + diesel	708.3	659.8	851.0	368.7
Local coal (hard coal + lignite + asphalt)	8515.2	8573.4	9013.4	9842.4
Imported coal	3912.6	6062.6	6064.2	7473.9
Natural gas + LNG	20269.9	21476.1	21222.1	22502.4
Biomass	236.9	288.1	344.7	464.8
Multi fueled solid + liquid	675.8	667.8	667.1	667.1
Multi fueled liquid + natural gas	4365.8	4074.0	3684.0	3719.0
Geothermal	310.8	404.9	623.9	775.1
Hydro (dammed)	16142.5	16606.9	19077.2	19408.5
Hydro (run of the river)	6146.6	7036.3	6790.6	7107.4
Wind	2759.6	3629.7	4498.4	5376.1
Solar	0	0	0	12.9
Thermic (non-licensed)	0	0	56.5	82.1
Wind (non-licensed)	0	0	4.8	11.5
Solar (non-licensed)	0	40.2	248.8	779.9
Summary	64,044	69,519	73,146	78,591

Sources TEİAŞ installed power capacity

Turkey's solar energy installed power has increased by 519%, leading it from an earlier position of 13th in Europe to current 7th place [35].

As of 31 November 2016, Turkey's total installed power of 78.591.8 MW has been made up of: Solar (12.9 MW licensed at 0.02%, and 779.9 MW non-licensed at 0.99%); hydroelectric (19,408.5 MW from dams at 24.7% and 7107.4 MW from running water at 904%); biomass (464.8 MW with 0.59%); wind (5376.1 MW licensed 6.84%, and 11.5 MW non-licensed at 0.01%); geothermal (775.1 MW with a contribution of 0.99%) [4].

Table 6 shows that in 2016 solar energy had a break-through year with licensed solar energy generating plants taking the lead. Energy generated from geothermal and biomass resources also increased during this same time period. Most likely the reason for the increase in biomass generated energy was due to the addition of this energy source to the national development plan, which transferred the management of waste to the municipalities, which in turn have developed their own ways to benefit from this resource. Thus we can see that Turkey's utilization of renewable energy resources has been making positive strides.

However, in November of 2016 the rapid rise in foreign exchange rates opened a debate on the payment of incentives. Three separate arguments have been placed on the table: (1) these incentives need to be lowered if the US dollar continues to be used as a basis; (2) if the dollar continues to be used as the basis for tariffs, its exchange rate should be stabilized and not allowed to flux; or (3) the incentives should be paid on a Turkish Lira basis. The competition scheduled to be held as per the Guidelines Governing Renewable Energy Resources Fields was postponed until 14 February, 2017, thus allowing a period of time for the continuation of discussions relative to the YEKDEM tariffs.

The Guidelines Governing the Certification and Support of Renewable Energy Resources includes the provision of 10-year guarantees. This implementation poses a risk for current investments after the expiration of the 10-year period and a risk for future investments as well. This cloud of uncertainty needs to be cleared so that long-term plans can be made.

In 2016 Turkey should modify its energy policy paradigms to direct the policy towards investments in electricity transmission infrastructures. Rather than focusing on non-licensed generation of less than 1 MW, attention should be turned towards renewable energy resource-based generation plants with high installed power rates. The Guidelines Governing Fields of Renewable Energy Resources should be modified to include those licensed solar energy and other resources generating plants with high installed power generating operations into competitions that should be held in 2017.

New alternative models should be established for YEKDEM and these models should use investment financing funds to assist investors in turning profits on their efforts.

To this end, those funds that are currently not being used and whose rates of profit are low, especially the unemployment insurance fund, which in 2015 reached a total of 94,117,471.1 TL [36] and the individual retirement insurance fund with its total of 51,996,814,532 TL [37] should be used to finance those plants that are generating electricity from renewable energy resources.

Internet-based public funding tools could also be utilized to support such plants. This is one of the new model proposals raised by this work.

All legal obstacles to the development and installation of rooftop electricity generating systems should be eliminated and the roofs of public buildings should be used as models in the advancement of such systems.

The Law on the Regeneration of Areas under Disaster Risk, which is popularly referred to as the Law on Urban Renewal, should be amended with articles [38] that provide incentives or other encouragements to the establishment of roof top energy systems.

The atlases on solar energy and wind energy potentials both need to be updated. A solar energy forecasting and monitoring center needs to be established.

Turkey's energy policy paradigms should be modified to encourage both the establishment of high installed energy capacity generating plants and the transfer of technologies and know-how. Adding requirements that companies entering licensing competitions must establish Turkish factories to manufacture the parts and equipment to be used in the energy generating plants to be set up will bring the future into clearer focus.

Separate incentive mechanisms and relative policies must be devised for systems based on non-licensed plants and solar rooftop generators that are directed to the use of immediate consumers.

References

1. http://www.irena.org/DocumentDownloads/Publications/IRENA_REmap_2016_edition_report.pdf. Accessed 28 Oct 2016
2. http://www.iea.org/topics/renewables/. Accessed 28 Oct 2016

3. https://www.worldenergy.org/wp-content/uploads/2016/10/World-Energy-Resources-Full-report-2016.10.03.pdf. Accessed 29 Oct 2016
4. http://www.teias.gov.tr/yukdagitim/kuruluguc.xls. Accessed 5 Dec 2016
5. http://www.eie.gov.tr/eie-web/turkce/YEK/ruzgar/ruzgar_en_hak.html. Accessed 29 Oct 2016
6. Adıyaman Ç (2012) Türkiye'nin Yenilenebilir Enerji Politikaları. Yüksek Lisans Tezi, Niğde Üniversitesi, Sosyal Bilimler Enstitüsü, Niğde
7. http://www.enerji.gov.tr/tr-TR/Sayfalar/Biyoyakit. Accessed 30 Oct 2016
8. http://www.eie.gov.tr/yenilenebilir/h_turkiye_potansiyel.aspx. Accessed 30 Oct 2016
9. http://www.eie.gov.tr/eie-web/turkce/YEK/gunes/tgunes.html. Accessed 29 Oct 2016
10. https://www.tbmm.gov.tr/kanunlar/k4628.html. Accessed 2 Nov 2016
11. www.epdk.org.tr/TR/Dokuman/2992. Accessed 2 Nov 2016
12. http://www.resmigazete.gov.tr/eskiler/2003/06/20030626.htm#2. Accessed 2 Nov 2016
13. http://www.resmigazete.gov.tr/eskiler/2005/05/20050518-1.htm. Accessed 2 Nov 2016
14. http://www.epdk.org.tr/TR/Dokuman/6777. Accessed 7 Nov 2016
15. http://www.resmigazete.gov.tr/eskiler/2013/03/20130330-14.htm. Accessed 7 Nov 2016
16. http://www.resmigazete.gov.tr/eskiler/2013/10/20131001-4.htm. Accessed 9 Nov 2016
17. http://www.resmigazete.gov.tr/eskiler/2013/12/20131206-11.htm. Accessed 9 Nov 2016
18. http://www.epdk.org.tr/TR/Dokuman/3044. Accessed 12 Nov 2016
19. http://www.epdk.org.tr/TR/Dokuman/7033. Accessed 12 Nov 2016
20. http://www.epdk.org.tr/TR/Dokuman/3157. Accessed 12 Nov 2016
21. http://www.epdk.org.tr/TR/Dokuman/2999. Accessed 13 Nov 2016
22. http://www.resmigazete.gov.tr/eskiler/2015/02/20150225-7.htm. Accessed 15 Nov 2016
23. http://www.resmigazete.gov.tr/eskiler/2016/10/20161009-1.htm. Accessed 15 Nov 2016
24. http://www.resmigazete.gov.tr/ilanlar/eskiilanlar/2016/10/20161020-4.htm#Ç03. Accessed 17 November 2016
25. http://www.resmigazete.gov.tr/ilanlar/eskiilanlar/2016/12/20161210-4.htm#Ç01. Accessed 17 Nov 2016
26. http://www.resmigazete.gov.tr/eskiler/2012/06/20120619-1.htm. Accessed 19 Nov 2016
27. http://www.resmigazete.gov.tr/eskiler/2016/08/20160827-7.htm. Accessed 19 Nov2016
28. http://www.adaso.org.tr/WebDosyalar/Yayinlar/Teşvik%20Belgeleri/Yatırım%20Teşvik%20Belgeleri%20Sunumu-Agustos%202016%20Adana%20Türkiye.pdf. Accessed 24 Nov 2016
29. http://www.kalkinma.gov.tr/Lists/Yaynlar/Attachments/722/Orta%20Vadeli%20Program%20(2017-2019).pdf. Accessed 26 Nov 2016
30. https://seffaflik.epias.com.tr/transparency/. Accessed 1 Dec 2016
31. http://www.resmigazete.gov.tr/eskiler/2016/04/20160429-8.htm. Accessed 1 Dec 2016
32. http://www.enerjigunlugu.net/icerik/19131/yeni-yekdem-bazi-reslerin-gelirini-15-dusurdu.html#.WEwYk_mLSUk. Accessed 1 Dec 2016
33. https://seffaflik.epias.com.tr/transparency/uretim/yekdem/yek-bedeli-yekbed.xhtml. Accessed 3 Dec 2016
34. http://www.epdk.org.tr/TR/Dokumanlar/Elektrik/Yekdem. Accessed 3 Dec 2016
35. http://www.solarpowereurope.org/fileadmin/user_upload/documents/2015_Market_Report/SPE16_Members_Directory_high_res.pdf. Accessed 4 Dec 2016
36. https://www.muhasebat.gov.tr/portal/file-download?id=173390&title=bilanco. Accessed 7 Dec 2016
37. http://web2.egm.org.tr/webegm2/chart/besgosterge/wg_sirketview_tablolu.asp?raportip=10. Accessed 9 Dec 2016
38. http://www.csb.gov.tr/db/altyapi/editordosya/6306%20SAYILI%20KANUN-degisiklik%20islenmis%20son%20hali.pdf. Accessed 10 Dec 2016

Green Wall Systems: A Literature Review

Yasemin Baran[1] and Arzuhan Burcu Gültekin[2(✉)]

[1] Department of Civil Engineering, Faculty of Engineering, Giresun University, Giresun, Turkey
yasemin.baran@giresun.edu.tr
[2] Department of Real Estate Development and Management, Faculty of Applied Sciences, Ankara University, Ankara, Turkey
abgultekin@ankara.edu.tr

Abstract. Sustainability today is creating an important and interesting approach combining the architecture and the environment. Such an approach is taking place in different forms and with different degrees of intensity. Within the challenges of energy crisis and climatic changes, architects started to develop new approaches to address the energy demands of buildings. One of these approaches involves green walls. Green walls are self-sufficient walls that are attached to a structure along a building's exterior making room for vegetation growth. During the last decades, several researches were conducted proving that green walls can contribute to enhance and restore the urban environment and improve performance of buildings. The aim of this paper is to review all types of green wall systems in order to identify and systematize their main characteristics and technologies involved, and the main differences between systems in terms of their composition and construction methods used. In this context, this paper focuses on the analysis of the green wall systems, and underlines a number of criteria that should be considered for successful applications in Turkey.

Keywords: Green wall · Green wall system · Green facade
Vertical garden energy

1 Introduction

One of the biggest problems of the twenty-first century is the rapid increase of concrete structures in parallel with urbanization. Urbanization, which has gained momentum in Turkey in recent years, causes environmental problems including even deforestation. With the increase of the amount of concrete structures, the open and green spaces decrease and the nature is pushed out of the city. It is almost impossible to see green lands in settlement areas. Urban life loses its quality day by day, adversely affecting the inhabitants of the city. We need more greenery in the working and living spaces where we spent the most of our daily lives. Spaces available outside of the building are mostly used for roads and parking lots. Landscape design around the buildings is not sufficient to improve the adverse effects and to provide for the needs of the people. Green walls both contribute to natural life and turn the facade of concrete structures green.

© Springer International Publishing AG, part of Springer Nature 2018
S. Fırat et al. (eds.), *Proceedings of 3rd International Sustainable Buildings Symposium (ISBS 2017)*, Lecture Notes in Civil Engineering 7,
https://doi.org/10.1007/978-3-319-64349-6_8

Green walls are the applications based on the principle of horticulture on the exterior of structures. It is a practice aimed at increasing the effect of green spaces in cities. There are positive environmental impacts of green walls for big cities where there is not enough space to have large gardens. Therefore, green wall applications are increasing rapidly all over the world, especially in European countries.

2 Green Walls

Green walls provide biological value as they create a healthful and qualified urban environment [1]. The green wall applications bring a new understanding into modern urban culture applying "horticulture" principles to wall surfaces as a concept [2].

2.1 History of Green Walls

Green wall applications, which are fairly common nowadays, were built in Babylon about 2500 years ago as the first example. Hanging Gardens of Babylon are called as the father of modern green walls and were built by King Nebuchadnezzar (Fig. 1) [3].

Fig. 1. Hanging Gardens of Babylon [3]

In 1920, climbing plants and fences on houses-model of vertical gardens—were encouraged in Britain and North America. In the late twentieth century, stainless steel cable systems were trained on green facades, and modular trellis panel and cable net systems governed the market in North America. In 1993 and 1994, universal city walk on California was the primary example of the trellis panel system [3].

In ancient Rome, people living in Pompeii grew grapes on their balconies, and Romans planted trees around the monuments of emperors. Vikings sodded roofs and the walls of houses to seek shelter from the rain and wind. During the Renaissance period, roof gardens and green walls became popular in Genova. In the sixteenth and seventeenth centuries, vertical gardens consisting of clinging-climbing plants were trained in Mexico, India and partially in Spain. Grape gardens came into prominence in Kremlin, Russia, in the early twentieth century; plant components were used on building walls and even in St. Petersburg Airport which was aspiring for adornment. In eighteenth century, snipped green walls were designed for decoration on open spaces. In 1930s, Le Corbusier developed the idea of attaching green and road roofs to urban areas, and also Frank Lloyd Wright used green roofs and vertical gardens in Midway gardens, Hollyhock House, Cheney House and Falling water house; two prominent names of the modern architecture [4].

Until the mid-twentieth century, green walls were used on traditional construction. However, green walls became popular as concept designs especially in northern Europe [5]. Green walls are visually enriched elements reviving the environment, besides they regulate the air indoors and outdoors. Green walls are decoratively easy on the eye at the same time improving the variety of diurnal species [3, 6].

2.2   Green Wall Applications

As a type of a green wall system, green facades can be illustrated as creeping plants and cascading vegetation. Green facades can be reconstructed on fences or columns, embedded in existing walls. Modular trellis panel, grid system, and a wire-rope net system are most commonly used as green facade systems [6, 7].

The rigid, light weight, three dimensional panel which is made from a powder coated galvanized and welded steel wire supporting plants with both a face grid and a panel depth, is the building block of this modular system. The reason behind not directly attaching the plantation components on structure is to the design a system to hold a green facade off the wall surface, thus helping the maintenance and integrity of the building membrane and ensuring a "captive" growing environment for the plant with supporting structures for tendrils. The characteristic of this panel is that it is recyclable and is made of recycled steel. Panels can be used to form shapes and curves, or can be piled up to create more green spaces. Panels can extend between structures and may be used for freestanding walls too, thanks to their rigidness [3].

Grids, which are used on green facades, assist climbing plants to grow faster allowing dense foliage. Meanwhile, plants which grow slower will need wire-net systems installed with smaller intervals. Anchors, high tensile steel and additional materials are used on both systems. Cable nets are often used to support slow-growing plants, which require additional support from a closer proximity. Both systems use high-tension steel cables, anchors and ancillary equipment. Flexible vertical and horizontal wire ropes can accommodate various sizes and patterns as they are attached with cross clamps [3, 6].

Green walls are also known as bio-walls or vertical gardens. These systems consist of pre-vegetated panels, vertical modules and erosion control blankets. These panels can be made of plastic, expanded polystyrene, synthetic fabric, clay, metal or concrete, and support a great variety and density of species. Green walls can be made of several materials and have a higher density, therefore, they need more protection than facades. Green walls consist of three components: metal frame, a PVC layer and air layer (soil is not necessary). Mixed vegetation, perennial flowers, low shrubs and ferns etc. are supported by this system. In order for easy maintenance of green walls and their adaptation to the prevailing climatic condition, suitable species should be selected.

Generally self-automated watering and nutrition systems are used to perform at various climatic conditions [3, 6].

Thanks to a series of technological advances, modules for green roof applications are designed and their usage is increased on modular green wall systems. Modular systems consist of square or rectangular panels which contain growing materials to assist plant growth [3].

Types of Green Walls. Green walls raise a new understanding of modern architecture. By means of green walls used for outdoors or indoors applications, the quality of green effect enhances and air pollution decreases. Besides being a decorative material, green walls actually have a refreshing effect on the buildings and save the structures from gloomy atmosphere with its green colour. Green facade systems are used in metal fence system, modular system and panel system.

Panel System Vegetation. This system can be implemented directly on facade or on lower construction. Green walls in which panel system is implemented on facade has the positive effect on decreasing the high temperature absorbing the heat [8]. A panel system which is flexible, modular and durable to heavy rain, earthquake and wind, can be used for all kinds of climate conditions.

Metal Fence System Vegetation. Using a metal fence system is the best way to constitute semi permeable green texture of which direction and dimension is previously defined. It is possible to design several kinds of green texture using various metal fences. In practice, the height of the metal fences changes depending on its shape; 450 cm in height and 2D or 3D fence applications are most common. The type of the plant used is also important [8]. The species of plant used should be a climber (creeper) species in this application. The plant can be placed in a pot, also it can be planted directly on the ground. Water need of the plants is supplied by drip irrigation system [9, 10].

Modular System Vegetation. Modular system is designed using pots of various shape and scale on facades or on the lower construction [8]. If the system is arranged modularly, fertilization and pruning processes can be easily performed on each potted plant. Liquid flow should be vertical using drained flowerpots which are situated by modular system in order to provide efficient us of the sprinkler system. Thus, liquid manure and nutrients can easily be applied to plants by the sprinkler system [11].

Plant modules are fixed by fastening link rods on the bearing profile. System equipment are located away from mechanic room which is connected to the units. Electric board, manure tank, water tank, water pumps and lime crusher are located in mechanic room [2].

Types of Plant Species for Green Walls. The choice of plant to be used on green walls should be suitable to the purpose and aim of the design. Plant species may have various effects on facades depending on the season or light need. It is crucial to use plant species which can grow under specific ecologic conditions when cultivating the wall surfaces. Plants used on Northward surfaces must be durable to cold and must be shade-tolerant, plants used on Southward surfaces must be selected according to their heat-resistance. As an example, on Southward facade wisteria florubinda, *Jasminum nudiflorum* and rosa sp. are suitable while hedera helix, hydreangea etc. are suitable for

Northward facades. *Humulus lupulus, Lonicera caprifolium* etc. should be preferred on eastward or westward wall facades [12]. There are three types of plants commonly used on green walls; self-clinging climbers, twining climbers and rambling shrubs.

Self-Clinging Climbers. Self-clinging Climbers generally do not need support but on plain walls support can be necessary; main species are *Hedera helix, Parthenocissus quinquefolia, Parthenocissus tricuspidata, Hydrangea petiolaris,* and *Euonymus fortunei* [13].

Twining Climbers/Creepers. Creepers need support; thin iron wire, rough plastic surfaces and lumber lath are generally used. Wire or wooden lattices are needed for some species. *Capsis radicans, Passiflora caerulea, Lathyrus odoratus, Polygonum baldschuanicum, Lonicera periclymenum,* Lonicera spp., *Clematis vitalba,* Clematis spp., *Humulus lupulus,* Aristolochia spp., *Jasminum officinale,* Vitis spp., Wisteria spp. are some examples of creepers [13].

Rambling Shrubs. Rambling shrubs are not climbers but they can prove creeper characteristics by holding on to latticed grids of the wall. *Rubus fruiticous, Jasminum nodiflorum, Rosa canina,* Rosa spp., *Forsythia suspensa,* Cotoneaster spp. and *Pyracantha atalantioides* are rambling shrubs commonly used for walls [13].

2.3 Advantages and Disadvantages of Green Wall Applications

Vertical vegetation systems offer a great variety of individual and social benefits. Individual benefits of green façades and living wall systems on a single outer shell scale are primarily related to energy savings from heating and cooling, increase of real estate value (or rent), and durability of facades. It is proved that vertical green layer create a more static air layer and contributes to the performance of building, reducing the energy demand by 40–60% in Mediterranean climate [14–16].

The benefits of green walls are variable depending on different building types, green wall technologies used, plant choices and plant coverage. Green walls provide economic, ecologic and aesthetic value. Depending on the density of the plants on the green wall, they provide shade for the building. Shade ensures the ambience and building are cooled relatively, so temperature reduction affects the building not to mention the urban environment. Plants can do it in two ways: direct shading and evapotranspiration. Trees are effective on absorbing gaseous and particulate pollutants from the atmosphere, improve the air quality by filtering out airborne particles in their leaves and branches [17, 18].

Nowadays, green wall applications are also seen on the urban areas with high-rise buildings and skyscrapers covering the building facades so neither limited space nor lack of land stands a problem nor the increase in air quality is inevitable [19]. Plants located around the building can improve construction stability reducing the effects of weather. Reduction of the climatic stress on building facades, prolonging the lifetime of the building, reduced costs, however insignificant, are the outcome of green wall applications [13].

The adverse effects of climate change and environmental pollution on urban life increase day by day. Spending most of our time between walls can cause psychological problems. But green zones are the places that provide the people the room to breathe. In daily life, people need green zones to freshen up. Green walls are one of the recommended options for developing environmentally sensitive and energy efficient designs. Green walls change the appearance of the cities. Since they have a positive effect on micro-climate, green walls have an important role in sustainability and green cities. Vegetation cover helps to stabilize temperature fluctuations, thus, provides protection against sunlight and wind while absorbing the noise [20].

Green walls have ecological, aesthetic, and economic impacts on residential areas, and these are as follows:

- They contribute to the conservation of habitat and biodiversity,
- They reduce the negative effects of urban heat islands,
- They absorb noise,
- They reduce the effects of the cooling winds by decreasing the heat losses due to wind,
- They ensure filtration of airborne particles,
- They contribute to the formation of a healthier environment absorbing pollutants and harmful substances such as dust available in the air,
- They are effective in decreasing air pollution and increasing the amount of oxygen,
- They substantially reduce the electromagnetic radiation buildings are exposed to,
- They provide significant aesthetic contributions to urban areas with numerous alternative design possibilities,
- They save energy by reducing the amount of heat lost through convection [20].

When it comes to sound level, a green wall ensures acoustic benefits. The improvement of the environmental conditions dense urban areas, reduced greenhouse gas emissions, adaptation to climate change, improved air quality, enriched urban wildlife, etc. are counted in the social benefits related to green surfaces [21]. Finally another benefit is that plants available on the green walls can absorb the sound acting like a sound barrier. Green walls which are one of the most eye-catching solutions designed against concrete urban invasion reduce the noise pollution in urban areas [19].

There are many positive aspects of the green walls as well as a few negative ones. When designing the building, it is necessary to consider the weight of the plant growth on the wall with respect to the effects of wind, snow, etc. loads. Green wall application costs more than the construction of a normal wall. Incorrect irrigation methods can also damage building walls. Damage occurs on the surface and also on vegetation on the walls. Additional maintenance costs are created due to green walls [22].

3 Green Wall Applications in Turkey

Population increase in the cities leads to construction of high-rise and densely located structures. With the change in the cities and therefore, in the park areas, green spaces are gradually diminishing and the citizens are spending most of their time between streets and the pavements moving further away from the nature day by day. In cities, the need to have more green spaces, the need to commune with nature in every opportunity and protection of natural resources have become even more important.

In recent years, awareness has been created emphasizing the importance of green living in every platform. Environmentalists and designers work together to create solutions for urban problems working on sustainable architecture and green buildings. Green wall applications are commonly implemented in the world. In Turkey, especially in the last years, there are many examples on both interior and exterior walls. Some of these examples from interior spaces are given in Table 1, and from exterior spaces in Table 2.

Table 1. Green wall applications in interior spaces in Turkey [23, 24]

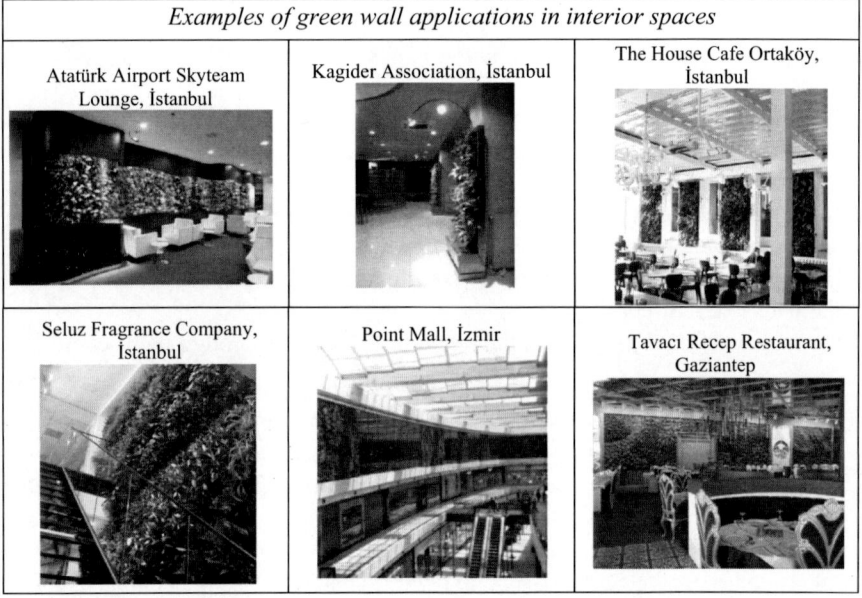

Green wall applications, which gain momentum in many Turkish cities in the recent years, are mostly found in Istanbul. The green wall applications in Turkey are mostly found on the external walls. As for the type of plant species, it is observed that all kinds of climbers, twining climbers and rambling shrub plant species are used, whereas the panel system is most commonly preferred. The types of green walls and plant species available in the examples given in Tables 1 and 2 are shown in Tables 3 and 4.

Table 2. Green wall applications in exterior spaces in Turkey [23, 24]

Examples Of Green Wall Applications In Exterior Spaces		
Siemens Shop, İstanbul	Sabiha Gökçen Airport, İstanbul	Atatürk Airport, İstanbul
Atatürk Airport Crossway, İstanbul	Antalya Airport, Antalya	İstanbul Metropolitan Municipality, Başakşehir
İstanbul Metropolitan Municipality, Edirnekapı	İstanbul Metropolitan Municipality, Merter	İstanbul Metropolitan Municipality, Altunizade
İstanbul Metropolitan Municipality, Topkapı	İstanbul Metropolitan Municipality, Güneşli	Esenler Municipality, İstanbul
Gaziosmanpaşa Municipality, İstanbul	İzmir Metropolitan Municipality, Narlıdere Crossroad, İzmir	Adana Metropolitan Municipality, Adana
Diyarbakır Metropolitan Municipality, Diyarbakır	Gebze Municipality, Kocaeli	Yeşilyurt Municipality, Malatya

Examples Of Green Wall Applications In Exterior Spaces

Körfez Municipality, Kocaeli	Beyşehir County Municipality, Konya	Sultangazi Municipality City Square, İstanbul
Üsküdar Municipality Jewelry Bazaar, İstanbul	Üsküdar Municipality Fishermen Bazaar, İstanbul	Swissotel Resort Bodrum Beach, Bodrum
Bodrum Mandarin Oriantel Hotel, Bodrum	*Hanna Boutique Otel, İstanbul*	Sinpaş GYO Ege Houses, Ankara
Aras Houses, İstanbul	Trio Houses, İstanbul	Erasta Mall, Antalya
Çağdaş Holding Duravit Shop, Bodrum	Point Mall, İzmir	Bağcılar Wedding Hall, İstanbul

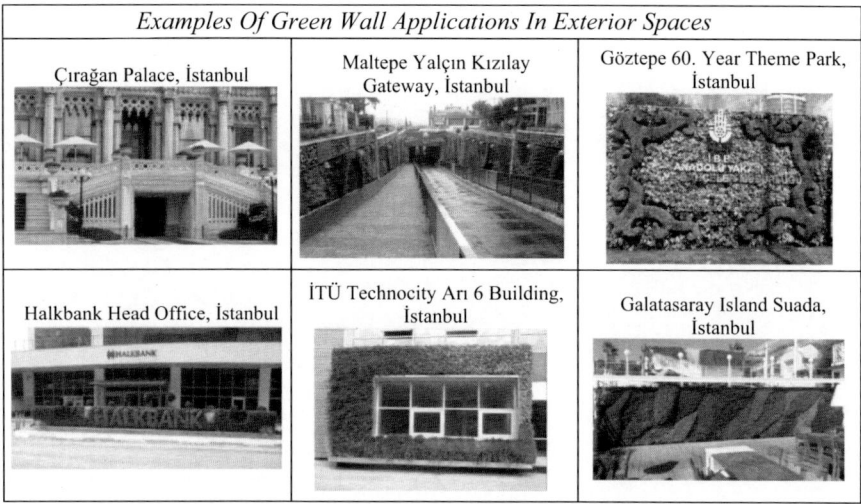

Examples Of Green Wall Applications In Exterior Spaces		
Çırağan Palace, İstanbul	Maltepe Yalçın Kızılay Gateway, İstanbul	Göztepe 60. Year Theme Park, İstanbul
Halkbank Head Office, İstanbul	İTÜ Technocity Arı 6 Building, İstanbul	Galatasaray Island Suada, İstanbul

Table 3. Green wall applications in interior spaces according to different types of green walls and plant species

Examples of interior spaces	Types of green walls			Plant species used		
	Panel system vegetation	Metal fence system vegetation	Modular system vegetation	Self-clinging climbers	Twining climbers creepers	Rambling shrubs
Atatürk Airport Skyteam Lounge, İstanbul	√			√	√	√
The House Cafe Ortaköy, İstanbul	√			√	√	
Seluz Fragrance Company, İstanbul	√				√	√
Point Mall, İzmir	√				√	√
Kagider Association, İstanbul			√		√	√
TavacıRecep Restaurant, Gaziantep	√			√	√	√
Siemens Shop, İstanbul	√			√	√	√
SabihaGökçen Airport, İstanbul	√				√	

Table 4. Green wall applications in exterior spaces according to different types of green walls and plant species

Examples of exterior spaces	Types of green walls			Plant species used		
	Panel system vegetation	Metal fence system vegetation	Modular system vegetation	Self-clinging climbers	Twining climbers creepers	Rambling shrubs
Atatürk Airport, İstanbul	✓				✓	
Atatürk Airport Crossway, İstanbul	✓				✓	✓
Antalya Airport, Antalya	✓				✓	
İstanbul Metropolitan Municipality, Başakşehir	✓				✓	
İstanbul Metropolitan Municipality, Edirnekapı	✓			✓	✓	✓
İstanbul Metropolitan Municipality, Merter	✓			✓	✓	✓
İstanbul Metropolitan Municipality, Altunizade	✓				✓	
İstanbul Metropolitan Municipality, Topkapı	✓			✓	✓	
İstanbul Metropolitan Municipality, Güneşli	✓					✓
Esenler Municipality, İstanbul	✓			✓	✓	✓
Gaziosmanpaşa Municipality, İstanbul	✓			✓	✓	✓
İzmir Metropolitan Municipality, Narlıdere Crossroad, İzmir	✓			✓	✓	✓
Adana Metropolitan Municipality, Adana	✓				✓	

(continued)

Table 4 (continued)

Examples of exterior spaces	Types of green walls			Plant species used		
	Panel system vegetation	Metal fence system vegetation	Modular system vegetation	Self-clinging climbers	Twining climbers creepers	Rambling shrubs
Diyarbakır Metropolitan Municipality, Diyarbakır	✓			✓		
Gebze Municipality, Kocaeli	✓			✓	✓	
Yeşilyurt Municipality, Malatya			✓		✓	
Körfez Municipality, Kocaeli	✓			✓		✓
Beyşehir County Municipality, Konya	✓					
Point Mall, İzmir	✓			✓	✓	✓
Sultangazi Municipality City Square, İstanbul	✓			✓	✓	✓
Üsküdar Municipality Jewelry Bazaar, İstanbul	✓			✓	✓	
Üsküdar Municipality Fishermen Bazaar, İstanbul	✓			✓	✓	✓
Swissotel Resort Bodrum Beach, Bodrum	✓			✓	✓	✓
Bodrum Mandarin Oriantel Hotel, Bodrum	✓			✓	✓	✓
Hanna Boutique Otel, İstanbul	✓			✓	✓	✓
Sinpaş GYO Ege Houses, Ankara	✓			✓	✓	
Aras Houses, İstanbul	✓					✓
Trio Houses, İstanbul	✓				✓	✓

(continued)

Table 4 (*continued*)

Examples of exterior spaces	Types of green walls			Plant species used		
	Panel system vegetation	Metal fence system vegetation	Modular system vegetation	Self-clinging climbers	Twining climbers creepers	Rambling shrubs
Erasta Mall, Antalya	✓				✓	
Çağdaş Holding Bodrum Duravit Shop, Bodrum	✓				✓	✓
Galatasaray Island Suada, İstanbul	✓			✓	✓	
Bağcılar Wedding Hall, İstanbul	✓			✓	✓	✓
Çırağan Palace, İstanbul			✓	✓	✓	
MaltepeYalçınKızılay Gateway, İstanbul	✓			✓	✓	✓
Göztepe 60. Year Theme Park, İstanbul	✓			✓	✓	
Halkbank Head Office, İstanbul	✓			✓	✓	✓
İTÜ Technocity Arı 6 Building, İstanbul	✓			✓	✓	✓

4 Conclusions and Recommendations

Increasing population of the cities leads to construction of high-rise and densely located structures. With this change in the cities, green areas like parks, gardens, etc. are gradually diminishing and citizens who are forced to spend most of their lives between streets and blocks of buildings are being pushed further away from nature every other day. The need to find room for green spaces in cities and protection of natural resources has become more and more important in every opportunity of reuniting with nature. Green areas can be emphasized in the roofs of buildings, in courtyards, inside and outside the walls, in short, horticulture applications on all building facades can be conducted without hindering the land use and development. Planted facade applications, which are becoming increasingly widespread in the world, have become an important industry in Europe, especially in terms of the infrastructure requirements of the system and their provision. However, such applications are generally not preferred for high-rise structures as they considerably increase both construction costs and expenditure.

References

1. Barış ME, Yazgan ME, Erdoğan E (2003) Roof gardens (In Turkish: Çatı Bahçeleri). Potted Ornamental Plants Publications, Ankara, p 67
2. Erdoğan E, Khabbazi PA (2013) Use of plants on building surfaces, vertical gardens and city ecology (In Turkish: Yapı Yüzeylerinde Bitki Kullanımı, Dikey Bahçeler ve Kent Ekolojisi). Turk J Sci Rev 6(1):23–27
3. Timur ÖB, Karaca E (2013) Vertical gardens. Adv Landsc. doi:10.5772/55763
4. Peck SW, Callaghan C, Kuhn ME, Bass B (1999) Greenbacks from green roofs: forging a new industry in Canada, CMHC, Toronto, p 78
5. Köhler M (2008) Green facades—a view back and some visions. Urban Ecosyst 11(4):423–436. doi:10.1007/s11252-008-0063-x
6. Yeh YP (2010) Green wall—the creative solution in response to the urban heat island effect. National Chung-Hsing University, Taiwan
7. Gonchar J (2009) Vertical and verdant, living wall systems sprout on two buildings. Paris and Vancouver, Architectural Record, McGraw-Hill Construction, Retrieved on Aug 2009, p 20
8. Yücel G, Elgin Ü (2010) Wall garden: vertical garden/green wall (In Turkish: Duvar Bahçesi: Dikey Bahçe/Yeşil Duvar). Blue Struc Mag Year 1:51–53
9. Lambertini A, Leenhardt J (2007) Vertical gardens: bringing the city of life. Thames & Hudson, London. ISBN: 978-0-500-51369-9
10. Van Uffelen C (2011) Façade greenery: contemporary landscaping. Braun Publishing, Salenstein
11. İpekçi CA, Yüksel E (2012) Planted building shell systems (In Turkish: Bitkilendirilmiş Yapı Kabuğu Sistemleri). 6. National roof and façade symposium, Uludag University, Faculty of Engineering and Architecture, Görükle Campus, 12–13 April, Bursa
12. Seçkin NP (2011) When perceiving green cover that approaches to the sun (In Turkish: Güneşe Yaklaşan Yeşil Örtüleri Algılarken). In: Material in architecture, UCTEA, TMMOB the chamber of architects Istanbul Büyükkent branch publication, year, 6, pp 42–50

13. Johnston J, Newton J (2004) Building green: a guide to using plants on roofs, walls and pavements. Greater London Authority. London, ISBN: 1 85261 637 7
14. Perini K, Ottelé M, Fraaij ALA, Haas EM, Raiteri R (2011) Vertical greening systems and the effect on air flow and temperature on the building envelope. Build Environ 46(11):2287–2294. doi:10.1016/j.buildenv.2011.05.009
Mazzali U, Peron F, Scarpa M (2012) Thermo-physical performances of living walls via field measurements and numerical analysis. In: Eco-architecture IV. Harmonization between architecture and nature, vol 165, pp 239–250. doi:10.2495/ARC120011
16. Alexandri E, Jones P (2008) Temperature decreases in an urban canyon due to green walls and green roofs in diverse climates. Build Environ 43(4):480–493
17. Loh S (2008) Living walls—a way to green the built environment. BEDP Environ Design Guide 1(TEC 26):1–7
18. Dwyer JF, Schroeder HW, Gobster PH (1994) The deep significance of urban trees and forests. In: the ecological city: preserving and restoring urban biodiversity. pp 137–150
19. Dunnett N, Kingsbury N (2008) Planting green roofs and living walls. Timber Press, Portland
20. Çelik A, Ender E, Zencirkiran M (2015) Vertical garden and related applications in turkey (In Turkish: Dikey Bahçe ve Türkïye'dekï Uygulamaları. J Agric Sci 8(1):67–70. ISSN: 1308-3945, E-ISSN: 1308-027X. www.nobel.gen.tr
Perini K, Rosasco P (2013) Cost Benefit Analysis for Green Façades and Living Wall Systems. Build Environ 70(11):110–121
22. Ottele M (2011) The green building envelope vertical greening. Delft Technical University, Delft. ISBN: 978-90-9026217-8
23. http://www.dikeybahcem.com/projelerimiz/
24. http://www.aktasplant.com/universal-grid-gallery/duvarbahce/

Achieving Sustainable Buildings via Energy Efficiency Retrofit: Case Study of a Hotel Building

Muhsin Kılıç$^{(\boxtimes)}$ and Ayşe Fidan Altun

Mechanical Engineering Department, Engineering Faculty, Uludag University,
Gorukle Campus, 16059 Bursa, Turkey
{mkilic, aysealtun}@uludag.edu.tr

Abstract. Recent studies show that demand for green buildings with minimal environmental impact is increasing. Among the buildings in anywhere, existing buildings consume more energy than new ones. Because of that, building energy efficiency retrofit plays a very important role for achieving sustainable building targets. In addition to creating possibilities to reduce energy consumption, retrofitting also improves occupants' health and comfort. Tourism sector is one of the largest energy consuming sectors in the world. However conducted energy retrofit studies mainly focused on domestic buildings so far. The focus of this paper is to make a literature review of energy efficiency retrofits and present an approach to find the most appropriate, energy efficient and cost effective way renovate a hotel building. In this paper, a hotel building is energy efficiently retrofitted according to the TSE Green Building Certificate standards. Results from the analysis ensure that energy efficiency retrofit, helped to reduce energy consumption of the building and decreased in operational costs and CO_2 emissions.

Keywords: Green buildings · Sustainable design · Energy efficiency retrofit

Nomenclature

BREEAM	Building Research Establishment Environmental Assessment Method
CASBEE	Comprehensive Assessment System for Built Environment Efficiency
DGNB	German Suitable Building Council
HVAC	Heating, Ventilating, and Air Conditioning
LEED	Leadership in Energy and Environmental Design
TOE	Tonne of Oil Equivalent
TSE	Turkish Standards Institution
PW	Net Present Worth
n	Service life (years)
i	Interest rate
A	Annual expenditure
C_o	Initial investment
RP	Repayment period
R_1, R_2, R_n	Cash flow (annual cost savings)
R	Annual energy cost savings

© Springer International Publishing AG, part of Springer Nature 2018
S. Fırat et al. (eds.), *Proceedings of 3rd International Sustainable Buildings Symposium (ISBS 2017)*, Lecture Notes in Civil Engineering 7,
https://doi.org/10.1007/978-3-319-64349-6_9

1 Introduction

Global warming, environmental pollution, extinction of fossil fuels and the price of energy rises are the biggest problems of the humanity. Construction industry consumes of the 40% of global resources, 12% of potable water reserves, 55% of wood products, 45–65% produced waste, 40% raw materials, and the emission of 48% of hazardous greenhouse emissions, which results to environmental pollution and global warming [1].

As a result, many countries have introduced green building rating systems for sustainable construction. These include but not limited to United States' Leadership in Energy and Environmental Design (LEED), United Kingdom's BREEAM and Turkey's TSE Green and Secure Building.

Over the past years, as being the 17th largest economy in the world, Turkey has experienced considerable growth and energy use has grown at an annual rate of about 4.5% from 2013 to 2015 [2]. According to the Turkish Statistical Institute's "2014 Energy Consumption Statistics of Turkey", service sectors consumed 14 million 597 thousand TOE energy in 2014 [3]. For this reason, energy efficiency is a critical issue for Turkey.

Among the buildings in anywhere, existing buildings consume more energy than the new ones. As a result, building energy efficiency refurbishment plays a mandatory role in order to achieve green building goals. The term "retrofit" has been used to describe a variety of improvements to an existing building or group of buildings [4]. According to Ibn-Mohammad et al., retrofitting existing building stock can decrease CO_2 emissions 15 times more by 2050 than their demolition and today's technological improvements offer promising refurbishment solutions [5]. The outcome of this paper is energy efficiency retrofit of an existing hotel building facility with optimal heating, ventilation and air conditioning (HVAC) system solutions which can meet the demands of the users. And eventually, prove that sustainably renovated buildings can generate numerous benefits in terms of energy and CO_2 reduction, cost savings and healthier environments.

Results from the assessment ensure that energy efficiency refurbishment, helped to decrease in energy consumption of the building and consequently reduced operational costs and CO_2 emissions in a very short payback period. Obtained results might be generalized to other similar realities and encourage investors and building owners to renovate.

1.1 Green Buildings

Over the past years, Turkey's economy has expanded and it became the 17th largest economy in the world. As a result, energy consumption increased considerably. Since, Turkey imports nearly all of its energy resources, it has to minimize its energy consumption. About 40% of whole energy consumption in Turkey belongs to building sector while it has a large saving potential, as a result, improvement of building energy performance seems inevitable [2]. Green building is a method which is environmentally responsible, supports resource efficiency throughout the buildings life cycle: during design, construction, maintenance, renovation and demolition [6]. In other words, a green building is one, whose construction and lifetime of operation assure the healthiest environment while minimizing resource utilization and greenhouse gas emissions [7]. There are numerous sustainable assessment schemes all around the world which aim to rate a building's environmental performance. Some of the widely used green building assessment schemes are GBTool (Canada), CASBEE (Japan), BREEAM (Britain), BEAM (Hong Kong), Green Star (Australia), LEED (USA), DNGB (Germany), and TSE Green and Secure Buildings (Turkey). Generally, they aim to assess the impact of the building on its environment. During the assessment, local conditions and regional priorities should take into consideration. For instance, as stated in a study of Suzer [8], the environmental concern priorities for buildings located in Middle East and Northern America may differ since both regions have different climatic conditions and natural resources. In order to assess a building's energy performance accurately, it is better to use the country's own assessment scheme. As a result of that, Turkish Standards Institution (TSE) prepared a national green building certification (TSE Green and Safe Building Certificate) according to Turkey's priorities and requirements. This certificate aims to measure sustainability, fire, earthquake and domestic safety of the buildings. However, other international certifications mostly do not take into account earthquake safety since it is not a threat for them. It is better to evaluate buildings with the certifications which prepared according to national conditions. Below, LEED and TSE Green Building Certificate categories and their weights are shown in Tables 1 and 2. Improving energy efficiency of existing buildings plays a vital role in order to achieve the goals of sustainable buildings. Energy retrofitting reduces CO_2 emissions, utility bills and maintenance costs and create jobs and career opportunities [9]. There are a number of approaches of building retrofitting such as installation of renewable energy systems, enhancing building envelope and upgrading HVAC systems [10]. HVAC system is the highest energy consuming component in a building, therefore improving HVAC system contributes in greater energy savings within the building [11]. In this study, significant attention has been paid to improve the energy efficiency of building's HVAC system.

Table 1. LEED NC, v.3 categories and their weights

LEED NC, v.3 categories	Category weights (points)	Percentage
Sustainable sites	26	23.6
Water efficiency	10	9.1
Energy and atmosphere	35	31.8
Materials and resources	14	12.7
Indoor environmental quality	15	13.6
Innovation	6	5.6
Regional priority	4	3.6
Total	110	100

Table 2. TSE Green Certificate categories and weightings

TSE Green Building Certificate categories	Category weights (points)	Percentage
Sustainable sites	30	9.77
Water efficiency	33	10.74
Energy and atmosphere	120	39.08
Materials and resources	33	10.74
Health, safety and comfort	34	11.07
Innovation	6	1.95
Carbon footprint	5	1.62
Facility management	13	4.23
Premium points	25	8.14
Land choice	8	2.60
Total	307	100

2 Methodology

As the 6th most popular tourist destination in the world, Turkey generates almost 30 billion USD from tourism sector, annually. By the end of 2015, there were 13,615 accommodation facilities in Turkey [12]. Since hotels are the most energy intensive of all building categories, their energy use and environmental impact can be quite large, especially in popular tourist destinations [13]. As a result, improving energy efficiency of existing hotel buildings is a necessary action in terms of reducing CO_2 reduction and improving thermal comfort levels for occupants. In order to identify the most suitable retrofitting actions and assess their effectiveness when implemented in a building, it is necessary to have available specific information on the energy characteristics, thermal performance, comfort conditions and existing problems [14]. The methodology of this paper is described below;

- Selection of the case study building
- Identify climate conditions
- Identify major energy losses (Survey of existing envelope, thermographic inspection, lighting etc.)
- Create an energy profile of the building by using available data on monthly electric and fossil fuel use
- List retrofit options according to Green and Secure Building Standards
- Make a cost/benefit analysis.

2.1 Selection of the Case Study Building

Thermal hotel facility which is located in Oylat, Bursa was selected in this study to serve the research purpose. Famous thermal water springs of Oylat has been using in the hotel's spa and baths. Hotel facility has 252 rooms and 5 floors above ground with a total floor area of 11,808 m^2 (Fig. 1). In particular, coal is used for the facility heating and thermal water is used in baths and spas.

2.2   Climate Conditions

The Turkish State Meteorological Service and TSE (Turkish Standards Institution) classified Turkish climate regions as "Thermal Insulation Regions" by using a degree-day method which was developed by the Turkish State Meteorological Service. In Turkey, there are four thermal zones and each zone demand different envelope characteristics. The limited U value of building envelope in different degree-day zones is shown in Table 3. Based on TS 825-2013, Bursa is the representative city for second climate zone.

Fig. 1. Hotel building

2.3   Thermographic Inspection

As means of uncovering inefficiencies in the HVAC system and building envelope, a thermographic inspection of the case study building was performed by using an infrared camera. Infrared images produced during the inspection were analyzed for heat loss. According to the visual inspection, building envelope appears to be in good condition and meet the current regulation demands. The inspection also revealed that mechanical installation is a significant source of heat loss and has a high potential for energy efficiency improvements. Due to TS-2164 thermal installation regulation, temperature difference between insulated surface and indoor air supposed to be of a maximum of 5 °C. However, thermographic inspection revealed that temperature difference is between 9 and 12 °C (Fig. 2). As a result of that, most of the pipelines have been replaced with pre-insulated geothermal pipes.

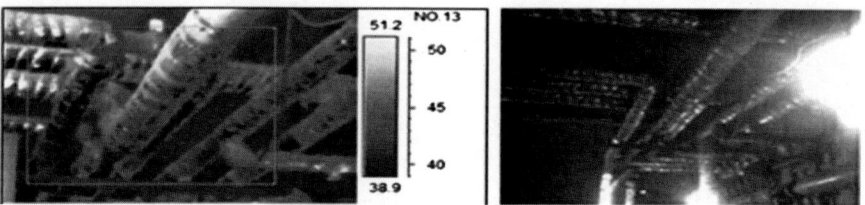

Fig. 2. Thermographic inspection of installation systems

Table 3. The limit U values different degree-day zones

	U exterior wall (W/m² K)	U ceiling (W/m² K)	U ground floor (W/m² K)	U windows (W/m² K)
1. DDZ	0.66	0.43	0.66	1.8
2. DDZ	0.57	0.38	0.57	1.8
3. DDZ	0.48	0.28	0.43	1.8
4. DDZ	0.38	0.23	0.38	1.8
5. DDZ	0.36	0.21	0.36	1.8

2.4   Lighting Audit

Energy conversation from artificial lighting is very important in hotels, since several areas of the building remain with the lights on throughout the night hours [14]. Almost 21.3% of the electricity that generated in Turkey is consumed for lighting, and if a good design is made, the total energy consumption for lighting purposes can be reduced by 20%. In order to investigate the efficiency of the lighting system, light measurement is made by using a lux meter. According to the measurements, lighting levels of the interior spaces need to be enhanced in terms of safety and comfort (Fig. 3).

Fig. 3. Hotel lobby lighting

2.5   Energy Audit Results Before the Retrofit

Based on the collected bills of 2015, annual energy consumption of the hotel is shown in Table 4. According to the investigation, the annual total electricity consumption is 1,140,685 kWh and annual consumption of coal is 669,210 kg. Data refer to the whole building consumption show a predominant use of coal with respect to electricity.

Table 4. Annual energy consumption of the hotel building

Energy	Annual consumption		Annual cost	
		TOE	TL	TL/TOE
Electricity	1,140,685 kWh	98.10	220,427	2247
Coal	669,210 kg	468.45	428,294	914.29

2.6   Proposed Retrofit Actions

There are a number of approaches of the building retrofitting and previous studies have predominantly focused on the technological aspects such as the installation of renewable energy systems, enhancing building envelope etc. [10]. The HVAC systems are one of the biggest energy consumers in hotel buildings. In this study, significant attention is paid to HVAC systems in order to transform an existing building to a green building.

The heat source of the building is coal and central heating system was used for the building. Annually, 670 t of standard coal is required for heating the building. Water source heat pumps (WSHP) have gained an increasing level of application in Turkey due to their higher COP values. Therefore WSHP with 3.8 COP was selected as the heat source of the retrofit. In addition to WSP, a condensing, fuel fired boiler was selected.

Since hotel building is located in Oylat, thermal water with a daily capacity of 4300 t can be reused by WSHP. Thermal water flow from the source at approximately 40 °C and it is discharged after using baths and spas of the hotel facility at 30 °C. Thermal water that discharged at 30 °C after being used in baths and spas, will be collected and pumped to WSHP in order to use the waste heat. Waste water will enter WSHP at 30 °C and exit at 70 °C. By this way energy consumption will be reduced and operational cost will be saved.

2.7   Energy Audit Results After the Retrofit

After retrofit implementations, the annual primary energy use and the cost of saved energy is presented in Table 5. According to the results, after retrofit, reduction of annual energy consumption is 528.57 TOE and reduction of annual energy cost is 294,000 TL.

Table 5. Energy consumption and energy cost after retrofit

	Energy	After retrofit		Annual energy cost (TL)
		Annual energy consumption		
Heat pump	Electricity	255,194 kWh/year	21.95 TOE	112,285
Boiler	Natural gas	19,430 m³/year	16.03 TOE	21,917

2.8 Evaluation of the Building According to TSE Green Building Certificate

In this part of the study, TSE Green Building Certificate energy category evaluation is applied to the hotel building. According to category requirements, renovated building is evaluated and scored. Scores after the evaluation is shown in Table 6.

For classification of buildings according to primary energy, CO_2 emissions, to set minimum energy performance requirements for major renovations of existing buildings, to evaluate feasibility of renewable energy sources, to provide inspection of heating and cooling systems, to limit greenhouse gas emissions, to determine building energy performance measures and to protect environment, Turkish Building Energy Performance Regulation came into force in December 2008 [2]. Eventually, obtaining an energy performance certificate for new and existing buildings is mandatory.

In order to be evaluated according to Energy category credits of TSE Green Building Certificate, Turkish Building Energy Performance Certificate document must be provided. After applying retrofit actions, hotel building's annual energy consumption and greenhouse gas emissions shown in Fig. 4. According to this certificate, the hotel building's energy performance class is B and CO_2 gas emission class is C.

2.9 Economic Analysis of the Retrofit

The economic problem of allocating limited resources to various needs often requires making cost-benefit analysis where costs and benefits over the lifetime of the project are evaluated and investments with positive net benefits are considered to be acceptable [15].

In this section, costs related to the investment is estimated and appropriateness of the investment is determined.

Before making an investment, present worth analysis (PW) should take into consideration (Eq. 1);

$$PW = -C_0 + \frac{R1}{(1+i)^1} + \frac{R2}{(1+i)^2} + \cdots + \frac{Rn}{(1+i)^n} \qquad (1)$$

Present worth analysis (PW) is a formula to calculate an estimate of how profitable the project or investment will be [16]. The $-C_0$ is the initial investment, which is a negative cash flow showing that money is going out as opposed to coming in [16]. Considering that the money going out is subtracted from the discounted sum of cash flows coming

Fig. 4. Energy performance certificate of the hotel building

in, the net present value would need to be positive in order to be considered a valuable investment [16].

In Table 7, Net present worth analysis of the investment is made and since the present net value is positive, the investment would be considered a valuable one. In addition to Present Worth Value analysis, calculating repayment period will be also beneficial in terms of investigating the feasibility of a project. Repayment period of a project can be calculated as stated below (Eq. 2);

$$RP = C_o/(R - A) \tag{2}$$

According to Eq. (2), RP value of the case study is 3.4 years.

Table 6. Evaluation of the building according to TSE certificate

Energy category criteria	Required documents	Category total score	Hotel building's score
Building class	−TSE Energy performance certificate	18	12
Energy efficiency		21	12
Renewable energy usage		48	33
Energy security and quality		3	2
Greenhouse gas emissions	−TSE Energy class document	5	3
Energy efficient appliances		4	3
Operating HVAC system		21	15
Total		120	80

Table 7. Net present worth analysis of the investment

(TL)	0	1	2	3	4	5	6	7
Initial investment	−1,000,000							
Energy cost savings		+294,000	+294,000	+294,000	+294,000	+294,000	+294,000	+294,000
Additional costs		−1000	−1000	−1000	−1000	−1000	−1000	−1000
Salvage								+1000
Cash flow (CF)	−1,000,000	+293,000	+293,000	+293,000	+293,000	+293,000	+293,000	+294,000
Present worth (PW)	$\frac{1}{(1+0.07)^0}$	$\frac{1}{(1+0.07)^1}$	$\frac{1}{(1+0.07)^2}$	$\frac{1}{(1+0.07)^3}$	$\frac{1}{(1+0.07)^4}$	$\frac{1}{(1+0.07)^5}$	$\frac{1}{(1+0.07)^6}$	$\frac{1}{(1+0.07)^7}$
CF × PW	−1,000,000	273,831	255,997	239,175	223,528	208,904	195,238	183,088
ε (CF × PW)	−1,000,000	−726,179	−470,182	−231,007	−7479	201,425	396,663	579,751

3   Conclusion

This paper provides a literature review about green building assessment schemes in particular LEED and Turkish Standards Institute, Green Building Certificate. Additionally building energy efficient retrofitting is examined via a case study of a hotel building. The experience of the case study of a hotel building has emphasized that how HVAC retrofitting according to green building standards, can increase energy efficiency and decrease operational costs. In this study, as a result of the retrofitting, the annual energy cost decreased by 50%.

The unpredictable nature of energy prices offers a compelling argument for improving energy efficiency of Turkey's building stock. Energy retrofitting according to green building assessment schemes energy related standards undoubtedly play a vital role in achieving energy efficiency goals of Turkey. As a result, comprehensive retrofit programs should be developed to achieve Turkey's energy targets.

References

1. Suzer O (2015) A comparative review of environmental concern prioritization: LEED vs. other major certification systems. J Environ Manag 154:266–283
2. Ferdos N (2015) An approach for cost optimum energy efficient retrofit of primary school buildings in Turkey. M.Sc. thesis. ITU Graduate School of Science, Department of Architecture. Istanbul
3. Sectoral energy consumption statistics (2014) http://www.turkstat.gov.tr/PreHaberBultenleri.do?id=21587. Accessed 20 Mar 2017
4. D'Agostino D et al (2017) Towards nearly zero energy buildings in Europe: a focus on retrofit in non-residential buildings. Energies 10:117. doi:10.3390/en10010117
5. Ibn-Mohammed T et al (2013) A decision support framework for evaluation of environmentally and economically optimal retrofit of non-domestic buildings. Sustain Energy Build 22:209–227
6. Azouz M, Kim J (2015) Examining contemporary issues for green buildings from contractors' perspectives. Procedia Comput Sci J 118:470–478
7. United States Environmental Protection Agency (EPA). https://www3.epa.gov/. Accessed 09 Nov 2016
8. Suzer O (2015) A comparative review of environmental concern prioritization: LEED vs. other major certification systems. J Environ Manage 154:266–283
9. Pengpeng X (2012) A model for sustainable building energy efficiency retrofit using energy performance contracting mechanism for hotel buildings in China. Ph.D. thesis, Hong Kong PolyTechnic University, Department of Architecture, Hong Kong
10. Zhou Z et al (2016) Achieving energy efficient buildings via retrofitting of existing buildings: a case study. J Clean Prod 112:3605–3615
11. Ruparathna R et al (2015) Improving the energy efficiency of the existing building stock: a critical review of commercial and institutional buildings. Renew Sustain Energy Rev 53:1032–1045
12. Investment Support and Promotion Agency of Turkey. http://www.invest.gov.tr/
13. Eang L et al (2009) A study on energy performance of hotel buildings in Singapore. Energy Build 41:1319–1324
14. Argiriou A et al (1996) Energy conservation and retrofitting potential in Hellenic Hotels. Energy Build 24:65–75
15. Bhattacharya S (2011) Energy economics, concepts, issues, markets and governance. Springer, New York. doi:10.1007/978-0-85729-268-1
16. Net Present Value, Finance Formulas. http://www.financeformulas.net/Net_Present_Value.html Accessed 01 Mar 2016

Application of Solid Wastes for the Production of Sustainable Concrete

Vasudha D. Katare[✉] and Mangesh V. Madurwar

Visvesvaraya National Institute of Technology, Nagpur, India
vasudhakatare@gmail.com, mangesh_bits@yahoo.com

Abstract. The production of concrete results in emissions of about 0.13 t of CO_2 per ton of concrete, equal to 1/9th the emissions of cement. The tremendous usage of conventional raw material creates a negative impact on the concrete industry which can be minimized by the application of solid wastes as a raw material. The present paper reviews various solid wastes in different compositions to develop sustainable concrete. In view of the utilization of agro-industrial wastes as a pozzolanic material, their physical and chemical characterizations were reviewed. The crushing strength of high strength concrete incorporated with various agro-industrial wastes as a pozzolanic material was studied. This paper also proposes a procedure for producing manufactured coarse aggregates made of solid waste (i.e. sediment dredged from dam reservoir). The crushing strength of concrete incorporating the manufactured coarse aggregates was experimentally measured. Along with it, the physico-mechanical properties of the artificial sand, produced using the industrial solid wastes (i.e. fly ash) were analyzed. Reusing of solid waste as a substitute raw material will not only solve its disposal problems but also serves as an economical option to develop the sustainable concrete. The present paper is useful for various researchers involved in using solid wastes to develop the sustainable concrete.

Keywords: Solid waste · Sustainable concrete · Reservoir sediment
Agro-industrial waste

1 Introduction

The continuous increase in demands of residential apartments and infrastructural facilities is because of the higher rate of increase in population as well as an improvement in living standards. This needs concrete for their construction. After water, concrete is the most consumed material in India, which ultimately requires large quantities of cement, sand, and aggregates for its production. The entire world is currently facing the most serious environmental issue of rapid depletion of natural resources due to their enormous increase in consumption rates. Cement production requires a tremendous amount of energy for its production and can generate nearly one ton of CO_2 for every ton of processed cement. Stringent new laws prohibit blasting of hills for obtaining aggregates. There is also ban on removal of natural sand from rivers. These are the crucially important materials for every new construction. These resources

© Springer International Publishing AG, part of Springer Nature 2018
S. Fırat et al. (eds.), *Proceedings of 3rd International Sustainable
Buildings Symposium (ISBS 2017)*, Lecture Notes in Civil Engineering 7,
https://doi.org/10.1007/978-3-319-64349-6_10

are exhausting very rapidly because of their tremendous usage. So it is a need of the time to find some substitute to these natural resources.

Disposal of a considerable quantity of agro-industrial wastes is currently one of the serious environmental issues. Utilization of such solid waste for the production of cement, sand, and aggregates will not only answer the disposal problem of solid wastes but also gives the sustainable solution to the depletion of natural resources. Another severe problem the world is facing is a water shortage. Dams built all over the world are filled with a large volume of sediment, which occupies from around 10% to more than 70% of reservoir volume. The only way out is the removal of this sediment and use it in the making of artificial aggregates. Finding an alternative raw material is a dire need of the time before it is too late.

2 Cementitious Materials

Figure 1 indicates the information on the status of waste produced from different agro-industrial sources in India [1]. A lot of waste produced from agro-industrial sources is wheat straw husk, sugarcane bagasse, paddy, groundnut shell, cotton stalk, coconut shell, jute fiber, wooden industry waste etc. [2]. Akram et al. [3] reported that the India is the second largest country in producing sugarcane i.e. 290 million tons yearly. Prasad et al. [4] reported that India alone annually generates 30 million tons of rice husk. Singh et al. [1] reported that the annual production of maize across India is about 18.5 and 5.55 million tons of maize residue i.e. corn cob is generated.

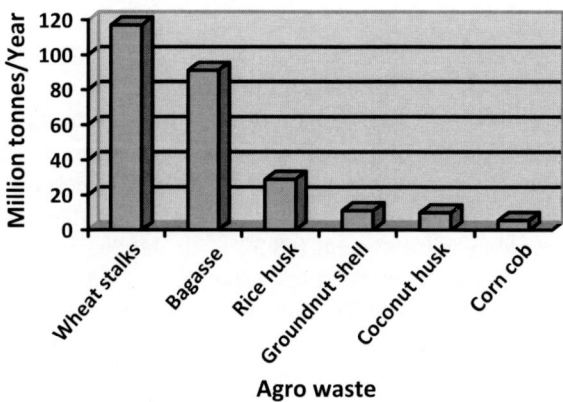

Fig. 1. Status of agro-industrial wastes produced in India (million tonnes/year) [1]

2.1 Physical Properties of Agro-Industrial Waste Ashes

From Table 1, it is seen that the particle size, specific gravity, and the density of all the agro-industrial waste ashes are lower than that of cement except for corn cob ash. The

Table 1. Physical properties of agro-industrial waste ashes

Property	Particle size (μm)	Particle size (%)	Specific gravity	Specific gravity (%)	Density (kg/cum)	Density (%)	Source
Cement	25	100	3.12	100	3140	100	[5–7]
Bagasse ash	5.4	21.60	1.85	57.81	90–140	2.87–4.46	[8, 9]
Rice husk ash	12.3	49.60	2.16	67.50	150–175	4.78–5.57	[8, 9]
Corn cob ash	29–45	116–180	2.5–3.6	78.13–112.50	300–330	9.55–10.51	[8, 10]

corn cob ash blended concrete doesn't show much improvement in the compressive strength of concrete compared to other agro-industrial waste ashes at early ages. So, larger particle size may be one of the reasons for not having any significant strength improvement. As the specific gravity and density of the agro-industrial waste ashes are lower than the cement, the concrete produced will be lighter in weight compared to the control concrete produced without agro-industrial waste ashes. It is examined that the variation in the results of the each property of different agro-industrial waste ashes is because of the disparity in the selection of suitable of incinerator/furnace, grinding method, burning time, burning temperature, and cooling time etc. of agro-industrial wastes.

2.2   Chemical Characterization of Agro-Industrial Waste Ashes

Table 2 shows that the residue after combustion of agro-industrial waste ashes gives the chemical characterization dominated by SiO_2 (silicon dioxide). More the percentage of amorphous silica in a material better is the pozzolanic activity. The review also shows that the agro-industrial waste ashes contain all the elemental oxides, which are present in cement. Nair et al. [11] analyzed that the large quantity of amorphous silica found in ashes burnt at the temperature of 500–700 °C. Habeeb et al. [12] examined that the rice husk ash is effective as a supplementary cementitious material i.e. it is dominated by amorphous silica content (88.32%). Jauberthie et al. [13] identified the origin of the amorphous silica in rice husk ash. The rice husk ash contains a strong concentration of silica occurring in an amorphous and crystalline (quartz) forms. The amorphous silica occurs mainly on the external face of the husk and to a lesser concentration on the inner surface. This amorphous silica explains the pozzolanic role. Adesanya et al. [14] investigated that, as the corncob ash percentage increases, additional silicate and alumina were available to react with the lime produced during hydration of cement to produce further cementitious products. The silica and alumina content are responsible for the development of cementitious compounds. Aprianti et al. [5] explained that when pozzolanic materials are combined with Portland cement, they react to form cementing particles whereas, by themselves, these ashes do not possess any cementitious properties. Thus, a cementitious material can exhibit a

Table 2. Chemical composition of agro-industrial waste ashes

Component	SiO$_2$	Al$_2$O$_3$	Fe$_2$O$_3$	CaO	MgO	K$_2$O	SO$_3$	LOI	Source
OPC	25.1	5.5	5.9	55	3.4	0.5	2.7	0.9	[7]
Bagasse ash	65	4.8	0.9	3.9	–	2	0.9	10.5	[7]
Rice husk ash	86.81	0.5	0.87	1.04	0.85	3.16	–	4.6	[15]
Corn cob ash	67.33	7.34	3.74	10.29	1.82	4.2	1.11	–	[16]

self-cementitious i.e. a hydraulic activity and contains quantities of CaO while a pozzolanic material requires Ca(OH)$_2$ to form strength.

2.3 Compressive Strength of Agro-Industrial Waste Ash Blended Concrete

Table 3 shows that researchers were utilized various agro-industrial wastes ashes in varied proportions blended with cement and also adopted various methodologies to produce ordinary, standard, and high strength concrete (As per IS 456 clause no. 6.1, 9.2.2, 15.1.1, and 36.1) of various grades. From the reviewed results of Table 3, it is observed that as the cement replacement (%) increases the (28 days) compressive strength decreases except for the rice husk ash blended high strength concrete. But still,

Table 3. Cement replacement and respective compressive strength of agro-industrial waste ash blended concrete

Agro waste	Cement replacement (%)	Comp strength 28 days (Mpa)	Control concrete strength (Mpa)	Source	Remarks as per IS 456 (clause 6.1, 9.2.2, 15.1.1 and 36.1) [17]
Bagasse ash	2	59.75	50.10	[18]	Standard/high-strength concrete
	6	57.25			
	10	56.20			
Bagasse ash	10	43	36	[19]	Standard concrete
	20	40			
	30	32			
Bagasse ash	5	29.50	13.80	[20]	Ordinary concrete
	15	19.32			
	25	17.73			
Rice husk ash	10	40.6	36.8	[21]	Standard concrete
	15	37.9			
	20	36.7			
Rice husk ash	10	81	76	[22]	High strength concrete
	20	86			
	30	80			
Corn cob ash	10	23	24	[23]	Ordinary concrete
	20	19			

the compressive strength values are more than the control concrete strength. Agro-industrial waste ashes when blended with cement, silica of the ashes react with $Ca(OH)_2$ and produces an additional calcium-silicate-hydrate gel. This gel greatly contributes towards the compressive strength of concrete. So, blending of cement with the pozzolana in concrete reduces the quantity of $Ca(OH)_2$ and increases the strength.

The 28 days compressive strength behavior of concrete is shown in Fig. 2. It explains that compressive strength decreases with the increase in percentage agro-industrial waste ashes at short term ages (28 days). Insignificant improvement in compressive strength of blended concrete is observed at lower ages (28 days). This can be attributed to the fact that the low rate of pozzolanic reaction at early ages. While Fig. 3 shows the increase in 90 days compressive strength of blended concrete as the percentage of agro-industrial waste ashes increases up to the certain limit. This is because the silica from pozzolan reacts with lime produced as the by-product of

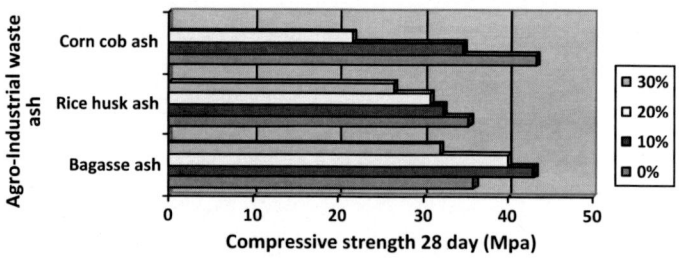

Fig. 2. Comparison of 28 days compressive strength of standard concrete with 10, 20, and 30% cement replacement [16, 24, 25]

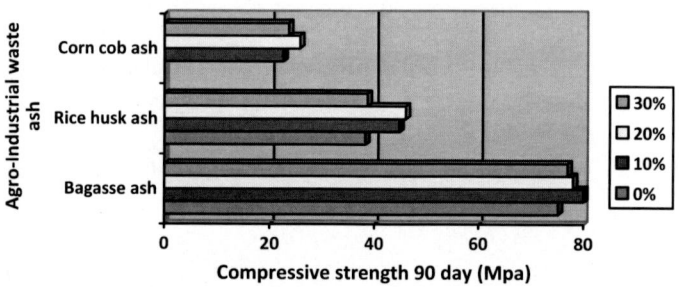

Fig. 3. Comparison of 90 days compressive strength of standard and high strength concrete with 10, 20, and 30% cement replacement [7, 26, 27]

hydration of ordinary Portland cement to form additional calcium-silicate-hydrate that increases the binder efficiency and corresponding strength values at later days of curing (90 days).

3 Manufactured Aggregates

The consumption of the concrete has been increasing in recent years in developing countries like India at a rate far exceeding that suggested by the economy growth rate of the construction industries. Concrete is an important part of society's infrastructure. Everyday life is greatly affected by concrete in numerous ways. The aggregates occupy 70–80% of the volume of concrete; their impact on various characteristics and properties of concrete is undoubtedly considerable. Blasting of hills and quarry mining for crushed aggregates will automatically come to a stop by using the manufactured aggregates instead of natural and crushed aggregates for making concrete, and all the subsequent adverse impacts on the environment will be avoided. The methodology for producing the manufactured aggregates out of reservoir sediment is proposed here.

3.1 Methodology Adopted

The fine sediment was collected from the Khadakwasla Dam Reservoir, Pune, Maharashtra, India. It was manually segregated to remove all unwanted materials such as shells, organic matter, and plastic bags. The fine sediment was then sun and air dried in order to reduce the moisture content from sediments. The sundried material was

Table 4. Physical test results of the fine sediments

Soil classification	Specific gravity	Liquid limit (%)	Plastic limit (%)	Plasticity index (%)	Ingredients (%)			
					Gravel (%)	Sand (%)	Silt (%)	Clay (%)
Highly organic	2.65	74.3	47.17	27.13	0	21.6	22.13	56.22

crushed into sufficiently fine size, nearly to a powder form. After crushing into a fine powder, the material was sieved manually through a 300μ sieve. The physical characteristic of the sediment is reported in Table 4. More than 50% particles were of clayey nature. It indicates that it was a highly organic clayey soil. Also, it was of highly plastic nature according to plasticity index value.

Trial mixes of dredged fine sediment and admixtures (Sodium hydroxide [NaOH], sodium silicate [Na_2SiO_3], Bottom ash and Cement) were prepared. Trial mix details are given in Table 5. Sieved dredged sediments and admixtures thoroughly mixed with optimum water content. Then the mixture was filled and compacted into the cylindrical molds. Then the molds were cut into angular shapes and random sizes by using trimming knife. These raw aggregate samples were sundried for 10 days. The dried aggregates were baked at the temperature of 1100 °C in a scientific oven by using Ramp and hold arrangement. All the six types of baked aggregates are as shown in Table 5. The manufactured aggregates were tested for determining the various engineering properties (crushing value, abrasion value, impact value, soundness, and water absorption) and to check whether they meet Indian Standards for aggregates to be used in construction. The results of engineering properties of the manufactured aggregates

Table 5. Trial mixes of fine sediment and admixture for manufacturing aggregates

Type 1	Type 2	Type 3	Type 4	Type 5	Type 6
Trial mix field sediment + 0% admixture	Field sediment + 2.5% NaOH	Field sediment + 2.5% Na_2SiO_3	Field Sediment 2.5% + bottom ash	Manufactured aggregates + coating of 10% Na_2SiO_3	Manufactured aggregates + coating of 10% cement

Table 6. Mechanical properties of manufactured aggregates

Sr. no	Experiment	Type 1	Type 2	Type 3	Type 4	Type 5	Type 6	Permissible limit (%)	IS 2386: 1963
1.	Crushing value (%)	29.19	25.2	20.3	27.5	21.8	19.0	<30	Part 4
2.	Impact value (dry) (%)	28.9	21.4	15.2	28.8	21.7	26.8	<30	Part 4
3.	Impact value (wet) (%)	31.9	23.4	19.7	30.0	22.0	27.0	<39	Part 4
4.	Water absorption (%)	11.3	5.0	3.1	11.3	6.3	6.7	<2	Part 3
5.	Abrasion value (%)	31.3	27.7	23.1	28.5	26.8	29.4	<30	Part 4
6.	Specific gravity (−)	1.4	1.3	1.2	1.4	1.4	1.7	–	Part 3
7.	Soundness value (%)	9.8	8.4	8.2	9.7	8.6	9.5	<18	Part 5

are shown in Table 6. All the six types of manufactured aggregates were tested for Crushing value, Impact value (Dry), Impact value (Wet), Water absorption, Abrasion value, Specific gravity, Soundness value in accordance with IS 2386 (Part 3)-1963, IS 2386 (Part 4)-1963, and IS 2386 (Part 5)-1963 Codes. The test results of manufactured aggregates were compared with the permissible limits given in respective IS code. The specific gravity of the produced aggregates ranging from 1.76 to 1.28 is significantly lower than the specific gravity of commercially available natural aggregates. By studying the test results it can be concluded that the produced aggregates can be used as Light Weight Aggregates (LWA) for structural concrete. Type 4 aggregate was found to have better strength amongst all six types, which confirms that manufactured aggregates are able to serve as structural aggregates.

4 Artificial Sand

One of the major challenges with the environmental awareness and scarcity of space for land-filling is the wastes/by-products utilization as an alternative to disposal. Throughout the industrial sector, including the concrete industry, the cost of environmental compliance is high. Use of industrial by-products such as foundry sand, fly ash, bottom ash, and slag as a substitute material for the natural river sand can result in significant improvements in overall industry energy efficiency and environmental performance. Use of natural river sand in concrete has a number of constraints pertaining to its availability, quality, cost, and environmental impacts. Waste foundry sand, bottom ash, and copper slag are reviewed in the present paper.

Table 7. Physical properties of artificial sand

Fine aggregates	Specific gravity	Unit weight (kg/cum)	Fineness modulus	Source
Sand	2.63	1890	3.03	[31]
Waste foundry sand	2.61	1638	1.78	[31]
Bottom ash	1.93	948	1.60	[31]
Copper slag	3.40–3.91	2080	3.47	[32, 33]

4.1   Physical Properties of Artificial Sand

Kim et al. [28] experimentally evaluated that the density of hardened concrete linearly decreases as the replacement ratio of bottom ash increases. As the bottom ash is having lowest specific gravity and the unit weight shown in Table 7. Kou [29] found that a slump value of fresh concrete gets affected by the fineness modulus of artificial sands. Substitution of artificial sands in concrete increases its slump value. Siddique et al. [30] reviewed that the water absorption capacity of the concrete decreases with increase in waste foundry sand content and so adversely affect the slump of the concrete. The effect of particle size of artificial sand on the strength properties of the concrete was also studied. Using waste foundry sand having the particle size in the range of clay and silt results in the decrease in the porosity of the concrete. Possibly, it can be the favorable effect on the strength properties of the concrete.

4.2   Chemical Characterization of Artificial Sand

The reviewed waste foundry sand exhibits pozzolanic properties since it contains low CaO content, highest silica content i.e. 83.8% and other oxides such as Al_2O_3, SiO_2, and Fe_2O_3 shown in Table 8. All other reviewed industrial wastes also have all the elements which are present in cement. Use of these wastes in the concrete industry can have the benefits of reducing the costs of disposal and helps to protect the environment.

4.3   Compressive Strength of Artificial Sand Blended Concrete

Aggarwal et al. [31] experimentally analyzed that the inclusion of waste foundry sand and bottom ash as fine aggregate does not affect the strength properties negatively as the strength remains within limits. The concrete was endowed with comparable mechanical properties and greater resistance to aggressive agents (chemical, physical and environmental). Table 9 shows the blending of artificial sands in the concrete in varied proportion. For every artificial sand different optimum percentage is obtained. The strength obtained by the replacement of industrial wastes is more than the control concrete strength except for the bottom ash blended concrete. Guney et al. [38] studied the reason for the decrease in strength property of concrete after excessive sand replacement. It is possibly due to the weakening in cement/aggregate adherence, an increase in the amount water required and retardation in cement hydration. In addition,

Table 8. Chemical composition of artificial sand (% by mass)

Component	SiO$_2$	Al$_2$O$_3$	Fe$_2$O$_3$	CaO	MgO	K$_2$O	SO$_3$	LOI	Source
Waste foundry sand	83.8	0.81	5.39	1.42	0.86	1.14	0.21	–	[34]
Bottom ash	56.4–57.9	29.2–22.6	8.44	0.75–2	0.4–3.2	1.29	0.24	0.89–1.67	[35, 36]
Copper slag	25.8–31.9	0.22–2.52	68.29–59.11	0.15–1.25	1.65	0.23–0.81	0.11–1.34	6.59	[33, 37]

Table 9. Sand replacement and respective compressive strength of artificial sand blended concrete

Artificial sand	Sand replacement (%)	Comp. strength 28 days (Mpa)	Control concrete strength (Mpa)	Source
Waste foundry sand	10	33.24	33.14	[39]
	20	32.58		
	30	31.24		
	40	29.48		
	50	25.23		
Bottom ash	10	21.41	39.52	[40]
	20	23.78		
	30	24.65		
	40	19.99		
	50	21.20		
Copper slag	10	46.00	45	[41]
	20	47.00		
	40	47.10		
	50	47.00		
	60	46.00		
	80	34.80		
	100	35.10		

water absorption during cement hydration and, then, water loss in the hardened concrete may cause an increase in the voids and, later, the occurrence of cracks in the concrete. The strength and durability of concrete decrease significantly under these unfavorable conditions.

5   Conclusions

The review shows that many agro-industrial waste ashes such as bagasse ash, rice husk ash, and the corn cob ash have the potential to be used as a supplementary cementitious material for producing sustainable concrete. The quality of highly reactive ash depends on the controlled burning conditions of agro-industrial waste i.e. selection of suitable of incinerator/furnace, burning time, burning temperature, cooling time, and grinding method etc. of agro-industrial wastes. Application of agro-industrial waste ashes in concrete enhances the compressive strength with the age of concrete. Better results were observed for 90 days age than the 28 days age in the present study. The use of less expensive agro-industrial waste ash is more desirable to decrease the overall production cost of concrete and to reduce the cement requirement. Which leads to decrease the environmental pollution and energy by cement factories thus providing economic and environmental benefits along with providing a way of disposing of this agricultural waste product which otherwise has a little alternative use. Using agro-industrial waste ashes as a supplementary material in concrete will be a valuable contribution and viable solution for sustainable construction. So, it can be concluded that the agro-industrial waste ash mineral is a promising pozzolanic material and can be fruitfully used as a supplementary material in Portland cement.

The reservoir sediments can be used as primary resource material for manufacturing lightweight aggregates that can achieve not only technical benefits but also can result in good social and ecological benefits. The Specific gravities of the produced manufactured aggregates ranged from 1.28 to 1.76, as against 2.67 for the natural aggregates. They also meet the requirements of relevant IS codes. Type 4 i.e. Sodium silicate admixture aggregates are having better properties amongst all six types of aggregates. Making artificial aggregates by using dam reservoir sediment will solve two problems with one solution, namely (1) Emptying the dam reservoirs and increasing its water storage capacity and (2) Offering an eco-friendly solution, which will not only provide an essential product for human use but also it will not create any new environmental problems. It is feasible to use the reviewed industrial waste ashes as fine aggregate in preparing concrete mixes. Replacement of fine aggregates with bottom ash can easily be equated to the strength development of normal concrete.

References

1. Singh J, Gu S (2010) Biomass conversion to energy in India-A critique. Renew Sustain Energy Rev 14(5):1367–1378. doi:10.1016/j.rser.2010.01.013
2. Pappu A, Saxena M, Asolekar SR (2007) Solid wastes generation in India and their recycling potential in building materials. Build Environ 42(6):2311–2320. doi:10.1016/j.buildenv.2006.04.015
3. Akram T, Memon SA, Obaid H (2009) Production of lowcost self compacting concrete using bagasse ash. Constr Build Mater 23:703–712. doi:10.1016/j.conbuildmat.2008.02.012
4. Prasad CS, Maiti KN, Venugopal R (2001) Effect of rice husk ash in whiteware compositions. Ceram Int 27:629–635. doi:10.1016/S0272-8842(01)00010-4
5. Aprianti E, Shafigh P, Bahri S, Farahani JN (2015) Supplementary cementitious materials origin from agricultural wastes—a review. Constr Build Mater 74:176–187. doi:10.1016/j.conbuildmat.2014.10.010
6. Ganesan K, Rajagopal K, Thangavel K (2007) Evaluation of bagasse ash as supplementary cementitious material. Cem Concr Compos 29(6):515–524. doi:10.1016/j.cemconcomp.2007.03.001
7. Rukzon S, Chindaprasirt P (2012) Utilization of bagasse ash in high-strength concrete. Mater Des 34:45–50. doi:10.1016/j.matdes.2011.07.045
8. Madurwar MV, Ralegaonkar RV, Mandavgane SA (2013) Application of agro-waste for sustainable construction materials: a review. Constr Build Mater 38:872–878. doi:10.1016/j.conbuildmat.2012.09.011
9. Singh D, Singh J (2016) Use of agrowaste in concrete construction. Int J Environ Ecol Family Urban Stud 6(1):119–130
10. Snellings R, Mertens G, Elsen J (2012) Supplementary cementitious materials. Rev Mineral Geochem 74:211–278. doi:10.2138/rmg.2012.74.6
11. Nair DG, Fraaij A, Klaassen AAK, Kentgens APM (2008) A structural investigation relating to the pozzolanic activity of rice husk ashes. Cem Concr Res 38(6):861–869. doi:10.1016/j.cemconres.2007.10.004

12. Habeeb GA, Fayyadh MM (2009) Rice husk ash concrete: the effect of RHA average particle size on mechanical properties and drying shrinkage. Aust J Basic Appl Sci 3(3):1616–1622
13. Jauberthie R, Rendell F, Tamba S, Cisse I (2000) Origin of the pozzolanic effect of rice husks. Constr Build Mater 14:419–423. doi:10.1016/S0950-0618(00)00045-3
14. Adesanya DA, Raheem AA (2009) Development of corn cob ash blended cement. Constr Build Mater 23:347–352. doi:10.1016/j.conbuildmat.2007.11.013
15. Gupta AI, Wayal AS (2015) Use of rice husk ash in concrete: a review. IOSR J Mech Civ Eng 12(4):29–31
16. Adesanya DA, Raheem AA (2009) A study of the workability and compressive strength characteristics of corn cob ash blended cement concrete. Constr Build Mater 23(1):311–317. doi:10.1016/j.conbuildmat.2007.12.004
17. IS456:2000. Plain and reinforced concrete—code of practice. Bureau of Indian Standard 1978 [Reaffirmed 2000], New Delhi
18. Otoko GR (2014) Use of bagasse ash as partial replacement of cement in concrete abstract. Int J Innov Res Dev 3(4):285–289
19. Amin N (2011) Use of bagasse ash in concrete and its impact on the strength and chloride resistivity. J Mater Civ Eng ASCE 23(5):717–720. doi:10.1061/(ASCE)MT.1943-5533. 0000227
20. Srinivasan R, Sathiya K (2010) Experimental study on bagasse ash in concrete. Int J Serv Learn Eng 5(2):60–66. doi:10.1017/CBO9781107415324.004
21. Givi AN, Rashid SA, Aziz FNA, Salleh MAM (2010) Assessment of the effects of rice husk ash particle size on strength, water permeability and workability of binary blended concrete. Constr Build Mater 24:2145–2150. doi:10.1016/j.conbuildmat.2010.04.045
22. Muthadhi A, Kothandaraman S (2013) Experimental investigations of performance characteristics of rice husk ash—blended concrete. J Mater Civ Eng 25:1115–1118. doi:10.1061/(ASCE)MT.1943-5533.0000656
23. Ettu LO, Arimanwa JI, Nwachukwu KC, Awodiji CTG, Amanze C (2013) Strength of ternary blended cement concrete containing corn cob ash and pawpaw leaf ash. Int J Eng Sci (IJES) 2(5):84–89
24. Amin N (2011) Use of bagasse ash in cement and its impact on the mechanical behaviour and chloride resistivity of mortar. Adv Cem Res 23(2):75–81. doi:10.1680/adcr.9.00023
25. Madandoust R, Mohammad M, Ahmadi H (2011) Mechanical properties and durability assessment of rice husk ash concrete. Biosyst Eng 110:144–152. doi:10.1016/j. biosystemseng.2011.07.009
26. Ganesan K, Rajagopal K, Thangavel K (2008) Rice husk ash blended cement: assessment of optimal level of replacement for strength and permeability properties of concrete. Constr Build Mater 22:1675–1683. doi:10.1016/j.conbuildmat.2007.06.011
27. Ettu LO, Anya UC, Arimanwa JI, Anyaogu L, Nwachukwu KC (2013) Strength of binary blended cement composites containing corn cob ash. Int J Eng Res Dev 6(10):77–82
28. Kim HK, Lee HK (2011) Use of power plant bottom ash as fine and coarse aggregates in high-strength concrete. Constr Build Mater 25:1115–1122. doi:10.1016/j.conbuildmat.2010. 06.065
29. Kou SC, Poon CS (2009) Properties of concrete prepared with crushed fine stone, furnace bottom ash and fine recycled aggregate as fine aggregates. Constr Build Mater 23:2877–2886. doi:10.1016/j.conbuildmat.2009.02.009
30. Siddique R, Singh G (2011) Utilization of waste foundry sand (WFS) in concrete manufacturing. Resour Conserv Recycl 55:885–892. doi:10.1016/j.resconrec.2011.05.001
31. Aggarwal Y, Siddique R (2014) Microstructure and properties of concrete using bottom ash and waste foundry sand as partial replacement of fine aggregates. Constr Build Mater 54:210–223. doi:10.1016/j.conbuildmat.2013.12.051

32. Al-Jabri KS, Hisada M, Al-Saidy AH, Al-Oraimi SK (2009) Performance of high strength concrete made with copper slag as a fine aggregate. Constr Build Mater 23:2132–2140. doi:10.1016/j.conbuildmat.2008.12.013

33. Brindha D, Nagan S (2011) Durability studies on copper slag admixed concrete. Asian J Civ Eng (Build Hous) 12:563–578

34. Singh G, Siddique R (2011) Effect of waste foundry sand (WFS) as partial replacement of sand on the strength, ultrasonic pulse velocity and permeability of concrete. Constr Build Mater 26:7. doi:10.1016/j.conbuildmat.2011.06.041

35. Singh M, Siddique R (2015) Properties of concrete containing high volumes of coal bottom ash as fine aggregate. J Clean Prod 91:269–278. doi:10.1016/j.jclepro.2014.12.026

36. Yüksel I, Bilir T, Özkan Ö (2007) Durability of concrete incorporating non-ground blast furnace slag and bottom ash as fine aggregate. Build Environ 42:2651–2659. doi:10.1016/j.buildenv.2006.07.003

37. Wu W, Zhang W, Ma G (2010) Optimum content of copper slag as a fine aggregate in high strength concrete. Mater Des 31:2878–2883. doi:10.1016/j.matdes.2009.12.037

38. Guney Y, Sari YD, Yalcin M, Tuncan A, Donmez S (2010) Re-usage of waste foundry sand in high-strength concrete. Waste Manag 30:1705–1713. doi:10.1016/j.wasman.2010.02.018

39. Ganesh Prabhu G, Hyun JH, Kim YY (2014) Effects of foundry sand as a fine aggregate in concrete production. Constr Build Mater 70:514–521. doi:10.1016/j.conbuildmat.2014.07.070

40. Syahrul M, Sani M, Muftah F, Muda Z (2010) The properties of special concrete using washed bottom ash (WBA) as partial sand replacement. Int J Sustain Constr Eng Technol 1:65–76

41. Al-Jabri KS, Al-Saidy AH, Taha R (2011) Effect of copper slag as a fine aggregate on the properties of cement mortars and concrete. Constr Build Mater 25:933–938. doi:10.1016/j.conbuildmat.2010.06.090

Sustainable Architecture Under the Timeline Frame: Case Study of Fujairah in UAE

Tahani Yousuf$^{(\boxtimes)}$ and Hanan Taleb

British University in Dubai, Dubai, UAE
Tahani.Yousuf@mopw.gov.ae, hanan.taleb@buid.ac.ae

Abstract. There is a significant link between thermal comfort and energy performance, since creating comfortable temperatures requires thermal energy. The tighter the tolerances of temperature demanded the more energy consumed. It is therefore important to understand the implications of relaxing those tolerances and what passive strategies are available as alternatives to mechanical temperature control. This study attempts to investigate the impact of passive environmental strategies on built form using a simulation of an existing residential building. The house is located in Al Bithna/Fujairah. The simulation was conducted using IES-VE software. The results show a reduction and savings when implementing different passive strategies. This begins with increasing window size and is followed by changing glazing type and finally adding a shading device to the base model. The aim of this study is to understand sustainability evolution in residential houses in the UAE across different time frames. The document will delve in the past to check the characteristics of old houses in the UAE and their attempts to maintain comfort and sustainability. Turning to the present, a case study is selected to which different passive design strategies are applied and a recommendation of what the future will look like in terms of sustainable residential houses is developed. The study concludes with a look at the impact of combined strategies to see the impact of different variables such as average day lighting, solar gain and cooling sensible load.

Keywords: Passive design · Arid · Design strategies · Timeline
UAE · IES

1 Introduction

In 2014, Ban Ki-moon stated that climate change was a hot issue for our times, affecting lives and producing high cost implications everywhere. The world is witnessing high levels of emission of CO_2 (Fig. 1), which is having an effect on climate change. These high emission levels are due to human activities which are mostly related to energy consumption. Hammad and Abu-Hijleh [7] stated that buildings consuming 40% of the total world energy and almost 70% of sulfur oxides and 50% of carbon dioxide emissions come from that sector. In the UAE, the second highest consuming sector for energy is the residential sector (Fig. 2). The country is heavily targeting sustainable development science which has proved that there is a strong relationship between energy performance and thermal comfort. This means that there

© Springer International Publishing AG, part of Springer Nature 2018
S. Fırat et al. (eds.), *Proceedings of 3rd International Sustainable
Buildings Symposium (ISBS 2017)*, Lecture Notes in Civil Engineering 7,
https://doi.org/10.1007/978-3-319-64349-6_11

are stricter demands on temperatures that relate to the consumption of energy. It is very important to know the consequences of those temperatures and related passive strategies are available as alternatives to mechanical temperature control.

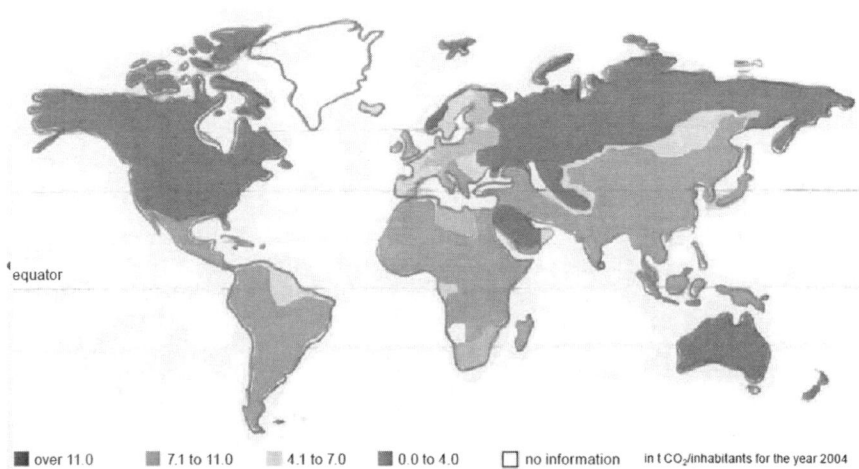

■ over 11.0   ■ 7.1 to 11.0   ■ 4.1 to 7.0   ■ 0.0 to 4.0   □ no information   in t CO₂/inhabitants for the year 2004

Fig. 1. CO_2 emissions distribution levels per capita. *Source* World Population for the Year 2004

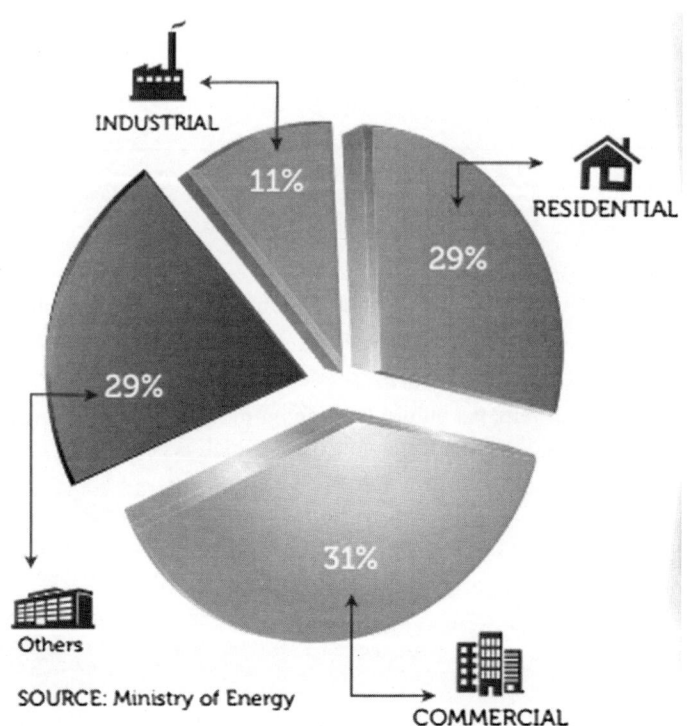

INDUSTRIAL

11%

RESIDENTIAL

29%

29%

Others

31%

SOURCE: Ministry of Energy

COMMERCIAL

Fig. 2. UAE electricity consumption 2013. *Source* Ministry of Energy, 2013

1.1   Climatic Data

Fujairah is located on the following coordinates: latitude 25.10 North and longitude 56.33 East. It has an area of approximately 1166 km^2 (Fig. 3).

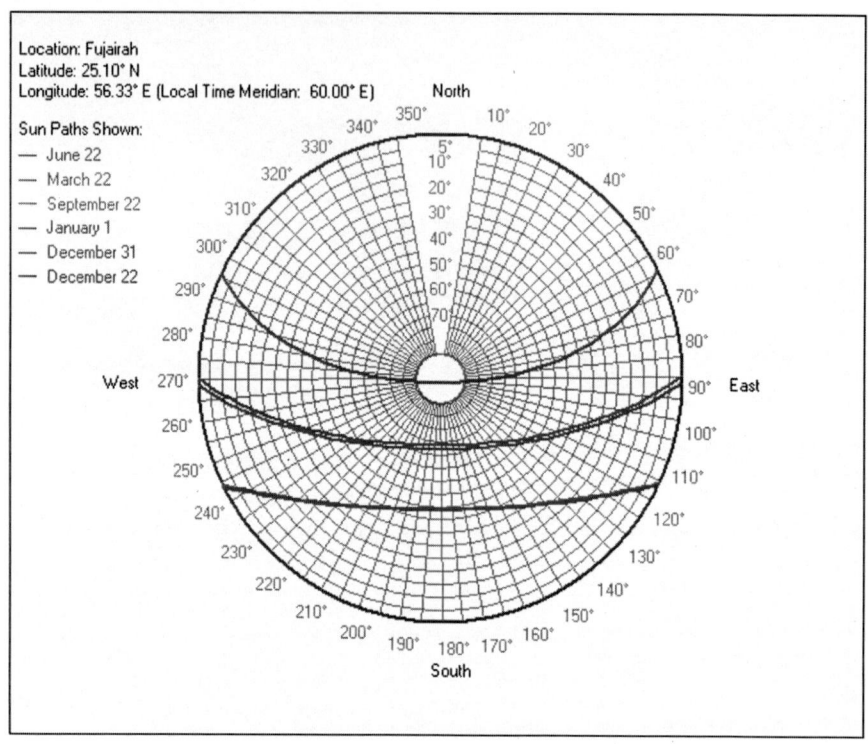

Fig. 3. Location of Fujairah. *Source* IES-VE software 2016 Daily/monthly

Temperature
Fujairah's weather is hot throughout the year. Summer is between April to October and temperatures gradually rise to an average of 43 °C between July and October (Fig. 4).

Relative Humidity
The weather in Fujairah remains humid throughout the year. Relative humidity ranges from 31% which is considered comfortable to 79% which is considered humid (Fig. 5).

March is considered very dry or very humid.

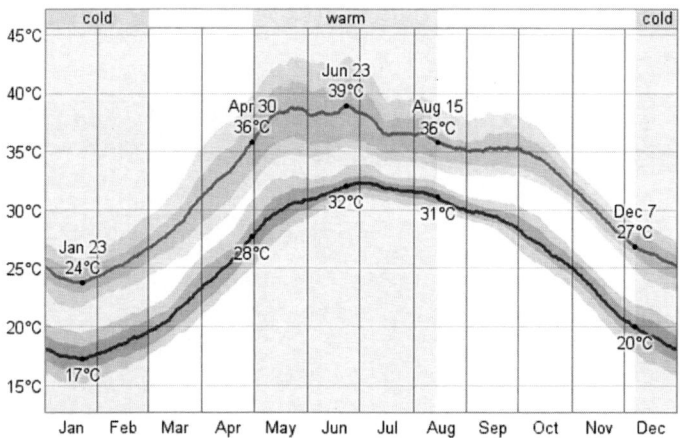

Fig. 4. Daily temperature in Fujairah. *Source* http://www.worldweatheronline.com/fujairah-weatheraverages/fujairah/ae.aspx

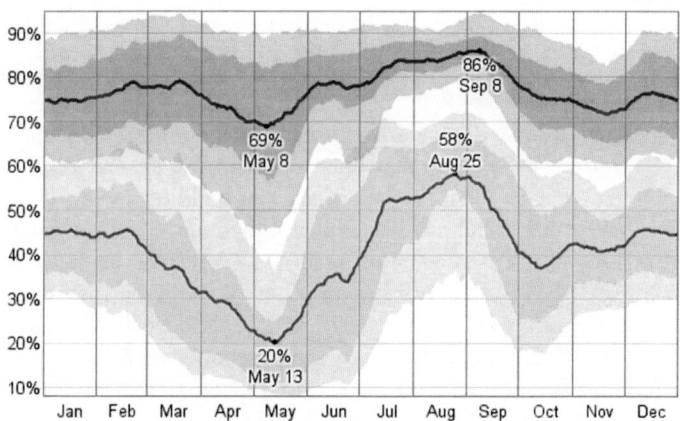

Fig. 5. Relative Humidity in Fujairah. *Source* http://www.worldweatheronline.com/fujairah-weatheraverages/Fujairah/ae.aspx

Dew Point

The most comfortable months of the year are December to February, with the rest of the year considered muggy (Fig. 6).

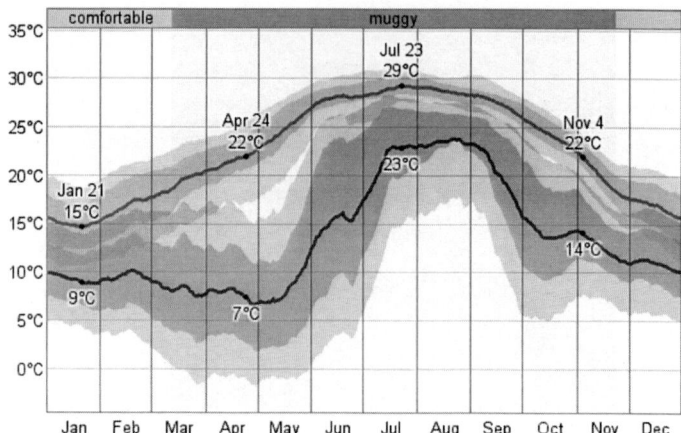

Fig. 6. Dew point in Fujairah. *Source* http://www.worldweatheronline.com/fujairah-weatheraverages/fujairah/ae.aspx

2 Literature Review

Olgyay [12] stated that there is a significant relationship between passive design strategies and the site and geographical location of a building. Designers need to be aware of orientation which will affect the amount of daylight entering the building and ventilation. It is important to begin the study with climatic data to check the shading, ventilation, thermal capacity and day lighting of the building.

Meyer [11] stated that in hot and dry climates, the main source of solar radiation is the roof, since it faces direct sun. Perusing in the literature, the following was discovered. Windows have a significant impact on daylight performance and visual comfort. Occupant comfort and performance will be associated with daylight performance, in addition to the enhancement of building energy performance.

Thermal comfort is subjective and related to glazing and shading devices characteristics. Tzempelikos et al. [15] state that it is possible to reduce the need for secondary air-conditioning by maintaining an operative temperature within the comfort zone in a good building envelope. In terms of window and glazing, the following findings are from the literature review: Wong and Istiadji [16] argue that the designer should place emphasis on window glass since this has the highest impact in terms of reducing annual consumption of energy. It is then necessary to start to add shading devices. Chi-Ming and Yao-Hong [4] show that window glass has a larger impact on energy consumption than the shading device effect does. Bessoudo et al. [3] examined the interior thermal atmosphere of a fully glazed office building with different types of shading. The use of roller and Venetian blinds showed an improvement in the inside thermal environment.

In 2010, Carmody et al. 2004 cited in Bessoudo et al., demonstrated that in a cold climate, a big window area which is facing south will have a substantial effect on enhancing interior thermal comfort for all type of windows except the double-glazed reflected type.

In terms of shading devices the following findings were taken from the literature review: Fiocchi et al. [6] state that to enhance the visual comfort and support occupant productivity, it is better to increase day light and decrease solar glare. Energy transmitted by windows varies depending on variables such as window type, side and overhang fins. Kensek et al. [8] state that few designers implement the shading devices in their projects although they offer many advantages such as minimising glare, reducing cooling load and monitoring radiation and light intensity. A study conducted by Tzempelikos and Athienitis [14], found that shading devices have great implications for the energy consumption of a building; they can make a saving of 50% of total energy consumed by cooling and can see solar and heat gain of at least 12%. In terms of shading device orientation, the following findings came from the literature review: Kim et al. [9] found that external shading devices which are horizontal have an impact on reducing cooling energy. Even a short one sees 11% less reduction in cooling energy. Chi-Ming and Yao-Hong [4] examined different types of shading devices; side fins, overhangs and box shades. They found that all these devices decrease daylight spread to interior spaces at different levels and that the maximum energy reduction effect was occurred with the box type. Palmero-Marrero and Oliveira [13] examined the position of louvers, and found that for the west and east facades vertical louvers are more suitable, while for south facades horizontal louvers have significant efficiency levels. The same results were found by Alzoubi and Al-Zoubi [1] when they studied south-facing facades with a vertical shading device.

Kim et al. [9] studied the sun path and effect of shading devices at different angles. He found that an angle of 0° has the greatest performance in terms of energy and this becomes the less at an angle of 60°. Lam and Miller [10] studied the effect of bio-shading for two years and found that overheating which occurs in the summer can be overreached from vertical living. The same impact would occur when placing a living plant within the void of double glazing to reduce energy consumption. Due to an absence of methodologies and measurements and methods to estimate the performance of the bio-shading, the study was not binding.

Vegetation can be an important influence in shading a building facade, particularly in a hot climate and this helps to improve the thermal effect and the air quality.

The American Institute of Architecture (AIA) [2] defines the best shading devices as those that reduce direct solar radiation and allow only diffused radiation to enter. A reduction in the amount of light entering a building will have the effect of reducing energy consumption. A lot of studies have looked into the effect of different design solutions for the building façade to reach an optimum design strategy.

These have included exterior and interior shading devices and different kinds of louver shading devices. The AIA also showed that in a low-rise building the natural environment can be used. Deciduous trees give great shade for the façade in all seasons and will transmit 20% solar radiation in summer. Koo et al. (2010) studied automated blind shading to gain the most advantage from daylight, reduce energy consumption from lighting and avoid glare. Chi-Ming and Yao-Hong [4] examined different kinds of shading devices—overhangs, side fins and box shades—finding that all of them reduced daylight penetration to interior spaces and that box shading has the optimum energy reduction effect.

In 2009, Palmero-Marrero conducted an investigation into the position of louvers and concludes that vertical louvers are much better for east and west façades while horizontal ones produce optimum efficiency on south façades. In addition, Ebrahim-pour and Maerefat [5] examined the use of exterior devices such as overhangs and side fins for single clear glazing. The conclusion was that it is more useful for any direction of window than advanced glazing like double clear pane or low-E pane glazing. Yu et al. [17] studied residential buildings in China, finding the best performances to be better wall insulation and window shading. This can save between 11.55 and 11.31% of energy consumption by air conditioners. Before the 1970s houses in the UAE were built from palm fronds called Al Areesh and usually used in the summer. In the mountains areas houses were constructed with irregular stone and internal walls were plastered with mud. The roofs were covered with palm fronds and were flat (Fig. 7).

Fig. 7. Areesh House. *Source* https://dreaminginarabic.wordpress.com/interestingsnippets/traditional-houses/

3   Research Objectives

The aim of this study is to understand the evolution of sustainability in a UAE residential house across different time frames. The document will look in to the past to find characteristics of old houses that maintained comfort and sustainability. In the present a case study will be selected with different passive design strategies applied. Finally a recommendation of what the future will look like in terms of sustainable residential houses will be provided. The main objectives are as follows:

- Study of an existing residential house constructed by the Ministry of Public Works to understand current conditions and performance of the house.
- Study the impact of changing window sizes on the average daylight factor.

- Study the influence of changing window glazing systems on solar heat gain, total electricity used and sensible cooling load.
- Study the impact of different types of exterior shading devices on solar heat gain, total electricity used and sensible cooling load.
- Study the impact of combined strategies as a final step to their effect on different variables such as average day lighting, solar gain and sensible cooling load.

4 Methodology

This study offers a simulation using EIS-VE software to test selected design strategies to reach an optimum scenario. It then combines each optimal scenario with each design strategy to see the results to compared with findings from the literature review.

4.1 Conceptual Framework

The conceptual framework used in this study is as follow (Fig. 8).

4.2 Case Study Description

Building description:

The residential house selected is in Al Bithna, Fujairah, with a 45° N orientation (Figure shows layout of house). The residential house selected is design number 763C with the following characteristics: One storey house, 2 bedrooms with 2 bathrooms attached, 4 window types, 4 A/C, 4 water heaters, 1 Majlis with a washroom and toilet and 1 kitchen, Total area: 143.35 m^2 (Figs. 9 and 10).

Building construction material:
Windows:

- W1 170 × 145, W2 100 × 110 and W3 60 × 80
- Glass used is double 24 mm thick consisting of: (6 mm Ref. Glass H.S + 12 mm A.S + 6 mm clear low E tempered glass
- Glass Specification: U-value (W/m^2 k) 2.8

Door sizes:

- D1 180 × 220 cm, D2–D3–D5 120 × 220 cm and D4 100 × 220

Window sizes:
Wall to window ratio = 9.153158, Total area of the walls = 173.91 m^2, Total area of the windows = 19 m^2
Bill for electricity and water consumption from January 2015 to January 2016.

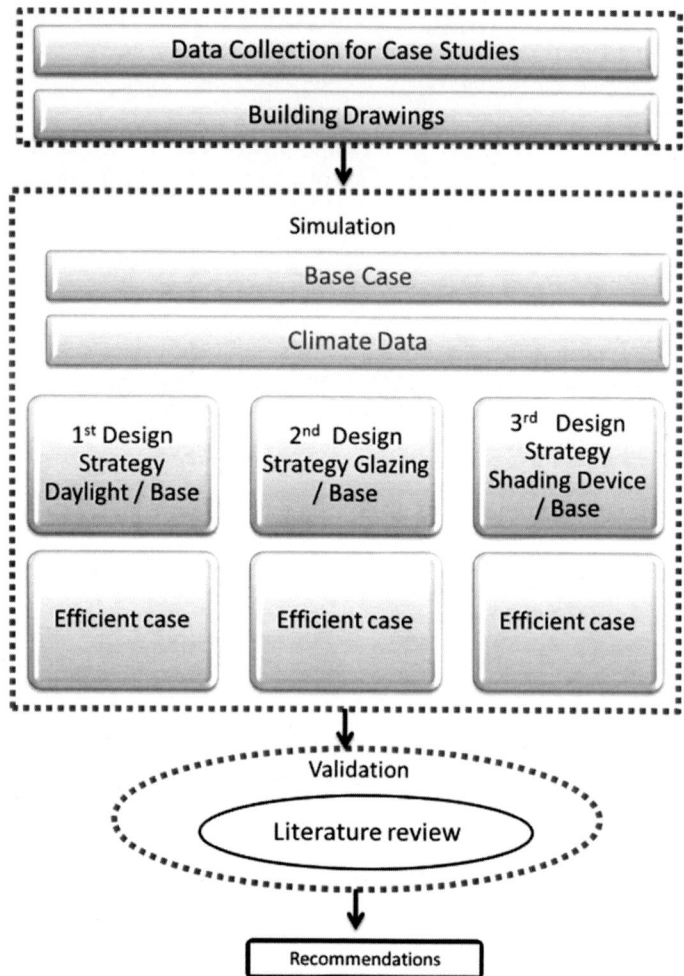

Fig. 8. Research process

Month	Electricity consumption	Water consumption
2015-01	580	11440
2015-02	595	9460
2015-03	352	6600
2015-04	905	16940
2015-05	975	12760
2015-06	1602	12760
2015-07	2190	12980
2015-08	2244	12540
2015-09	1735	12540

(*continued*)

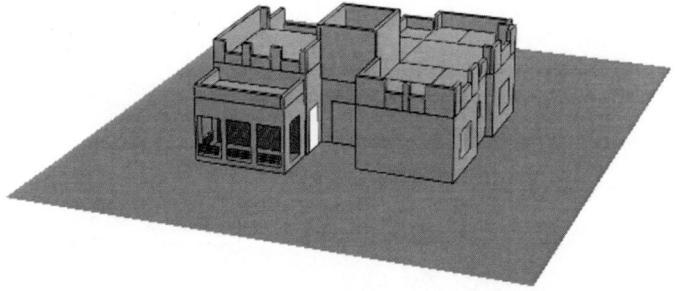

Fig. 9. Front elevation of the model

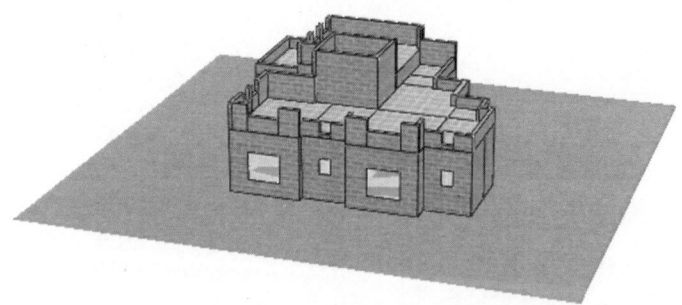

Fig. 10. Right elevation of the model

(continued)

Month	Electricity consumption	Water consumption
2015-10	1625	13860
2015-11	1237	10120
2015-12	595	12100
2016-01	519	8800

4.3 Assigned Scenarios

In this paper the selected strategies are:

1. Window size: the simulation will include the base model and increase window size to 15 and 20% respectively.
2. Glazing: the simulation will conducted for the base then for double and finally triple glazing.
3. Shading device: the simulation will include the base model without shading device, then with horizontal, vertical and finally combined shading device. Finally a combination of all of the above strategies will be adopted to find the optimum design.

5 Results

See Fig. 11.

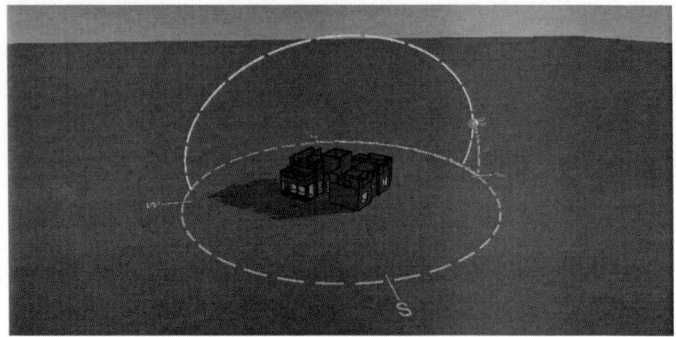

Fig. 11. Base Sun cast simulation

5.1 First Design Strategy: Increasing Window Area

This strategy involves an increase (15–20%) of window area for the Majlis, Bedroom 1 and Bedroom 2 and we will see the impact of the daylight factor. Note that the Majlis is on the northern side of the house and the bedrooms are located on the south side (Figs. 12, 13 and 14).

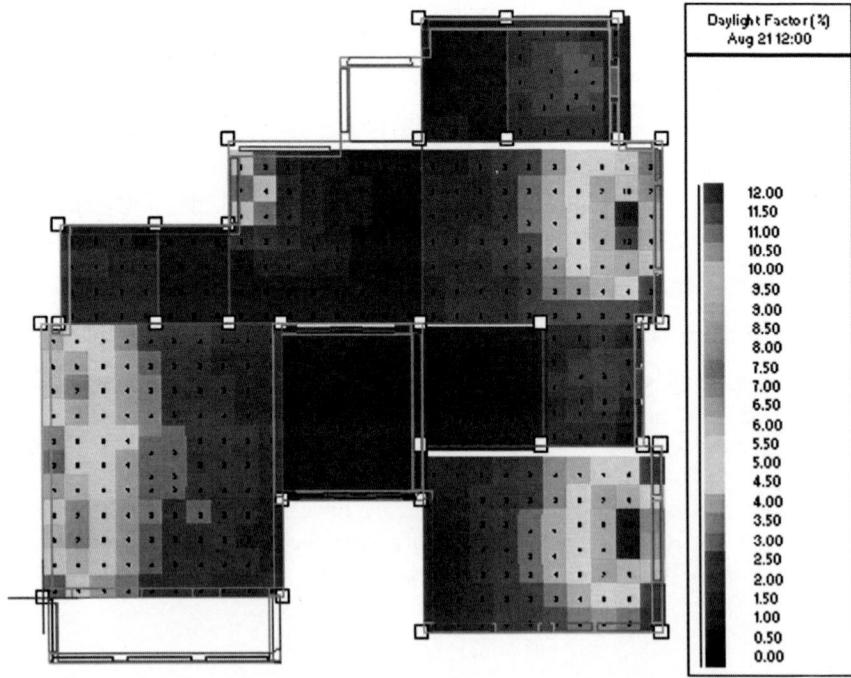

Fig. 12. Average daylight factor for the base model in August

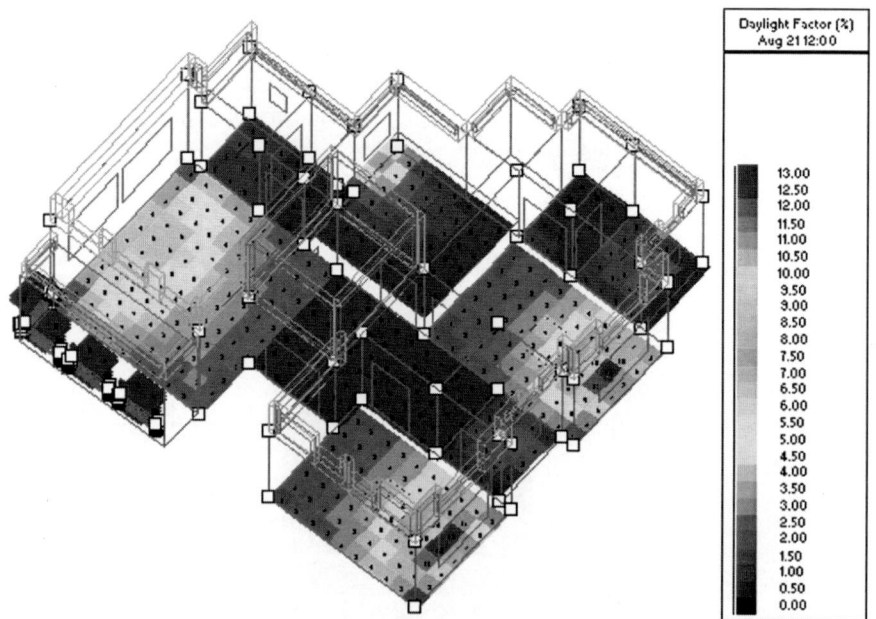

Fig. 13. Average daylight factor after increasing the window size up to 15%

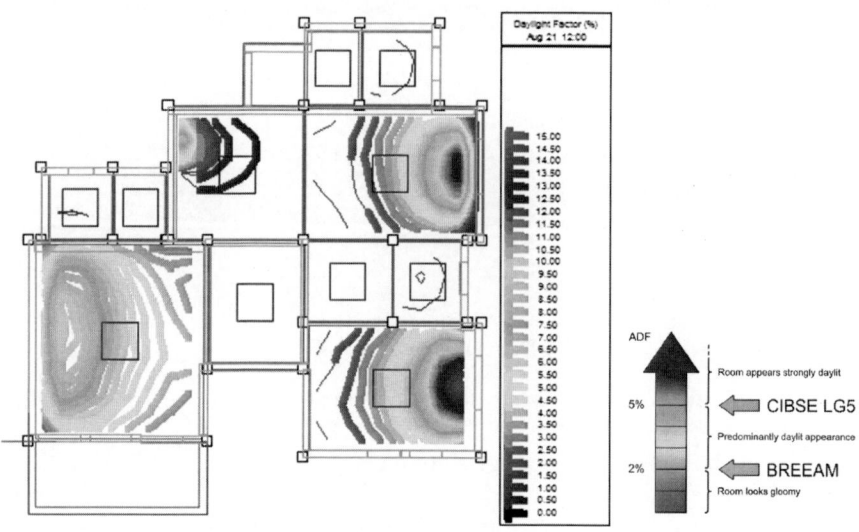

Fig. 14. Average daylight factor after increasing the window size to 20%

According to an assessment method for sustainable buildings (BREEAM): 2.08 is the ratio between DFmin/DFmean (uniformity) and should at least be 0.4 or the minimum point daylight factor should be at least 0.8%. (ADF in kitchen 2% and all living rooms should achieve a minimum ADF of at least 1.5%). According to the pearl design system used in Abu Dhabi, a residential unit should have a 2% minimum daylight factor or 200 lx daylight level for 20% of the living space floor area. Increased window size increases the average daylight factor and, in both cases, complied with BREEAM and Pearl specifications (Table 1).

Table 1. Results from simulation of daylight

	Room	Values			Uniformity	Diversity
		Min.	Ave.	Max.	(Min./Ave.)	(Min./Max.)
Base Model	(Majlis)	1.20%	3.00%	7.00%	0.39	0.17
	(BDRM1)	1.00%	3.40%	12.20%	0.3	0.08
	(BDRM2)	0.90%	3.30%	12.10%	0.26	0.07
increasing 15%	(Majlis)	1.60%	3.40%	6.10%	0.45	0.26
	(BDRM1)	1.40%	4.30%	13.20%	0.34	0.11
	(BDRM2)	1.20%	4.00%	12.30%	0.3	0.1
increasing 20%	(Majlis)	2.20%	5.50%	10.10%	0.4	0.22
	(BDRM1)	1.20%	5.00%	15.10%	0.25	0.08
	(BDRM2)	0.20%	3.50%	12.80%	0.05	0.01

Appearance	Uniform energy implications
Room looks gloomy	Electric light needed most of the day
Predominantly daylight appearance but supplementary artificial lighting is needed	Good balance between lighting and thermal aspects
Room appears strongly daylight	Daytime electric lighting rarely needed, but potential thermal problems due to overheating in summer and heat losses in winter

5.2   Second Design Strategy: Different Window Glazing

The base model used single glazing; these scenarios double and triple clear glazing with the following U-value (Figs. 15, 16, 17 and 18).

As shown above using different glazing offers a reduction in cooling load by 8.63 and 13.22% for double and triple glazing respectively and in terms of saving in total electricity, 5.065 and 7.72% savings are made for double and triple glazing respectively (Table 2).

5.3   Third Design Strategy: Different Shading Devices

The existing model has no shading device, so the scenarios will be horizontal, vertical and a combined device (Figs. 19, 20, 21 and 22).

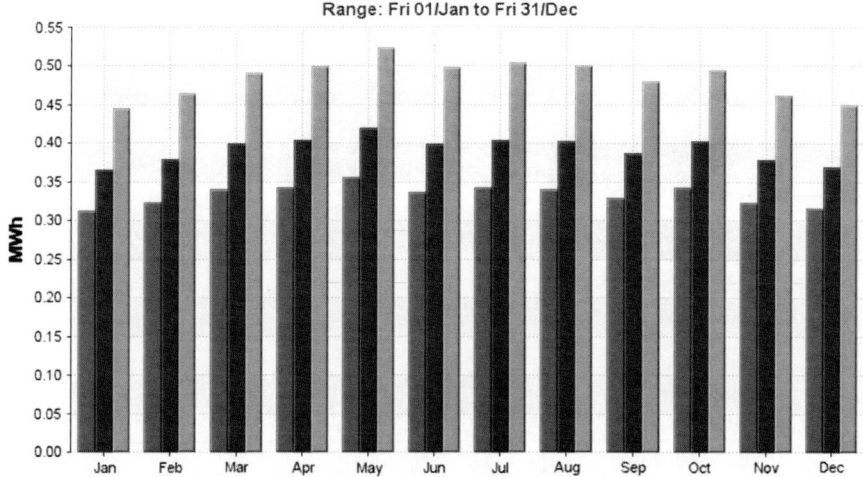

Fig. 15. Solar gain for three rooms with different glazing

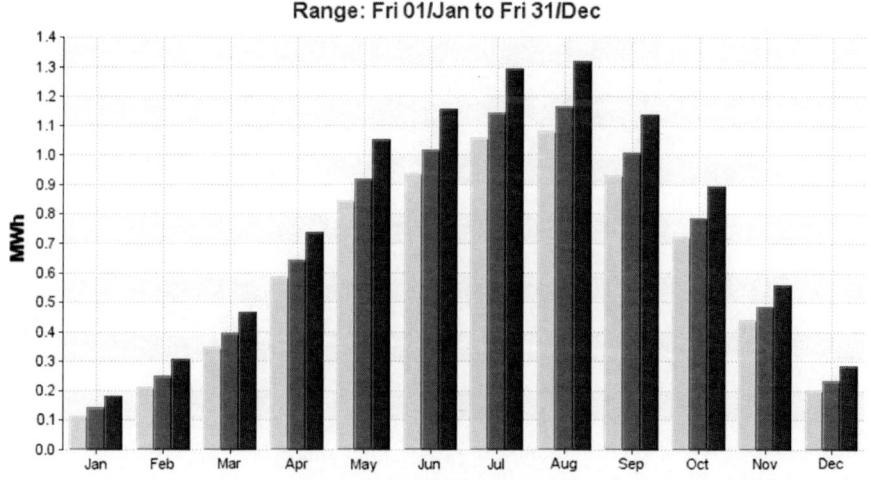

Fig. 16. Cooling plant sensible load for three rooms with different glazing

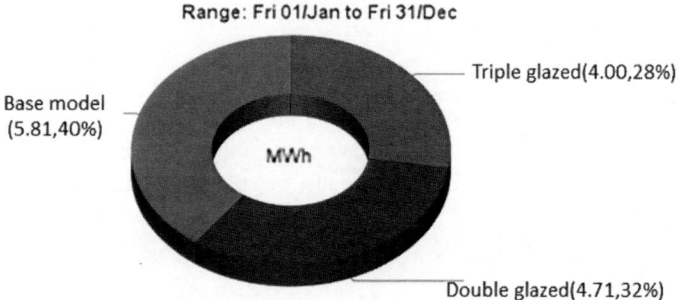

Fig. 17. Solar gain comparison for different glazing

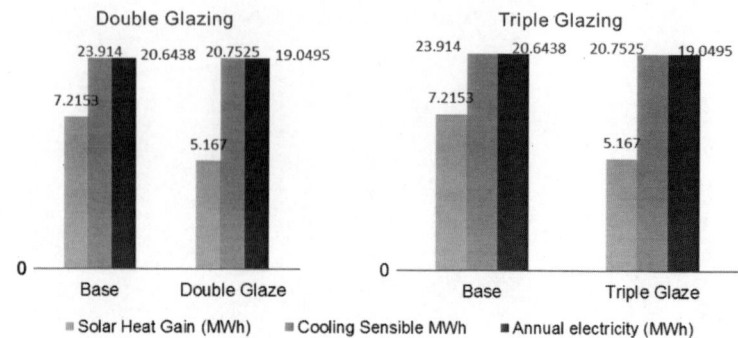

Savings / Reduction With	Solar Heat Gain	Cooling Load	Total electricity
double glazing	16.7	8.6	5.1
triple glazing	28.3	13.2	7.7

Fig. 18. Savings and reduction for different glazing types

Table 2. The U-value of different glazing

Scenari	Variables	U-value (Glass)	U-value (net)	LT
1	6mm, clear float (Reference scenario)	5.6928	5.5979	0.89
2	Double-glazing,	2.6505	2.8598	0.89
3	Triple-glazing, clear	1.8761	2.1629	0.89

No big difference between base model and adding the horizontal device. Using a vertical device gives a predominantly daylight appearance but supplementary artificial lighting is needed. It offers a good balance between the lighting and thermal aspect. When the combined device was used, it has the same effect as the base model in March but better effects in August (Table 3).

5.4   Optimal Scenario

In this scenario, the optimum for each situation was applied to the base model and the following results were found (Figs. 23, 24, 25 and 26).

From the above figures and tables, it can be seen that the combined passive design strategies have a great impact on cooling loads, on solar heat gain and offer savings of 19% on total electricity costs (Table 4).

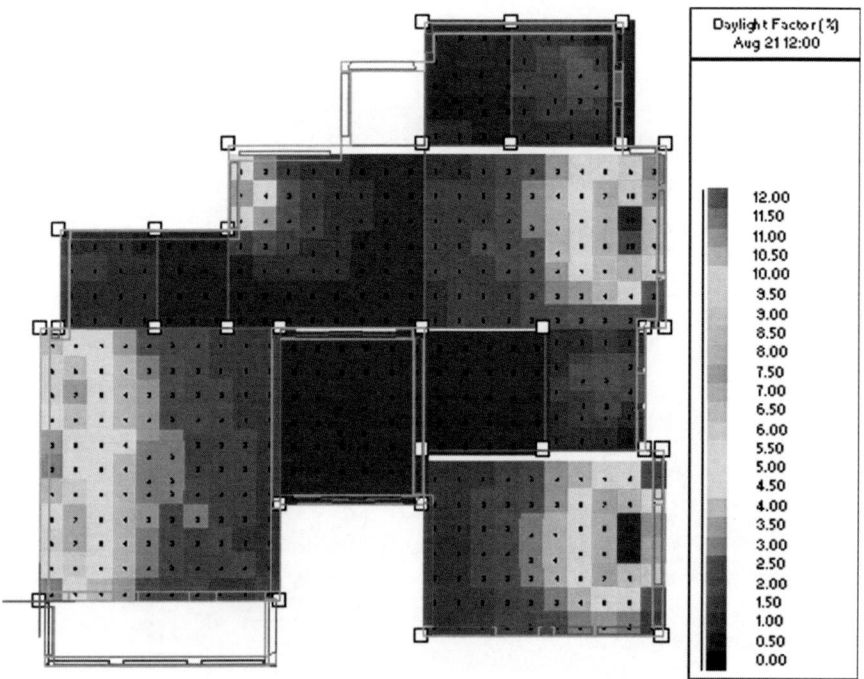

Fig. 19. Average daylight factor for the base model

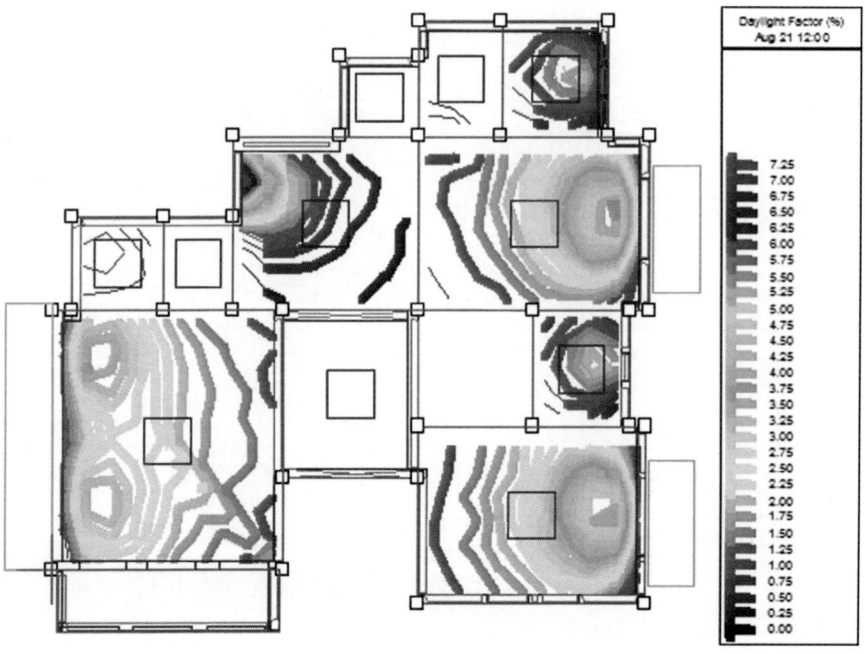

Fig. 20. Average daylight factor with a horizontal device

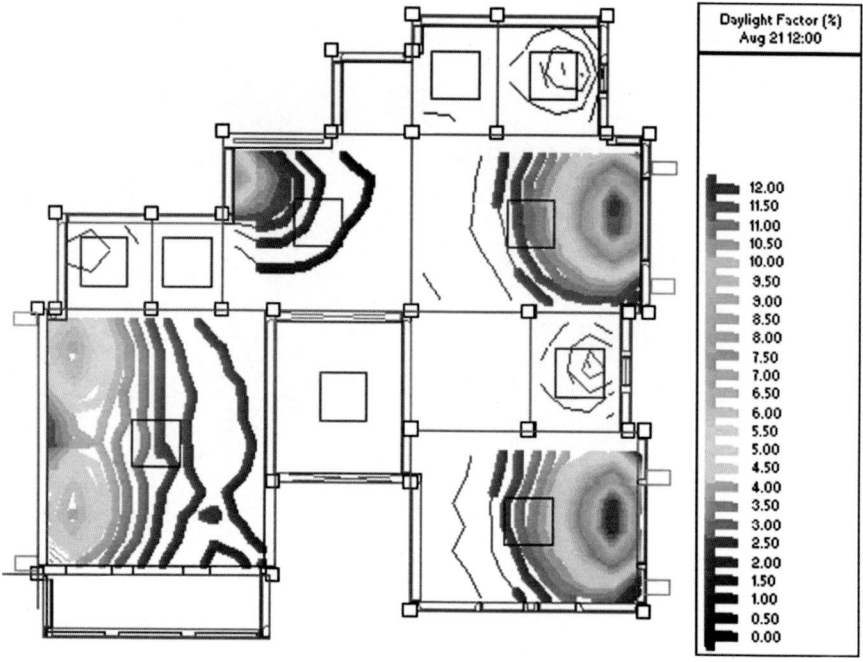

Fig. 21. Average daylight factor with a vertical device

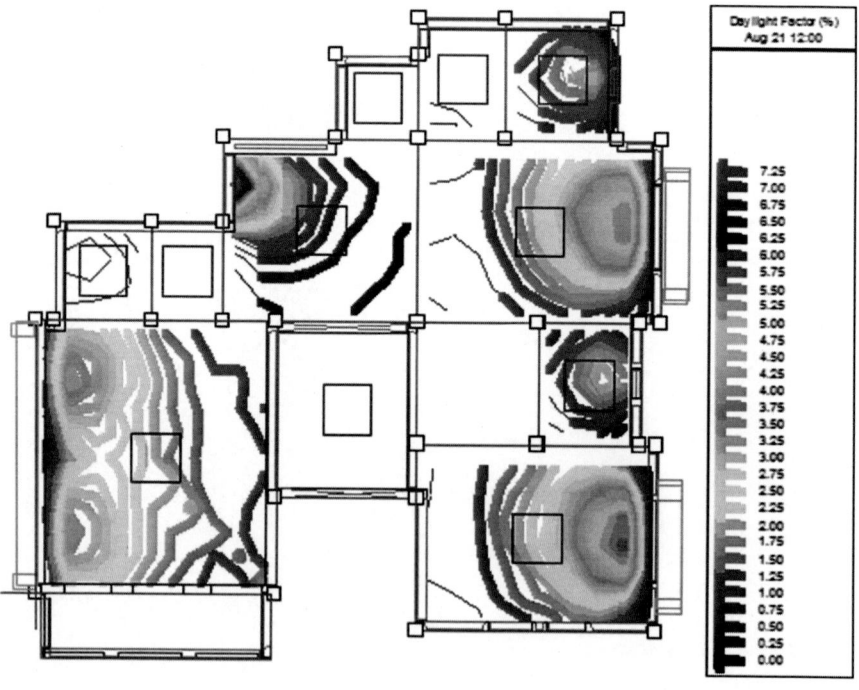

Fig. 22. Average daylight factor with a combined device

Table 3. Daylight factor for different shading devices

		August				
	Room	Values			Uniformity	Diversity
		Min.	Ave.	Max.	(Min./Ave.)	(Min./Max.)
Base Model	(Majlis)	1.2 %	3.0 %	7.0 %	0.39	0.17
	(BDRM1)	1.0 %	3.4 %	12.2 %	0.30	0.08
	(BDRM2)	0.9%	3.3 %	12.1 %	0.26	0.07
Horizontal SDs Overhang 60 cm	(Majlis)	1.6 %	3.4 %	6.1 %	0.45	0.26
	(BDRM1)	1.4 %	4.3 %	13.2 %	0.34	0.11
	(BDRM2)	1.2 %	4.0 %	12.3 %	0.30	0.10
Vertical SDs Side Fine	(Majlis)	2.2 %	5.5 %	10.1 %	0.40	0.22
	(BDRM1)	1.2 %	5.0 %	15.1 %	0.25	0.08
	(BDRM2)	0.2 %	3.5 %	12.8 %	0.05	0.01
Combined SDs	(Majlis)	0.4%	1.9 %	4.1 %	0.22	0.10
	(BDRM1)	0.2 %	1.9 %	5.8 %	0.10	0.05
	(BDRM2)	0.3 %	1.8 %	5.5 %	0.15	0.05

Appearance	**Uniformity** Energy implication
Room looks gloomy	Electric lighting needed most of the day
Predominantly daylight appearance but supplementary artificial lighting is needed	Good balance between lighting and thermal aspects
Room appears strongly dalight	Daytime electric lighting rarely needed , but potential thermal problems due to overheating in summer and heat losses in winter

6 Discussions

Passive design is proven to be related to orientation, day-light and geographical location. The results can contribute to the selection of economically efficient glass as there is a relationship between windows type, glass properties and shading device performance. The total electricity savings are largest for triple-glazing, followed by double and single.

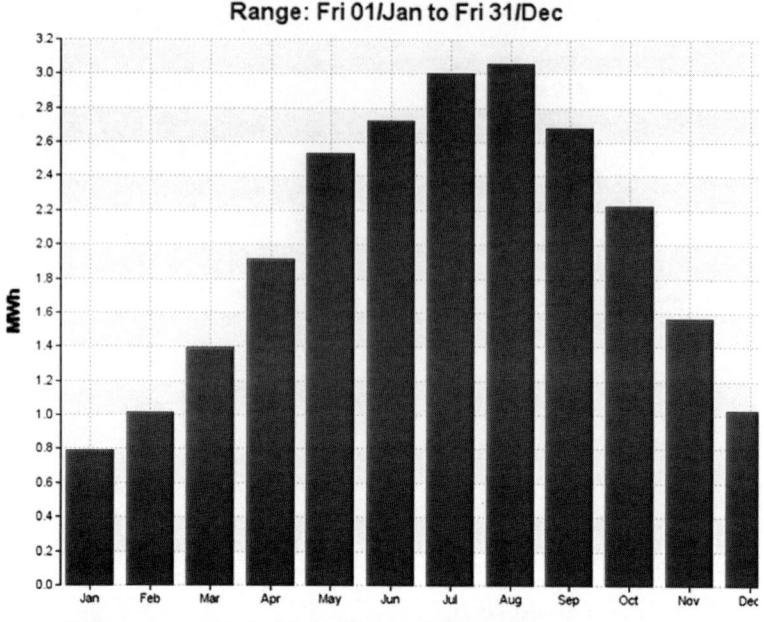

■ Cooling plant sensible load: 22 rooms (763BasScenario.aps)

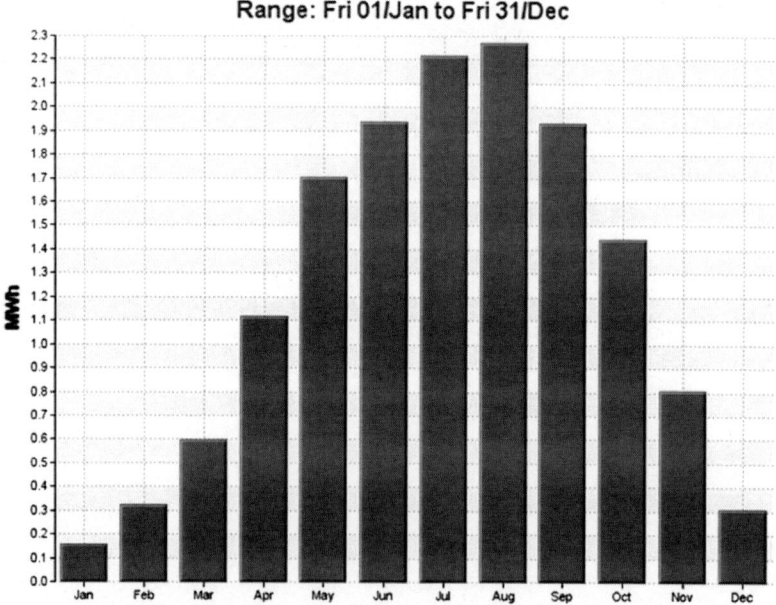

■ Cooling plant sensible load: 68 rooms (optimumScenario.aps)

Fig. 23. Cooling plant load (base and optimum model)

Base scenario

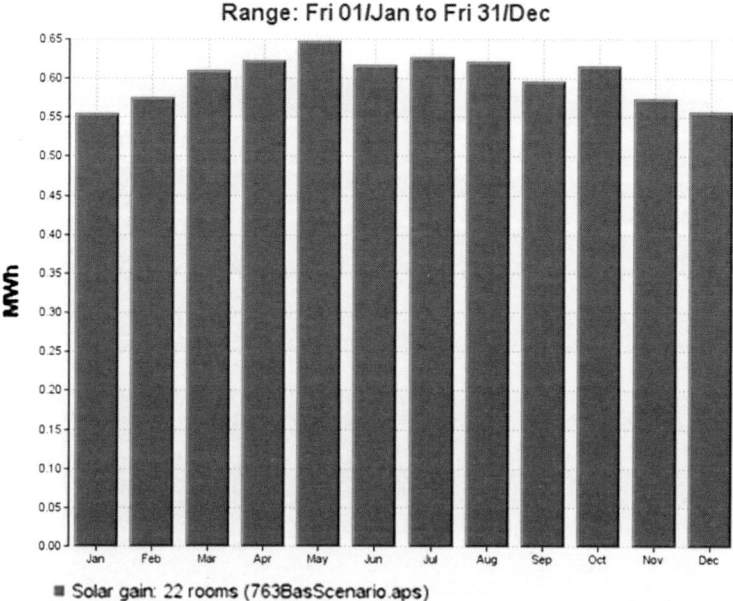

Optimum scenario

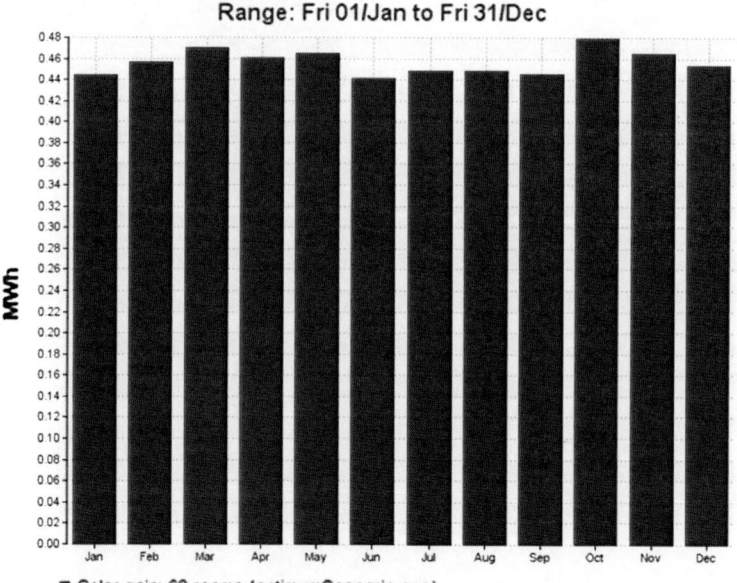

Fig. 24. Solar gain (base and optimum model)

Base scenario

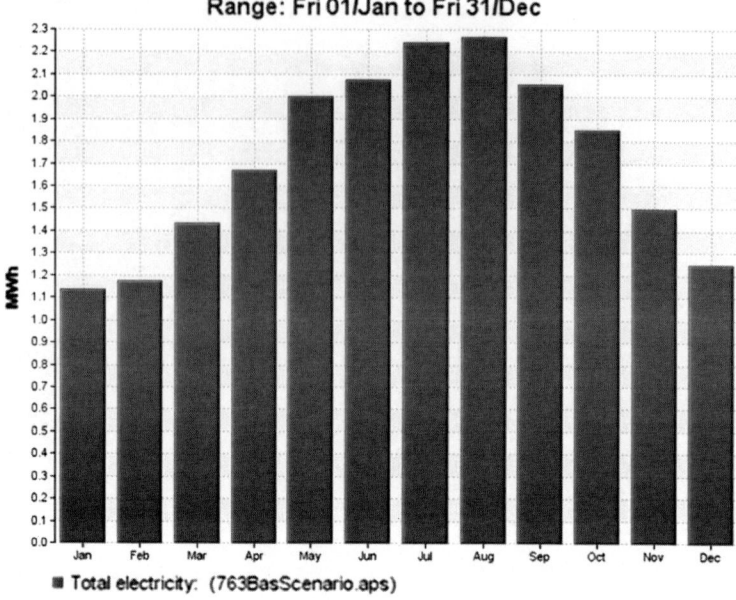

Optimum scenario

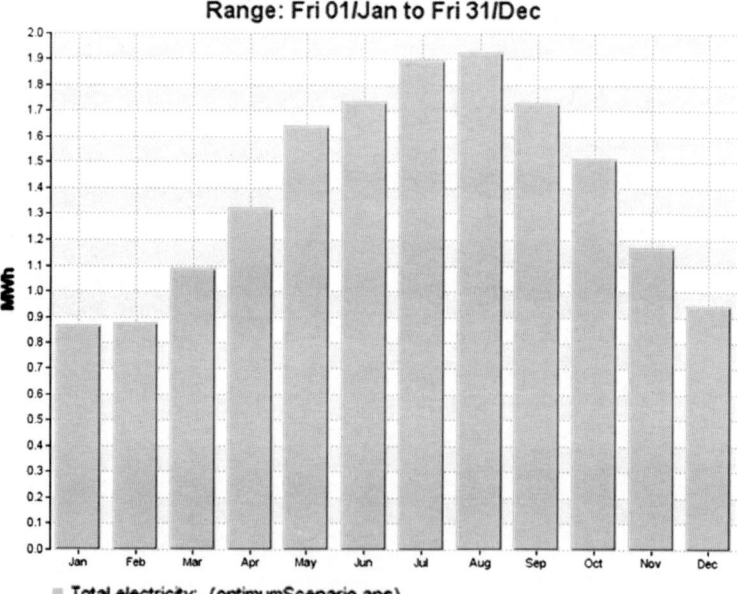

Fig. 25. Total electricity (base and optimum model)

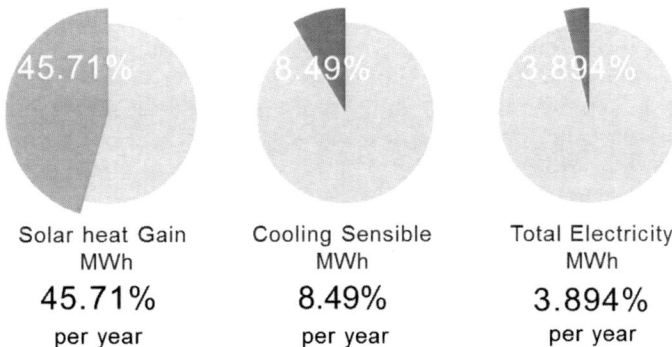

Fig. 26. Savings and reduction in solar heat gain, cooling load and total electricity (base model, optimum model)

Table 4. Average daylight factor for base and optimum scenario

Var. Name	Location	Min	Ave (%)	Max (%)	Uniformity Min/Ave	Diversity Min/Max
Base model	Majlis	Load (kW) (%)	1.7	3.5	0.22	0.11
	BDRM1	0.4	2.4	7.6	0.06	0.03
	BDRM2	0.2	2.2	7.4	0.12	0.06
Optimum scenario	Majlis	0.3	2.0	3.3	0.15	0.09
	BDRM1	0.3	3.5	10.1	0.09	0.03
	BDRM2	0.6	3.5	11.2	0.18	0.06

7   Conclusions

This study has proven that the different passive strategies which were tested in this project have huge implications for total electricity costs, solar gain and cooling loads. This is a path to achieving reductions in CO_2, leading to sustainability. As a future recommendation, different strategies could be implemented to achieve more reductions in total electricity costs. The Ministry of Public Works is targeting low emission houses to maintain sustainability.

References

1. Alzoubi H, Al-Zoubi A (2009) Assessment of building façade performance in terms of daylighting and the associated energy consumption in architectural spaces: vertical and horizontal shading devices for southern exposure facades. Energy Convers Manag 51 (2010):1592–1599
2. American Institute of Architecture (AIA) (2008) Environmental study. Available from: http://www.aia.org

3. Bessoudo M et al (2010) Indoor thermal environmental conditions near glazed facades with shading devices—Part I: experiments and building thermal model. Build Environ 45 (11):2506–2516

4. Chi-Ming L, Yao-Hong W (2011) Energy-saving potential of building envelope designs in residential houses in Taiwan. Energies 4:2061–2076

5. Ebrahimpour A, Maerefat M (2010) Application of advanced glazing and overhangs in residential buildings. Energy Convers Manag 52(2011):212–219

6. Fiocchi C, Hoque S, Shahadat M (2011) Climate responsive design and the Milam Residence. Sustainability 3(2011):2289–2306

7. Hammad B, Abu-hijleh (2010) The energy savings potential of using dynamic external louvers in an office building. Energy Build 42:1888–1895

8. Kensek K, Noble D, Schiler M, Setiadarma E (1996) Shading mask: a teaching tool for sun shading devices. Autom Constr 5:219–231

9. Kim G, Lim H, Lim T, Schaefer L, Kim J (2012) Comparative advantage of an exterior shading device in thermal performance for residential buildings. Energy Build 46:105–111

10. Lam M, Miller A (2009) Shading performance of vertical deciduous climbing plant canopy. Build Environ 45:81–88. Accessed 5 Feb 2016. Available from: http://www.sciencedirect.com

11. Meyer WT (1982) Energy economics and building design. McGraw-Hill, New York, London

12. Olgyay V (1963) Design with climate: bioclimatic approach to architectural regionalism. Princeton University Press, Princeton

13. Palmero-Marrero A, Oliveira A (2009) Effect of louver shading devices on building energy requirements. Appl Energy 87:2040–2049. Accessed 5 Feb 2016. Available from: http://www.sciencedirect.com

14. Tzempelikos A, Athienitis A (2006) The impact of shading design and control on building cooling and lighting demand. Sol Energy 81(2007):369–382

15. Tzempelikos A, Bessoudo M, Athienitis A, Zmeureanu R (2010) Indoor thermal environmental conditions near glazed facades with shading devices e Part II: thermal comfort simulation and impact of glazing and shading properties. Build Environ 45:2517–2525. Accessed 5 Feb 2016. Available from: http://www.sciencedirect.com

16. Wong NH, Istiadji AD (2004) Effect of external shading devices on daylighting penetration in residential buildings. Light Res Technol 36(4):317–333

17. Yu L, Watanabe T, Hiroshi Y, Gao W (2008) Research on energy consumption of urban apartment buildings in China. J Environ Eng 73:183–190

Determination of Optimum Insulation Thickness on Different Wall Orientations in a Hot Climate

Erhan Arslan and Irfan Karagoz[✉]

Department of Mechanical Engineering, Uludag University, Bursa, Turkey
arslan.erhan92@gmail.com, karagoz@uludag.edu.tr

Abstract. In recent years, alternative and renewable energy sources have become essential and their use in this direction is increasing. When these energy sources are used, energy saving is also assessed. The main objective of this work is to determine the optimum insulation thickness according to the cooling requirement of a building in a hot climate. During the summer period different wall directions has been evaluated in Bursa. For the optimum insulation thickness in all wall directions, firstly total solar radiation, cooling transmission load and then cost analysis are performed. While solar radiation was calculated, factors such as declination angle, geometric factor, sun clock angle, altitude, ground reflectance and latitude angle were taken into account. The differences in cooling transmission load for insulated and uninsulated walls were observed. The heat transfer coefficient was determined according to inside and outside air-film thermal resistances and total thermal resistance of the wall without the insulation. In addition to these, the volumes, thicknesses, thermal conductivities and costs of the thermal insulation materials are also considered. When cost analysis is performed, PWF is calculated and also interest rate, inflation rate, cost of insulation materials and electrical cost were taken into account. Depending on the location of the building, insulation material which should be used for minimum cost and maximum efficiency and the required properties of this insulation material have been determined.

Keywords: Optimum thickness of insulation · Cooling transmission load Solar radiation · Cost analysis

1 Introduction

Energy is an essential factor of production in production and one of the main indicators reflecting the economic and social development potential of a country. Rapid depletion of natural resources, increased environmental pollution and the payment of high wages for existing energy require efficient use of energy [1]. It is a fact that Turkey is not very rich in terms of energy resources. 60–65% of your energy needs are imported from abroad [2]. The environmental pollution, which is a consequence of the limitations of energy resources and the consumption of energy, has become compulsory. Looking at the sectorial distribution of energy consumption, it is seen that housing and conversion

© Springer International Publishing AG, part of Springer Nature 2018
S. Fırat et al. (eds.), *Proceedings of 3rd International Sustainable Buildings Symposium (ISBS 2017)*, Lecture Notes in Civil Engineering 7, https://doi.org/10.1007/978-3-319-64349-6_12

sector (electricity generation) account for 26% [3]. Energy requirements, especially of the residential building sector, are an important part of the total energy consumption in many countries [4]. For example, in Turkey, the building sector was the second largest consumer of energy with 25.793 million tons of equivalent energy (MTOE) in 2001, and its demand is estimated to reach 41.7 MTOE by 2020 [5].

Energy conservation can be provided by reducing energy consumption in buildings [2]. One way to achieve this is to apply heat insulation to the exterior walls of the building [2]. External insulation system can be easily applied to new buildings and existing buildings. In the buildings being used, all application have been realizing outside the building during the application [6].

While the thickness of the thermal insulation increases, the cost of the insulation material increases, energy costs are also reduced because of the reduction in heat loss from the building. The fuel cost and the energy cost in the building were evaluated and the optimum insulation thickness was determined from the total cost-insulation thickness axis according to these cost values.

The TS 825 thermal insulation rules in buildings have been made compulsory since 2000 and have begun to be implemented in new buildings [7]. TS 825 has set new limits on heating expenditures according to area and volume ratios of buildings aims to save energy [7]. It is aimed to provide energy savings in buildings [2].

There are many studies in the literature on optimizing the thermal insulation thickness. However, in most of the studies, there is no analysis according to different directions (east, west, north and south). Ucar [8] used four different insulation materials for determination of optimum insulation thickness on the exterior wall but did not work for wall orientation. Ozel and Pihtili [9] determined the optimum insulation thickness for different wall directions, but they didn't use the degree-day method.

In this study, the determination of the optimum insulation thickness value to be applied to the exterior walls is focused. Considering the effect of solar radiation, it has been determined how the insulation thickness changes for vertical surfaces (outside walls of the building) facing in different directions. For this purpose, in order to calculate the values of instantaneous solar radiation falling on the surfaces, the outside air temperature data of Bursa is taken into account. After determining DG values due to direction, optimum insulation thicknesses, which minimize the sum of energy and insulation costs, are calculated.

2 Mathematical Formulation

2.1 Calculation of Solar Radiation

In order to calculate the values of solar radiation falling on the horizontal surface, firstly, the daily solar radiation on a horizontal surface is determined [10], which is given by:

$$\frac{I_h}{I_{o,h}} = \left(a + b\frac{S}{S_o}\right) \tag{1}$$

where I_h is the monthly average daily global solar radiation, $I_{o,h}$ is the monthly average daily extraterrestrial radiation, S the day length, S_o the maximum possible sunshine duration, a and b are empirical coefficients relating with the region [10].

a and b are a function of solar declination angle (δ) and latitude angle (ϕ) and altitude (Z), as given by the equations [10]:

$$a = 0.103 + 0.000017Z + 0.198\cos(\phi - \delta) \tag{2}$$

$$b = 0.533 - 0.165\cos(\phi - \delta) \tag{3}$$

The monthly average daily extraterrestrial radiation on per unit of horizontal surface can be computed as follows:

$$I_{o,h} = \frac{G_{sc}}{\pi}\left[1 + 0.033\cos\left(n\frac{360}{365}\right)\right]\left[\cos\phi\cos\delta\sin\omega_s + \frac{\pi}{180}\omega_s\sin\phi\sin\delta\right] \tag{4}$$

where G_{sc} is the solar constant, ω_s is the sunset hour angle for the month, and n is the day of the year. The solar constant is given as 1367 W/m^2 [10]. The declination angle according to the number of days of the year is calculated using Eq. (5) [11].

$$\delta = 23.45\sin\left(360\frac{n+284}{365}\right) \tag{5}$$

The clock angle is calculated using Eq. (6) [11]:

$$\omega_s = \text{arc}\cos[-\tan(\phi)\tan(\delta)] \tag{6}$$

T_e is the sol–air temperature including the effect of solar radiation on the outdoor temperatures and is expressed as follows [12]:

$$T_e = T_o + \frac{aI_T}{h_o} - \frac{\varepsilon\Delta R}{h_o} \tag{7}$$

where T_o is the outdoor air temperature which is taken as 15 °C. I_T and a denote the total solar radiation and solar absorptivity of the outdoor wall surface, respectively. $\varepsilon\Delta R$ is the correction factor and is assumed to be 4 °C for horizontal surfaces and 0 for vertical surfaces from ASHRAE [12]. The summer design values of the ratio a/h_o are determined to be 0.052 m^2 K/W [13].

The total solar radiation (I_T) is calculated as [10, 12]

$$I_T = I_h\left(1 - \frac{I_d}{I_h}\right)R_b + I_d\left(1 + \frac{\cos\beta}{2}\right) + I_h\rho\left(1 - \frac{\cos\beta}{2}\right) \tag{8}$$

Ground reflectance ρ is usually taken as 0.2. The geometric factor R_b is the ratio of beam radiation on the tilted surface to that on a horizontal surface at any time and is calculated as [12]:

$$R_b = \frac{\cos\theta}{\cos\theta_z} \tag{9}$$

The equation of mean daily diffuse solar radiation on a horizontal surface (I_d), which is based on approximately 10-year measurements, is developed by Tiris et al. (1995), as follows [10]

$$I_d = I_h(0.703 - 0.414K_T - 0.428K_T^2) \tag{10}$$

$$\begin{aligned}\theta = {} &\sin\delta\sin\phi\cos\beta - \sin\delta\cos\phi\sin\beta\cos Y + \cos\delta\cos\phi\cos\beta\cos\omega \\ &+ \cos\delta\sin\phi\sin\beta\cos\omega\cos Y + \cos\delta\sin\beta\sin Y\sin\omega\end{aligned} \tag{11}$$

$$\theta_z = \cos\delta\cos\phi\cos\omega + \sin\delta\sin\phi \tag{12}$$

where γ is surface azimuth angle. γ is zero for an inclined plane facing south. It is taken as negative from the south to the east, and the north, and positive from south to west, and the north, i.e. −180° < γ < + 180° [12].

2.2   The Structure of the External Walls and Parameters

The characteristics of the wall in the study are given in Table 1 [12] and some financial and cost parameters are given in Table 2. This year's (2015) data is used for interest and inflation rates [14, 15].

Table 1. Wall materials, thermal conductivities and thermal resistances

Wall materials	Thickness (m)	Thermal conductivities 'k' (W/mK)	Thermal resistances 'R' (m² K/W)
External plaster	0.025	1	0.025
Brick block	0.19	0.45	0.422
Internal plaster	0.025	1.4	0.0178
XPS	u	0.029	u/0.029
EPS	u	0.038	u/0.038

While optimal insulation thickness is calculated, lifetime cost analysis is one of the methods used. The present value factor depends on the inflation rate (g) and the interest rate (i) [16].

$$PWF = \frac{(1+r)^N - 1}{r(1+r)^N}, \quad \begin{pmatrix} i > g & r = \frac{i-g}{i+g} \\ i > g & r = \frac{g-i}{1+i} \end{pmatrix} \tag{13}$$

Table 2. Fuel and financial parameters

Parameters	Values
Inflation rate (g)	7.16%
Interest rate (i)	7.50%
Present worth factor (PWF)	11.113
Cost of electricity (TL/kWh)	0.42
Coefficient of performance (COP)	2.5

$$PWF = \frac{N}{1+i}, \quad i = g \tag{14}$$

The N lifetime in the equation is taken as 13 years in the calculations (Table 2).

2.3 Cooling Transmission Load Calculations

Yearly cooling load is separately calculated from daily transmission loads which are added over summer period [12]. In the degree-days for CDD and the degree-hours methods for CDH, the yearly transmission load per unit of wall area is estimated (in J/m^2) by the following expression [12]:

$$Q_c = 86400 \cdot CDD \cdot U \tag{15}$$

$$Q_c = 3600 \cdot CDH \cdot U \tag{16}$$

where CDD and CDH are cooling degree-days and cooling degree-hours, respectively. These values for the climate of Bursa are calculated from the meteorological data depend on wall orientation. The overall heat transfer coefficient of the wall can be expressed as follows [18].

$$U = \frac{1}{R_i + R_w + R_{ins} + R_o} \tag{17}$$

where R_i and R_o are the inside and outside air-film thermal resistances and R_w is total thermal resistance of the wall without the insulation. The thermal resistance of the insulation layer R_{ins} is given by [5]:

$$R_{ins} = \frac{x}{k} \tag{18}$$

where x and k are the thickness and thermal conductivity of the insulation material, respectively.

2.4   Determination of Optimum Insulation Thickness

The use of the thermal insulation reduces the load of air-conditioning, so, the energy cost of cooling in summer. However, the prices of materials of insulation increase the initial costs of construction. Consequently, an economic analysis should be carried out in order to estimate the optimum insulation thickness which minimizes the total cost including the insulation and the energy consumption costs. The total costs per unit of area of wall are given by [17]:

$$C_t = C_{enr} \cdot PWF + C_i \cdot L_i \tag{19}$$

where C_{enr} is the cost of energy consumption (TL/m^2), PWF is Present Worth Factor and C_i is the cost of insulation per unit of area (TL/m^2). The annual cost of energy per unit of area of the wall for cooling (C_{enr}) is given by [17]:

$$C_{enr} = \frac{Q_c \cdot C_{el}}{COP} \tag{20}$$

where Q_c is yearly cooling transmission load (kWh/m^2), COP is coefficient of performance of air-conditioning system and is C_{el} cost of electricity (TL/kWh) [18].

3   Result and Discussion

3.1   Environmental Conditions

The investigation is carried out for all wall orientations at the climatic conditions of Bursa (latitude: 40.23° N, longitude: 29.01° E). The outdoor air temperatures are obtained by averaging hourly measurements recorded in meteorological data [19]. The summer design values of the ratio a/h_o are determined to be 0.052 m^2 K/W for dark-colored surfaces [13]. Hourly variation of incident solar radiation and sol–air temperature for all wall orientations in July 15 are shown in Fig. 1a, b, respectively.

The incident solar radiation is highest at 10:00 for east orientation and at 14:00 for west orientation while it is maximum at noon (12:00) for south and north orientations. It is seen that maximum peak value of sol–air temperatures of south oriented wall is appears for July 15.

3.2   Cooling Transmission Loads

Daily and yearly total cooling transmission loads of uninsulated and insulated walls for the all wall orientations are shown in Figs. 2 and 3. June gives the lowest cooling loads while August gives the highest cooling loads. The maximum cooling load is obtained for east, north and south wall in August while it is obtained for west walls in July.

The yearly cooling loads of uninsulated wall are obtained as 61.51, 32.78, 90.56 and 77.66 (MJ/m^2 year) for south, north, east and west oriented walls, respectively (Fig. 3). The lowest cooling load is provided for north wall while the highest cooling

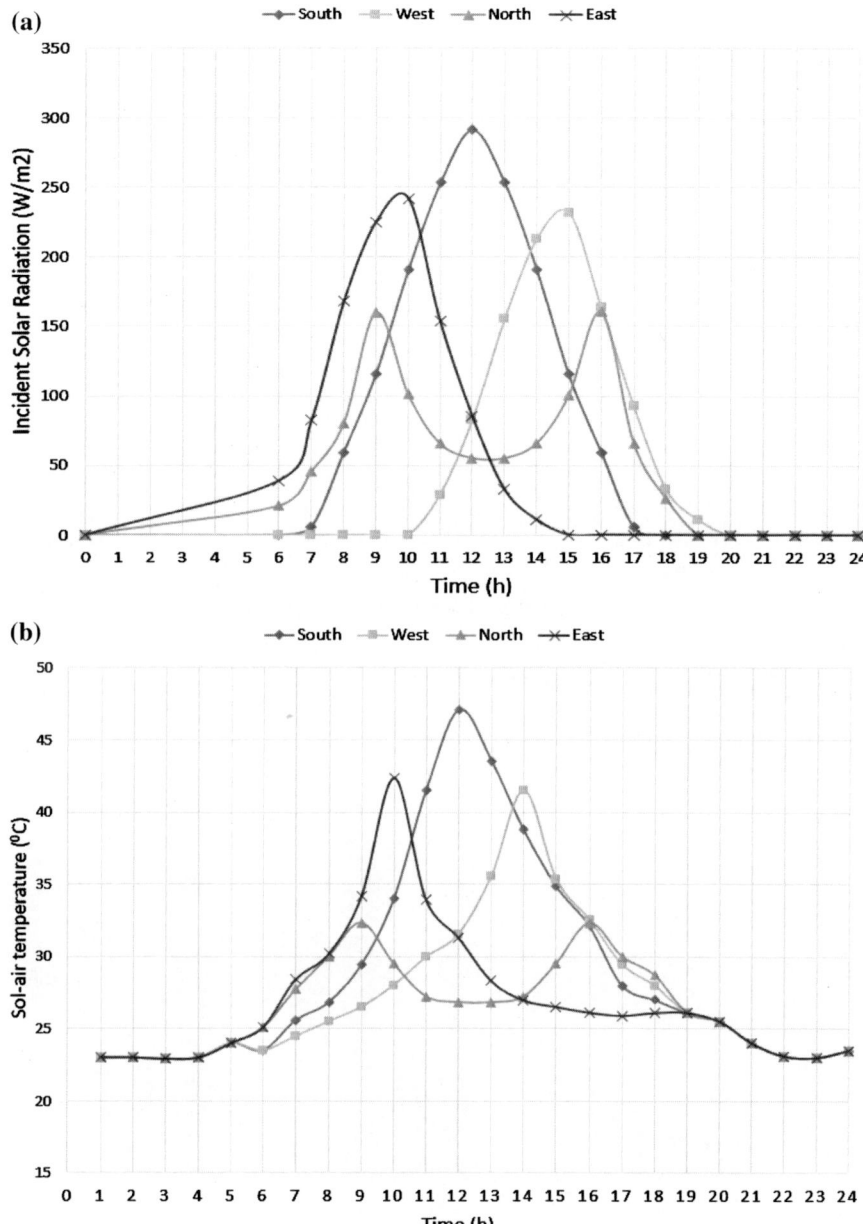

Fig. 1. (a) Hourly variation of incident solar radiation and (b) sol–air temperature for July 15

load is obtained for east walls. It is seen that the yearly cooling loads decrease when the wall is insulated. It is also seen that EPS carries out lower cooling loads than XPS.

Figure 4 indicates variation of yearly cooling transmission loads due to increasing insulation thickness for all wall orientations. So, transmission loads decrease as the

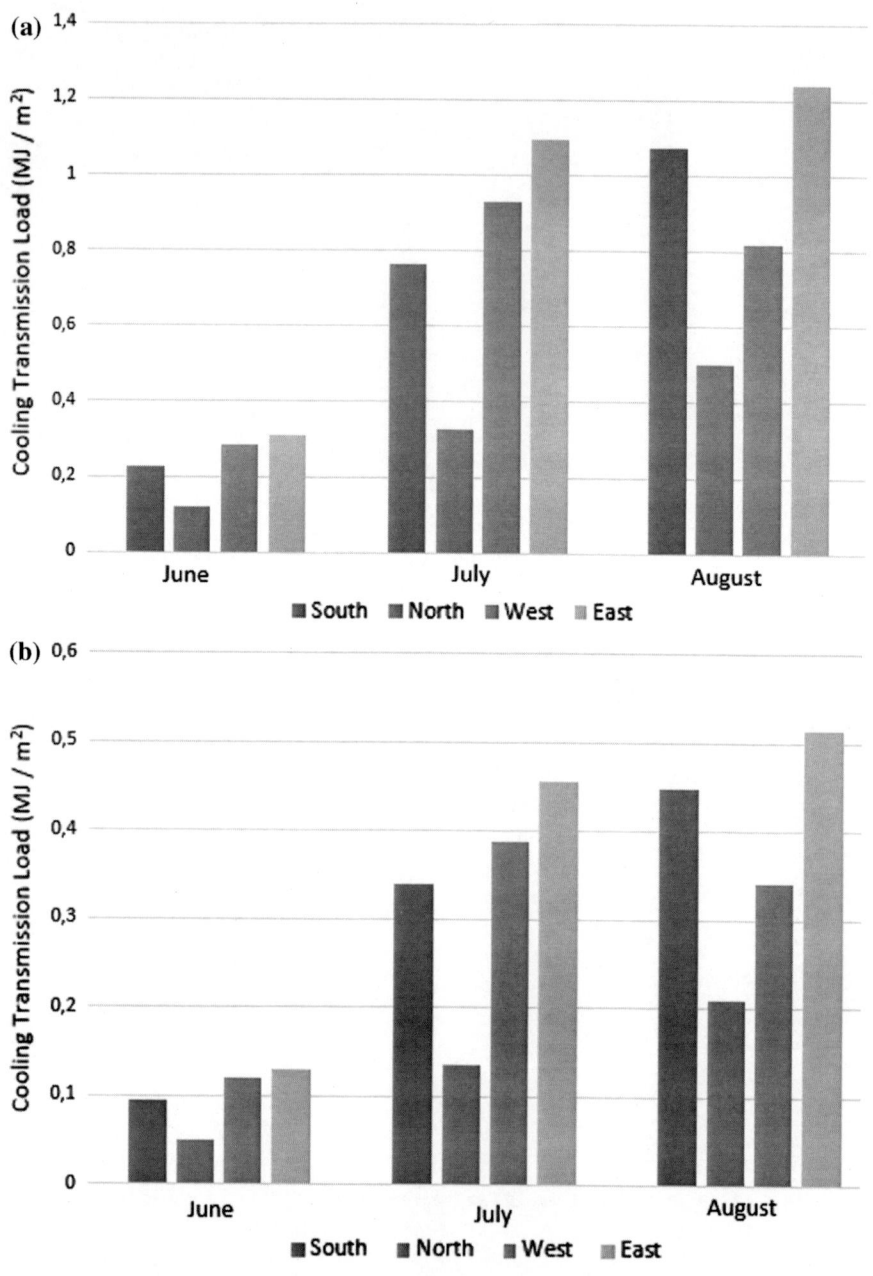

Fig. 2. (a) Daily cooling transmission load of uninsulated and (b) 4 cm insulated walls with XPS for the representative days of each month of summer for the all wall orientations

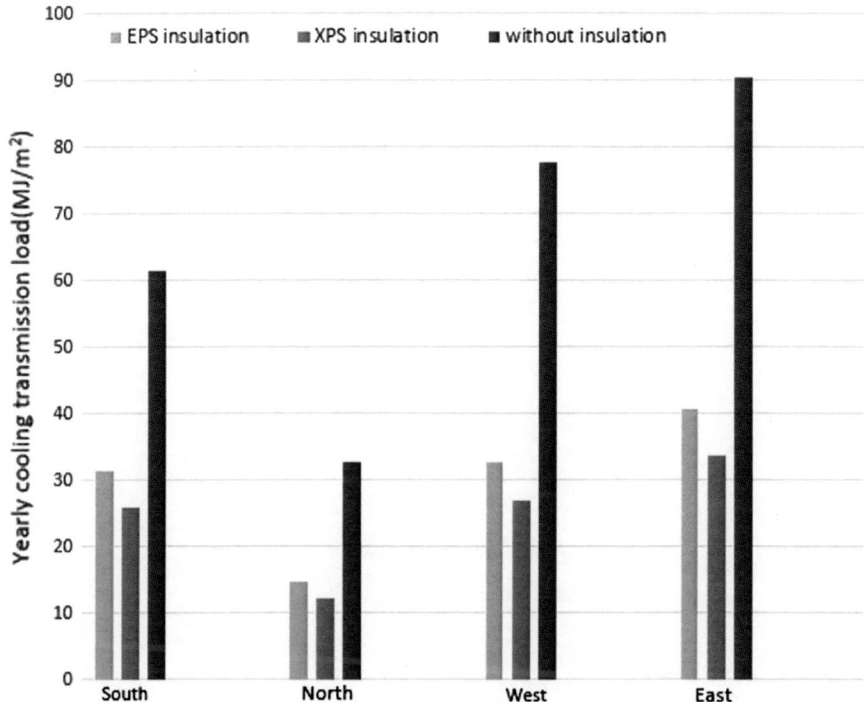

Fig. 3. Yearly cooling transmission load of uninsulated and 3 cm insulated walls for the all wall orientations

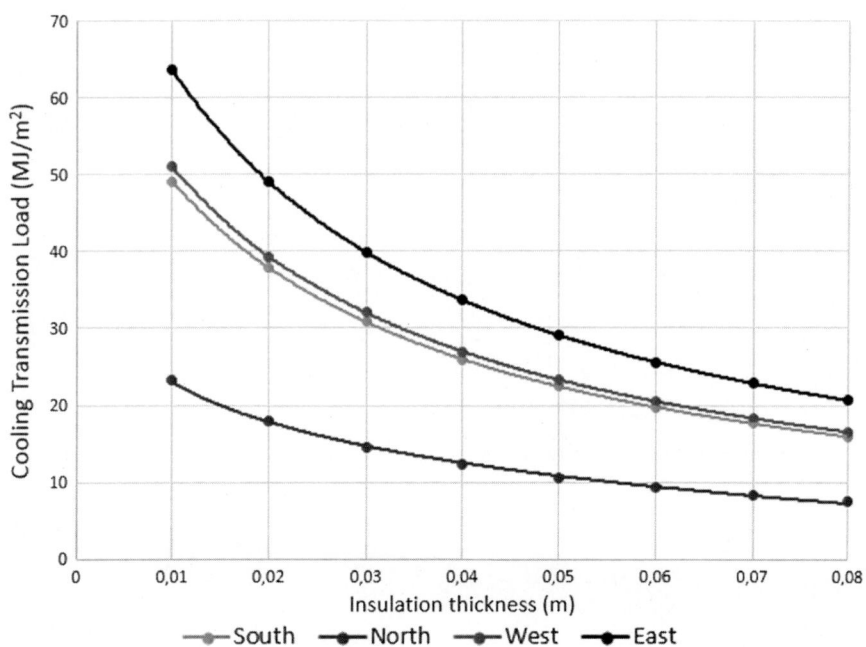

Fig. 4. Variation of yearly cooling transmission load versus insulation thickness for the all wall orientations

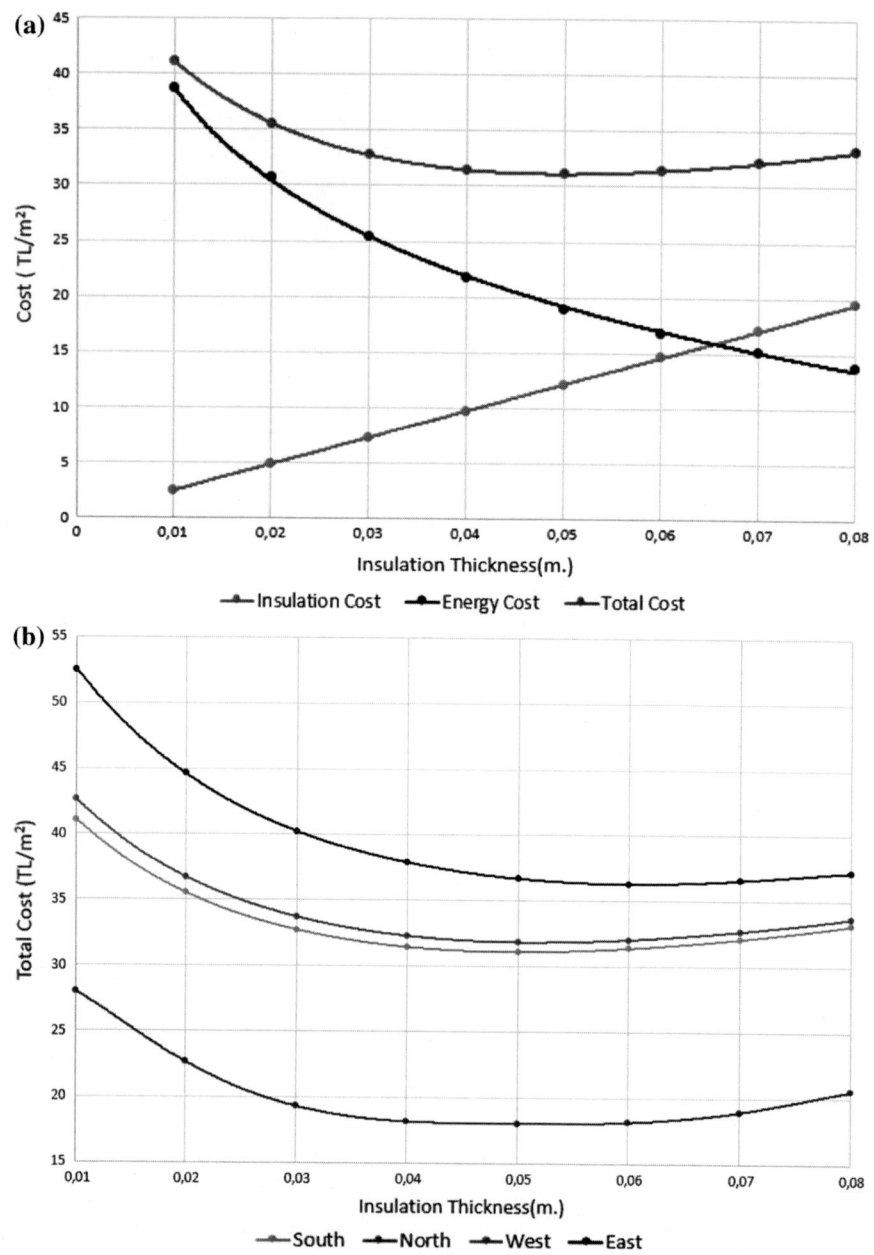

Fig. 5. Variation of cost with insulation thickness: (**a**) for a south facing wall, and (**b**) for all wall orientations

insulation thickness increases. As the insulation thickness increases, cooling trans-mission loads for all wall orientations become closer.

3.3   Optimization of Insulation Thickness

To determine the optimum insulation thickness, the cooling transmission load obtained is used as a model including the cost of insulation material and the present value of energy consumption cost over lifetime of 13 years of the building. Figure 5a shows variation of cost to insulation thickness for a south oriented wall. As expected, while the energy cost decreases with increasing insulation thickness, the insulation cost increases with increasing insulation thickness. Total cost is sum of energy and insu-lation costs. The insulation thickness at which the total cost is the minimum as shown in Fig. 5b is taken as the optimum insulation thickness. It is seen that optimum insulation thicknesses for south, north, east and west orientations are obtained as 5.2, 4.5, 6.1 and 5.4 cm, respectively.

The results of the other studies related to effect of wall orientation on optimum insulation thickness are shown in Table 3.

Table 3. The optimum insulation thickness and minimum total cost, for all wall orientations

Orientation	Optimum insulation thickness (m)	Minimum total cost (TL/m^2)
South	0.052	31.12
North	0.045	18.16
East	0.061	36.78
West	0.054	32.16

4   Conclusion

In this study, thermal and economic parameters for insulated building walls in a hot climate are investigated. The results show that the lowest cooling load is obtained as 32.78 (MJ/m^2 year) for north wall while the highest cooling load is obtained as 90.56 (MJ/m^2 year) for east wall. It is seen that the yearly cooling loads decrease when the wall is insulated. This decrement in all wall orientations is 37.86% for 4 cm XPS insulation and 45.78% for 4 cm EPS insulation. It is seen that optimum insulation thicknesses for south, north, east and west orientations are obtained as 5.2, 4.5, 6.1 and 5.4 cm, respectively. While the lowest value of optimum insulation thickness is obtained for the north-facing wall which has minimum cooling load, highest value of optimum insulation thickness is obtained for the east-facing wall which has maximum cooling load.

The most economical orientation is north with an optimum insulation thickness of 4.5 cm. for cooling season. To get highly accurate results, heat transmission loads in the building walls and therefore, economic parameters must be determined under the thermal conditions.

In this study, it is concluded that the cooling transmission load is an important factor to determine the insulation thickness. Therefore, calculations were made only for summer months. However, it can be said that the heating transmission load should also be taken into account by considering the winter months, in order to achieve more reliable results. If this article is taken into consideration, it can be suggested that the heating transmission load can be added to the account and the exact results can be obtained by finding the insulation thickness and the total cost.

References

1. Erkoc E, Senkal MC (2013) The state of energy in world and Turkey. Muhendis ve Makine 54:32–44
2. Kaynaklı O, Sevcan O, Karamangil MI (2012) Determination of optimum thermal insulation thickness considering solar radiation and wall orientation. J Fac Eng Archit Gazi Univ 27:367–374
3. Ministry of Foreign Affairs, Republic of Turkey 2016. http://www.mfa.gov.tr
4. Fertelli A (2013) Determination of optimum insulation thickness for different building walls in Turkey. Transaction of Famena XXXVII-2, Department of Mechanical Engineering, Cumhuriyet University, Sivas, Turkey. ISSN 1333-1124 103-113
5. Oğulata TR (2012) Sectoral energy consumption in Turkey. Renew Sustain Energy Rev 6:471–480
6. Sezer FS (2005) Progress of thermal insulation systems in Turkey and exterior wall insulation systems in dwellings. J Uludağ Univ Fac Eng Archit 10:79–85
7. Compulsory Standard Document TS 825 (1999) Thermal insulation rules in buildings. Ankara, Turkey
8. Ucar A, Balo F (2009) Determination of the energy savings and the optimum insulation thickness in the four different insulated exterior walls. Renew Energy 35:88–94
9. Ozel M, Pihtili K (2006) Optimum location and distribution of insulation layers on building walls with various orientations. Build Environ 42:3051–3059
10. Kaynakli O, Kaynakli F (2016) Determination of optimum thermal insulation thickness for external walls considering the heating, cooling and annual energy requirements. Uludağ Univ J Fac Eng 21:227–242
11. Külcü R (2015) Modelling of solar radiation reaching the earth to Isparta province. Süleyman Demirel Univ J Fac Agric 10(1):19–26
12. Ozel M (2015) Determination of optimum insulation thickness based on cooling transmission load for building walls in a hot climate. Energy Convers Manag 66:106–114
13. Bolattürk A (2007) Optimum insulation thicknesses for building walls with respect to cooling and heating degree-hours in the warmest zone of Turkey. Build Environ 43:1055–1064
14. Turkish Statistical Institute, Republic of Turkey 2016. http://www.tuik.gov.tr
15. Ozel M (2014) Effect of insulation location on dynamic heat-transfer characteristics of building external walls and optimization of insulation thickness. Energy Build 72:288–295
16. Kameni MN, Raminosoa RR, Mamiharijaona R, Rene T, Orosa JA, Elvis W, Meukam P (2015) Study of the economical and optimum thermal insulation thickness for buildings in a wet and hot tropical climate: case of Cameroon. Renew Sustain Energy Rev 50:1192–1202

17. Ghedamsi R, Settou N, Saifi N, Dokkar B (2014) Contribution on buildings design with low consumption of energy incorporated PCMs. In: The international conference on technologies and materials for renewable energy, environment and sustainability, vol 50, pp 322–332
18. Daouas N (2011) A study on optimum insulation thickness in walls and energy savings in Tunisian buildings based on analytical calculation of cooling and heating transmission loads. Appl Energy 88:156–164
19. State Meteorological Station, Records for weather data, Turkey. https://www.mgm.gov.tr

An Examination of the Energy-Efficient High-Rise Building Design

H. Handan Yücel Yıldırım⬤, Arzuhan Burcu Gültekin$^{(\boxtimes)}$⬤, and Harun Tanrıvermiş

Faculty of Applied Sciences, Department of Real Estate Development and Management, Ankara University, Ankara, Turkey
handany@aydiner.com.tr, arzuhanburcu@yahoo.com,
tanrivermis@gmail.com

Abstract. The need for sheltering that started with the existence caused to the concept of housing and has gone through various phases with sedentary life and urbanization. Due to urbanization, a process experienced in parallel with industrialization and economic development, the need for energy has increased. With the increasing need of energy, a great majority of natural resources are being used in the construction industry, particularly at high-rise buildings. It becomes increasingly impossible to meet energy needs of high-rise buildings with nonrenewable resources. Nevertheless, under today's conditions and with current technology, use of renewable energy sources is a quite expensive method. In order to meet the increasing demand in energy, it is required to use existing energy in an efficient and productive manner. For this purpose, this study suggests a conceptual framework including principles, strategies and methods related to energy efficient building design while examples of energy-efficient high-rise buildings applied worldwide are examined within the suggested framework.

Keywords: Energy · Energy-efficient building · High-rise building
Design parameters

1 Introduction

About 11,000 years ago, our primitive ancestors, who lived in the region between the rivers Euphrates and Tigris in stone age when pottery was unknown (Neolithic age without pottery), had architectural skills and they had gathered in regular intervals (at Göbekli Tepe) for religious rituals. Gatherings due to these rituals had led to sedentary life [1] in time. The need for sheltering that started with the existence resulted with the concept of housing and has gone through various phases in parallel with sedentary life and urbanization. Urbanization is a process experienced in parallel with industrialization and economic development. Due to rapid population growth and industrialization, migration from rural areas to urban areas has increased and resulted in rapid urbanization accompanied by unhealthy life conditions in cities.

© Springer International Publishing AG, part of Springer Nature 2018
S. Fırat et al. (eds.), *Proceedings of 3rd International Sustainable Buildings Symposium (ISBS 2017)*, Lecture Notes in Civil Engineering 7,
https://doi.org/10.1007/978-3-319-64349-6_13

Rapid urbanization resulted with certain problems. Housing problem caused by the migration to cities following industrialization has led to unhealthy and irregular building and need for energy has increased. With the increasing energy demand, a great majority of natural resources are being used in the construction industry, particularly at high-rise buildings. It becomes increasingly impossible to meet energy needs of high-rise buildings with nonrenewable resources. Nevertheless, under today's conditions and with current technology, use of renewable energy sources is a very expensive method. In order to meet increasing energy demand, it is required to use existing energy in an efficient and productive way. For this purpose, this study suggests a conceptual framework including principles, strategies and methods related to energy efficient building design while examples of energy-efficient high-rise buildings applied worldwide are examined within the suggested framework.

2　An Examination of Energy-Efficient Building Design Parameters

Due to the increasing need of energy in cities, studies on energy-city relationship in urban planning processes have gained importance. Principles, strategies and methods for energy efficient urban planning of eligible and habitable residential and natural environment are proposed in the paper by Yücel Yıldırım et al. [2] are presented in Table 1.

Table 1. Principles, strategies and methods of energy-efficient urban planning

Principles (P)	Strategies (S)	Methods (M)
Energy Conservation (EC)	Reducing utilization of nonrenewable energy resources (EC1)	Reduction in energy consumption (EC1.1)
		Integration of energy technologies to city, elimination of the deficiency of renewable energy systems (EC1.2)
		Considering local climates at building design (EC1.3)
	Generation and utilization of renewable energy resources (EC2)	Enforcement of the regulations of implementation for renewable energy generation in settlements (EC2.1)
		Creation of aids and incentives for utilization of renewable energy sources (EC2.2)
		Arrangement of spatial areas containing renewable energy utilization (EC2.3)
		Development of social awareness and training on renewable energy (EC2.4)
	Determination of policies and basic principles for compliance and preventive actions for climate change (EC3)	Legislating and enforcement of the law on climate change (EC3.1)

(*continued*)

Table 1. (*continued*)

Principles (P)	Strategies (S)	Methods (M)
		Regulations for increase of energy efficiency and savings for controlling and reducing greenhouse gas emissions (EC3.2)
		Preparation of climate maps of settlements, and keeping them updated (EC3.3)
	Reduction of pollution (EC4)	Balanced distribution, preservation, and enhancement of green spaces within settlements (EC4.1)
		Connection of existing outdoor and green spaces to each other and to pasture area (EC4.2)
		Utilization of local vegetation suitable for climate (EC4.3)
		Development of urban forestry (EC4.4)
		Implementation of green wall and roof systems (EC4.5)
Land Conservation (LC)	Conservation of topographic structure of land (LC1)	Provision of harmony between land usage and topographic structure (LC1.1)
	Conservation of habitat (LC2)	Formation of inventory for natural resources, using values as basis of spatial planning (LC2.1)
		Preservation and growth of agricultural areas (LC2.2)
	Development of settlement plans by energy efficient development form and structure (LC3)	Selection of right location for upper-scale decisions based on climatic properties (LC3.1)
		Reducing heat island impact (LC3.2)
		Minimization of infrastructure and superstructure problems arising from land (LC3.3)
Water Conservation (WC)	Increasing utilization efficiency of water resources (WC1)	Utilization of systems allowing efficient usage of water
		Taking legal measures for efficient utilization and management of water resources and enforcement of the law on water management (WC1.1)
		Reduction in water consumption (WC1.2)
		Unpolluted utilization of water resources (WC1.3)

(*continued*)

Table 1. (*continued*)

Principles (P)	Strategies (S)	Methods (M)
Waste Reduction (WR)	Formation of waste and recycling systems (WR1)	Promotions to local administrations for waste systems and recycling (WR1.1)
		Increasing public sector supervision in waste management (WR1.2)
		Sorting of wastes on site, use of recycling technologies (WR1.3)
Ensuring Accessibility (EA)	Generation of environment-friendly urban transportation policies and plans (EA1)	Drawing plans of transportation and land usage suitable for public transportation (EA1.1)
		Developing energy-efficient transportation means and systems (EA1.2)
		Minimization of private vehicle ownership (EA1.3)

Today, a great majority of natural resources utilized in all areas of life, including transportation, industry and building, are being used in the construction industry, which deteriorates ecological balance and makes environment having harmful impact on human health. Solution of environment and energy problems relies on increasing use of renewable energy sources and efficient use of energy. Design parameters on efficient use of energy in construction industry may be listed as follows [3, 4]:

- Selection of location,
- Topography,
- Position of the building and its distance to other buildings,
- Direction of the building,
- Form of the building,
- Building facade's physical properties that affect heat transmission,
- Outdoor brightness level,
- Non-building obstacles that may affect climate and visual comfort,
- Physical properties of the building's indoor spaces,
- Dimensions and structure of building components such as windows and glasses,
- Properties of components constituting artificial lighting system,
- Solar control and natural ventilation systems.

In consideration of methodologies and data presented in articles by Gültekin and Yavaşbatmaz [5], Koç and Gültekin [6], Gültekin [7], Yılmaz and Hotunluoğlu [8], Barış [9], and Bashiri and Begeç [10], RoT Ministry of Energy and Natural Resources General Directorate of Renewable Energy year 2012 Activity Report [11] and Urbanisation Council Commission Report [12], book composed by Atabay et al. [13], MA thesis of Zinzade [14], "Regulation on Principles and Procedures Pertinent to Increasing Energy Efficiency in Transport" issued by the Ministry of Transportation [15], principles (P), strategies (S) and methods (M) on energy-efficient urban planning were assessed with design parameters related to energy-efficient building design. In conclusion to this paper, methods relevant to energy-efficient building design were presented from Tables 2, 3, 4, 5 and 6 and a conceptual framework was suggested.

Table 2. A conceptual framework on energy-efficient building design (energy conservation)

Energy-efficient urban planning			Methods of energy-efficient building design
P	S	M	
EC	EC1	EC1.1	Selecting appropriate location for building
			Appropriate position of the building and appropriate distance to other buildings
			Orientating the structure in accordance with physical environment data
			Shaping form of structure in accordance with physical environment data
			Making use of the daylight at lighting
			Selecting energy-efficient construction materials
			On exterior surfaces, using construction materials of colors compatible with the climate
			Using high-performance joinery and glasses
			Reducing the building crust surface
			Introducing energy saving by efficient insulation systems
		EC1.2	Using solar batteries at power generation
			Making use of solar collectors at water heating
			Considering position relevant to the sun at energy-efficient building design
			Using wind turbines at power generation
			Making use of wind energy at ventilation and cooling
			Providing lighting with renewable energy sources
		EC1.3	Application of architecture compatible with the local climate
			Selecting local construction materials
	EC2	EC2.1	Enactment of regulations aiming at increasing energy efficiency at buildings
			Preparing a Enactment of Regulations aiming at increasing efficiency of energy resources and energy use at buildings
		EC2.2	Supporting energy-efficient projects to be implemented
			Providing incentives to reduce energy density
			Operation of a fixed-rate guarantee incentive mechanism on the condition of not being equal for each renewable energy source
			Supporting the right of unlicensed production for integration of small-scale enterprises to the national economy and to assure their efficient use
			Providing financial incentives such as VAT exemption, customs duty exemption
EC	EC2	EC2.3	Making use of renewable energy sources at building design
			Designing exemplary energy-efficient projects
			Implementation of pilot applications
		EC2.4	Efficient use of media

(continued)

Table 2. (*continued*)

Energy-efficient urban planning			Methods of energy-efficient building design
P	S	M	
			Supplying training and certification services jointly with the energy management for energy surveys and efficiency enhancing projects
			Concerning energy efficiency, improvement of effective and productive cooperation with public bodies and agencies, universities, private enterprises and non-governmental organizations
			Preparing training and awareness videos
			Holding competitions
	EC3	EC3.1	Implementation of policies preventing climate change
			Preventing misuse of land
			Development and use of green technologies
			Following international meetings on climate change and implementation of decisions taken
		EC3.2	Development of renewable recovery techniques for elimination of greenhouse gas emissions
			Protection of natural resources with their physical, biological and ecological features
			Use of standardized construction materials that pose no problem of health and pollution
		EC3.3	Employing climate change maps as basis at building design
			Assessment of biological data
	EC4	EC4.1	Improving green area standards set forth in the zoning legislation
			Preserving current green areas during design process
			Introduction of a green mass standard that will eliminate greenhouse emission
			Constant improvement of the settlement's green value through standards, incentives and certification
		EC4.2	Arranging at least 40% of the land surface in current transformation zones as green areas or areas containing green elements
			Creating a network of green and open areas to improve urban flora and fauna, reduce greenhouse gas emission and to enable air circulation
		EC4.3	At settlements, using vegetation compatible with local climate conditions
			Using construction materials compatible with local climate conditions
			Preserving current vegetation as possible
		EC4.4	Preserving current forest areas
			Increasing the number of urban forests
		EC4.5	Selecting the plants in accordance with dominant wind direction
			Introducing vegetation compatible with climate conditions for each direction

Table 3. A conceptual framework on energy-efficient building design (land conservation)

Energy-efficient urban planning			Methods of energy-efficient building design
P	S	M	
LC	LC1	LC1.1	Constructing the building in compliance with topography
			Protection of natural topography
			Introduction of a design that is compatible with natural features
			Selecting location in compliance with density of settlement
	LC2	LC2.1	Preserving existing natural resources
			Using natural resources inventory as a basis in design and its integration
			Protecting fertile land
		LC2.2	Preventing use of agricultural land for other purposes, preventing build-up
			Improvement of agricultural land lost due to misuse and their reclaiming for agriculture
	LC3	LC3.1	Forming construction areas according to climate data
			Efficient use of climate areas
		LC3.2	Preserving current tree cover
			Increasing areas to be forested
			Choosing right locations for the trees
			Selecting appropriate locations for plants in immediate vicinity of the buildings
			Implementation of green area works for improving climate of the settlement
			Application of green wall systems
			Application of green roof systems
		LC3.3	Identification of infrastructure and superstructure problems
			Ensuring spontaneous, mutual and reliable exchange of information with local agencies that implement infrastructure and superstructure works
			Creating sources for infrastructure and superstructure investments
			Identification of material, safety, positioning and excavation standards
			Coordinated implementation of planning and repair-maintenance works
			Carrying fertile land left inside the build-up area to green areas and making use of them

Table 4. A conceptual framework on energy-efficient building design (water conservation)

Energy-efficient urban planning			Methods of energy-efficient building design
P	S	M	
WC	WC1	WC1.1	Use of instalments and equipment that consume water efficiently
			Installation of a waste water and grey water plant
			Using rainwater collection and storing systems
		WC1.2	Planning of water resources use
			Increasing efforts of awareness and consciousness in all segments of the society
			Compliance with standards on water retention in buildings

(continued)

Table 4. (*continued*)

Energy-efficient urban planning			Methods of energy-efficient building design
P	S	M	
			Efficient and effective use of water
			Landscape arrangements that use water efficiently and require little maintenance
			Choosing in the landscape design plants that require little water and maintenance
			Increasing areas that have vegetation resistant to drought
			Treatment and reuse of waste water
			Reuse of rainwater after collecting at appropriate areas
			Treatment and reuse of grey water
		WC1.4	Renovation of sewage systems to prevent contamination of water resources
			Controlling polluting elements from sewage and storage areas by use of necessary technologies
			Reducing use of toxic pesticides

Table 5. A conceptual framework on energy-efficient building design (waste reduction)

Energy-efficient urban planning			Methods of energy-efficient building design
P	S	M	
WR	WR1	WR1.1	Enhancing cooperation between local governments, public and private enterprises
			Implementation of training programs to raise public awareness and behavioral change on waste management
			Ensuring communication between public and public bodies authorized for collection, transport and disposal of wastes
			Implementation of the deposit system by local governments
		WR1.2	Developing the material management plan to prevent loss in source and production of wastes
			Identifying locations of waste and recovery plants in relevant plans
			Disposal of wastes without harming the habitat and topography
			Disposal of wastes without causing pollution in soil and water
			Raising awareness in issues of waste management and recycling
		WR1.3	Use of recyclable and reusable construction materials
			Use of construction materials that are capable of rapid self-renewal
			Sorting, storage and classification of reusable materials

Table 6. A conceptual framework on energy-efficient building design (ensuring accessibility)

Energy-efficient urban planning			Methods of energy-efficient building design
P	S	M	
EA	EA1	EA1.1	Preparing a transport plan that keeps the demand of transportation at a minimum level
			Development of pedestrian/bicycle transport systems and processing on physical plan
			Improvement of methods to enable use of public transport from the regional parking lots to urban centers
			Enhancing rail transport systems in urban transport
		EA1.2	More common use of clean fuels in transport
			More common use of vehicles with less fuel consumption
			Raising awareness of the consumer and making low-emission vehicles attractive
			More common use of smart traffic practices and smart transport systems that make use of information and communication technologies
			Raising efficiency standards in vehicles
		EA1.3	Planning relations between sheltering, work and instalment areas in a way that requires minimal energy
			Extending public transport network
			Creating pedestrian lanes
			Introduction of local parking lots

3   An Examination of Worldwide Energy-Efficient Building Design on High-Rise Buildings

Importance of efficient use of energy has been perceived in different manners in each country and in this context; different solutions and recommendations have been suggested. United States of America (USA) became one of leading countries to realize importance of energy efficiency and starting from 1970s, when an oil crisis was experienced, studies on energy efficiency were continued with an increasing momentum.

In consequence of the energy efficiency studies conducted in USA, achievements were obtained between the years 1973 and 2005, such as significant contribution to protection of environment, improving energy efficiency of household appliances, and avoiding construction of further power plants. Furthermore, USA did not find those achievements sufficient and a national action plan called "Vision 2025" was prepared in 2008 due to reasons such as energy efficiency is an untouched and low cost energy source which enhances energy supply security and reduces future risks of carbon policies that are already ruled by uncertainty [16].

On the other hand, European Union (EU) member states, starting from the beginning of 1970s, have implemented studies in order to reduce dependency on oil, enhance energy supply security, support competition by decreasing energy costs, reduce unemployment, protect the environment and minimize emission of greenhouse gases [16]. EU Commission, with its indirect taxation study in the year 2007, revised the Directive on Energy Taxation, attempted to introduce an encouraging energy taxation system, has examined the benefits of tax reductions and other incentives to increase the production of high energy efficient certified equipment and devices [17]. Furthermore, the EU, concerning energy savings, has enacted the Directive 2002/91/EC of the European Parliament and of the Council of 16 December 2002 on the Energy Performance of Buildings [18], Council Directive 2003/96/EC of 27 October 2003 Restructuring the Community Framework for the Taxation of Energy Products and Electricity [19], Commission Directive 2003/66/EC of 3 July 2003 Amending Directive 94/2/EC Implementing Council Directive 92/75/EEC with Regard to Energy Labelling of Household Electric Refrigerators, Freezers and Their Combinations [20], Communication from the Commission on the Implementation of the Energy Star Programme in the European Union in the Period 2006–2010 [21], Evaluation of the Energy Labelling and Eco-design Directives [22] and Directive 2006/32/EC of the European Parliament and of the Council of 5 April 2006 on Energy End-Use Efficiency and Energy Services and Repealing Council Directive 93/76/EEC [17, 23].

In Japan, where negative impact of the 1970s oil crisis was experienced, "Energy Conservation Laws" [24] was revised in 1999. In Japan, where energy efficiency studies are supported by the government through financial models that include tax incentives, long-term reimbursement credits, industrial corporations and the public support the studies on a voluntary basis and city governments are implementing from time to time various efficiency programs within their boundaries [25, 26].

In the process of design and application procedures of buildings, which have a major share in worldwide energy consumption, efficient use of energy shall be enabled when the legislative regulations and methods presented in the Table 2 evaluated together.

Due to decreasing energy resources and increasing environmental problems in the world, design and application of energy-efficient high-rise buildings are on the rise. Because initial investment costs of these buildings are high, office building applications are more commonly observed. In this paper, it has been examined whether the high-rise building examples in the literature are constructed according to the methods related to energy-efficient building design and whether the methods are fully implemented. Visionaire Building, Solaire Building and Helena Building in USA, Burj Mohammed Bin Rashid Tower in the United Arab Emirates, Telus Garden Building in Canada and Sky Terrace @Dawson Building in Singapore are some of the energy-efficient high-rise building projects. These examples, in consideration of methods of energy-efficient design listed in Tables 2, 3, 4, 5 and 6, are examined in the Table 7.

When it comes to energy-efficient building design approach, we failed to obtain sufficient information on international examples of high-rise buildings from literature. Therefore, examples of housing projects presented in the Table 7 were examined on limited information. It was found out that methods listed in the suggested conceptual framework are being partially implemented in the examples of high-rise building projects.

Table 7. High-rise building examples constructed in accordance with energy-efficient building design approach

High-rise buildings	Energy-efficient building design approach
Visionaire Building [5, 27–29]  *Country*: New York City, USA *Date of the Project*: 2006–2008 *Height*: 109.73 m *Number of Floors*: 35 *Intended Use*: Housing	• LEED Platinum certificate • Custom made BIPV panels (solar batteries) applied on western and eastern facades of the building generate electric power • Building has a rainwater storage system • Rainwater collected on the roof is used for irrigation of green areas • About 35% of the building's electric load is supplied from renewable energy sources • Materials used at construction of the building had been supplied from an area of a diameter of 800 km and from recyclable resources • Construction materials of self-renewing nature, such as bamboo were used indoor areas • Green roof application contributes in decreasing heat island effect • Durable and recyclable construction materials were used and efficient use of sources was ensured in the building • Materials used in the building were supplied locally and economically sustainable design was provided • Non-toxic construction materials were used during construction of the building • Insulating glass is used • Full length floor-to-ceiling glasses makes possible better use of natural light • Building has a waste water treatment system that enables reuse of waste water in the building • Air filtering system allows fresh air entry inside the building • In order to reduce significantly the power demand of building the heating and cooling systems are operated with natural gas

(continued)

Table 7. (*continued*)

High-rise buildings		Energy-efficient building design approach
Solaire Building [5, 30–33]	 *Country*: New York City, USA *Date of the Project*: 2001–2003 *Height*: 85.29 m *Number of Floors*: 27 *Intended Use*: Housing	• LEED Golden certificate • Green roof application designed to collect almost 70% of the rainwater contributes in decreasing urban heat island effect • Rainwater collected on roofs is used at irrigation of green areas • Excess water not absorbed by vegetation is collected with grey water in a cistern and treated; it is used at irrigation of the green roof and parks in the vicinity • Waste water from the building is recycled • Solar batteries supply energy to meet the building's energy need • Supplying energy by solar batteries reduces energy consumption costs • Two thirds of materials used at construction of the building were supplied from intimate environs • 93% of waste materials produced at construction was recycled • Selecting recyclable construction materials increased source efficiency of the building • Tax deductions reduced construction costs and encouraged investor for construction of more energy-efficient buildings
Helena Building [5, 34–38]		• LEED Golden certificate • Green roof application contributes in decreasing urban heat island effect • Rainwater is collected and used at appropriate locations • Solar batteries generate energy • 20% of materials had been supplied from an area of a diameter of 800 km and use of local materials was supported • High-performance facade components reduce harmful effects of sun on indoors • Supplying construction materials from recyclable and local sources, energy and source efficiency was enhanced, costs

(*continued*)

Table 7. (*continued*)

High-rise buildings	Energy-efficient building design approach
 *Country*: New York City, USA *Date of the Project*: 2002–2005 *Height*: 122.2 m *Number of Floors*: 37 *Intended Use*: Housing	were decreased and environmental pollution was reduced • Supplying the building's energy demand from renewable energy sources such as wind and sun minimizes energy consumption • High-performance double glass application reduced the harmful effect of ultraviolet rays on furniture • Glass reinforced concrete used on the exterior allows better use of natural light • Ventilation culverts on windows enable ventilation of rooms without opening windows • Wet surfaces, cupboards and doors were constructed of panels that are composed of wheat stalks that are easily renewable, recyclable and non-toxic, which contributed in reduction of wastes • Energy-saving power switches were used • Waste water obtained from the black water treatment plant was used in toilets, ventilation and air-conditioning systems (HVAC) and cooling towers of ¾ of buildings and at garden irrigation and efficient use of water was thus ensured • Devices with 'Energy Star' sign were used to reduce energy consumption
Burj Mohammed Bin Rashid Tower [39–41]   *Country*: Abu Dhabi, United Arab Emirates *Date of the Project*: 2007–2014	• Energy efficiency was improved by means of triple-wall facade system • With the facade lining of high reflectivity that requires minimum-level maintenance, heat island effect was decreased, compatible conditions of comfort were supplied indoors • Energy efficiency was improved by solar collectors • Shading components were used on the facade to provide heat and visual comfort indoors • Towers were directed in positions that decrease sun effect • Tower roofs were designed to allow installation of further solar panels • Since power is supplied from a regional center, natural ventilation is supported by windows that can be opened

(*continued*)

Table 7. (*continued*)

High-rise buildings		Energy-efficient building design approach
	Height: 381.2 m *Number of Floors*: 88 *Intended Use*: Housing	• High-efficiency water fittings were used for water saving • Local plant species were grown to decrease garden irrigation and for water saving • Local materials were used to decrease costs of transporting imported materials at construction
Telus Garden Building [42–45]	 *Country*: Vancouver BC, Canada *Date of the Project*: 2012–2016 *Height*: 167 m, 88 m *Number of Floors*: 53 Condo, 24 Office *Intended Use*: Condo, Office	• Office building has LEED Platinum certificate. Condo has LEED Golden certificate • About 300 solar panels placed on the roof are generating 65000 kWh of power each year • Green roof application contributes in reducing heat island effect • Motion-sensitive, energy efficient lighting system reduces energy consumption • Building hosts a fully integrated and smart building system that controls all systems including energy-efficient lighting, heating, cooling and fire alarms. This system allows utilizing fresh air instead of recycled air • Rainwater is collected and used at appropriate locations • Building, which is included in the regional power system, contains a system that converts indoors heat for heating of air and water and approximately 1.000 tons kg of CO_2 gas emission is prevented per year • Since the building is located in proximity of the Sky Train, it hosts an electric vehicle charging station and bike facilities • Green mortgage bond was used in funding of the office buildings

(*continued*)

Table 7. (*continued*)

High-rise buildings	Energy-efficient building design approach
Sky Terrace *@Dawson Building* [46–48]  *Country*: Soo Khian Chan, Singapore *Date of the Project*: 2008–2015 *Height*: 142 m *Number of Floors*: 43 (5 towers) *Intended Use*: Housing	• BCA Green Mark Platinum certificate • Rainwater is collected • Drip irrigation system is used for irrigation • Shallow depressions where rainwater is directly diverted without any processing and natural and foreign plants are grown host a 'rainwater garden', in other words, a bio-retention system [49]. Water collected in these areas is used for garden irrigation • Energy is generated by solar panels located on the roofs and generated energy is used for lighting of common areas and operation of lifts • High-performance double glass application enables decrease in heat loss • Water-efficient devices are being used • Motion-sensitive sensors are used to decrease power consumption at common areas and staircases to decrease energy consumption • Since they are equipped with variable voltage, frequency adjuster and sleep mode operation costs of the lifts have been decreased • Wastes are collected in the waste chute

It is observed that methods such as utilization of renewable energy sources, application of rainwater collection and treatment systems, reuse of treated waste water at appropriate areas, application of green roof systems to reduce the heat island effect, use of devices with high efficiency for water saving, use of local and recyclable construction materials, making use of natural lighting and ventilation to reduce energy consumption and application of high-performance facade and glass systems to decrease heat loss have come into prominence. We failed to get access to information such as positions, building directions, topographic compatibility and preservation of natural resources at buildings.

4 Conclusion and Suggestions

In today's world, where most of the world population lives in cities, it is a rather expensive method to provide all of increasing energy need from renewable energy sources with today's technology and conditions. It is therefore necessary to use existing energy efficiently and productively, which is a cheaper application to meet the increased energy need.

The conceptual framework suggested in this study may be adopted as a guideline in energy efficient building design, and consequently, may be used as a guide for different disciplines. In energy-efficient building design, although application of systems and studies that enhance energy performance increase initial investment costs, operation and maintenance costs required throughout the life of these buildings may be 5–10 times bigger than these application costs [50]. Accordingly, efficient and productive utilization of energy and renewable energy resources must be adopted as a government policy, and the public awareness must be increased. People shall be more selective as public awareness increases on efficient and productive utilization of energy, and investments of lesser energy consumption for less money shall be taken into consideration. In today's world, where economic growth and social welfare lead to more energy consumption, decrease in energy consumption, preservation of the environment, and reduction the burden of energy costs on the economy shall be provided with the enforcement of laws and regulations for efficient and productive utilization of energy.

Since energy is one of the elements that affect production costs at the highest level, generating the same quantity of output by using technologies that require less energy consumption for energy efficiency may result in decrease in costs. This will enhance competitive capacities of countries. Furthermore, by efficient use of energy, consumption will be lessened and accordingly, the countries' dependency on foreign energy sources will be reduced.

References

1. Haber Bilgi. http://www.haberbilgi.com/bilim/arkeoloji/b-anadolu-v.html
2. Yücel Yıldırım HH, Gültekin AB, Tanrıvermiş H (2016) Evaluation of energy efficient urban planning approach. In: SBE16 İstanbul—international conference on sustainable built environment, smart metropoles proceedings book, pp 228–229
3. Dikmen ÇB (2011) Exemplifying energy-efficient building design criteria. Politek J 14 (2):121–134
4. Manioğlu G (2011) Assessment of energy-efficient building design and renewal studies with examples. Install Eng J Issue 126:35–47
5. Gültekin AB, Yavaşbatmaz S (2013) Sustainable design of tall buildings. J Croat Assoc Civ Eng Gradevinar 65(5):449–461
6. Koç Y, Gültekin AB (2010) Green roofs and applications thereof in Turkey. In: 5th National symposium on roof and facade, İzmir
7. Gültekin AB (2014) Proposal of solutions within the context of sustainable architectural design guidelines. In: 19 International congress of building life: future of architecture, Architecture for Future Bursa

8. Yılmaz O, Hotunluoğlu H (2015) Incentives for renewable energy and Turkey. Adnan Menderes Univ Soc Sci Inst J 2(2):74–97
9. Barış E (2005) Urban planning, urban ecosystem, and trees. TMMOB Chamber Urban Planners Plann J :156–163 (Ankara)
10. Bashiri DH, Begeç H (2015) Examination of energy-efficient design in high buildings. In: 2th National Building Congress and fair, building construction, use, and conservation processes, TMMOB Chamber of Architects, Ankara Branch
11. RoT (2013) Ministry of Energy and Natural Resources Directorate General of Renewable Energy, Year 2012 Activity Report
12. Ministry of Public Works and Settlement, Urbanization Council, Spatial Planning System and Institutional Structuring Committee Reports, p 780, (2009)
13. Atabay S, Karasu M, Koca C (2014) Climate change and our future. Yıldız Technical University Faculty of Architecture, İstanbul
14. Zinzade D (2010) Scrutinizing sustainability in design of high structures. Postgraduate Thesis, İstanbul Technical University, Science Institute, İstanbul, Turkey, pp 10–12. (Consultant: Y. Demir)
15. Bylaw on principles and procedures on increasing energy-efficiency in transportation. Ministry of Transportation, Official Journal, Date: 09.06.2008 No: 26901
16. 42th Period Energy Study Group (2012) TMMOB Chamber of electrical engineers, energy efficiency report. EMO Publications, Ankara, 27–30
17. Energy efficiency strategies in buildings and what to do in Turkey, 2010–2023 heat insulation planning report. İZODER—association of heat, water, and noise insulators, İstanbul, 11, (2010)
18. Official Journal of the European Communities. http://www.mo.org.tr/UIKDocs/energyperformancedirective.pdf
19. Official Journal of the European Union. http://www.ebb-eu.org/legis/OJ%20taxation%20EN.pdf
20. Official Journal of the European Union. http://eur-lex.europa.eu/legal-content/EN/TXT/PDF/?uri=CELEX:32003L0066&rid=6
21. European Commission. http://eur-lex.europa.eu/legal-content/EN/TXT/PDF/?uri=CELEX:52011DC0337&rid=7
22. European Commission. http://eur-lex.europa.eu/legal-content/EN/TXT/?qid=1484905495246&uri=CELEX:52015SC0143
23. Official Journal of the European Union. http://eur-lex.europa.eu/legal-content/EN/TXT/PDF/?uri=CELEX:32006L0032&rid=23
24. Asia Energy Efficiency and Conservation Collaboration Center. http://www.asiaeec-col.eccj.or.jp/chronicle/index.html
25. Enerji Gazetesi.ist. http://www.enerjigazetesi.ist/kose-yazisi-dunyada-enerji-verimliligi-japonya-abd-ve-ab/
26. Kavak K (2005) Energy efficiency in Turkey and world and examination of energy efficiency in Turkish industry. DPT—Directorate General of Economical Sectors and Coordination. Specialty Thesis, Ankara, pp 68–71
27. Aluminum Siding. http://blog.bisam.com.tr/2014/09/kendi-elektrigini-ureten-cevreci-binalar.html
28. Atelier Ten. http://www.atelierten.com/2011/projects/visionaire-residential-tower/
29. Condopedia. http://www.condopedia.com/wiki/Visionaire
30. Altpower. http://www.altpower.com/projects/bipv/thesolaire/
31. http://www.balmori.com/portfolio/the-solaire
32. CTBUH Research paper. http://global.ctbuh.org/resources/papers/download/1308-overview-of-sustainable-design-factors-in-high-rise-buildings.pdf

33. The Greenroof and Greenwall Projects Database. http://www.greenroofs.com/projects/pview.php?id=464
34. Helena 57 West. http://www.helena57west.com/#/gallery/pure/
35. A View on Cities. http://www.aviewoncities.com/buildings/nyc/thehelena.htm
36. Durst Organization. https://www.change.org/p/durst-organization-helena-building-nyc-new-unreasonable-pet-policy-revision-request
37. Sustainable Design Resources. http://sustainabledesignresources.pbworks.com/w/page/6050482/Case%20Studies
38. http://global.ctbuh.org/resources/papers/download/1209-sustainable-design-in-high-rise-residential.pdf
39. CTBUH. Tall Buildings. http://ctbuh.org/TallBuildings/FeaturedTallBuildings/FeaturedTallBuildingArchive2015/BurjMohammedBinRashidAbuDhabi/tabid/7090/language/en-US/Default.Aspx
40. CTBUH Annual Awards. http://awards.ctbuh.org/media/ctbuh-names-best-tall-buildings-for-2015/
41. The Urban Developer. https://www.theurbandeveloper.com/worlds-four-best-tall-buildings-named/
42. CTBUH Research Paper. http://www.vancitybuzz.com/2015/09/telus-garden-office-tower-vancouver/
43. TELUS. http://about.telus.com/community/english/news_centre/news_releases/blog/2015/07/24/telus-garden-office-tower-partnership-issues-225-million-in-green-bonds-to-retire-construction-financing
44. Sustainability Report 2015. https://sustainability.telus.com/en/sustainability-strategy/introducing-telus-garden/
45. Integral Group. http://www.integralgroup.com/projects/telus-garden/
46. Archello. http://www.archello.com/en/project/skyterrace-dawson
47. The Magazine of the Institution of Engineers, Singapore. http://citeseerx.ist.psu.edu/viewdoc/download;jsessionid=487898EBBF130A19FFB5ADFA38323F19?doi=10.1.1.174.877&rep=rep1&type=pdf
48. Inhabitat. http://inhabitat.com/singapores-solar-powered-sky-terrace-residential-towers-combine-all-the-best-of-green-living/
49. Müftüoğlu V, Perçin H (2015) Rain garden within scope of sustainable rain water management. İnönü Univ J Art Des. ISSN: 1309-9884. 5/11:27-37
50. Çakmanus İ, Kaş İ, Künar A, Gülbeden A (2010) An evaluation on high-performance sustainable buildings. Turkey Engineering News 461–462:38–46 (55/2010/3-4)

On the Feasibility of Using Photovoltaic Panels to Produce the Electricity Required at a Chemical Treatment Plant on Industrial Company

Akın Yalçın[1(✉)], Zafer Demir[2], and Nesrin Çolak[3]

[1] Institute of Science and Technology, Energy Resources and Management, Eskişehir, Turkey
aknyalcn@gmail.com
[2] Anadolu University Porsuk MYO, Eskişehir, Turkey
zaferdemir@anadolu.edu.tr
[3] Anadolu University Ulaştırma MYO, Eskişehir, Turkey
nesrincola@anadolu.edu.tr

Abstract. This work consists of a study focusing on the feasibility of using photovoltaic panels to produce the electricity required by plant operations at a chemical treatment plant belonging to an industrial company. Using available statistical data, it provides estimates relative to the amount of solar radiation received by the Eskişehir region, a large proportion of which is exposed to sunlight every day. The study reveals in detail the operating periods and power of the pumps used to pump the wastewater and the chemicals, and performs daily and instantaneous measurements of the power used at the plant over a one week time period. Following these evaluations, the study determines the instantaneous maximum power used by the system and provides the calculations necessary for photovoltaic panels and equipment.

Keywords: Photovoltaic · Panel · Solar · Sun

1 Introduction

When we look at the International Energy Agency (IEA), US Energy Information Administration (EIA) and the European Environment Agency resources, we estimate that human need will be around 400,000 TWh by 2016. Below, renewable energy potentials are given

*Solar energy 1575 EJ (438,000 TWh),
*Wind energy 640 EJ (180,000 TWh),
*Geothermal energy 5000 EJ (1,400,000 TWh),
*Biomasses 276 EJ (77,000 TWh),
*Hydropower 50 EJ (14,000 TWh)
*Ocean energy 1 EJ (280 TWh).

© Springer International Publishing AG, part of Springer Nature 2018
S. Fırat et al. (eds.), *Proceedings of 3rd International Sustainable Buildings Symposium (ISBS 2017)*, Lecture Notes in Civil Engineering 7,
https://doi.org/10.1007/978-3-319-64349-6_14

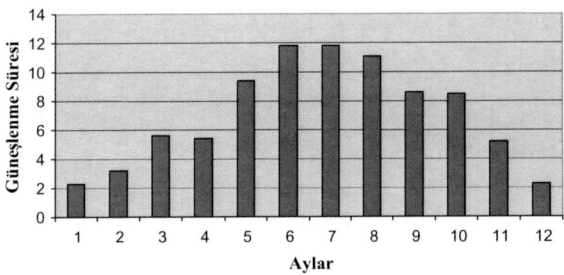

Fig. 1. Average hourly sunlight periods of the province. (*Source* EYEKPA, p. 7)

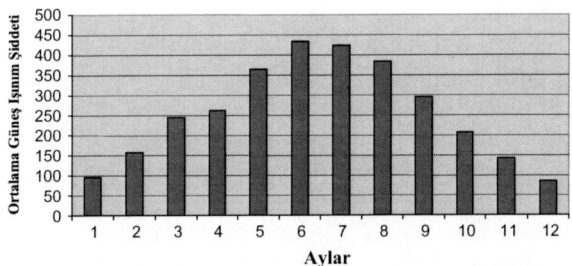

Fig. 2. Intensity of solar radiation in the Eskişehir province. (*Source* EYEKPA, p. 7)

Solar energy potentials of Eskişehir are displayed in Fig. 1, which provides the average hourly sunlight periods of Eskişehir and Fig. 2, which provides the monthly solar radiation intensity values.

As seen in Fig. 1, the months of June and July provide the longest periods of solar energy availability, followed by May, August, September, and October. From these data which show us that the region receives a monthly average of 300–400 km/m^2 of solar radiation and an average daily sunlight period of 8 h, we may conclude that Eskişehir will be an amenable location for the harnessing of solar energy.

The radiation intensity values and daily average of hours of sunlight of the Eskişehir region are seen to exceed national averages [1].

2   Available Technologies for the Production of Electricity from Solar Energy

This section of the paper reviews the methods and technologies used in the conversion of solar energy into electricity, with special focus on photovoltaic cells and solar radiation power generation systems. This section also provides general technical information and capacity volumes of these systems.

2.1   Photovoltaic (PV) Systems

2.1.1   Photovoltaic (PV) Panels [2, 3]

Photovoltaic Cells (PV batteries) are composed of semi-conductor materials that can directly convert surface solar rays into electric energy. In accordance with photovoltaic principles, when direct sunlight strikes the cell, an electric charge occurs on its terminals. The photovoltaic process is that in which two dissimilar materials in close contact produce an electric potential in the space between them when struck by photon rays. The photovoltaic cells are interconnected in a parallel series and formed into panels protected from the elements with glass, polymers or other kinds of other kind of surface materials to form panels. Today there are many different kinds of PV cells. These have been summarized below:

(a) **Crystallized silicon**: Mono-crystallized silicon blocks are first grown and then sliced into 200 micron thick wafers. Under laboratory conditions it has been determined that PV cells made out of blocks of these fine wafers have efficiency of 24%, while the commercial models have efficiency rates of 15%.

(b) **Gallium Arsenide (GaAs)**: This material provides 25% efficiency under laboratory conditions and 28% as an optical concentrator.

(c) **Amorphous Silicon**: SI (thin-film) photovoltaic cells made up of this non-crystalline form of silicone have efficiency rates of approximately 10%, while the commercial models rate at around 5–7%.

(d) **Cadmium Telluride (CdTe)**: CdTE is a poly-crystalline material that is expected to significantly lower the costs of photovoltaic cells. Under laboratory conditions small CdTe cells provide efficiency rates of 16%, and approximately 7% in commercial applications.

(e) **Copper Indium Gallium Selenide (CuInSe2)**: Under laboratory conditions cells made of this poly-crystalline materials provide efficiency rates of 17.7% and 10.2% efficiency of prototypes built as energy converters.

(f) **Optic Concentrator Cells**: Models of devices with lenses or mirrors that can concentrate light at 10–500 magnitudes have model efficiency of 17% and cell efficiency that can attain as much as 30%. These concentrators are made up of simple and inexpensive plastic materials.

2.1.2   Other Equipment Included in PV Systems [4]

(a) **Inverter**: inverters, which are referred to as the heart of the systems, are used to convert direct current outputs of PC cells into utility frequency alternating current. The 12 or 24 DC current produced by the panels are converted into 24 volts of alternating current. It does so by taking the constant DC voltage and changing it to a sine wave curve (or a curve that approximates a sine wave). The power of the inverter used has to accord with the type of system established.

(b) **Charge Regulator**: A charge regulator or controller is the device used to lower or limit to desired levels the rate of the current derived from the solar energy. Generally used in off-grid systems, the most important criterium in the selection of this device is the efficiency value.

(c) **Accumulator**: The accumulator is used in off-grid systems to store the energy from the system in a chemical environment. When desired, the device releases the energy from its terminals in the form of electricity. The batteries used are lead-acid permanently placed and can withstand several emptying and filling processes; however, the location of their placement is important as these batteries may emit dangerous gases. However, if they are to be used indoors dry batteries have to be used.

(d) **Peak Power Monitor**: The power derived from a PV cell is in direct ratio to the amount of solar radiation to which it is exposed; i.e.,as the solar radiation increases so does the wattage power. This is the maximum power that can be produced by the cell or panel, and is termed the "peak power." It is measured in WP (watt peak).

(e) **Installation Set**: These sets are needed to install the PV panels on roofs, in fields, or in other applications. Currently there are two types of such sets, permanent and trackers.

2.1.3 Off-Grid and on-Grid Systems [5]

Off-grid systems are used to provide electricity to isolated farms and mountain facilities that are not connected to the electrical power network, to run well motors, transmission antenna, and boats. These are also referred to as island systems. The electric energy generated by these PV panels is stored in batteries and then converted into alternating current with inverters. The generated energy can also be used as direct current (Fig. 3).

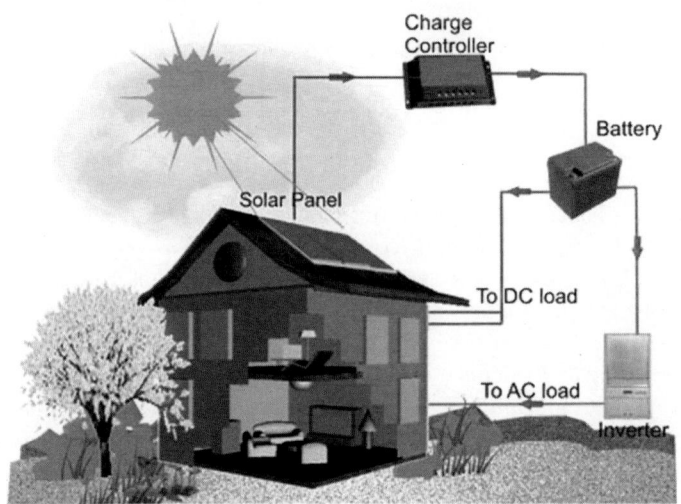

Fig. 3. Off-Grid PV applications. (*Source* https://solenturk.com)

2.1.4 On Grid PV Systems [5]

On-Grid systems are used in connection with municipal electrical distribution stations. The most significant difference between on-grid and off-grid systems is that the on-grid systems do not use groups of accumulators. Additionally, the inverters used in on-grid systems do not have the same technical features as the off-grid accumulators. As known, inverters are used to convert DC current to the kind of alternative current used by household appliances and devices. However, the inverters used with on-grid systems are synchronized so as to accord with the power networks and just as these inverters can be connected to the household distribution panel boxes, their paired meter counters allow them to be tied into the power network itself (Fig. 4).

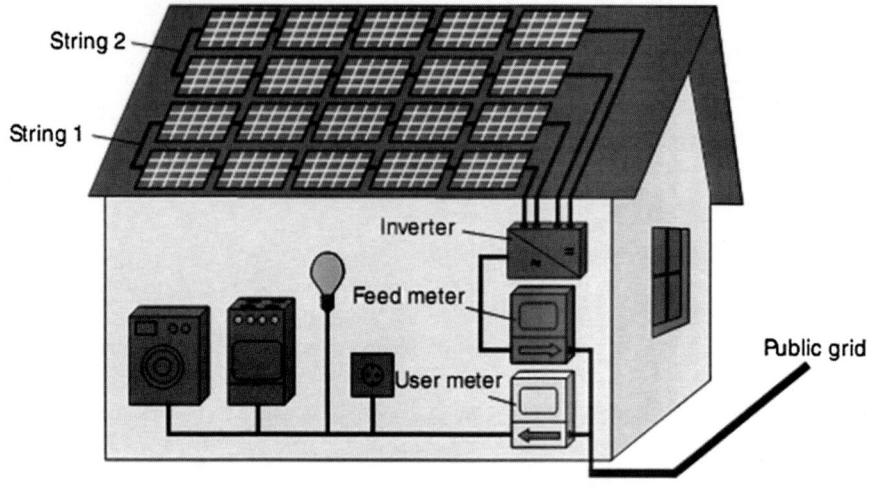

Fig. 4. On-grid system. (*Source* https://solenturk.com)

2.2   Concentrated Solar Power (CSP) Systems [6]

2.2.1   Dish Systems

Dish systems concentrate on the solar focalization point by tracking the sun from two axises. Thermal energy is harnessed by using a suitable working fluid at the concentrated point on the dish. Once the energy has been converted to heat, it is sent to the generator via thermodynamic heat transfer, or a Stirling engine is positioned at the focalization point to convert the solar energy to electricity, providing an efficiency rate of approximately 30%.

2.2.2   Power Tower Systems

Single-focused flat, movable mirrors (also called heliostats) are mounted onto a solar-energy receiver called a tower. Solar heat is reflected onto the heat exchanger and is concentrated. Mounted on the tower is a collection of pipes filled with a viscous fluid that absorbs the solar energy in a three-dimensional mass. This fluid is then pumped to the Rankin machine where it is converted to electricity.

2.2.3   Solar Chimney

A tower containing soil and air and covered with solar-exposed transparent material is allowed to heat to a temperature that exceeds the outside temperature. Because hot air rises it forms a roof slope and if this air is directed towards a high chimney an air flow is created within the chimney. A horizontal wind turbine placed at the entrance to the chimney converts this wind to electricity.

2.2.4   Solar Pool Systems

The black colored floors of these 5–6 m deep pools absorb solar rays and provide water heated to 90 °C. Just as this hot water can be pumped with a heat exchanger and used for heat, it can also be used in the production of electricity when a Rakın inverter is used.

2.2.5   Parabolic Trough Systems

This system is one of the most commonly used linear concentrator thermal systems. The collectors are composed of series of parabolic cross-sections. The inner parts of the collector have mirrored, reflective surfaces. The solar energy focuses on an absorbent pipe that runs the length of the inner surface. Generally speaking, the collectors are situated in along a single axis that tracks the sun as it moves from east to west. The collected heat is sent to a power station where it is converted into electricity.

3   Turkey's Legal Regulations and Current Situation [7]

Turkey's energy policies call for the amount of renewable energy to make up 30% of all of Turkey's energy needs by the year 2023. In line with these targets, beginning in 2010 various legal modifications and reforms have been made to the country's energy sectors and energy sector investors are being provided with certain incentives.

Legal Provisions and Laws

- Number 4628: Law Governing Electricity Markets
- Number 5346: Law Governing the Use of Renewable Energy Resources in the Production of Energy
- Number 5627: Law Governing Energy Efficiency
- Number 6446: Law Governing Electricity Markets

Regulations

- Regulations Governing the Licensing of Electricity Markets
- Regulations Governing the Rules and Procedures Relative to the Provision of Renewal Energy Resource
- Regulations Regarding the Manufacture of Parts and Accessories Used at Facilities Producing Electric Energy from Renewable Energy Resources,
- Regulations Governing Solar Energy Production Plants
- Regulations Governing the Production of Non-Licensed Electricity Plants in the Electricity Market

The laws and regulations mentioned above have been investigated so as to illuminate the work carried out at the trial study of an industrial plant located in Eskişehir with emphasis on the subjects discussed below:

3.1 Non-licensed Electricity Production

Limited to the meeting of personal energy requirements: Co-generation Plant—Micro Cogeneration Plant—Those actual individuals or corporations that establish a energy production plant producing a minimum of 500 kw of energy to be used for personal requirements and obtained from renewable energy resources are excused from the requirement to obtain a license. Any person (actual individual or corporation) that is a subscriber to the electricity network may establish a non-licensed electricity production plant. These individuals must have at least one consumption entity registered in their own name, meaning they must be subscribers to the state's network. Those individuals who are not subscribers are not permitted to establish a non-licensed electricity production plant. Those who are not actual individual or corporate subscribers (such as apartment managers) are not permitted to establish a non-licensed electricity production plant.

Non-licensed electricity production plants must be established so as to accord with the relative laws and regulations. Those individuals and/or corporations who apply and meet the above stipulations are then given the right to produce the electricity needed to meet their own energy requirements. Of these, actual individuals are only authorized to establish micro-cogeneration plants, while actual individuals and/or corporations are allowed to establish cogeneration plants, but both of these are only authorized to produce the electricity needed to meet their own needs. The excess energy generated by those individuals who establish micro-cogeneration facilities and those actual individuals and corporations that establish facilities based on renewable energy resources are given ten-year long rights within the framework of the YEK (renewal energy) Support Mechanism to sell this excess energy to a retail-licensed energy distribution company. Appendix I of the YEK Law provides a scaled listing of the proposed prices of such energy as per resource employed. Those individuals who establish micro-generation facilities use the lowest price on the scale. **What this means is that the government only guarantees the oversight of the system, but does not guarantee sales or purchase of the electricity.**

3.2 Sale of Produced Electricity

The excess energy produced from an electricity producing facility based on renewable energy resources is evaluated via the YEK Support Mechanism for sale to a local retail-sales licensed energy distribution company. While there is no limit on volume of sales, there has to be an ongoing consumption process that proceeds between the subscribed energy generation facility and the consumer supplier company to which it has been attached.

And while the YEK Support Mechanism operated by TEİAŞ (Turkish Electricity Transmission Corporation) is a market-based purchasing mechanism, it does not engender sales to the state, but only rather guarantees that it will oversee the operating of the government monitored system.

According to Appendix I of the YEK Law, excess electricity produced by renewable energy based systems will be purchased for a period of ten years according to the scaled price list. Following this ten year period new modifications to the Law will clarify how the process will be implemented in the future.

3.3 Sample Calculation

To benefit from the regulations legislating non-licensed electrical generation, 132 m^2, of a 180 m^2 roof will be exposed to the sun and it is in this area that a solar energy harnessing facility will be installed. Assuming that we will need 6.6 m^2 to obtain 1 kW, a 20 kw power photovoltaic solar panel will need to be established.

Now, assuming that the facility has been built and is operable:

Also assuming that the plant will receive, on average, 7 h of sunshine daily, we may conclude that the plant can produce 7 × 20 kW = 140 kWh of electric energy daily. Assuming that the plant consumes 70 kWh on a single day, we can calculate that each day the system will generate 140–70 = 70 kWh of excess energy. Let us now assume that the next day is cloudy and the plant receives low levels of radiation. In this kind of instance, if the plant receives 4 h of solar radiation it will produce 4 × 20 = 80 kWh of electricity and will thus produce 80–70 = 10 kWh of energy that surpasses the system's requirements.

Accordingly, let us calculate the amount of 13.3 cents (USD) of support incentives to be received per two days of solar energy.

Let us assume that each case outlined above occurred 15 days per month.

(15 × 70 = 1050) + (15 × 10 = 150) = 1200 × 0.133 × 1.85 = 159.6 TL (not including dskb/energy distribution utilization charges and other legal obligations).

Added to this calculation is this plant's electricity consumption which equals 15 days of 10 kWh and 15 days of 15 kWh. According to this calculation, (30 × 70) or 2100 kWh of electricity is consumed monthly. If this individual had not built this generating plant, electric consumption would have to be paid by subscription rate, which would mean that 2100 × 26 Kuruş = 546 TL of energy charges would have had to be remitted. In addition, the individual would also have to have paid dskb and other legal obligations. In conclusion, this plant:

– Setting aside costs due to feasibility, investments, and operating, the facility will receive 159.6 TL of support (not including dskb and other legal obligations) and will also be exempted from paying an electric bill of 546 TL (dskb and other obligations not added).

4 An Operation's Chemical Waste Water Treatment Plant in Eskişehir

Table 1

Table 1. Installation power consumption operating hours

Equipment	P	Tot.	Time	Energy	P (% 80)	Daily total comp.
	kW	kW	Saat	kW	kW	kW
Mixer	0.75	0.75	1	0.75	0.6	0.6
Oily wastewater pump	2.2	4.4	20	2.2	1.76	35.2
Wastewater equalization pumps	2.2	4.4	12	2.2	1.76	21.12
Acidification tank mixer	0.37	0.37	20	0.37	0.296	5.92
Daf tank stripper	0.25	0.25	12	0.25	0.20	2.4
Daf tank pumps	2.2	4.4	24	2.2	1.76	42.24
Coagulation tank mixer	0.75	0.75	20	0.75	0.6	12
Neutralization tank mixer	0.75	0.75	20	0.75	0.6	12
Flocculation tank mixer	0.55	0.55	20	0.55	0.44	8.8
Sludge removal pumps	1.1	2.2	1	1.1	0.88	0.88
Finned settling tank sludge pumps	1.1	2.2	1	1.1	0.88	0.88
Filter press feeding pumps	3	6	3	3	2.4	7.2
P. tank mixer	0.37	0.74	1	0.74	0.592	0.592
Anionic pump	0.25	0.5	20	0.25	0.2	4
Lime tank mixer	0.55	0.55	1	0.55	0.44	0.44
Clean water discharging mixer	2.2	2.2	10	2.2	1.76	17.6
Installed power	kW-Saat	1		23.1936	**19.328**	
Total consumption	kW-Gün					**171.272**

Source An industrial company
Significance of bold type to specify the total amount

4.1   Expected Developments

A treatment plant was designed to treat the waste water of a 200 m³/daily capacity, four furnace plant. If three of the furnaces are working, no extra work on the treatment center will be required, even if there is an increase in capacity. Quite the opposite, in coming periods some of the pumps and motors will be exchanged with new and more efficient models, thus further decreasing energy requirements.

The panels to be mounted on the roof have an average weight of two tons. The roof weight bearing capacities will be recalculated and any necessary supports will be added (Fig. 5).

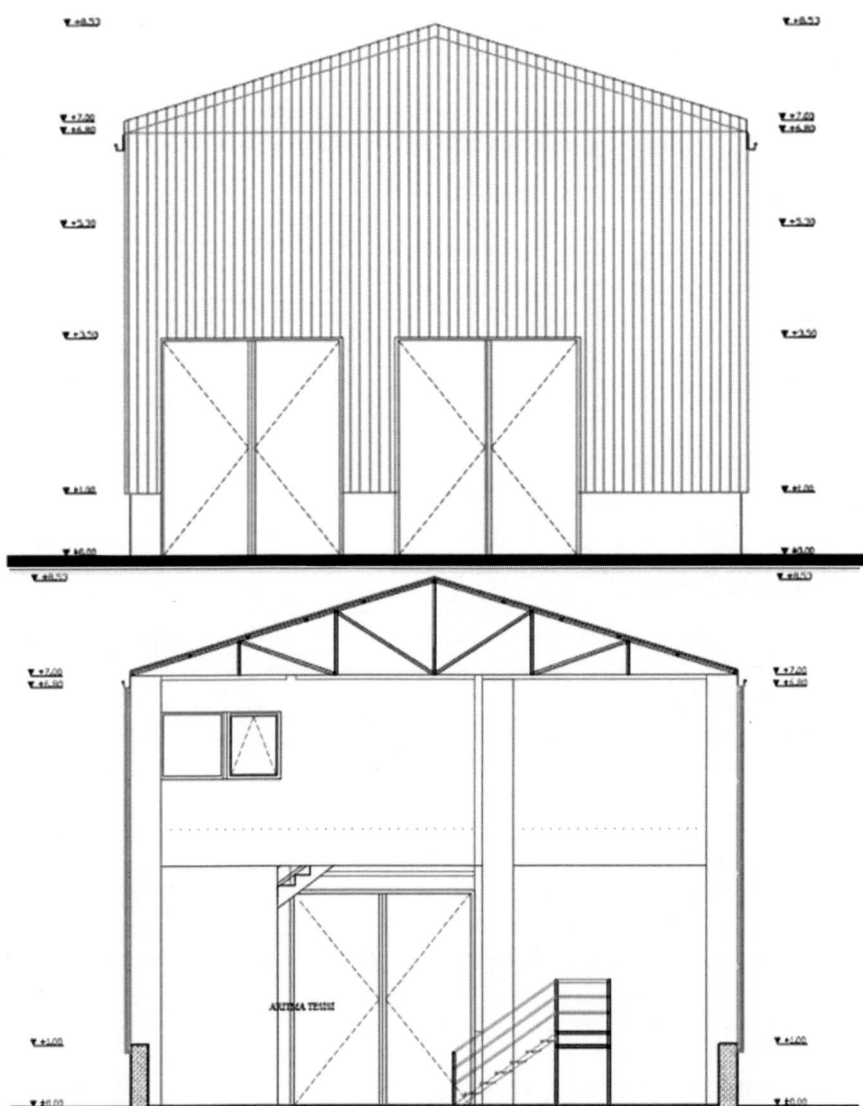

Fig. 5. Design of treatment plant. (*Source* An industrial company)

4.2   Pitch of Design and Roof

4.3   Actual Measurements

One-week long measurements were carried out of instantaneous and 24 h long electricity consumption rates as provided by the electric panel box belonging to the treatment facility and located in the factory's power supply room. The instantaneous consumption during the day were found to be at an average level of 14 kw, while the 24 h total average consumption was approximately 160 Kwh (Table 2).

Table 2. Actual instantaneous and daily measurements

Date	28.02	29.02	01.03	02.03	03.03
24 h	158 kw	160 kw	178 kw	150 kw	158 kw
Ins. cons.	14 kwh	8 kwh	4 kwh	12 kwh	14 kwh

Source An industrial company

5 Feasibility and Equipment Selection

Daily Consumption = 172 kW = 172,000 W
  1 h Maximum Expended Power = 20 kW = 20,000 W

5.1 Panel Selection

Because the calculations are done according to the highest generating power of the facility, a 250 W generating power photovoltaic cell will be selected. Each panel is 1.65 m^2 and weighs 20 kg.

Eighty panels will be needed to meet daily consumption requirements. These 80 panels will be situated on the roof, as shown below, one a space of 13.2 × 10 m, thus covering a 132 m^2 area.

The roof covers a 180 m^2 area and is in the shape of an 18 m × 10 m rectangle. The 10 m area will be completely covered by panels, while there will be a 1.6 m wide overlay. Panels will also be installed on the 9 m slope to the east with a 0.5 overlay.

Offers for 250 W capacity panels were received from 5 separate companies. The Y firm was selected as it has a good reputation and also gave the lowest offer.

X company: 74,000 TL
Y company: 53,760 TL
Z company: 52,800 TL
W company: 57,600 TL
U company: 47,360 TL

5.2 Battery Selection

A: W/V
  172,000/12 V = 14,333 Ah
  1 day storage = 72 batteries

200 Ah 12 V Jel Akü
X company: 1500 TL
Y company: 1750 TL
Z company: 1800 TL

Considering that Eskişehir is located in the continental (cold winter) meteorological region with low nighttime temperatures, the facility will need batteries in which to store the energy. Also considering the size of the area and number of panels, calculations called for the use of 72, 200 Ah 12 V gel batteries. The battery group will be interconnected on the roof of the 8 story building both parallel to each other and in a series. The batteries will be stored in the command room located on the second floor of the treatment center.

Total Costs of the Battery Group: 108,000 TL

5.3   Inverter Selection

Calculations call for a one hour total drawn power (20,000 W) and thus a total of four inverters will be needed.

X company: 5000 TL 5000 W/24 V
Y company: 7500 TL 5000 W/24 V

İnverter Costs: **20,000 TL**

5.4   Charge Regulator Solar Cable (6 mm)

MPPT	50 m
350 TL	**500 TL**

Other miscellaneous expenditures: **17,000 TL**
Total Investment Expenditures:
53,760 + 108,000 + 20,000 + 500 + 350 + 17,000 = 199,610 TL
Amortization Period:
Electricity Unit Cost: **0.39 TL/kWh**
Annual Electricity Consumption: 172 * 365 days = 62.780 Kwh
Annual Electricity Costs: **25,000 TL/year Amortization will thus be achieved in a period of 8 years.**

6   Conclusion

This feasibility study suggests that the operation of a waste water treatment plant with solar panels may be considered as a workable alternative. Such a system with a one-day storage capacity will cost approximately 200,000 TL to establish. Another alternative to consider would be a system that does not include a battery group. On days that the solar radiation is insufficient, this system could work with back-up electricity supplied by the central distribution network. In this case, the plant will utilize energy supplied from two different systems. In such a case, even if the hour maximum of kw electric energy is drawn from the system (feasibility used 20 kW/h), but still sufficient energy is not generated from the system, the system will automatically begin to pull electricity from the central energy distribution system. If such a system is utilized, the total investment costs will decrease by almost one half, to approximately 100,000 TL. And, since the system will be shut off during the nights and during long and cold winter periods, there will be no change to the amortization period. One of the biggest advantages of such a system is that because batteries only have a working life of 5 years, the owners will not have to deal with batteries, their operations and their replacements.

The calculations are based on assumptions that the panels will work at 100% efficiency, but in actuality, work carried out thus far has shown that with today's technology and conditions, these panels can not yet work at total efficiency. Taking current unit electricity prices into consideration, the amortization period for all solar energy systems range between 7 and 8 years. This kind of long period is due to the fact that investment costs remain very high and that the equipment and accessories used do not all supply the same 24 h efficiency rates. Still, it must be remembered that the demands for electricity in Turkey are continuously rising and the prices are sure to also rise as other generation resources are depleted. In these cases, the amortization period will decrease in a few years time.

References

1. Kurban M, Başaran Filik Ü, Hocaoğlu F. O. ve Mert Kantar Y (2007) Eskişehir Bölgesi İçin Güneş Işınım Şiddeti, Güneşlenme Period Ve Sıcaklık Arasındaki İlişkinin Analizi
2. www.enegylan.sandia.gov
3. Karapınar İlçesi'nde Güneş Enerjisine Dayalı Electricity Üretim Tesisi Yatırımları için Enerji İhtisas Endüstri Bölgesi Kalmasına Yönelik Fizibilite Çalışması Raporu
4. Kalogiru SA (2009) Solar energy engineering process and systems
5. Solimpeks Photovoltaic Katalog (2006)
6. Filik ÜB, Kurban M, Aydin G, Hocaoğlu FO, Anadolu Üniversitesi İki Eylül Kampüsü, Eskişehirdeki Yenilenebilir Enerji Kaynaklarının Potansiyel Analizi
7. Güneş Enerjisi Sanayicileri ve Endüstrisi Derneği

Behavior of Mortar Samples with Waste Brick and Ceramic Under Freeze-Thaw Effect

Selçuk Memiş[1]([✉]), İ.G. Mütevelli Özkan[1], M.U. Yılmazoğlu[1],
Gökhan Kaplan[2], and Hasbi Yaprak[1]

[1] Department of Civil Engineering, Faculty of Engineering and Architecture,
Kastamonu University, Kastamonu, Turkey
smemis@kastamonu.edu.tr
[2] Kastamonu University, Kastamonu Vocational Schools, Kastamonu, Turkey

Abstract. Increasing number of industrial facilities and population concentration in particular regions along with overconsumption are main reasons of the increased environmental pollution. It is a necessity to preserve available resources and to keep the waste in control in order to achieve a sustainable development goal. In recent years, concepts of waste management, recycling and sustainability have gained importance with regards to the construction industry. Today, approximately 35 billion tons of concrete is produced worldwide and 80% of this amount consists of aggregated manufactured using natural resources. A significant environmental impact is the case even for the production of cement, a binding agent for concrete, which accounts for 1 ton CO_2 emission in order to produce 1 ton of cement. The main subject of this study is the production of a sustainable construction material with the use of ceramic instead of both aggregate and cement. Clay is defined as a common natural material with fine-grains, with layers and a high water absorption capacity. Ceramic products are construction materials which can replace cement generally in the form of artificial puzzolana. The main subject of this study is the use of the waste obtained from ceramic plants which produces ceramic products in mortars. Taguchi L9 array design was used as part of this experimental study. Water-binder ratio was set to 0.50 in the preparation of the mixes and natural aggregates were used. Aggregates were then replaced by pieces ceramic and brick at a percentage between 20 and 60% while cement was replaced by ceramic and brick powder at a percentage between 10 and 30%. The mixes were then subjected to freeze and thaw tests at 30, 60 and 90 cycles in accordance with ASTM C 666 standard. Dynamic modulus and mechanical properties of the mortars subjected to f-t effect were then identified. When the results of the tests were examined, it was found that the compressive strength at 7th and 28th days were decreased with the increase of the volume of ceramic and brick powder used while it was found that the use of ceramic and brick powder did not have a significant effect on the compressive strength at the 90th day. The use of ceramic and brick aggregate led to favorable results in terms of freeze-thaw resistance. Especially the use of 10% ceramic [(whiteware) CA] and 20% other ceramic [(brick) BA] aggregate in mortars subjected to 90 f-t cycles increased the dynamic modulus while similar results were found for the use of 5% ceramic powder and 10% brick powder in mortars subjected to 90 days of f-t cycles. This

© Springer International Publishing AG, part of Springer Nature 2018
S. Fırat et al. (eds.), *Proceedings of 3rd International Sustainable Buildings Symposium (ISBS 2017)*, Lecture Notes in Civil Engineering 7,
https://doi.org/10.1007/978-3-319-64349-6_15

study shows that waste material obtained from ceramic and bricks industry can be repurposed in the construction industry.

Keywords: Sustainability · Waste management · Durability
Freeze-thaw resistance · Mortar

1   Introduction

Consumption of energy and natural resources has step up proportionately to civilization development and humanity population growth is one of the biggest environmental concerns today. In addition to the increasing emission of greenhouse effect gases, unbalanced consumption of natural resources will eventually lead to their exhaustion, as in the case of natural resources [1].

The exploitation of natural resources for construction purposes, in particular non-renewable resources, leads to millions of tons of construction and demolition waste (CDW) every year. Since most countries have no specific processing plan for these materials, they are sent to landfill instead of being reused and recycled in new construction. According to Eurostat, the total amount of waste generated in the European Union in 2010 was over 2.5 billion tones, of which almost 35% (860 million tons) derived from construction and demolition activities (Fig. 1). Of the total waste generated by the construction and demolition activities, and other activities, 97% was mineral waste or soils such as excavated earth, road construction waste, demolition waste, rock waste and others [2, 3].

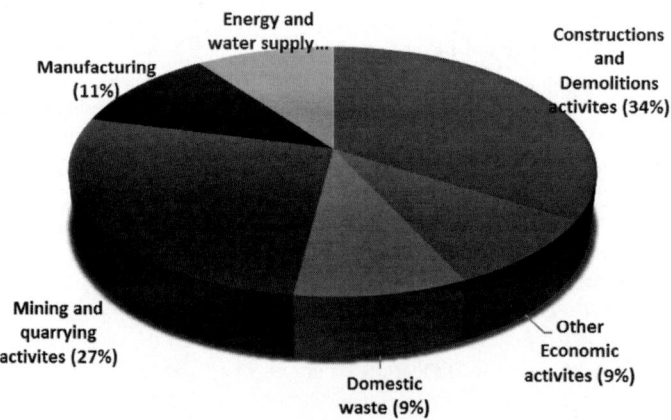

Fig. 1. Total waste generated in European Union [2]

This waste problem is becoming increasingly problem because of the growing quantity of waste (like industrial, construction and demolition) cause in spite of the measures which have been taken in recent years at European Union and Turkey, national and regional levels aimed at controlling and organizing waste management.

The necessity to manage these wastes is become one of the most urge upon issues of nowadays, requiring certain actions aimed at preventing waste generation like reuse, recycling and waste to energy systems. This necessity is promotion of resource recovery systems as a means of using the natural resources contained within waste and reducing environmental impact. Use of such waste, in addition to helping protect the environment, offers a series of advantages such as a reduction in the use of other raw materials, contributing to an economy of natural resources. Moreover, reuse also offers benefits in terms of energy, thus recovering the energy previously can be incorporated during production [4]. For instance, Portland cement clinker production consumes of energy that 850 kcal per kg of clinker and has a considerable environmental impact. This involves solid quarrying for raw materials, as it takes 1.7 tones to produce 1 tons of clinker, as well as the emission of greenhouse and other gases (NOx, SO_2, CO_2) into the atmosphere. Around 850 kg of CO_2 are spread per tons of clinker produced [5]. In this way, both the lost energy is recovered and pollution is prevented.

As many developing countries have all over the world, Turkey has also been generating a huge amount of CDW waste, which generates serious environmental problems to deal with. Because of the Urban Renewal Law at Turkey, it is estimated that demolition and maintenance of the structures at the end of their economic span result 4–5 million ton/year of concrete and demolish waste (CDW). As recycling and reuse are alternatives to minimize the impact of energy and raw material consumption on the environment, waste that can be potentially used for concrete production can be recycled aggregate, which obtained via concrete and demolish waste (CDW) [6].

Concrete is the most common and useful material, which has contributed to the progress of civilizations throughout last century, in the construction industry. However, construction activities demand a significant amount of natural materials, which significantly modifies the natural sources and creates environmental problems, in order to produce cement and aggregate [6, 7]. The increasing and unsustainable consumption of natural resources is the cause of great concern for the environment and economy with the excessive production of construction and demolition waste (CDW). Construction industry is a big sector, which can include several efforts to promote ecological efficiency. Thanks to this feature, waste can be considered that the reusable of CDW at new construction [8]. In other words, construction and demolition waste (CDW) can be use for possible feasible alternative as aggregates [as named recycled aggregates (RA)] into the production of new concrete, mortar or plaster. Recycled aggregate has become one of the sustainable technologies in concrete industry because of great environmental advantages in past decades [9–11]. Reuse of construction and demolition wastes (CDW) may be provide a beneficial way which leads sustainable engineering approaches to mortar and concrete mix design [6, 7]. In other words, we can say that fine or coarse aggregates and cement could be successfully replaced by different kind of solid waste. This waste was mainly composed of concrete and masonry waste, both components with a high potential for being re-used or recycled as unbound materials or cement-treated materials in road construction, recycled aggregates (RA) for the production of concrete, mortar, drainage materials and also for asphalt materials [12].

When investigations involving recycled aggregates (RA) and supplementary cementing materials are analyzed, aggregates with pozzolanic properties are generally seen to be used more frequently as recycled aggregates (RA) or supplementary cementing materials such as blast furnace slag, pulverized soda lime glass, residuals of glass separate collection, treated bottom ashes and biomass fly ash, activated slag, rice husk ash, metakaolin and calcined clays, ceramic residues [13] for cement and concrete, bricks, tile, wood, etc. for aggregate.

Pozzolan is defined as a siliceous material which, in itself, possesses little or no cementing property but which will, in finely divided form and in the presence of moisture react chemically with calcium hydroxide at ordinary temperatures to form compounds possessing cementitious properties [14]. Pozzolans and other materials with similar characteristics, when used in mortars or concrete, allow reducing the cement content for a pre-determined mechanical strength, thus replacing a material with very high-energy consumption in its manufacture [8].

Ceramic material, which is defined as an inorganic, non-metallic, generally crystalline oxide, nitrile or carbide material, can withstand very high temperatures and raw materials include clay minerals such as kaolinit [15]. Heat treated clays from ceramic products, such as bricks and tiles that were milled and incorporated in mortar, was used materials in ancient times, when natural pozzolans were not available [16, 17].

Researchers is revealed that ceramics (Fig. 2) might have potential as pozzolans or as aggregates when incorporated into mortars or concrete along with being compatible with buildings masonry. Apart from these environmental advantages, there are other benefits from using ceramic waste. When used to partially replace the constituents of mortars, ceramic particles help to reduce the consumption of natural aggregates or binder [17].

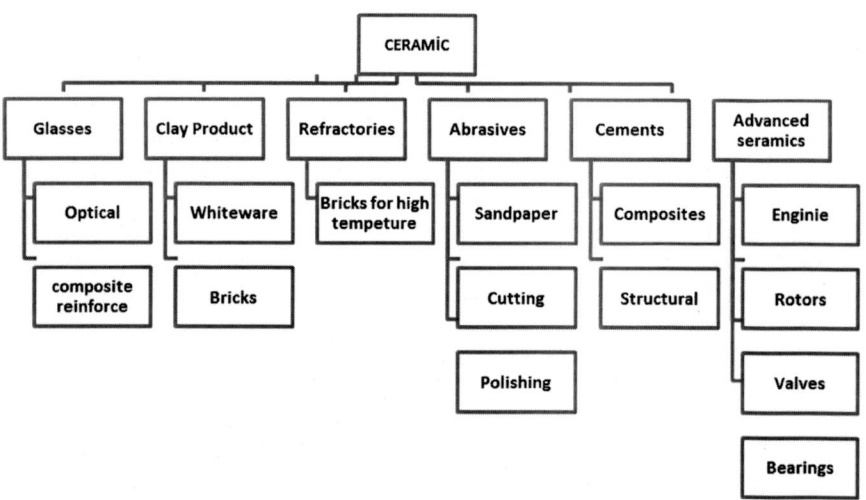

Fig. 2. Classification of ceramic by type [19]

Ceramic waste may originate from two sources, which is the ceramics industry and is associated with construction and demolition. It has been estimated that about 30% of the daily production in the ceramic industry goes to waste. Most of the previous studies have investigated the use of ceramic waste in concrete or mortar as sand or coarse aggregate. Some researches were also done in which the use of the materials in concrete as a substitute for cement were investigated [4, 18].

Various authors have researched the possible use of ceramic waste. Most of these studies do not include ceramic waste from CDW, but rather from ceramic industry waste: clay roof tiles, ceramic sanitary ware and brick. Finely crushed ceramic waste has been used for making cement, as a substitute of cement for mortar production [20] and as an addition to mortar. The coarse fraction of ceramic waste has been used as recycled aggregate in concrete production and fine fraction as recycled sand in mortars [12].

Ceramic powder and grains (Fig. 3) have been widely used in mortars. The variety of these materials was observed that when they were combined with certain substances the resulting mortars have improved characteristics. Over time, it has been demonstrated that the addition of small ceramic particles confers improved characteristics on mortars and that pozzolanic reactions might occur [16].

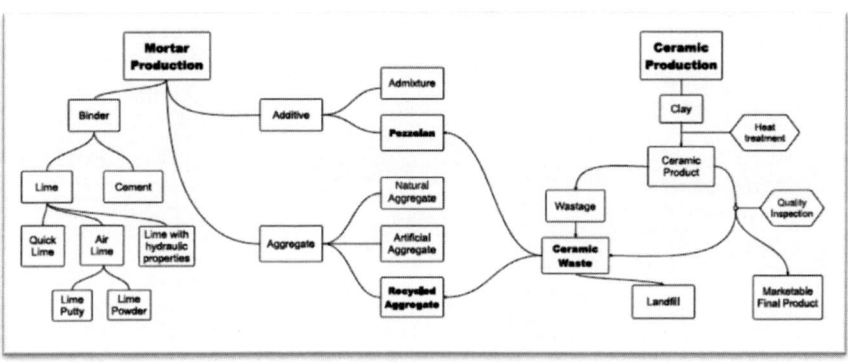

Fig. 3. Type of aggregate used in this study [17]

During the freeze-thaw cycles, the concrete or mortar will expand and contract alternately caused by conversion between solid state and liquid state of water in the microstructure. These expansion and contraction of the water can fatigue the concrete microstructure over time and may lead to the disintegration and failure of the concrete material. And then concrete or mortar will be caused after the action of freeze-thaw cycles [21].

In this study, the recycling rate of baked earth products obtained from brick and ceramic industry was increased with their use in mortar production. In this context, this study used 20–60% ceramic products instead of natural aggregate and 10–30% ceramic products instead of cement. Some of the waste ceramic materials were used into mortars mixtures in different proportions, partially replacing the natural sand or binder at, is to determine the influence of ceramic aggregate and ceramic powder on the properties and freeze-thaw resistance of cement mortar.

2   Materials and Method

2.1   Materials

In the mortar production process, washed stream aggregate at 0–4.75 mm sieve size obtained from the city of Kastamonu was used. Ceramic waste (brick and ceramic aggregate) was used in combination with the aggregate replacing the natural aggregate at a ratio between 20 and 60%. The particle distribution of waste aggregate was then reduced to 0–4.75 mm in order for it to match natural aggregate size. CEM I 42.5 R cement was used in mortar production in accordance with TS EN 197-1 standard. Ceramic waste (CA and BA powder) was used in combination with cement replacing the cement at a ratio between 10 and 30%. Chemical and physical properties of the binding agents used in mortar production are shown in Table 1.

Table 1.   Chemical and physical properties of the binding agents

Component (%)	CEM I 42.5 R	Ceramic powder	Brick powder
CaO	64.70	1.73	8.81
SiO_2	19.12	55.73	54.20
Al_2O_3	4.75	29.76	14.90
Fe_2O_3	3.53	5.41	5.75
MgO	0.94	3.30	6.56
Na_2O	0.21	1.96	2.03
K_2O	0.88	3.11	2.22
SO_3	2.85	0.29	0.59
Cl^-	0.009	–	
Free lime	1.41	–	
Loss on Ign.	2.36	6.89	
Insolubles	0.23		
C_3S	58.79		
C_2S	12.69		
C_3A	7.74		
C_4AF	9.83		
Specific W.	3.13	2.55	2.95
Specific surface (cm^2/g)	3400	3520	3330

Granulometric properties of the aggregate used in mortar production are shown in Fig. 4. Physical properties of the aggregates are shown in Table 2.

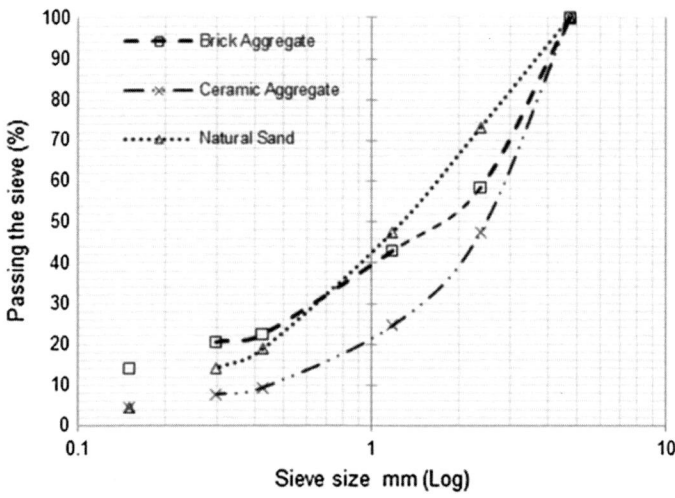

Fig. 4. Granulometric properties of aggregates

Table 2. Physical properties of aggregates

	Brick aggregate	Ceramic aggregate	Natural aggregate
Specific weight (SSD)	2.52	2.40	2.63
Water absorption (%)	4.6	8.3	2.3

2.2 Method

In this study, ceramic (CA) and brick(BA) powder was used in combination with the cement replacing the cement at the ratios of 5, 10 and 15%. Ceramic powder was obtained grounding the ceramic waste for 120 min using a ball mill. Due to the fact that the material has the slope of the cement, Due to the fact that the material, which is the factory waste, has the slope of the cement (Table 1), no grounding process was necessary for brick powder. Ceramic and brick aggregate was used in combination with the natural aggregate replacing the natural aggregate at the ratios of 5, 10 and 15%. Ceramic and brick waste was reduced to a fine-grain aggregate size using a rock crusher. Taguchi L9 assay matrix was used in the preparation of mixs. The variables used in the experimental study are given in Table 3.

Table 3. Dependent variables of Taguchi L9 assay matrix

Taguchi L9 assay matrix		Level 1 (%)	Level 2 (%)	Level 3 (%)
Aggregate replacement	Ceramic aggregate	10	20	30
	Brick aggregate	10	20	30
Cement replacement	Ceramic powder	5	10	15
	Brick powder	5	10	15

10 different mix, 9 with aggregate and cement replacements and 1 reference, were prepared as part of mortar preparation process. Mix ratios of the mortars are given in Table 4.

Table 4. Aggregate and cement replacement ratios

Mix #	Aggregate replacement		Cement replacement	
	Ceramic aggregate (%)	Brick aggregate (%)	Ceramic powder (%)	Brick powder (%)
1	10	10	5	5
2	10	20	10	10
3	10	30	15	15
4	20	10	10	15
5	20	20	15	5
6	20	30	5	10
7	30	10	15	10
8	30	20	5	15
9	30	30	10	5
10 (R)	0	0	0	0

W/B (water/binder) ratio of the mortar mix was set at 0.50 while their a/c (aggregate/cement) ratio was set at 3. Mortars were produced in prismatic samples of $4 \times 4 \times 16$ cm in size. Mortars were removed from the mold after 24 h and preserved in water saturated with lime at 20 ± 2 °C until the experiment day. Uniaxial compression tests and 3-point flexural tests were conducted on the mortar samples at 7th, 28th and 90th day. Freeze-thaw resistance of the mortars produced using ceramic products was explored in order to determine their durability properties. Having been cured in drinking water for 28 days, the mortar samples were then subjected to f-t tests.

Once the curing process was completed, mortars were subjected to freeze-thaw tests at 30, 60 and 90 cycles. Freeze-thaw temperatures and durations were defined in accordance with the ASTM C 666 standard in the f-t tests. Durability factors of the mortars were identified having measured their relative dynamic elasticity modulus. Relative dynamic modulus was calculated using Eq. (1). Durability factor, on the other hand, was calculated using Eq. (2).

$$P_c = \frac{n_c^2}{n^2} \times 100 \tag{1}$$

where c is the number of cycles of freezing and thawing, nc is the resonant frequency after c cycles, and n is the initial resonant frequency (at zero cycles).

$$DF = \frac{N}{M} \times P_c \tag{2}$$

where Pc is the relative dynamic modulus, N is the number of cycles completed, and M is the planned duration of testing. Testing is usually halted when the relative dynamic modulus falls below 50–60% of its initial value.

3 Results and Discussion

3.1 Mechanical Properties of Mortars

Compressive strengths of the mortars at 7th, 28th and 90th days were calculated relatively. The effect of replacement ceramic aggregate ratio on compressive strength is shown in Fig. 5.

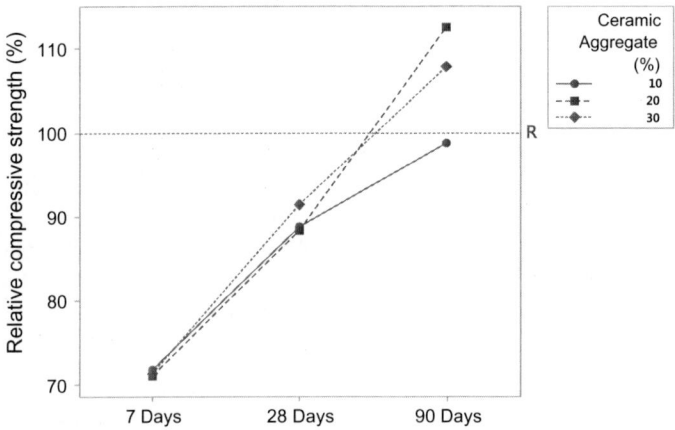

Fig. 5. The effect of replacement ceramic aggregate ratio on relative compressive strength

As shown in Fig. 5, increasing the curing time increases compressive strength. The compressive strength results obtained from the 7th day and 28th day shows that the use of ceramic aggregate decreases the compressive strength by approximately 10–30%. When the compressive strength results obtained from the 90th day are examined, it was found that the use of ceramic aggregate at ratios of 20 and 30% increases the compressive strength by approximately 10%. This situation can be explained that the material under passing the sieve, which is kept under 25% sieve, was showed a puzolonic feature and increased its strength. The use of 10% ceramic aggregate has an unfavorable effect on the compressive strength at the 90th day. The use of 30% ceramic aggregate has a favorable effect on the compressive strength at the 28th day. However, considering the compressive strength results for 90th day, it can be said that the use of 20% ceramic aggregate is more suitable. The effect of replacement brick aggregate ratio on compressive strength is shown in Fig. 6.

As shown in Fig. 6, increasing the curing time increases compressive strength of the mortars with brick aggregate additive. The use of brick aggregates at the ratios of 20 and 30% gave similar results for all experiment days. Compressive strength of the

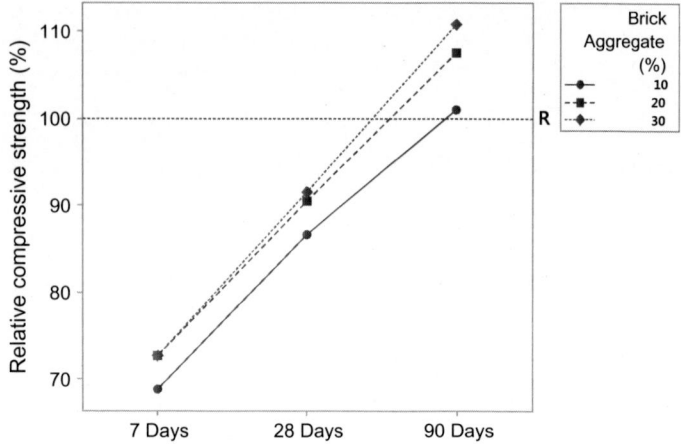

Fig. 6. The effect of replacement brick aggregate ratio on relative compressive strength

mortars at the 28th day is reduced by 10–15% when compared to the reference mix. However, Fig. 6 shows the favorable effects of the use of brick aggregate at the 90th day. The use of brick aggregate, especially at a ratio between 20 and 30%, increases the compressive strength by approximately 10%. As it was the case for ceramic aggregate, the use of 10% brick aggregate is not suitable enough. The effect of replacement ceramic powder ratio on compressive strength is shown in Fig. 7.

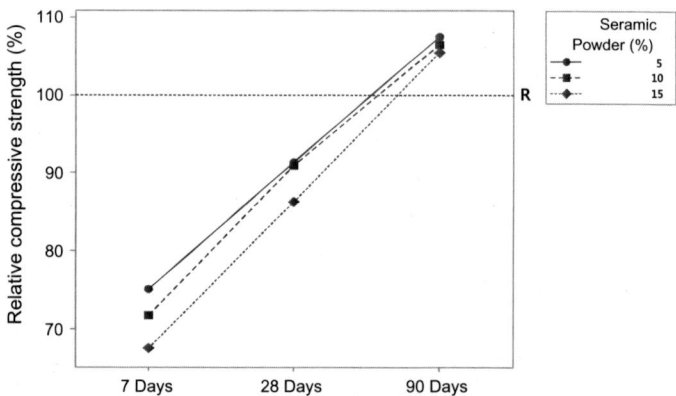

Fig. 7. The effect of replacement ceramic powder ratio on relative compressive strength

In terms of compressive strength at the 7th day, increasing the ceramic powder ratio leads to decreases in the compressive strength. The use of ceramic powder at a ratio between 5 and 10% decreases the compressive strength by approximately 10% at the 28th day. Exploring the compressive strength at the 90th day, a distinctive difference was found between mortars. The use of ceramic powder increases the compressive strength by approximately 5% at the 90th day. The effect of replacement brick powder ratio on compressive strength is shown in Fig. 8.

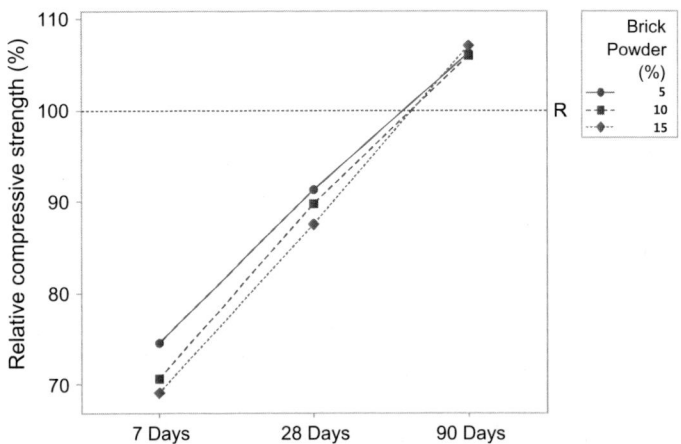

Fig. 8. The effect of replacement brick powder ratio on relative compressive strength

As shown in Fig. 8, increasing the brick powder ratio has an unfavorable effect on the compressive strength at 7th and 28th days. However, this unfavorable effect is eliminated at the 90th day. The use of brick powder increases the compressive strength by approximately 5% at the 90th day, as it was the case for ceramic powder.

3.2   Freeze-Thaw Resistance Properties of the Mortars

Properties such as relative dynamic modulus (Eq. 1) and durability factor (Eq. 2) were identified for the freeze-thaw resistance of the mortars. Figure 9 shows the relative dynamic modulus of the mortars. The graph in Table 4 shows the mixes with their relevant numbers and their mix properties. Mix #10 represents the reference mortar. Increasing the f-t cycles in the freeze-thaw tests applied to the mortars decreases the relative dynamic modulus. The optimal mix ratios were defined with Taguchi optimization conducted on the f-t tests with 30, 60 and 90 cycles. Table 5 shows the optimized and estimated values for relative dynamic modulus calculated using maximized function.

Durability factors of the mortars are shown in Fig. 10 with respect to their mix numbers. It was found that mortars show similar properties at the end of 30 and 60 cycles. However, mortars show distinct differences in terms of their durability factors at the end of 90 cycles. An optimization was made using the maximized function for the durability factors of mortars. Optimization results are shown in Table 6.

The optimization conducted on the relative dynamic modulus and durability factors showed that the optimal ceramic aggregate ratio is between 10 and 20% while the optimized brick aggregate ratio is between 20 and 30%. It was found that a ceramic

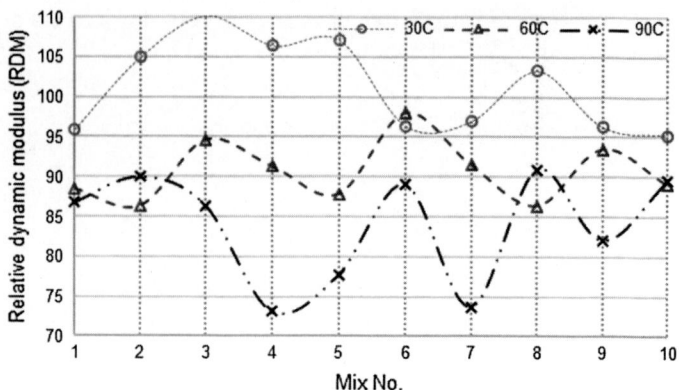

Fig. 9. Relative dynamic modulus of mortars

Table 5. Optimization results for the relative dynamic modulus

Cycle	Ceramic aggregate (%)	Brick aggregate (%)	Ceramic powder (%)	Brick powder (%)	Estimated RDM
30C	10	20	15	15	114.4
60C	20	30	15	10	98.4
90C	10	20	5	10	97.2

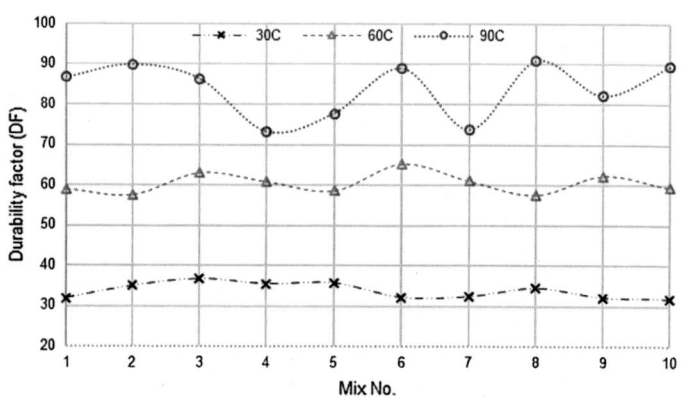

Fig. 10. Durability factors of mortars

Table 6. Optimization results for durability factor

Cycle	Ceramic aggregate (%)	Brick aggregate (%)	Ceramic powder (%)	Brick powder (%)	Estimated RDF
30C	10	20	15	15	38.2
60C	20	30	15	10	65.6
90C	10	20	5	10	97.2

powder ratio of 5 or 15% and a brick powder ratio between 10 and 15% contribute to the freeze-thaw strength.

4   Conclusions

The following conclusions can be drawn from this experimental study;

- The use of ceramic aggregate at ratios of 20 and 30% increases compressive strength.
- The use of brick aggregate at ratios of 20 and 30% has a favorable impact on the compressive strength.
- The use of brick and ceramic aggregate at a ratio of 10% has an unfavorable impact on the compressive strength.
- Increasing the amount of ceramic and brick powder used results in an unfavorable effect on compressive strength at 7th day and 28th day. However, this unfavorable effect is eliminated at the 90th day due to pozzolanic activity.
- The use of ceramic and brick aggregate led to favorable results in terms of freeze-thaw resistance. Especially the use of 10% ceramic and 20% brick aggregate increased the relative dynamic modulus of mortars subjected to 90 f-t cycles.
- The use of ceramic and brick powder improves freeze-thaw resistance of the mortars. The use of 5% ceramic and 10% brick powder increased the relative dynamic modulus of mortars subjected to 90 f-t cycles.
- This study shows that waste material obtained from ceramic and bricks industry can be repurposed in the construction industry. It will be possible to produce sustainable construction materials as a result of utilization of such waste material.
- The use of industrial waste which leads to environmental problems in the construction industry is important in terms of ecology, sustainability, durability and recycling.

References

1. Alves AV, Vieira TF, Brito J, Correia JR (2014) Mechanical properties of structural concrete with fine recycled ceramic aggregates. Constr Build Mater 64:103–113
2. Silva RV, Brito J, Dhir RK (2014) Properties and composition of recycled aggregates from construction and demolition waste suitable for concrete production. Constr Build Mater 65:201–217

3. Silva RV, Brito J, Dhir RK (2015) Comparative analysis of existing prediction models on the creep behaviour of recycled aggregate concrete. Eng Struct 100:31–42
4. Khalil NM (2014) Exploitation of the ceramic wastes for the extraction of nano aluminum oxide powder. J Ind Eng Chem 20:3663–3666
5. Puertas F, García-Día I, Barba A, Gazulla MF, Palacios M, Gómez MP, Martínez-Ramírez S (2008) Ceramic wastes as alternative raw materials for Portland cement clinker production. Cement Concr Compos 30(9):798–805
6. Çakır Ö (2014) Experimental analysis of properties of recycled coarse aggregate (RCA) concrete with mineral additives. Constr Build Mater 68:17–25
7. Pacheco-Torgal F, Jalali S (2010) Reusing ceramic wastes in concrete. Constr Build Mater 24(5):832–838
8. Silva J, Brito J, Veiga R (2008) Fine ceramics replacing cement in mortars partial replacement of cement with fine ceramics in rendering mortars. Mater Struct 41:1333–1344
9. Ann KY, Moon HY, Kim YB, Ryou J (2008) Durability of recycled aggregate concrete using pozzolanic materials. Waste Manag 28:993–999
10. Khaldoum RK (2007) Mechanical properties of concrete with recycled coarse aggregate. Build Environ 42(1):407–415
11. Mueller A, Sokolova SN, Vereshagin VI (2008) Characteristics of lightweight aggregates from primary and recycled raw materials. Constr Build Mater 22(4):703–712
12. Ledesma EF, Jimenez JR, Ayuso J, Fernandez JM, Brito J (2015) Maximum feasible use of recycled sand from construction and demolition waste for eco-mortar production e Part-I: ceramic masonry waste. J Clean Prod 87:692–706
13. Bignozzi MC, Saccani A (2012) Ceramic waste as aggregate and supplementary cementing material: a combined action to contrast alkali silica reaction (ASR)
14. Mehta PK (1987) Natural pozzolans: supplementary cementing materials in concrete. CANMET Spec Publ 86:1–33
15. Carter CB, Norton MG (2007) Ceramic materials: science and engineering. Springer, Berlin, pp 3–4. ISBN: 978-0-387-46271-4
16. Matias G, Faria P, Torres I (2014) Lime mortars with ceramic wastes: characterization of components and their influence on the mechanical behaviour. Constr Build Mater 73:523–534
17. Matias G, Faria P, Torres I (2014) Lime mortars with heat treated clays and ceramic waste: a review. Constr Build Mater 73:125–136
18. Andrejkovicova S, Velosa AL, Ferraz E, Rocha F (2014) Influence of clay minerals addition on mechanical properties of air lime–metakaolin mortars. Constr Build Mater 65:132–139
19. Materials and Manufactoring processes 20 (2017) http://slideplayer.com/slide/4616756/
20. Naceri A, Hamina M (2009) Use of waste brick as partial replacement of cement in mortar. Waste Manag 29:2378–2384
21. Shang H, Zhao T, Cao W (2015) Bond behavior between steel bar and recycled aggregate concrete after freeze-thaw cycles. Cold Reg Sci Technol 118:38–44

Part II
Environmental Policies and Practices

Sustainable Design Approaches on Packaging Design

Ceyda Özgen(⊠)

Gebze Technical University, Gebze, Kocaeli, Turkey
cozgen@gtu.edu.tr

Abstract. In this study, 'Concept of Sustainability' effects on company management decisions and these decisions impacts on sustainable product design will be emphasized. Impact of sustainable design criteria on company's organizational structure has been determined by a case study in the packaging industry. Examined the structure of the general condition of the Turkish packaging industry for answer to the question of "What is sustainable packaging?" is discussed within the framework of literature study and definitions of packaging unions. During the study, sustainable packaging assessment tools are examined and the evaluation criterias are identified. The concept of the sustainability effects on sustainable management approaches, strategies and the decision-makers investigated by case study methods of the companies operating in the packaging sector. This study focuses on sustainability issues of two leading firm's practices and studies in the Turkish packaging industry. Both companies are examined to assess the sustainability of the current stages of the product design and production, as well as applications that include sustainable design and manufacturing innovations aimed to explore and investigate.

Keywords: Sustainability · Product design · Packaging design

1 Introduction

The concept of sustainability refers to sustainable development or sustainable life. In the Bruntland Report it is defined, as "Sustainable Development is development that meets the needs of the present without compromising the ability of future generations to meet their own needs". Later on, during the environment and development conference (UNCED, The Earth Summit), organized by United Nations, significant decisions were taken and the terms "sustainable consumption" and "development" began to gain more importance [1].

Within the scope of this study, the effects of the concept of sustainability on administrative decisions of companies and the subsequent effects of these decisions on sustainable product design are covered. The aim of this study includes "denoting important points to producers to consider during the design process of sustainable products and putting these points into use". With the case study carried out in packaging sector, the aim is to determine the effects of sustainable design criteria of packing products on the organizational structure of companies. For the analysis of selected

© Springer International Publishing AG, part of Springer Nature 2018
S. Fırat et al. (eds.), *Proceedings of 3rd International Sustainable Buildings Symposium (ISBS 2017)*, Lecture Notes in Civil Engineering 7,
https://doi.org/10.1007/978-3-319-64349-6_16

evaluation units, in the life-cycle analysis phase, the sustainability of energy and materials is researched. This study was conducted with two members of Packaging Industries Association: Anadolu Glass Inc. and Plaş Plastics Inc.

2 The Concept of Sustainability

According to Bruntland Report published in 1987, the source of interest in sustainable development lies in the belief that existing human actions harm the environment and these actions will lead to serious negative results [2]. As World Commission on Development and Environment published the report entitled *Our Common Future* in 1987, the concept of sustainable development began to be discussed in public domain. In the report, a series of social and environmental obstacles were defined including the unsustainable industrial development requiring global attendance. To remove these obstacles, the following were advised; efficient utilization of resources in the industry and activities related to industry, less pollution and less waste production, utilization of renewable resources instead of non-renewable resources, reducing side effects on human health and the environment [3].

Later in 1992, during United Nations' Conference on Environment and Development (UNCED, The Earth Summit) organized in Rio, following significant decisions taken, the terms "sustainable consumption" and "development" began to gain more importance. The aim of the concept of sustainable development includes rational use of natural resources by protecting environmental values, reducing the use of resources and elevating quality of life [4].

2.1 Sustainable Product Design

The word "to design" is defined as "the creative act of stating a problem—consisting of decisions made to achieve the aims of the design—and solving it during various stages of the design act" [5]. Design is described as a creative activity incorporating the aim of establishing versatile features of objects, processes, services and systems through their overall lifecycles [6]. Otto, defines sustainable product design as follows: "Sustainable design should deliver the best (social, environmental, and economic) performance or result for the least (social, environmental and economic) cost" [7]. This strategic approach is defined as the design of products, processes, services and systems. It is concerned with the instability between the demands of the society, environment and the economy. The system defined here is a system that aims to dissolve this instability and it incorporates an overall thinking process about the effects of these three areas now and in the future [7].

Papanek states [8], "Industrial designers, industries and administrations should all together give an answer to the question of how much social and ecological harm is done to our society". In his book entitled *Design for the Real World*, Papanek claims industrial designers, industries and administers are all subject to environmental responsibility.

Tischner and Charter [4], argue that sustainable design means more than eco-design or design for the environment. As for sustainable design, social and ethical components are juxtaposed together with economic and environmental ideas into the life-cycle of the product. Sustainable design may also be defined as environmentally responsible product design and development, which incorporates a product life-cycle perspective together with approaches integrating work, culture and organizational skills [9].

"The eco-design of energy-using products" directive (EUP) defines sustainable design as "environmental features should be integrated into product design to improve performance, sustainable design criteria should be provided all throughout the life-cycle, it should include purchase of raw materials, manufacturing, packaging, transportation and distribution, installation and maintenance, use and end-of-life stages based on sustainable design needs" [10–12]. Sustainable design should be compatible with the current corporate culture in order to increase economic value and reduce environmental damage. The word sustainability refers to ecology and economics that are inextricably adhered to good design in design practices [4].

Design actions for sustainability are different from general design activities with their environmental assessments, solution-seeking methodologies and strategy definitions. Moreover, the designer's expertise in sustainable design is more influential than sustainable design tools in defining environmental assessment and strategy [13].

3   Overview of Packaging Sector

3.1   Sustainable Packaging Design

What is meant by commercial packaging today was developed in the 1700s. With the developments taken place in those years, the focus was on the practical function of the package rather than on its aesthetic quality. In the 19th century, exceptional advancements in transportation occurred. When transcontinental travel and shipping started, the travel time was quite long so a need to pack and ship products safely without spoilage arose [14].

As Robert Opei states [15]: "the main function of packaging is to protect the product (as we expend its existence and provide distribution) make the product meet other products but avoid drawing the product away from its own standing position". Opei points to packaging design as an element of marketing and underlines its consequence for today. Paul Southgate, the author of *Total Branding By Design* [16], and James Pilditch, the author of *The Silent Salesman* [17] both wrote about the significance of packaging as and instrument of marketing.

Every industrial product–produced by an industry—is packaged somehow and is sent to its final consumer. Because of this reason, the main function of packaging is to preserve and protect the product it carries in. However packaging plays an intermediary role in communication taking place between the product and the consumer in the retail area [18].

Results of research presented in International Association of Packaging Research Institutes (IAPRI) conference in 2004 show that it is difficult for most of the participants to define packaging and sustainability as complementary [19]. The organization argues that for sustainable packaging, the products should comply with the following four qualities; Efficient: meet social and economic targets; Effective: meet targets of effective use of materials, energy and water as much as possible; Cyclic: recyclable within industrial and environmental systems, and Safe: non-toxic and non-polluting.

Another organization Sustainable Packaging Coalition (SPC) provided a 7-way definition [20]. According to SPC (2006), sustainable packaging should obtain the following criteria [21]:

1. All throughout the product life-cycle, it should be safe and healthy for the consumers of the product and for the whole society.
2. It should provide for the needs of the market in terms of performance and price.
3. Sustainable energy should be used in production, distribution and recycled material sources… etc.
4. Maximum amount of recycled material should be used.
5. Production techniques and applications that are respectful to the environment should be chosen.
6. Product life-cycle analysis should be conducted and materials used should be a part of healthy solutions all throughout the process.
7. It should be designed physically with an aim to balance material and energy use.
8. It should incorporate reuse and renewable qualities and it should have an industrial cradle-to-cradle cycle.

3.2   Turkish Packaging Industry

For marketing various agriculture and food products and other non-food products produced in present and lately developing industrial branches and specifically for export goods, packaging has an increasingly vital role. Turkish packaging industry is growing by 10% each year and becoming an important competitor in the global market. Approximately 5000 manufacturers are producing packaging products in Turkey. Most of these manufacturers are located in Istanbul, Izmir, Kocaeli, Gaziantep, Adana, Ankara, Konya and Balıkesir [22].

In the last decade, significant changes have been taking place in life standards as a result of increase in per capita income. Growing tendency for urbanization, increase in average life span, changes in consumption habits and consumer expectations all lead to development of self-service methods in consumption centers. The organized retail trade percentage, which was a little above 20% in 2000s, rose above 40%. Since this percentage is above 80% in economically developed European Union countries and countries in North America, it may be argued that the demand for packaged goods will rise in Turkey too [23]. Gross production capacity of packaging industries in Turkey is presented in Table 1.

Table 1. Packaging industry production capacity of 2007–2015 [23]

Production types	2007 (tons)	2015 (tons)
Paper	60,000	85,000
Cardboard	415,000	577,000
Corr. cardb.	1,370,000	2,031,000
Plastic	1,470,000	2,988,000
Metal	299,500	428,000
Glass	659,000	1,153,000
Wood	385,000	510,000
Total	4,658,500	7,772,000

3.3   Conclusion of Section

For the present and still developing industry in Turkey, packaging design has become a significant issue. As it will be discussed in Sect. 3 of this article, according to the research on Turkish packaging industry, Turkish packaging industry is growing by 10% each year and becoming an important competitor in the global market. In 2011, Turkish packaging industry had reached a capacity of $12 billion approximately. For the period between 2001 and 2016, volume of plastic packaging production had tripled while glass packaging production had doubled. In the light of all these acquired information, it is seen that packaging design is quite significant for Turkish industry. Further studies in sustainable packaging design are crucial for both packaging sector and the local economy.

According to data collected between 2007 and 2016, it can be seen that glass packaging production had increased from 659,000 tons to 1,153,000 tons. In the case of plastic packaging production, the production capacity had increased from 1,470,000 tons to 2,988,000 tons. In the Fig. 1 there is Turkish packaging sector production rates [23].

Fig. 1. Turkish packaging sector production rates [23]

4   Research Method

4.1   Framework of Method

Bayazıt [24] define research conducted in the field of design as follows: "Design research is a systematical research and information acquisition on design and design action". The aim of design research includes the discovery of man-made artifacts and their involvement in both academic research and manufacturing sector. The scope of this research includes packaging industry applications in Turkey.

In the context of real life research, descriptive and revealing case studies are designed to collect research findings [25]. This research consists of two parts: 1. Background studies. 2. Case studies.

Background studies consist of screening of publications on sustainable design in academic literature, printed sectorial information and discussions of the issue with academics and professionals in the sector. These background studies constituted the basic knowledge for interviews conducted within the scope of the case study.

The case study is considered as the most effective method for a research in which a researcher cannot control or predict the experiment outcomes in real life context [25]. The two cases in which the case study method can be used optimally are described as follows. One of these is the search for a direct descriptive question "What happened?" or an explanatory question "Why and how?". This research aims to produce information through design and production models in today's packaging manufacturers.

Packages designed by manufacturers Anadolu Glass Packaging Industry Inc. and Plaş Plastics Inc. are analyzed in accordance with the criteria defining sustainable packaging, and product-centered environmental management systems. For the products selected as evaluation units, life-cycle analysis method (energy consumption, material usage) is taken as an example to determine the effects of sustainable design.

Three types of information resources were used: interview, documentation and physical objects. According to the model explained by Chung [26], interviews were held at three levels. For this research, both open-ended and close-ended interviews were carried out at strategic, tactical and operational levels. The information was recorded with a voice recorder. An evaluation unit is needed to determine data collection and evaluation techniques [25]. Glass and PET packaging were identified as the evaluation units. Three types of evidence sources have been used for these evaluation units. These are: the product itself as a physical object, the company documents-reports of the products as documents and interviews.

Langrish [27] notes six basic models of case studies. According to Langrish models, this study aims to obtain representative information on "sustainable product design and production application" through the two packaging companies that carry out the best applications.

5 Findings

This study is based on the data presented in the thesis entitled "Analysis of the Concept of Sustainability on Package Design in Terms of Company Strategy" [28]. All data used in the findings section are obtained from closed-ended and open-ended interviews and the reports of production facilities presented in this thesis.

5.1 General Evaluation

5.1.1 Awareness of Sustainable Design

Closed-ended interview participants indicate that decisions on sustainable design are made mainly by the "Design Department" and the "Marketing Department". In response to the participants' answers to interview questions, the state and producers have a major role in the evaluation of responsibilities towards environmental issues and sustainability.

5.1.2 Main Routers

Within the scope of the closed-ended interviews, companies were asked to specify the legislations related to their products and activities. These legislations were pointed out as Packaging Waste Regulation, ISO 1440, Turkish Food Codex, EU Packaging and Packaging Waste Regulation and REACH. Participants were asked to evaluate the products produced by their companies in terms of compliance with the environmental legislations specified.

5.1.3 Importance of Environmental Issues

In the closed-ended interview, companies were asked to evaluate the impact of environmental factors on their markets, taking into account their own products. Anadolu Glass Inc. stated that the impact was extremely effective, whereas Plaş Plastic Inc. stated that it was less effective. Manufacturers were asked to rate the market pressures on the business in case of environmental issues. Both companies underlined that there was little pressure.

The companies were asked to evaluate how the development of environmental issues would provide a market advantage for their companies. It is seen that all employees of the both companies think that the development of environmental issues will increase brand reputation, effect corporate identity positively and this will provide competition power.

5.1.4 Obstacles

Company employees have stated that environmental issues are very important to them, but they underline that there is no competition in this sector about environmental issues. The difficulties faced by companies in implementing environmental issues seem to be expressed in terms of lack of understanding about environmental benefits, lack of financial resources, lack of customer demand, system and culture.

5.1.5   Product Development Process

The employees of the companies surveyed within the scope of this research were asked about the actions they performed during the product development process and were asked to evaluate upon them. Selecting harmless materials, reducing weight and volume, using materials efficiently and reducing energy consumption are seen as the most important applications.

5.1.6   Application of Sustainable Design

Participants of the companies surveyed in the research were asked to evaluate the sustainable design criteria applied to their products. Reducing material usage and diversity, reducing energy consumption, reusing the product, reducing the weight and volume of the product are the main issues.

5.2   Analysis of Evaluation Unit

As the evaluation units, olive oil bottles were selected as the most preferred and sold bottles of both packaging manufacturers. For this analysis, 2 products with similar forms produced by the two companies were examined within the scope of this research.

5.2.1   Comparison of Glass and PET Packaging

No detailed material and energy consumption reports have been generated for each product in the production facilities under review. In this context, the anticipated value of recycling energy for analysis is unknown. In accordance with the Turkish Food Codex, the use of recycled material is taken as "0" for PET packaging. The rate of use of recycled glassware is assumed to be 30%. Energy items spent for recycling, reuse, improvement, waste, transportation and service for the packages produced from both materials are not included in the production reports.

Analysis of packages made from glass and PET materials presented in Table 2 are shown graphically in Figs. 2, 3, 4, 5 and 6.

Table 2. Glass and PET packaging

Product	Company	Weight (g)	Production energy (kWh)	Product images
Olive oil bottle 1000 cc 123,399	Anadolu Cam Sanayi A. Ş. Glass Bottle	610	1.0467	
Olive oil bottle 1000 cc Efes	Plaş Plastik A.Ş. PET bottle	50	0.043	

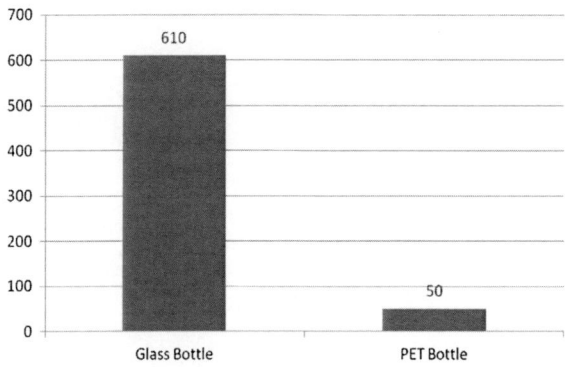

Fig. 2. Glass and PET bottles weight (g)

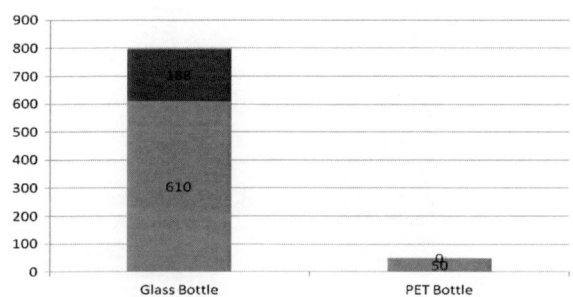

Fig. 3. Recycled materials using of glass and PET packaging (g)

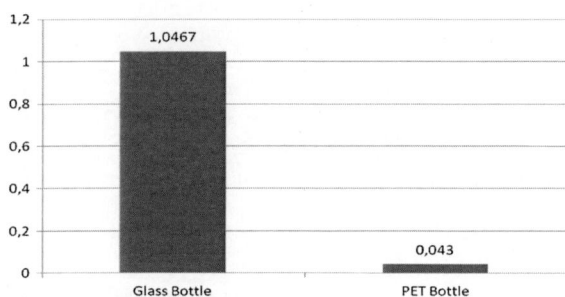

Fig. 4. Energy using of glass and PET packaging (kWh)

The automatic calculators in Url-1 [29] are used for the calculation of CO_2 emitted in transporting glass and PET packages. The parameters used in these calculations are given in Table 3. A simulated-road of 1302 km has been applied in the calculations carried out with these parameters. Table 4 shows the CO_2 releases. Figure 7 shows CO_2 emission figures for trucks carrying glass and PET packaging. Figure 8 shows the fuel consumption of glass carrying trucks.

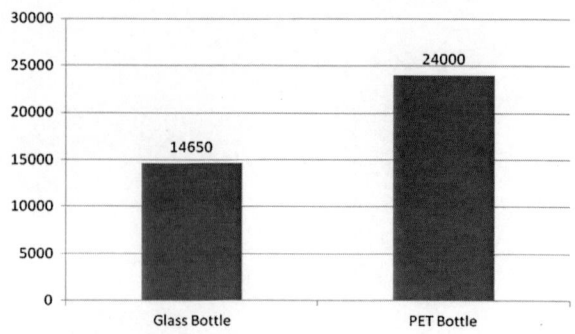

Fig. 5. Transportation of glass and PET packaging (pieces)

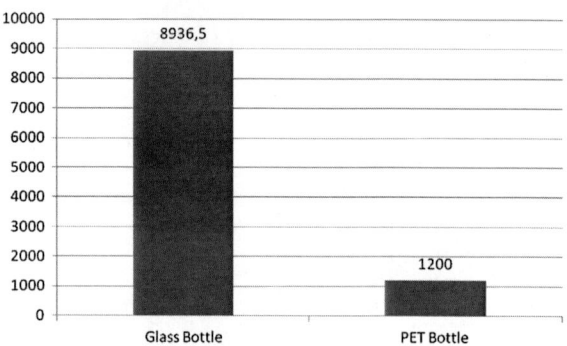

Fig. 6. Transportation of glass and PET packaging (kg)

Table 3. CO_2 calculation parameters

Truck size	16 Ton
Fuel consumption	0.40 l/km
Truck efficiency factor	80%
CO_2 emission/fuel consump.	2.630 kg/l

Table 4. CO_2 emission (kg)

Transportation truck (16 tones)	CO_2 emission (kg)	Efficiency g CO_2/t km	Weight (kg)
Truck with glass olive oil bottle	1000	82	8936
Truck with PET olive oil bottle	100	82	1200

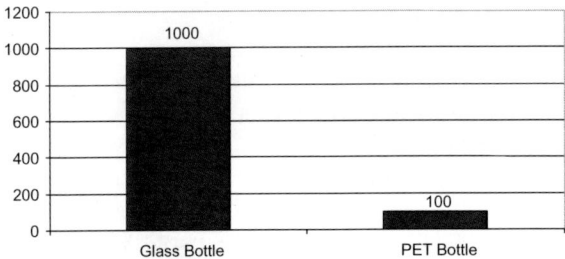

Fig. 7. CO_2 emission figures for trucks carrying glass and PET packaging

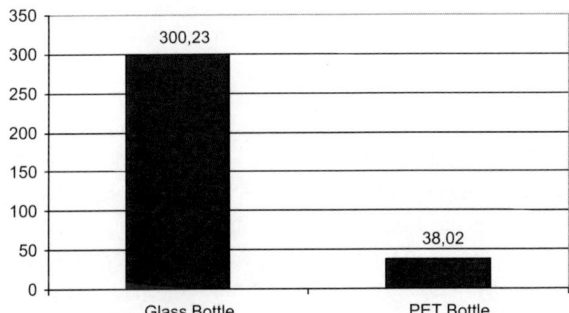

Fig. 8. Fuel consumption of trucks carrying glass and PET packaging

6   Conclusion and Recommendations

6.1   Conclusion

Brundtland Report (1987), which redefined the development approach, proposed "Sustainable Development" to the agenda and with the Rio Declaration (1992) the issue has begun to be widely discussed. According to Porter and Van Der Linde [30] this issue has led companies to make new environmental innovations that improve competition, and as a result, less polluting production processes have been developed.

Within the scope of the research, it is seen that in the applications of the companies surveyed, the idea of designing sustainable products is not targeted directly but because of the laws and regulations, competition creation and economic reasons, sustainable product criteria are applied partially. It is seen that the work carried out within the framework of company strategic plans generally includes reduction of materials at the product level, applications for easy and rapid production at the production stages, and reduction of the energy consumed in production. In order to be able to sustain their assets in the packaging industry and to be able to respond to the competition, it is stated that the relation of packaging with the environment is taken into consideration in product design and presentation to the market.

In the findings section, it is observed that according to the common view of the companies, the government and producers both have vital roles about environmental issues and responsibilities towards sustainability. On the issue of being environmentally conscious in the designs they produce, manufacturers are mostly influenced by industries and other producers; and the customer factor comes as secondary. A further influential factor is determined as the European Union and consumers. It is seen that the most significant inducement about the environmental consciousness in the companies' work on sustainability can be listed as being in compliance with the legislation and regulations, consumer needs and market opportunities. In this framework, it has been pointed out that the regulations of "Packing Waste Regulation", "Turkish Food Codex", "European Union Packing and Packaging Waste Directive" are well known and practiced by the companies.

It is seen that the companies have a consensus on the necessity of being more active about environmental issues and actions in their organizations. Further observations show that all the employees of both of the companies think that the development of environmental issues will increase their brand reputation and positively affect corporate identity.

6.1.1   Result of Evaluation Units Analysis

Information on the amount of energy used to produce the product, the amount of material used to produce the product, the amount of recycled material usage, and the shipping conditions of the product from the production facility have been collected in the production, assembly, use-service and retirement phases. The production stages of glass and PET packaging in production facilities were examined and it was determined that the production stage for both packaging types was formed by one step.

When packaging is considered in the context of recycling, it is seen that packages made from glass materials can be recycled, but it has been determined that packages made from PET materials cannot be used in food packaging as recycled PET materials.

For the analysis of the products examined within the scope of the research, the following material-energy use analysis model was carried out in accordance with the current production stages and the information obtained. Figure 9 shows the model of evaluation analysis used for the packages examined in the research.

Analysis based on packaging clearly shows that PET packaging is produced using much less material and energy than glass packaging, CO_2 emissions are less in transportation of PET and in parallel to that less fuel is consumed during the transportation. These evaluations can be seen in Table 5.

Similar results have also appeared in the results of a study of baby food jars produced from glass and PET bottles. Humbert et al. [31] reported that plastic containers produced using 14–24% less energy, caused 28–31% less input on global warming, contained 31–34% less respiratory inorganic materials, caused 28–31% less terrestrial acidity and pollution.

As a result of the findings obtained, how realistic the discourse "glass is a nature-friendly material" has been open to debate. Although petroleum-derived PET material is seen as the enemy of the environment because of its source material, in accordance with the findings of this study and due to the efficiency of the production methods and transportation of PET, this acceptance turns out to be not true.

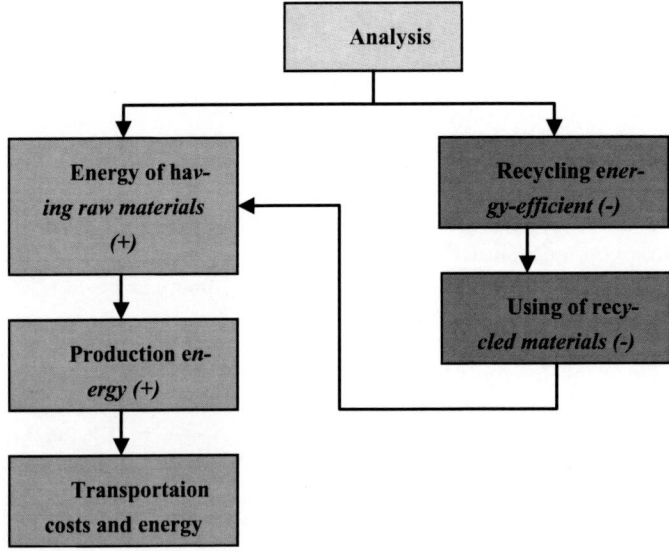

Fig. 9. The model of evaluation analysis used for the packages

Table 5. Evaluation of glass and PET packaging (glass bottle/PET bottle)

	Weight	Production energy	Transportation of piece	Transportation piece for 1 unit kg CO_2
Glass bottle/PET bottle	12.2 pieces	24 pieces	1.6 pieces	16.4 pieces

6.2 Recommendations

It is clear that sustainable production design and methods are important contributors to the solution of environmental problems such as waste reduction and natural resource conservation, reducing production costs and thus reducing product prices and operating profits. In order for sustainable production systems to be implemented in a healthy and effective way, it is necessary for the term sustainability to be in the management decisions of the company, and starting from the design stage, the products should be dealt with and developed in line with these criteria.

How much the concept of sustainability is present in design and production in Turkey is an issue of debate. Turkey's institutional capacity in terms of sustainability has been reported as limited compared to examples from the European Union [32, 33]. Küçüksayraç [34] further states that for the first stage of innovation support, sustainability-related services and tools need to be developed more quickly for sustainable design.

The European Union Compliance Criteria Framework, Kyoto Protocol guidelines and related laws and regulations point to the fact that, manufacturers have to use sustainable product design and manufacturing methods and inevitably have to be informed and implement necessary applications.

References

1. Bruntland G (1987) Our common future, the world commission of environment & development. Oxford University Press, Oxford
2. Bhamra T, Evans Zwan SF, Cook M (2001) Moving from eco-products to eco-services. J Des Res 1:2
3. Gertsakis J, Lewis H (2003) Sustainability and the waste management hierarchy. In: A discussion paper on the waste management hierarchy and its relationship to sustainability. EcoRecycle Victoria, Melbourne
4. Tischner U, Charter M (2001) Sustainable product design. In: Charter M, Tischner U (eds) Sustainable solutions: developing products and services for the future. Greenleef Publishing, Sheffield, pp 118–138
5. Bayazıt N (1994) Endüstri Ürünlerinde ve Mimarlıkta Tasarlama Metotlarına Giriş. Literatür Yayıncılık, İstanbul
6. ICSID (2014) International council of societies of industrial design. http://www.icsid.org/about/about/articles31.htm. Accessed Mar 2014
7. Otto BK (2005) About: sustainability. http://www.designcouncil.org. Accessed 19 Sept 2006
8. Papanek V (1984) Design for the real world. Van Nostrand Reinhold Co, London
9. Kim NK (2008) A model of component-based product-oriented environmental management system (C-POEMS) for small and medium-sized enterprises. Doctoral thesis, Brunel University School of Engineering and Design, London
10. EC (2003) Proposal for a directive of the European Parliament and of the council on establishing a framework for the setting of eco-design requirements for energy-using products and amending council directive 92/42/EEC. European Commission COM(2003) 453 final, Brussels
11. EU (2005) European Union directive 2005/32/EC of the European Parliament and of the council, establishing a framework for the setting of ecodesign requirements for energy-using products and amending council directive 92/42/EEC and Directive 96/57/EC and 2000/55/EC of the European Parliament and of the council. http://ec.europa.eu/environment/waste/directive2005:32:D09242EEC_index.htm. Accessed 6 July 2005
12. Goosey M (2004) End-of-life electronics legislation—an industry perspective. Circuit World 30(2):41–45
13. Vallet F, Eynard B, Millet D, Mahut SG, Tyl B, Bertoluci G (2013) Using eco-design tools: an overview of experts' practices. Des Stud 34(3):345–377
14. Ertem H (1999) Endüstri Ürünleri Tasarımı Açısından Ambalajın İncelenmesi ve Ambalaj Tasarım Yönetimi için bir Model Önerisi (Sanatta Yeterlik Tezi). Marmara Üniversitesi, Sosyal Bilimler Enstitüsü, İstanbul
15. Opie R (1991) Packaging source book, a visual guide to a century of packaging design. Brown and Company, Little
16. Southgate P (1994) Total branding by design: using design to create distinctive brand identities. Kogan Page, London
17. Pilditch J (1973) The silent salesman: how to develop packaging that sells. Brookfield Publishing Co, Philadelphia

18. Bayazıt N (2004) Investigating design: a review of forty years of design research. Des Issues 20(1):16–29
19. Lewis H, Sonneveld K (2004) Unwrapping the discourse: product stewardship and sustainability in the Australian packaging industry. In: 14th IAPRI world conference on packaging, Lidingö, Sweden, June 13–16
20. Lewis H, Fitzpatrick L, Verghese K, Sonneveld K, Jordon R (2007) Sustainable packaging redefined. Sustainable Packaging Alliance, Melbourne
21. SPC (2006) Sustainable packaging coalition report. Institutional Report, Melbourne, March 3
22. Bektaşoğlu S, Esen B (2007) Ambalaj Sanayi. İGEME, Ankara
23. ASSR (2007–2016) Ambalaj Sanayi Sektör Raporu. Ambalaj Sanayicileri Derneği, İstanbul
24. Bayazıt N (2005) Endüstriyel Tasarım Olarak Ambalaj. *Yapı Dergisi*, Sayı 284, Temmuz 2005, 101–105
25. Yin R (2003) Case study research: design and methods. Sage, Thousand Oaks
26. Chung KW (1992) The meaning of design management and its strategic value. In: Design management institute fourth international design management research and education forum, London Business School
27. Langrish J (1993) Case studies as a biological research process. Des Stud 4(14):357–364
28. Özgen C (2013) Sürdürülebilirlik Kavramının Firma Stratejisi Açısından Ambalaj Tasarımına Etkilerinin İrdelenmesi. İstanbul Teknik Üniversitesi, Fen Bilimleri Enstitüsü, İstanbul (Doktora Tezi)
29. Url-1 (2013) http://www.tschudilogistics.com/page/121. CO_2 Calculater. Accessed 12 Mar 2013
30. Porter M, Van der Linde C (1996) Green and competitive: ending the statement. In: Welford R, Starkey R (eds) The Earthscan reader in business and the environment. Earthscan, London
31. Humbert S, Rossi V, Margni M, Jolliet O, Loerincik Y (2009) Life cycle assessment of two baby food packaging alternatives: glass jars vs. plastic pots. Int J Life Cycle Assess 14 (2):95–106
32. TC ÇOB, TTGV (2010) TC Çevre ve Orman Bakanlığı & Türkiye Teknoloji Geliştirme Vakfı, Türkiye'de Temiz Üretim Uygulamalarının Yaygınlaştırılması için Çerçeve Koşulların ve Ar-Ge İhtiyacının Belirlenmesi Projesi. TC ÇOB ve TTGV, Türkiye, Ankara
33. Küçüksayraç E, Keskin D, Brezet H (2015) Intermediaries and innovation support in the design for sustainability field: cases from the Netherlands, Turkey and the United Kingdom. J Clean Prod 101(2015):38–48
34. Küçüksayraç E (2015) Design for sustainability in companies: strategies, drivers and needs of Turkey's best performing businesses. J Clean Prod 106:455–465

The Aim of Sustainable Urban Development and Climate Change Policies

Ayse Nur Albayrak[1(✉)] and Seda H. Bostanci[2]

[1] Gebze Technical University, Kocaeli, Turkey
nur4134@gmail.com
[2] Namik Kemal University, Tekirdag, Turkey
shbostanci@nku.edu.tr

Abstract. Although sustainability has been the dominant discourse in environmental and development policies for a long time, today the world is in an environmental crisis originated from climate change. Awareness of climate change rising on a global scale leads policymakers to play a more active role in order to find a solution to climate change. On the one hand, Kyoto Protocol and following Paris Agreement provide a course of action about how to bring the international consensus to national level; on the other hand, they form a frame in order the public to evaluate local policies and strategies easily. In this regard, the subjects of how the agreement reached at an international level are interpreted at a national level and how national policies and strategies are formed become the points to be considered. Policies of climate change are taken shaped in two main paths. These can be defined as eliminating the factors causing climate change, mitigation, and adapting to the negative effects of climate change. While forming policies, countries concentrate on these paths in accordance with their characteristics and priorities and announce their national strategies and action plans. In this study, first of all, the literature of climate change is reviewed, and then policies at local level are discussed and the examples discussed are examined comparatively. Thereafter, it is focused on the example of Turkey and recommendations of policies are developed. It is expected the results of the study to contribute in order to develop more efficient policies for sustainable urban development.

Keywords: Sustainable urban development · Climate change · Turkey

1   Introduction

Being able to form a sustainable development strategy has been the dominant course since 1970s. Countries within the changing global system and under the competitive pressure have focused on developing, improving their living conditions and urbanization. Environmental problems increasing with urbanization and decreasing natural resources have become precursors of fragility of earth. In the uncertain future which climate change drives the world is now an urgent matter which has to be considered. The definition "meeting today's requirements without making any concessions of the possibility of meeting next generations' requirements" [1] made for sustainability literally expresses the main conception which environmental policies have to include today.

© Springer International Publishing AG, part of Springer Nature 2018
S. Fırat et al. (eds.), *Proceedings of 3rd International Sustainable Buildings Symposium (ISBS 2017)*, Lecture Notes in Civil Engineering 7,
https://doi.org/10.1007/978-3-319-64349-6_17

Difficulties expected to be confronted put cities into the center of environmental problems. Increase in the rate of urbanization, the large population living in cities and the largeness of the disadvantageous and low-income groups piling up in mega-cities especially in developing countries requires solution seeking on a global scale. "Sustainable urban development" emphasized by Habitat III could both be the key to an urbanization which is less harmful to the environment and facilitates the results arising depending on climate change.

Material flows, engineering-stressed approaches focusing on the impacts of consumption and production on local and global environment as well as subjects like social rights and equality, organizational capacity, participation and financial sustainability are considered for a sustainable city [2]. In this context, vision, economy and society, biodiversity, ecological footprints, modeling cities on ecosystems, sense of place, empowerment, partnerships, sustainable production—consumption and governance are the principles of planning a sustainable city [3]. In this sense, not only environment but also social and economic subjects should be put on the agenda as well.

Within this framework, it is essential that many subjects at urban level should be discussed; the meaning and the content of the current green-eco approaches should be questioned from the point of view of sustainability. Many projects prepared with the eco label and launched to the market are criticized because their themes related to sustainable development have unrelated usages for marketing [4]. What to be reached through the strategies to be developed is a stage that "urban development is in a harmony with nature and consumes the minimum resource." In this regard, it is predicted that sustainable urban development should be formed by three pivots. These pivots can be summarized as spatial development, socio-economic structure and ecology-environment [5].

The stage where environmental problems are shows that the most important drawback to the aim of sustainable urban development is climate change. Many studies on planning focus on the interaction between urbanization and climate change, try to research thoroughly this interaction in terms of both developing of policy and implications. With this object, many subjects like "population dynamics, economic structure, cultural shift, and social dynamics" are discussed on different scales as "rural-to-mega cities" and urban policies are developed to adaptation and mitigation [6].

In this study, this accumulation on the field of planning is tracked, strategies and action plans developed directed to implementation on country and city-region scale are examined, and climate policies of Turkey are evaluated. For this purpose, in the second section following the prologue climate change is discussed in terms of policy development, these policies are exemplified internationally in the third section. Therefore, policies developed at local level are examined comparatively and an evaluation framework is created for the example of Turkey. In the fourth section, it is focused on the example of Turkey. Turkey is one of the countries that are open to ecologic and economic threats occurring depending on climate changes. Although Turkey has signed climate change agreements within sui generis conditions, the country is on the brink of significant environmental problems with its economy high-in development aims. In the conclusion part, it is expected that policy recommendations developed considering the example of Turkey make a contribution to developing sustainable urban development policies of other countries which share the similar features with Turkey and enrich the literature.

2 Climate Change

The starting point of climate change is global warming which is being caused by the increase of GHG emissions in the atmosphere. A long-term shift in the statistics is described as the definition of climate change. Depending on global warming, too much changing has been observed in the world and also tropical cyclones- storm surge, extreme rainfall, riverine floods, heat or cold waves, drought, and also changing temperature, precipitation and sea-level rise [7] have been expected to realize. In this picture, both biological diversity and agricultural product pattern might be affected negatively. These incidents not only affect on natural lands or rural regions. Because of climate change, it has been expected that there will be important problems in the cities. Not only traumas and instantaneous loss of lives related to natural disasters, socio-economic losses are also in question, especially, decreasing employment, damaged infrastructure and neighborhoods. Another risk of loss of agricultural production, increasing poverty and also migration from vulnerable regions to the cities in developed countries and consequently both ecological risks and socio-economic changes are being mentioned. In recent years, extreme climate events have been observed and thus, cities are face with increasing risks. Especially in developing countries, cities face to allocate their limited financial resources to avoid the negative impacts of climate change.

There is almost a consensus on anthropogenic reasons of climate change but it is difficult to describe a definite policy framework for the solution. Each geographic region has different priorities in the context with environmental policies and also economic development according to their own characteristics and vulnerabilities. Basically, two set of policy have been applied: Mitigation and adaptation. Mitigation is defined as "human interventions to reduce the sources or enhance the sinks of greenhouse gases" [8]. Today, many of mitigation strategies are related to energy policies and also urbanizations process. Thereby, in order to reduce GHG emissions, a holistic policy approach has been needed. Mitigation strategies are mostly related to national level. On the other hand "adaptation strategies protect local communities from sudden and immediate dangers. Shifting to wind, solar or wave energy is also a key adaptation strategy as smaller, more decentralized forms of power generation reduce the risks associated with widespread power loss through severe storm event, or from peak power loads under temperature extremes" [9]. "Adaptation policies" relatively focus on regional-local level. Especially many of policies and strategies at municipal management level aim for adaptation.

3 Local Cases

It is seen that Nordic countries take a leading part in adaptation to climate change and developing climate change strategies as well as in sustainability. Particularly, Finland becomes prominent among other countries with how thoroughly it deals with the subject and its participative process. Within the scope of Finnish Government Report on National Climate Strategy in 2001, adaptation strategies in coordination with ministries and then climate strategies were formed. While reducing greenhouse gas emissions was suggested as the most efficient way in order to decrease the impacts of climate change, climate change was discussed not only as a subject related to environment but also it was considered with its economic-social and cultural development aspects. An integration of adaptation strategies with industry-specific development as well as industry-specific approach was adopted as a prerequisite. Potential impacts of climate change include unexpected natural events and disasters. In this regard, anticipatory and preliminary midterm-long term policies are recommended [10]. Today in Finland, the legal framework on climate change includes law and regulation on many subjects. CC Act (2015) has defined that a greenhouse gas emissions reduction target of at least 80% by 2050, compared to 1990 [11]. Similar emission reduction targets have been declared by EU and also other countries. After 20/20/20 climate targets for EU [12], new climate targets was declared in 2014. According to new document, 2030 climate targets of EU include 40% cut in greenhouse gas emissions (from 1990 levels), 27% of EU energy from renewable energy resources and 27% improvement in energy efficiency [13].

Energy is the first subject to be focused on in order to reduce greenhouse gas emissions. Getting rid of carbon-based energy resources and heading to renewable energy resources, increasing energy productivity and supporting researches on energy are primary subjects to deal with [14–16]. Mitigation strategies are a subject that primarily developed countries discuss. On contrary, non-industrial countries don't have usages causing carbon emissions and moreover; they experience a dilemma of the necessity of becoming rich in order to put the climate change adaptation strategies into practice and industrialization in turn. "Carbon-trade" used in developed countries such as Canada, Japan as a part of mitigation strategies have been developed as a method which uses the limitedness in carbon emissions of developing countries for the benefit of developed countries. The contradiction here is that carbon emissions are reduced "being unfair to undeveloped countries."

Protecting nature and the strategy of using agriculture-forest resources in order to stock up carbon are tried to put into practice under pressure of urbanization. The problem of being unable to protect natural vegetation impacts food production and socio-economic structure depending on geography of country. One of the facts that make country borders unclear against climate change is the wave of migration experienced from rural-to-urban, from developing-to-developed countries as a result of climate change.

Rise of sea level has impacts on cultivated areas and forests at coastal zones, tourism and harbor activities as well as island countries. Countries which are in danger of rise of sea level develop solutions depending on their necessities. Flood problem onshore in Bangladesh is resolved by protecting natural vegetation. It is aimed to extend greenbelt onshore by protecting it and to take precautions against flood [17].

Cities reserve both activities causing climate change and population that climate change probably affects. That's why cities are key actors of fight against climate change. Within this period, it is seen that participative processes are planned [18] and also actions that give priority to awareness, education and research are supported [10]. Cities of developing countries particularly focus on adaptation strategies whereas mitigation strategies which are parallel to national goals are defined for cities of developed countries. In this sense, the most important subject is reducing C emissions.

Except for Canberra and Copenhagen which aim to zero carbon emissions, other cities aim to reduce it dramatically.

Transportation and building sectors are the primary effective sectors in reducing carbon emissions in cities. Especially dwellings are the most important emission resources in cities. Raising awareness in public increases the demand of "green/certificated" building in construction sector. A green building is "the practice of increasing the efficiency with which buildings and their sites use energy, water, and materials, and reducing building impacts on human health and the environment, through better siting, design, construction, operation, maintenance, and removal—the complete building life cycle" [19]. "The process of capital formation of many companies supporting green building production must be discussed in terms of how environment friendly they are. Today, one of the criticisms of the concept of green building is that it has become a trend and a marketing tool. Some enterprisers feature brand value of certification system. Since each certification system has application prerequisites and they have various ratings, it is seen that enterprises get points easily by ignoring some of certification criteria" [20].

Electric cars as well as encouraging public transportation, rail systems and bicycle-passenger based solutions are recommended in order to reduce emission in transportation [14, 21]. It is certain that these recommendations are closely related to city's macro-form. Transportation-infrastructure recommendations which will be effective in fight against climate change within urban sprawl-especially widely observed in the USA- are difficult to develop. Instead, European cities where pedestrianism and bicycle-riding are common are relatively advantageous. In this regard, compact city strategies [14] are suggested in order to fight against climate change. City form itself which limits development in nature and uses infrastructure in a more effective way becomes an urban sustainability indicator [22, 23]. Apart from that, approaches such as smart growth [16], low carbon community [21], low carbon economy [24] and sustainable land use [24] are in action plans for a compatible with climate and more sustainable urban development.

It is seen that subjects of adaptation to climate change and fight against natural disasters are prioritized in cities. In this regard, the main goal is to increase resilience of cities [17, 25]. Strategies developed for London grade the possibilities of flood/inundation and describe the social groups which will be affected from natural disaster. It is suggested that green areas should be expanded and biological diversity corridors should be formed in order to reduce the impact of heat island [18]. Even if developing countries mostly don't have economic activities causing climate change, but they have to face with disaster caused depending on changing climate conditions. Fragility in cities of these countries rises especially because their population consists of low-income families, they have lack of infrastructure and they don't have technological opportunities like early-warning systems. Moreover, disadvantageous and low-income groups have been affected more in cities of developed countries. Concordantly, local governments should describe weaknesses occurring depending on climate change and take significant precautions in order to find solutions [26]. In this sense, subjects like disaster management, protecting nature are common topics for cities-regions in developed and developing countries. On that note, action plans and strategy documents include how disadvantageous groups in city will be reached in the process of adaptation to climate change, and what precautions will be taken. The London Climate Change Adaptations Strategy explains that "adaptation should be seen as a journey, rather than a destination" [18].

4 Current Approaches and Climate Change Policies in Turkey

The concept of "sustainable development", which were brought to the agenda and easily accepted by 1980s' Turkish society in parallel with the global developments, have found a place for itself in a wide range of fields at both local and national levels. Despite unsatisfying level of practices of sustainability in the field of urban development, sustainable urban development has been considered as a common target both within the planning processes and policy documents. The acceptance of climate change, on the other hand, was not realized that easily. Although climate change was handled by different committees [27] in 1990s, it took a long time for this issue to be a part of policy making processes. Turkey was one of the countries that signed the climate change framework convention, however with a purpose of not falling behind the global agenda and its concerns relevant to future collaborations with OECD members and European countries. Hence, it is impossible to claim that either developing issues at international level were followed sufficiently or the emergent approaches were interiorized within this process. In accordance with this claim it is also stated that Turkey only focused on the issues relevant to its legal status within the framework of the convention and could not develop efficient policies during the period between the years 1990 and 2004 [28]. This period can also be mentioned as a period when Turkey intensified its attempt to comprehend the issue of climate change and become a part of the process.

According to official reports on climate change, Turkey is one of the risky countries in terms of the effects of climate change. As a component of climate change governance, urban climate governance in Turkey is of great importance [29]. When the sustainable urban development targets and climate change policies are considered together, Ministry of Environment and Urbanization comes to the forefront as Turkey's most effective governmental body. However, MEU (Ministry of Environment and Urbanization) has also been criticized as the Ministry became the only authorized body in urbanization policies and public works procedures of local administrations. On the other hand, efforts of the Ministry that systematize different public works, sustainability action plans and urban development process are regarded as the benefits of this body. The Ministry defines a system of values and fundamental principles in terms of sustainable urbanization with its publication titled as "KENTGES (Comprehensive Urban Development Strategies and Action Plan) 2010–2023". This publication states that climate change has already been discussed in various fields and it is aimed by the report to identify adjustment and reduction policies relevant to the climate change issue as well as procedure and principles for the planning and structuring processes carried out in local environments [30].

Climate Change Action Plan (2011–2023), which was prepared by the MEU, is a document that lays the foundation of climate change policies of Turkey. Various sectors are discussed within the scope of this document, such as energy, constructions, industry, transportation, waste, agriculture, land use and forestry, cross-sectoral issues and harmonization. The document gives priority to identify strategies, targets, specific periods and actions relevant to water resources management, agricultural sector and food security, ecosystem services, biodiversity and forestry, risk management in case of natural disasters, human health, common cross-sectoral issues and adjustment [31].

Within the period between 2012 and 2014, following the publication of CCAP (Climate Change Action Plan) document, investments in energy sector and the efforts for development came into rather more prominence. However, this situation resulted in a regression in terms of climate change and obscured the implementation process of the strategies identified by CCAP document. In the national report on Habitat III Conference which was held in Quito, Ecuador in October, 2016, Turkey decided to discuss climate change as an issue within the new agenda of urbanization [32]. The report considers both energy-efficiency and combating climate change and emphasizes ecological housing standards and the need for change in consumption patterns. The report also outlines the content of National Climate Change Strategy Document.

Since 2014, climate change has gained a new momentum at a national level and within this scope some local action plans and (green) energy strategies have been identified and described. And also it is observed that municipalities of cities and even districts have launched specific projects including reports, strategies and practices relevant to climate change. The projects are generally considered at an infancy phase in terms of reporting and strategies but the regular measurement of carbon emission levels can be considered as a pioneer implementation within this field. Municipalities of Gaziantep, Istanbul and Bursa cities and districts of Nilufer (Bursa), Cankaya (Ankara) and Kadikoy (Istanbul) can be given as examples for the local administrations that have specific climate change action plans.

Climate Change Action Plan of the Metropolitan Municipality of Gaziantep has been conducted by ICE, Mavi Consultants and University of Gaziantep and funded by AFD (French Development Agency) and the MM (Metropolitan Municipality) of Gaziantep. Energy issues and greenhouse gas emissions were analyzed within the scope of this action plan and specific targets were identified to reduce greenhouse gas emissions. Establishment of eco-environments, "energy information points" for the local residents and a "regional-local energy agency" for various institutions, improvements in public transportation by using the current infrastructure, transition to green technologies and reducing water consumption and loss by raising awareness and developing resources and network management are among the outstanding projects within this context [33].

Metropolitan Municipality of Istanbul has commenced preparations for the Istanbul Climate Change Action Plan and has completed the feasibility study in 2016 [34]. Information about this plan can be obtained through the official web site of the Municipality and the television channel of Istanbul's subway system that broadcast relevant developments. This practice can be regarded as a unique strategy to inform the public about recent developments and projects.

The priority of Bursa Metropolitan Municipality Climate Change Action Plan is the issue of greenhouse gases in Bursa city. This plan aims energy saving and reduction in emissions through certain fields such as urban development—built environment, transportation, renewable energy sources, management of solid waste and waste water, industrial actions and services, agriculture, animal husbandry, forestry, awareness raising campaigns and efficiency in natural energy sources etc. The Municipality has organized training programs and workshops to ensure a participatory approach within the implementation process [35]. When municipalities' climate change reports are analyzed, it is seen that they have common contents but differ in presentation, interpretation and local dynamics.

Several district municipalities, particularly those of metropolitan areas, have launched various projects on the climate change. When projects of Nilufer (Bursa), Cankaya (Ankara) and Kadikoy (Istanbul) Districts are analyzed in general: As a member of Healthy Cities Association, Municipality of Nilufer District (Bursa) has outstanding projects in environment and health issues as well as developing "eco-cities". Sustainable Energy Action Plan prepared by the Municipality of Nilufer (Bursa) presents a specific schedule on carbon emission and its levels based on the data from 2013 [36]. Municipality of Kadikoy District (Istanbul) has a "Transformation and Coordination" unit under the body of Environmental Protection and Control Department. This unit can be given as a significant development among the projects carried out by local administrations [37]. Municipality of Cankaya District (Ankara) have categorized its climate and energy projects under specific titles in the official website of the Municipality. Climate policy events and climate resilience through rain harvesting are two of these titles [38]. Considering the prospective key roles of district municipalities in developing participatory sustainable community models, climate change projects of these local administrations have a significant potency of effect.

5 Conclusions and Policy Recommendations

Efforts in combating climate change has yet to be at an optimum level, while a significant progress has occurred in terms of awareness. Climate change has been discussed in policies of various countries with different geographical, economic and social characteristics. These policies do not only deal with environmental challenges but also lay the groundwork for a social transformation. However, other social problems and needs pose as an obstruction to developing new policies needed in combating climate change and implementing these policies effectively. Developed countries give higher priority to mitigation policies, while developing countries pay more attention to adaptation. However, the current level of climate change requires all the states in the world to work together regardless of their levels of development.

Sustainable urban development is related with spatial development and socio-economic structure as well as the concept of ecology. All these policies should be handled in combating climate change in company with holistic strategies. In this context, the subjects drawing attention in the international examples are;

- Zero (or lower) C/GHG emissions, energy efficiency buildings, industries, vehicles etc.
- Innovative and technological solutions for human settlements,
- Sustainable urban development, compact urban form and sustainable infrastructure,

- Sustainable public transportation, electric vehicles, cycling, walking etc.
- Food security, supporting agricultural product pattern,
- Supporting resilience of the community, defining vulnerable social groups, developing socio-economic conditions,
- Sustainable (long-term) planning,
- Technological and organizational monitoring systems for disaster risks,
- Increasing green areas—protecting natural lands.

Turkey, which has an active economy agenda in parallel with its development targets, is considered as a developing country that tries to keep the balance of preservation and use of its natural sources. The main challenge of the country is to develop long-term environmental policies at a national level and integrate these policies with those in other fields.

It is seen that Turkey has already emphasized the importance of climate change and identified relevant strategies in its policy documents. The next step should be the discussions about realization of these policies. Most of the targets determined at a national level are postponed due to the economic problems and other issues related with the national development. It is also seen that climate change is yet to find a place in local environment and communities around the country. Communities that are not aware of climate change, neither pay attention to combating this problem. Moreover, most of the environmentalist projects and practices considered as a part of municipal visions are far from yielding significant results. Therefore, it is significant to evaluate effectiveness and appropriateness of these projects within this context.

The first step in this framework should be identifying common targets and conflicts that occur at the intersection of environmental issues and urbanization, industrialization and development fields. It is significant to prepare a guideline to on which the whole society would compromise, include key actors that could be effective within the process and develop a participatory approach. Awareness and knowledge of policy makers relevant to climate change should be another priority in addition to those of communities within this process. In order to support sustainable urban development, innovative and environment-oriented approaches are needed in policies in different fields in addition to current policies directly related with environmental issues. Within this context, various approaches ranging from eco-municipalism to green city and green accounting approaches, from industrial ecology to industrial symbiosis should be adapted in the implementation process.

Different approaches of the cases which analyzed within the study give us clues about the effectiveness of social characteristics within the implementation of climate change policies. Combating climate change is only possible through a participatory and sustainable social structure. The achievements in this context are highly connected firstly with the efforts to objectively analyze behavioral, sociologic and psychological structures and lifestyles of societies and secondly with practices that could change conventional attitudes and customs. The reasons that make certain countries more inclined to have sustainable social structures and the methods that drive the progress within this context should also other issues to be analyzed. This kind of a behavioral analysis in Turkey would be useful in future studies.

References

1. WCED (1991) Our common future. TCS Foundation (in Turkish)
2. Hoornweg D, Freire M (2013) Building sustainability in an urbanizing world, a partnership report. Urban Development Series Knowledge Papers No. 17, World Bank—Urban Development and Resilience Unit. http://www.globalurban.org/WorldBankSustainableCitiesReport.pdf
3. Newman P, Jennings I (2008) Cities as sustainable ecosystems: principles and practices. Island Press, USA
4. UN (2012) Back to our common future. Sustainable development in the 21st century (SD21) project. https://sustainabledevelopment.un.org
5. Albayrak AN, Senlier N (2015) Industrial ecology for sustainable cities. In: Environment and ecology at the beginning of 21st century. St. Kliment Ohridski University Press, Sofia, pp 190–202. ISBN:978-954-07-3999-1
6. Romero-Lankao P et al (2014) A critical knowledge pathway to low-carbon, sustainable futures: Integrated understanding of urbanization, urban areas, and carbon. Earth's Future 2:515–532. doi:10.1002/2014EF000258
7. IPCC (2007) Fourth assessment report: climate change 2007 (AR4). UNEP-WMO. http://www.ipcc.ch
8. IPCC (2014) Fifth assessment report: climate change 2014 (AR5). UNEP-WMO. http://www.ipcc.ch
9. Hamin EM, Gurran N (2009) Urban form and climate change: balancing adaptation and mitigation in the US and Australia. Habitat Int 33(3):238–245
10. Marttila V et al (2005) Finland's national strategy for adaptation to climate change. Ministry of Agriculture and Forestry of Finland, Finland
11. http://www.ym.fi/en-us/the_environment
12. http://ec.europa.eu/clima/policies/strategies/2020/index_en.htm
13. https://ec.europa.eu/energy/en/topics/energy-strategy/2030energystrategy
14. ACT (2012) AP2, a new climate change strategy and action plan for the Australian Capital Territory. ACT Government-Environment and Sustainable Development Directorate, Canberra
15. Nova Scotia Environment (2009) Toward a greener future, Nova Scotia's Climate Change Action Plan. ISBN: 978-1-55457-261-8
16. New Brunswick (2014) New Brunswick Climate Change Action Plan 2014–2020
17. MoEF (2009) Bangladesh Climate Change Strategy and Action Plan 2009. Ministry of Environment and Forests, Government of the People's Republic of Bangladesh, Dhaka, Bangladesh. ISBN:984-8574-25-5
18. GLA (2011) Managing risks and increasing resilience, the Mayor's Climate Change Adaptation Strategy, London. ISBN:978-1-84781-469-2
19. https://www.sustainability.gov/Resources/Guidance_reports/Guidance_reports_archives/fgb_report.pdf
20. Celik E (2009) Yesil bina sertifika sistemlerinin incelenmesi Turkiye'de uygulanabilirliklerinin degerlendirilmesi. Istanbul Technical University, Science Institute, Master Thesis, Istanbul (in Turkish)
21. Ontario (2016) Ontario's Climate Change Action Plan 2016–2020
22. Banister D (2008) The sustainable mobility paradigm. Transp Policy 15:73–80
23. Jabaren YR (2006) Sustainable urban forms: their typologies, models and concepts. J Plan Educ Res 26:38–52
24. Quebec (2012) Quebec in action greener by 2020, 2013–2020 Climate Change Action Plan. ISBN:978-2-550-65192-5

25. GoK (Government of Kenya) (2013) National Climate Change Action Plan 2013–2017
26. Shaw K, Theobald K (2011) Resilient local government and climate change interventions in the UK. Local Environ Int J Just Sustain 16(1):1–15
27. Turkes M, Sumer UM, Kilic G (1992) Atmosferin korunmasi ve iklim degisikligi. UIKG/AKID Report, Devlet Meteoroloji Isleri Genel Mudurlugu, Ankara (in Turkish)
28. Sahin U (2014) Turkiye'nin iklim politikalarinda actor haritasi. IPM—Istanbul Politikalar Merkezi-Sabanci Universitesi—Stietung Mercator Girisimi (in Turkish)
29. Demirci M (2015) Kentsel iklim degisikligi yonetisimi. Erciyes Universitesi Iktisadi ve Idari Bilimler Fakultesi Dergisi 46:75–100 (in Turkish)
30. MEU (2010) KENTGES EylemPlani (2010–2023) (in Turkish). http://www.kentges.gov.tr/dosyalar/kentges_tr.pdf
31. MEU (2012) Turkiye Cumhuriyeti Iklim Degisikligi Eylem Plani (2011–2023) (in Turkish). http://iklim.cob.gov.tr/iklim/Files/IDEP/%C4%B0DEP_TR.pdf
32. MEU (2014) Turkiye Habitat III Ulusal Raporu (in Turkish). http://www.csb.gov.tr/db/mpgm/editordosya/file/HABITAT/HABITAT%20III%20ULUSAL%20RAPOR_TURKCE.pdf
33. Gaziantep Greater Municipality (2011) Gaziantep Iklim Degisikligi EylemPlani: Enerji ve Sera Gazi Emisyon Profili On Eylem Plani ve UygulamaStratejisi Final Raporu (in Turkish). http://www.afd.fr/webdav/shared/PORTAILS/PAYS/TURQUIE/PAGE%20D'ACCUEIL/Gaziantep-CCAP-TR-final-20111102.pdf
34. Istanbul Greater Municipality (2016) Istanbul Iklim Degisikligi Eylem Plani Hazirlanmasi Isi (in Turkish). http://www.ibb.istanbul/tr-TR/Pages/IhaleIlani.aspx?IhaleID=15105
35. Bursa Greater Municipality (2015) Bursa Buyuksehir Belediyesi Iklim Degisikligi Eylem Plani (in Turkish). http://www.bursa.bel.tr/dosyalar/BBB_IDEP_Kas%C4%B1m2015.pdf
36. Nilufer Municipality (2016) Bursa Nilufer Belediyesi Surdurulebilir Enerji Eylem Plani (in Turkish). http://www.nilufer.bel.tr/dosya_yoneticisi/enerjieylem.pdf
37. Kadikoy Municipality (2016) Cevre Koruma ve Kontrol Mudurlugu (in Turkish). http://www.kadikoy.bel.tr/Kurumsal/Mudurlukler/cevre-koruma-ve-kontrol-mudurlugu
38. Cankaya Municipality (2015) Iklim ve Enerji Calismalari (in Turkish). http://www.cankaya.bel.tr/pages/3392/Iklim-ve-Enerji-Calismalari/

Acoustics Treatment for University Halls

Ammar Abdulkareem$^{(\boxtimes)}$, Mahmood Abu Ali, and Emad Mushtaha

Department of Architectural Engineering, College of Engineering, University of
Sharjah, Sharjah, UAE
U00033952@sharjah.ac.ae

Abstract. This paper aims to solve one important problem while designing
halls with domes in universities such as a central lobby of Architectural Engineering Department at the University of Sharjah. Most architects consider
aspects/factors of esthetics, form, function, and to some extent sustainability
besides illumination, but in general they drop issues of acoustics from their
designing phases. The absence of sound quality and clarity of speech in the
central hall of AE Department has led to non-comfortable sound environment
for conducting some activities such as juries and exhibitions. Listening to a
person giving his/her presentation or talk could be difficult due to the large
volume of the space, finishing materials inside the building, and the effect of the
dome that causes an excess amount of reverberation time (RT). This paper
discusses various factors that affect sound environment such as building shape,
volume, and materials. Using REVIT and ECOTECT model in the simulation,
the study could suggest alternatives to the existing building scenario. Furthermore, an ideal combination of various techniques has been proposed to achieve
economical and acoustical aspects.

Keywords: University halls · Acoustics · Reverberation time · ECOTECT

1 Introduction

A critical factor that plays a vital role in sound quality is the reverberation time.
Reverberation time is the persistence of sound after sound source has ceased [1].
Reverberation time is the single most important parameter for judging the acoustical
properties of a room and its suitability for various uses [2]. The Architectural Engineering building in the University of Sharjah is built square with a semi-spherical dome
placed on top of a central atrium, in which many activities are conducted. Finishing
materials used include marble for flooring, plaster for walls, and PVC panels for false
ceiling. This type of design is commonly applied in educational institutes in Sharjah.
The low absorption characteristics of marble and plaster foster the effect of creating a
highly reverberant space that is unsuitable for activities such as juries, or exhibitions,
food events, etc. Thus, lowering the RT value would significantly enhance sound
environment, which is our aim in this regard.

© Springer International Publishing AG, part of Springer Nature 2018
S. Fırat et al. (eds.), *Proceedings of 3rd International Sustainable
Buildings Symposium (ISBS 2017)*, Lecture Notes in Civil Engineering 7,
https://doi.org/10.1007/978-3-319-64349-6_18

2   Methodology

In order to explore evaluate the sonic environment in AE Dept., the study started by investigating sound environment in the building using sound meter for two different days. Decibels were recorded for the targeted area as shown in Figs. 1 and 2. Revit and Ecotect were used for the simulation.

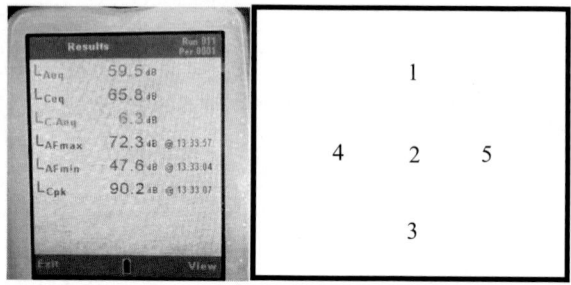

Fig. 1 Sound meter measuring in five different locations in the central lobby

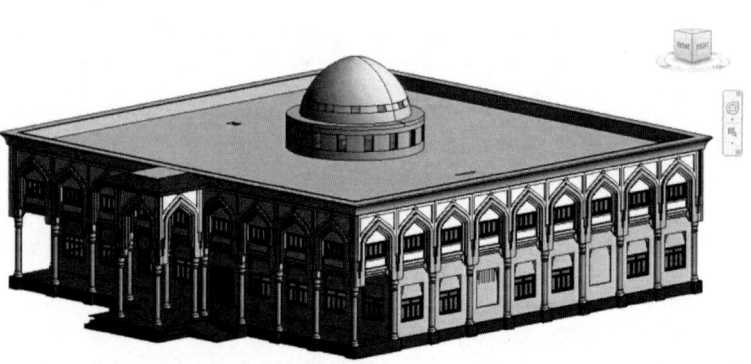

Fig. 2 Revit model for M8 building

2.1   Measuring Sound Levels in the Architectural Eng. Dep. Central Lobby

Sound meter was used to measure sound levels with HVAC System and mechanical equipment in the building for five different locations inside the central lobby.

The measurement was taken in weekends on Saturday, 28 Nov. 2015, at, 7:00 AM. The readings were as follows:

Point 1: 50.2 dB
Point 2: 46.5 dB
Point 3: 44.6 dB

Point 4: 45.6 dB
Point 5: 46.8 dB

Another experiment was taken during weekdays for ground
And first floors when it is occupied. The measurements and readings showed noisy
spaces that cannot be suitable for academic environments. Results in Fig. 2 demonstrate
sound levels received during working days/h (13:00–16:00). As shown in Table 1.

Table 1. Sound levels

Floor	Time			
	10:30 am (dB)	12:30 pm (dB)	2:30 pm (peak hour) (dB)	4:30 pm (dB)
Ground	60.8	57.7	68.3	59.5
First	66.3	67.6	74.6	50.2

2.2   Factors Affecting the Acoustical Behavior in University Halls

According to Sabine's equation in calculating the reverberation time, there are two
important factors that alter the acoustical behavior in any building [3]. The main factor is
the volume, so reducing the volume of the department will efficiently lower the reverberation time and consequently decrease sound levels in that area. As to be clarified in the
upcoming discussions, a method was sought to reduce the volume without changing the
shape of the building or dome. The second factor is related to finishing materials applied
on building surfaces. Adding materials that have higher absorption factor will also be
effective to reduce both the reverberation time and sound level. Initially, the existing
reverberation time is calculated to assess the current state of sound perception inside.

2.3   Reverberation Time Calculations

An equation for calculating the "Optimum" Reverberation Time, according to Stephens
and Bate in 1950, is applied as follows:

$$T60 = K\left[0.0118 \times \frac{V}{3} + 0.1070\right]. \tag{1}$$

With a volume of 3055.7 m^3 and K = 4 for speech, then:

$$T60 = 4\left[0.0118 \times \frac{3055.7}{3} + 0.1070\right] = 1.11\,s$$

With Sabine formula,

$$T60 = 0.161\left(\frac{V}{A}\right) \tag{2}$$

The existing RT value could be obtained. Taken the condition that there are no seats or people within the space:

$$T60 = 0.161 \left(\frac{3055.7}{49.03} \right) = 10.03 \, s$$

A reasonable explanation for such a large RT value is the extremely low absorption coefficients of existing finishing materials. Because of this, the total absorption area was found to be 49.03 m² Sabines. This small magnitude has severely reduced the quality of the acoustical behavior of the interior surfaces.

2.4   Computer Model Simulation

Given the nature and complexity of the interior geometry of the university hall examined, implementing a feasible computer model theory is important to obtain good quality data. In this work, a REVIT model was exported to ECOTECT to obtain the exact surface areas and volume taking into consideration the pervious calculations. After Simulating the existing case, different kinds of actions were applied to the model.

3   Existing Reverberation Time Simulation

After running the simulation, it is found that the reverberation time (RT) is 10.03 s at 500 Hz (Fig. 3). The RT value has exceeded the optimum value. Thus, the university halls are generally poorly suited for delivering speeches because of the excess in reverberation time. As shown below in Fig. 4.2, Frequency and RT correlation, the estimated decay curves at each frequency have revealed that additional absorption was needed at mid-frequency ranges between 250 Hz and 1 kHz.

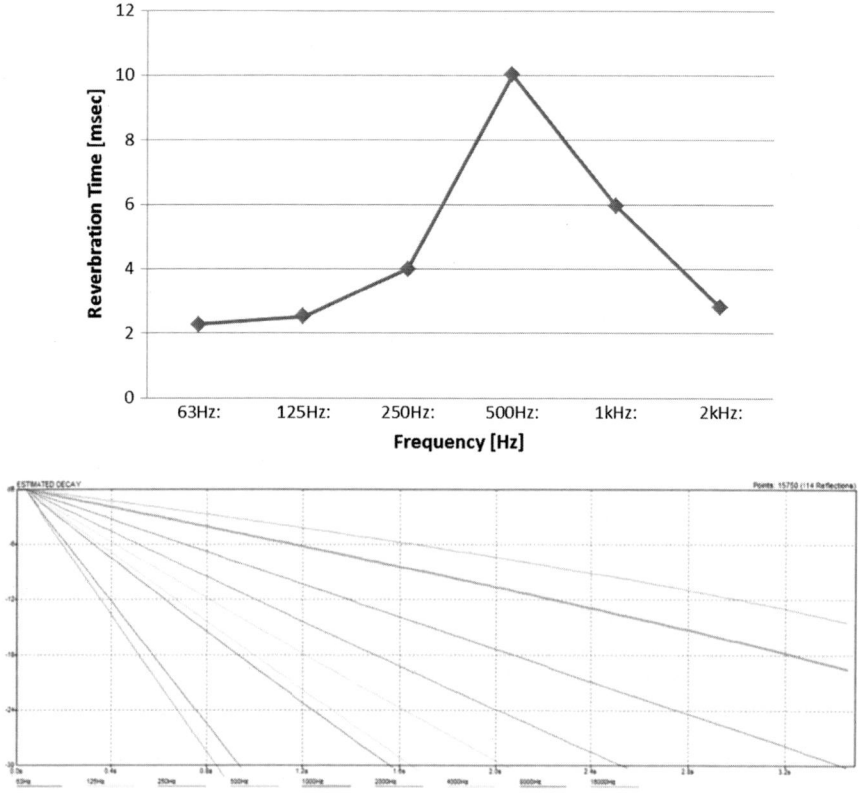

Fig. 3 (Hz–ms) graph and estimated decay

4   Acoustical Treatments

As outlined previously, the central hall requires more absorption surfaces to include in the building in order to dramatically enhance the acoustical environment. The dome form is one of the most inconvenient forms in acoustics design because of the concave curvatures. The propagating sound energy does not escape without reflecting several times inside the dome shell. Given this effect, the reflected sound energy from the dome returns to the main space with a time delay, resulting in echoes or noise in the hall and the reduction in the percentage of intelligibility [4]. Therefore, alternatives that could be considered to reduce the RT and increase the clarity of speech are as follows.

4.1   Reducing the Volume by Adding Mesh Between the Ground Floor and the First Floor

Mesh fabric was inserted in the model between ground and first floor to improve the acoustical environment. This mesh has also taken into consideration the penetration of natural light into space. Integrating the mesh fabric could help in eliminating the reflectance of sound rays as RT was reduced to 5.45 ms at 500 Hz as shown in Fig. 4.

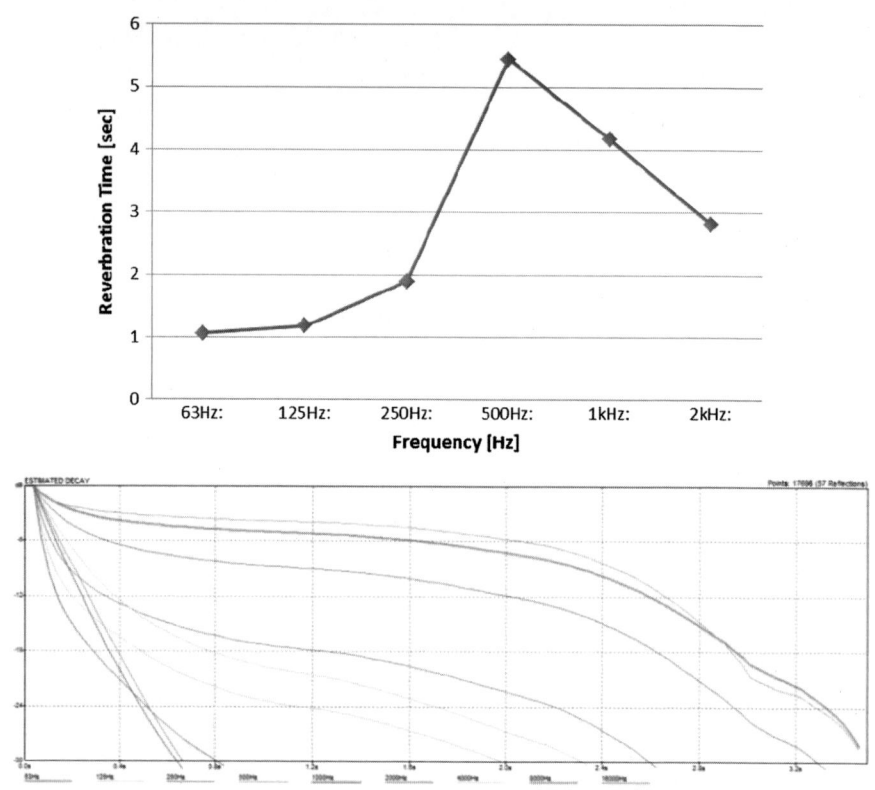

Fig. 4 (Hz–ms) and estimated decay graphs after applying mesh fabric

4.2   Adding Absorption Panels

As indicated earlier, more absorption was required for the mid-frequency range shown in Fig. 5. Thus, utilizing an appropriate type of acoustical panels is critical. Installations of porous absorption panels have reduced RT to 5.5 ms at 500 Hz. This result is almost similar to past trial, which is a good alternative.

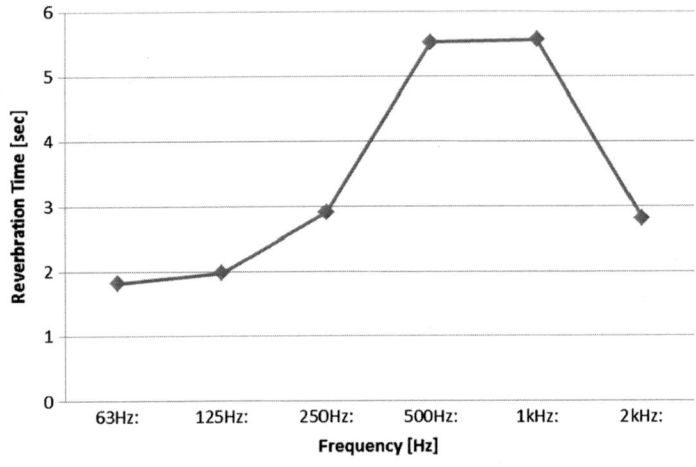

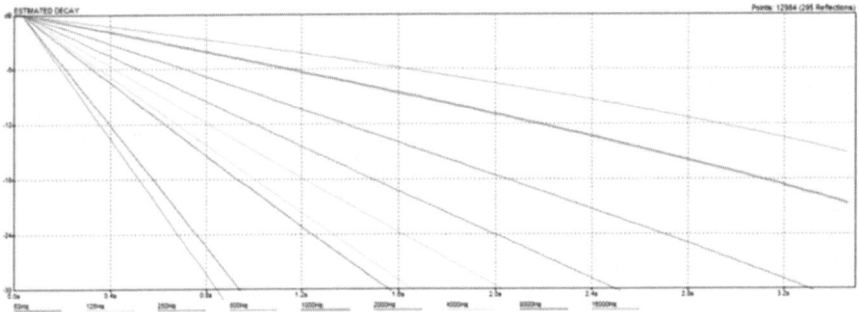

Fig. 5 (Hz–ms) and estimated decay graphs after applying absorption panels

4.3 Adding Carpet

In general, carpets have a high sound absorption that reaches up to 50% when used over large areas. Carpets were laid on the ground and first floors during the simulation and it was noticed that the RT was reduced to 3.79 ms at 500 Hz (Fig. 6). The proposal has performed better than earlier options of absorption panels and reduction of volumes.

4.4 Adding Wood Cladding

Wood is considered a desired material in acoustical treatments for its warm and natural appearance. Wood has an excellent sound absorption compared to previous alternatives. The authors have integrated this wood classing on internal walls of the both ground and first floors within the central hall area. From the simulation, the RT was decreased to 2.76 ms at 500 Hz (Fig. 7).

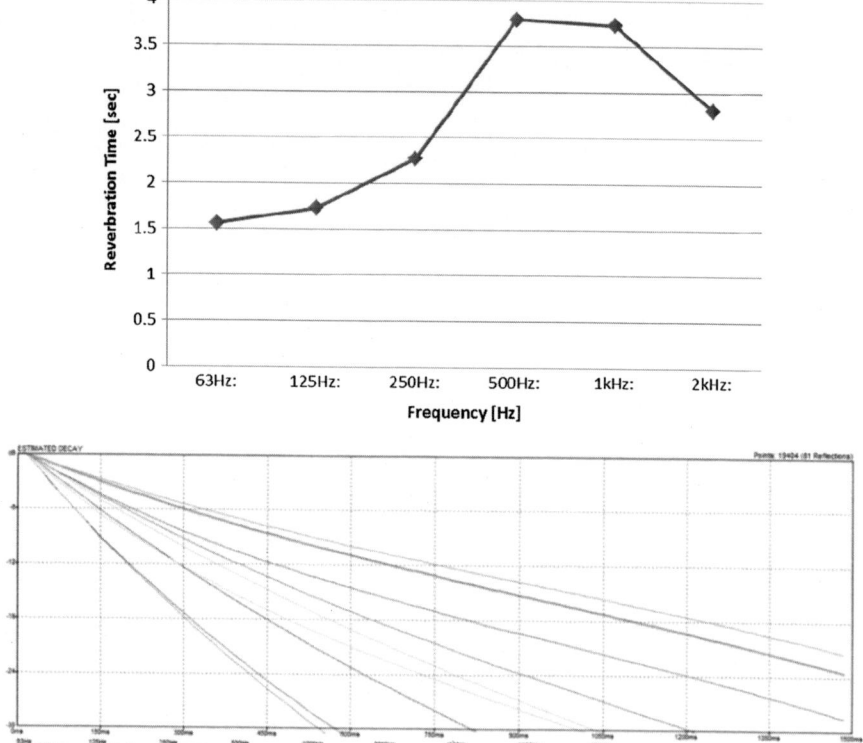

Fig. 6 (Hz–ms) and estimated decay graphs after adding carpet

4.5   Adding All Methods at Once

Although different alternatives were applied individually to enhance the acoustical behavior and reduce the RT levels, combining them together will probably magnify their effect and therefore achieve better satisfactory results. This combination has reached to an acceptable RT reduction value of approximate 0.7 from 10 ms, which is about 90% improvement (Fig. 8).

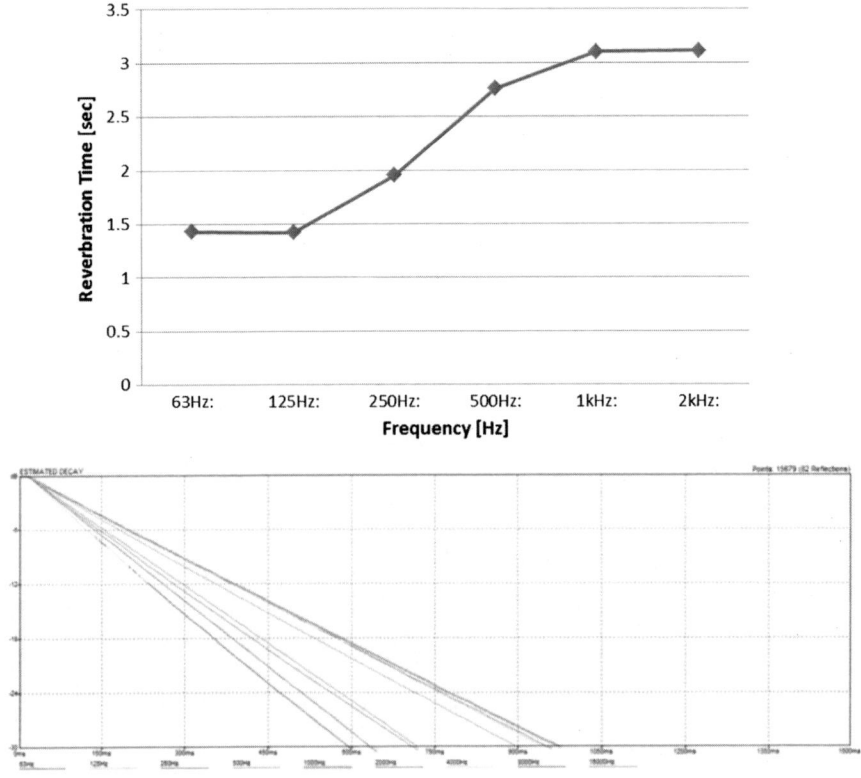

Fig. 7 (Hz–ms) and estimated decay graphs after adding carpet

5 Cost of Applied Materials

Earlier simulation showed that adding all methods at once was certainly the most efficient technique to have a maximum reduction of the reverberation time. However, adding all materials would cost to some extent in terms of money. Thus, it is very important to roughly estimate the cost of each proposal. Therefore, a strategy that satisfies the objective in terms of cost and absorption would be as shown below:

1. Mesh Fabric cost $=$ Area $\times$ Cost (per square meter) $= 141.7\,\text{m}^2 \times \$2 = \$243.4$
2. Absorption panels cost $=$ Area $\times$ Cost (per square meter) $= 88.9\,\text{m}^2 \times \$80.8 = \$7183.1$
3. Carpet cost $=$ Area $\times$ Cost (per square meter) $= 405.7\,\text{m}^2 \times \$5.3 = \$2139$
4. Wood cladding cost $=$ Area $\times$ Cost (per square meter) $= 456\text{m}^2 \times \$32 = \$14,592$
5. All methods cost $= \$24,157.5$

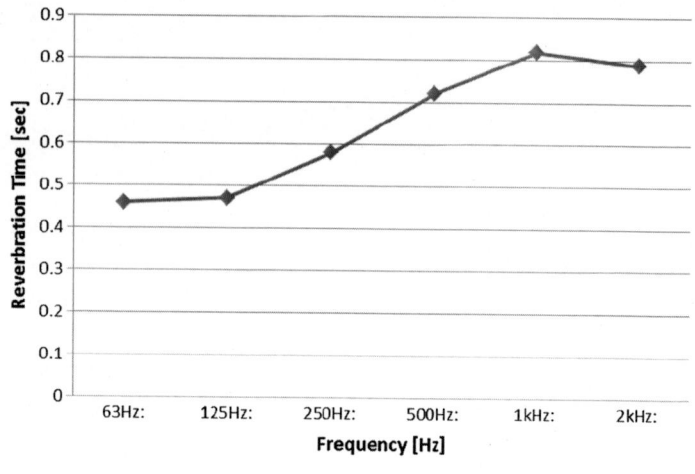

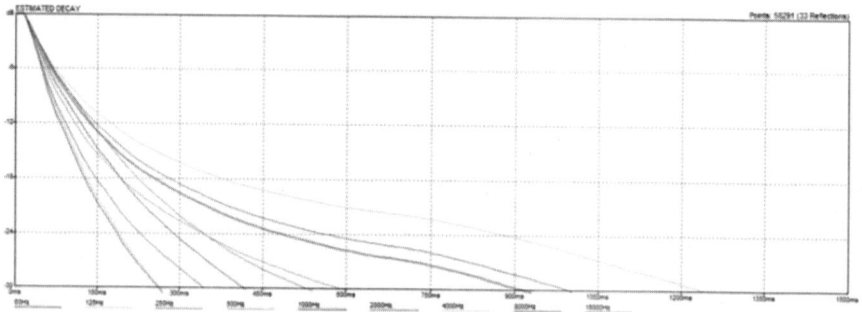

Fig. 8 (Hz–ms) and estimated decay graphs after applying all methods

6 Conclusion

In this work, the acoustic environment of common Sharjah university halls was evaluated. These low absorption coefficients of materials in halls resulted in a reduction of the total absorption area of the space. The computer simulation for the hall revealed that the materials commonly used for the interior needed acoustic treatment. This treatment will establish more absorption to the space and reduce any unfavorable acoustic occurrence that might happen from having surfaces that reflect large amounts of acoustic energy within the space. The decay curves showed that further absorption was needed at mid frequency range. The installation of porous panels and the addition of absorption by changing the different surface materials characteristics improved acoustic quality, indicating that these techniques are effective. A decrease in RT was observed. Another outcome is the increase in clarity and definition of speech. Using some materials and changing some parameters caused to reduce the reverberation time from 10.03 s to the comfortable range. Two aspects were taken under consideration to achieve the most efficient solution that can be applied to the department. Taking acoustic absorption into account by applying all the methods will ultimately produce the maximum sound quality. Moreover, adding mesh fabric between the two floors to cover the void was found to be the most convenient option in terms of cost. However, Using mes h and carpet, or absorption panels and carpet together, could satisfy both of these aspects in the most effective and economic approach.

References

1. McMullan R (2007) Environmental science in building. Palgrave Macmillan, Basingstoke
2. Lamancusa SJ (2000) Engineering noise control. Pennsylvania
3. Szokolay S (2003) Introduction to architectural science: the basis of sustainable, design. Elsevier, Britain
4. Ismail M (2012) Parametric investigation of the acoustical performance of contemporary mosques. Elsevier, Cairo

Kuznets's Inverted U Hypothesis: The Relationship Between Economic Development and Ecological Footprint

Ceyda Erden Özsoy[(⊠)]

Department of Economics, Faculty of Economics and Administrative Science,
Anadolu University, Eskişehir, Turkey
ceydae@anadolu.edu.tr

Abstract. Sustainable development is defined as "development that meets the needs of the present without compromising the ability of future generations to meet their own needs". The concept of sustainable development contains both environmental sustainability and economic development. One simple way to assess sustainable development is to use the Ecological Footprint and Human Development Index (HDI). HDI measures a country's average achievements in the areas of health, knowledge, and standard of living. The Ecological Footprint measures a country's demand on nature and can be compared to available bio capacity. The HDI-Footprint, using simple indicators, prominently reveals how far removed the world is from achieving sustainable development. For all countries, the goal should be high human development and a low ecological footprint per capita. Environmental Kuznets Curve is located in the sustainable economic development literature puts forward that the inverse U shape relationship between the level of economic development and environmental degradation. In this study, the ecological footprints of countries are compared with the level of human development and the validity of Kuznets inverted U hypothesis being investigated. Measuring these two variables reveals that very few countries come close to achieving sustainable development.

Keywords: Economic development · Human development index
Ecological footprint · Biological capacity · Environmental kuznets curve

1 Introduction

Sustainable development is a commitment improving the quality of human life while living within the carrying the capacity of supporting ecosystems. "Development" is shorthand for committing to well-being for all. "Sustainable" implies that such development must come at no cost to future generations. In other words, development is required to occur within what the planet's ecosystems are able to provide season after season, year after year. It needs to be enabled within the means of nature [1]. For all countries, the goal should be high human development and a low ecological footprint per capita. However only a few countries come close to creating such a globally reproducible high level of human development without exerting unsustainable pressure on the planet's ecological resources [2]. Unfortunately, human beings cut trees faster

© Springer International Publishing AG, part of Springer Nature 2018
S. Fırat et al. (eds.), *Proceedings of 3rd International Sustainable Buildings Symposium (ISBS 2017)*, Lecture Notes in Civil Engineering 7,
https://doi.org/10.1007/978-3-319-64349-6_19

than they mature, harvest more fish than the oceans can replenish, or emit more carbon into the atmosphere than the forests and oceans can absorb. The sum of all human demands no longer fits within what nature can renew. The consequences are diminished resource stocks and waste accumulating faster than it can be absorbed or recycled, such as with the growing carbon concentration in the atmosphere [3].

One simple way to assess sustainable development is using the Ecological Footprint and Human Development Index (HDI). The United Nations HDI is an indicator of human development that measures a country's achievements in the areas of life expectancy, education, and income. The Ecological Footprint measures a people's demand on nature and can be compared to available bio capacity. The HDI-Footprint, using simple indicators, prominently reveals how far removed the world is from achieving sustainable development [2].

An Ecological Footprint less than 1.7 global hectares per person makes those resource demands globally replicable. The United Nations considers an HDI over 0.8 to be "very high human development". These two thresholds define two minimum criteria for global sustainable development—an average Footprint (significantly) lower than 1.7 gha per person and an HDI of at least 0.8. Measuring these two variables reveals that very few countries (Cuba, Sri Lanka etc.) come close to achieving sustainable development [1].

The general proposition that economic growth is good for the environment has been justified by the claim that there exists an empirical relation between per capita income and some measures of environmental quality. It has been observed that as income goes up there is increasing environmental degradation up to a point, after which environmental quality improves. (The relationship has an "Inverted-U" shape.) [4]. In this paper we investigate the relationship between economic development (by using HDI) and environmental degradation (by using ecological footprint). According to Environmental Kuznets Hypothesis (EKH), as economies developed, reduces ecological footprint. But unfortunately, data says developed countries have high ecological footprint scores.

2 Kuznets's Inverted U Hypothesis

The Environmental Kuznets Curve (EKC) hypothesis postulates an inverted U shaped relationship between different pollutants and per capita income. Environmental pressure increases up to a certain level as income goes up; after that, it decreases. Environmental quality deteriorates at the early stages of economic development and subsequently improves at the later stages. In other words, environmental pressure increases faster than income at early stages of development and slows down relative to Gross Domestic Product (GDP) growth at higher income levels. This systematic relationship between income change and environmental quality has been called the Environmental Kuznets Curve [5]. The EKC is named for Simon Kuznets (1955) who hypothesized that income inequality first rises and then falls as economic development proceeds [6].

In the first stage of industrialization, pollution grows rapidly because high priority is given to increase material output, and people are more interested in jobs and income than clean air and water. People are too poor to pay for abatement, and/or disregard environmental consequences of growth [7]. The rapid growth inevitably results in greater use of natural resources and emission of pollutants, which in turn put more pressure on environment. Environmental degradation increases with growing income up to a threshold level beyond which environmental quality improves with higher income per capita. This relationship can be shown by an inverted U shaped curve. Figure 1 demonstrates EKC [8]. As can be seen on figure, the EKC proposes that indicators of environmental degradation first rise, and then fall with increasing income per capita.

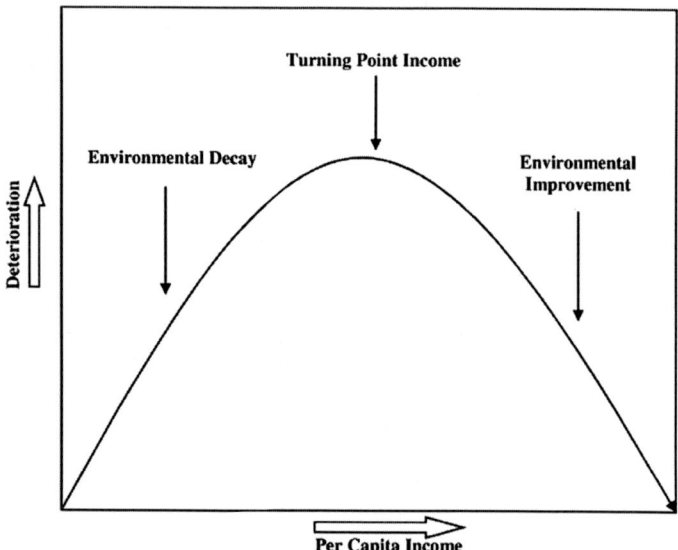

Fig. 1 Environmental Kuznets curve

As economic development accelerates with the intensification of agriculture and other resource extraction, at the take-off stage, the rate of resource depletion begins to exceed the rate of resource regeneration, and waste generation increases in quantity and toxicity. At higher levels of development, structural change towards information-intensive industries and services, coupled with increased environmental awareness, enforcement of environmental regulations, better technology and higher environmental expenditures, results in levelling off and gradual decline of environmental degradation. As income moves beyond the EKC turning point, it is assumed that transition to improving environmental quality starts. Thus, it could be a depiction of the natural process of economic development from a clean agrarian economy to a polluting industrial economy, and, finally, to a clean service economy [5].

3 Methods

Sustainable development is defined as "development that meets the needs of the present without compromising the ability of future generations to meet their own needs". The concept of "sustainable development" contains both environmental sustainability and economic development. One simple way to assess sustainable development is to use the Ecological Footprint and Human Development Index (HDI).

We examine sustainable development in terms of its two dimensions. We assess progress in development with the HDI because it is one of the most widely used overall measures of human well-being. The other dimension of sustainable development is the commitment to develop within the ecological capacity of planet Earth. This can be measured with the Ecological Footprint, a resource accounting tool that assesses how much of the regenerative capacity of the biosphere is occupied by human activities.

3.1 Human Development Index

HDI measures a country's average achievements in the areas of health, knowledge, and standard of living. Since 1990, the United Nations Development Program (UNDP) has used the Human Development Index in its annual Human Development Report. The purpose of the report is to show how well the management of economic growth and human development is actually improving human well-being in the nations of the world. The report defines human development as the "process of enlarging people's choices to live a long and healthy life, to be educated, have access to resources needed for a decent standard of living. So the index focuses on "health, education and the standard of living" as proxies for people's ability to live long and prosperous lives [9].

The health dimension is measured using life expectancy at birth. This also serves as a proxy for other aspects of well-being such as adequate nutrition and good health.

The education dimension is measured by mean of years of schooling for adults aged 25 years and more and expected years of schooling for children of school entering age. These two separate indicators are intended to reflect the level of knowledge of the adult population as well as the investment in the youth.

A the standard of living dimension is measured by gross national income per capita adjusted to reflect purchasing power parity. The HDI uses the logarithm of income, to reflect the diminishing importance of income with increasing Gross National Income [10]. The HDI is the geometric mean of normalized indices for each of the three dimensions. Figure 2 demonstrates dimensions and indicators of HDI [11].

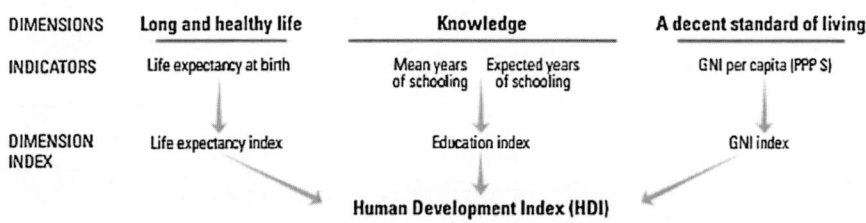

Fig. 2 Human development index

An HDI value of 1.0 implies that a country has achieved the maximum value for each sub-index, and a value of zero implies that the country is at or below the minimum value for all sub-indices. UNDP defines an HDI score of 0.8 as the limit between high and very high human development.

Figure 3 shows human development map prepared for 2015 [12]. According to the map the North America (USA and Canada), Argentina and Chile in the Latin America,

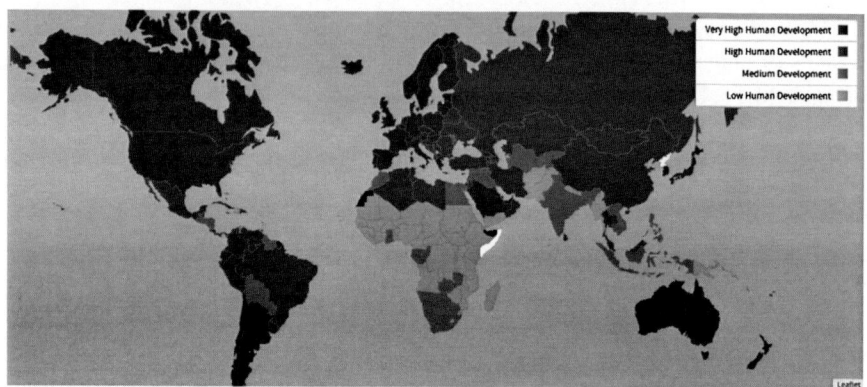

Fig. 3 Human development map (2015)

all of the East European countries and Australia and Nez Zealand have high human development. On the other hand, most African countries have low human development.

3.2 Ecological Footprint

Ecological footprint is a natural resource accounting tool that measures the ecological sustainability. The simplest way to define ecological footprint would be to call it the impact of human activities measured in terms of the area of biologically productive land and water required to produce the goods consumed and to assimilate the wastes generated.

The Ecological Footprint (EF) was developed by Mathis Wackernagel and William Rees as a way to account for flows of energy and matter into and out of the human economy and convert those flows into a measure of the area of productive land and water required to support those flows. The EF is intended to be used as a resource management tool for assessing whether and to what extent an individual, city, or nation is using available ecological assets faster than the supporting ecosystems can regenerate those assets [10].

The Ecological Footprint measures how much of the regenerative capacity of the biosphere is used by human activities. It does so by calculating the amount of biologically productive land and water area required to support a given population at its current level of consumption and resource efficiency. A country's Footprint is the total area required to produce the food, fibre and timber that it consumes, absorb the waste it generates, and provide area for its infrastructure [13]. Because trade is global, an individual or country's Footprint includes land or sea from all over the world. Without further specification, Ecological Footprint generally refers to the Ecological Footprint of consumption. The most commonly reported type of Ecological Footprint is Ecological Footprint of consumption (EFC). It is defined as the area used to support a defined population's consumption.

The consumption Footprint (in gha) includes the area needed to produce the materials consumed and the area needed to absorb the carbon dioxide emissions. The consumption Footprint of a nation is calculated in the National Footprint Accounts as a nation's primary production Footprint plus the Footprint of imports minus the Footprint of exports, and is thus, strictly speaking, a Footprint of apparent consumption. The national average of per capita Consumption Footprint is equal to a country's Consumption Footprint divided by its population.

EF always compares bio capacity. Bio capacity serves as a lens, showing the capacity of biosphere to regenerate and provide for life. It allows researchers to add up the competing human demands, which include natural resources, waste absorption, water renewal, and productive areas dedicated to urban uses. As an aggregate, bio capacity allows us to determine how large the material metabolism of human economies is compared to what nature can renew [13].

In contrast to the Footprint, which addresses demand on ecosystems, bio capacity describes the supply side—the productive capacity of the biosphere and its ability to provide a flux of biological resources and services useful to humanity. Both Footprint and bio capacity are measured in global hectares (gha). Global hectare represents a hectare of land with world average bio productivity. In 2015, the global per capita Footprint was 2.7 gha, and the per capita Footprint of nations with available data ranged from 0.4 gha/cap in Eritrea to 15.8 gha/cap in Luxembourg. In 2015, globally available bio capacity was 1.7 gha/cap.

Figure 4 shows Ecological Deficit/Reserve Map prepared by Global Footprint Network. While greens are biocapacity creditors, reds are bio capacity debtors [14]. If a

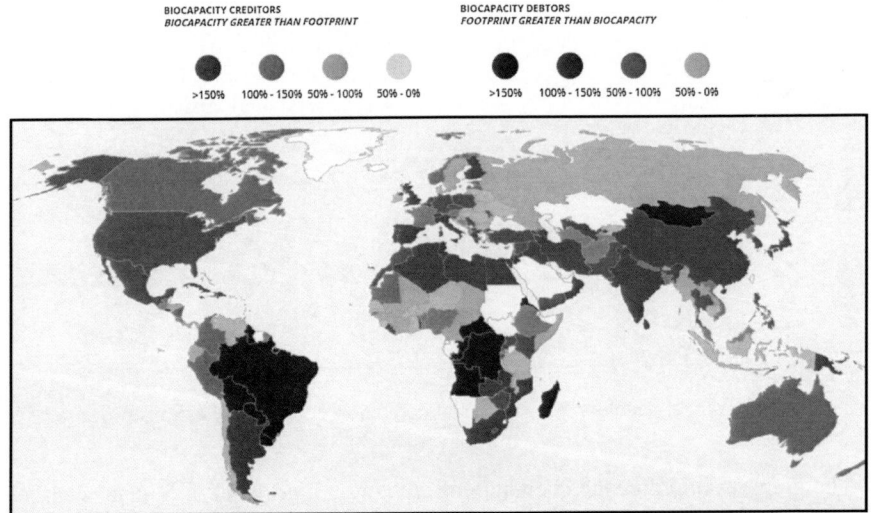

Fig. 4 Ecological deficit/reserve map

country Ecological Footprint is bigger than bio capacity, it is bio capacity debtors or vice versa. Note that developed countries are debtors, developing countries are creditors.

4 Measuring Sustainable Development

Sustainable development is a commitment improving the quality of human life while living within the carrying the capacity of supporting ecosystems.

SUSTAINABLE: Living within the means of planet Earth requires an average Ecological Footprint per person of less than 1.7 global average hectares (the supply of biologically productive planetary surface area that exists per person). The Ecological Footprint measures how much of the planet's surface people demand from nature for food, fiber, timber, and waste absorption (particularly for CO_2 from fossil fuel). Currently, the Footprint of humanity is 2.7 global average hectares per person.

DEVELOPMENT: Human Development Index On a scale of zero to one, 0.8 is considered the threshold for a very high level of development.

Successful sustainable development requires that the world, on average meets at a minimum these two criteria. These two thresholds define two minimum criteria for sustainable development. We argue that an HDI of no less than 0.8 and a per capita Ecological Footprint less than the globally available bio capacity per person (less than 1.7 global average hectares) represent minimum requirements for sustainable development that is globally replicable. For all countries, the sustainable development goal should be high human development and a low ecological footprint per capita.

Figure 5 illustrates combined the Human Development Index and Ecological Footprint of Nations in a graph [1].

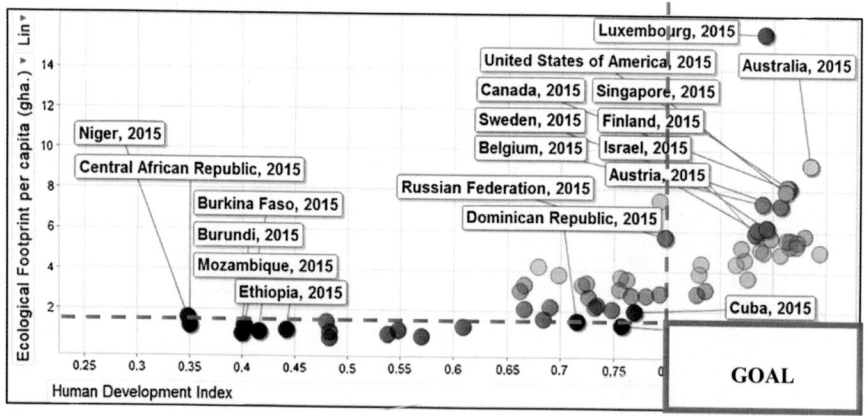

Fig. 5 Human development index and ecological footprint of nations (2015)

The graph exemplifies the challenge of creating a high level of human well-being without depleting the planet's or a region's ecological resource base. This indicates that

Table 1 Human development index and ecological footprint values of selected countries

Countries	HDI		Ecological footprint per capita (gha)	
	Ranking	Value	Ranking	Value
Luxembourg	19	0.892	1	15.8
Australia	2	0.935	2	9.3
USA	8	0.915	3	8.2
Canada	9	0.913	4	8.2
Singapore	11	0.912	5	8.0
Trinidad and Tobago	64	0.772	6	7.9
Oman	52	0.793	7	7.5
Belgium	21	0.890	8	7.4
Sweden	14	0.907	9	7.3
Estonia	30	0.861	10	6.9
Latvia	46	0.819	11	6.3
Israel	18	0.894	12	6.2
Mongolia	90	0.727	13	6.1
Austria	23	0.885	14	6.1
Finland	24	0.883	15	5.9
Lithuania	37	0.839	16	5.8
Slovenia	25	0.880	17	5.8
Switzerland	3	0.930	18	5.8
South Korea	17	0.898	19	5.7
Russia	50	0.798	20	5.7
New Zealand	9	0.913	21	5.6
Ireland	6	0.916	22	5.6
Denmark	4	0.923	23	5.5
Turkmenistan	109	0.688	24	5.5
Germany	6	0.916	25	5.3
Netherlands	5	0.922	26	5.3
Czech Rep.	28	0.870	27	5.2
France	22	0.888	28	5.1
Belarus	50	0.798	29	5.1
Japan	20	0.891	30	5.0
Norway	1	0.944	31	5.0
UK	14	0.907	32	4.9

countries in Europe and North America have very high Ecological Footprints and acceptable Human Development Indexes (above 0.8), while countries in Africa have unacceptably low Human Development Indexes (below 0.8) but have Ecological Footprints within the biosphere's allowable capacity per person. The lower right quadrant represents the goal of sustainable development, i.e., high human development, within levels of resource consumption that can be extended globally. Measuring these two variables reveals that very few countries (Cuba, Sri Lanka etc.) come close to achieving sustainable development. Despite growing global adoption of sustainable development as an explicit policy goal, we find that in the year 2015 (latest available) any countries surveyed met both of these minimum requirements.

Table 1 shows Human development index and ecological footprint values of some countries. Data shows high human development countries have high ecological footprint scores. An overall trend in high-income countries over the past twenty-five years that improvements to HDI come with disproportionately larger increases in Ecological Footprint, showing a movement away from sustainability.

5 Conclusions

Sustainable development represents a commitment to advancing human well-being, with the added constraint that this development needs to take place within the ecological limits of the biosphere. Progress in both these dimensions of sustainable development can be assessed: we use the HDI as an indicator of development and the Ecological Footprint as an indicator of human demand on the biosphere. The Ecological Footprint and HDI represent strict, yet widely accepted, metrics for ecological sustainability and human development.

We examine sustainable development in terms of its two dimensions. We assess progress in development with the HDI because it is one of the most widely used overall measures of human well-being. The other dimension of sustainable development is the commitment to develop within the ecological capacity of planet Earth. This can be measured with the Ecological Footprint, a resource accounting tool that assesses how much of the regenerative capacity of the biosphere is occupied by human activities.

Environmental Kuznets Curve is located in the sustainable economic development literature puts forward that the inverse U shape relationship between the level of economic development and environmental degradation. In the early stages of economic growth degradation and pollution increase, but beyond some level of income per capita, which will vary for different indicators, the trend reverses, so that at high income levels economic growth leads to environmental improvement. This implies that the environmental impact indicator is an inverted U shaped function of income per capita.

In this study, the ecological footprints of countries are compared with the level of human development and the validity of Kuznets inverted U hypothesis being investigated.

We argue that an HDI of no less than 0.8 and a per capita Ecological Footprint less than the globally available bio capacity per person (less than 1.7 global average hectares) represent minimum requirements for sustainable development that is globally replicable. Despite growing global adoption of sustainable development as an explicit policy goal, we find that in the year 2015 any countries surveyed met both of these minimum requirements. We also find an overall trend in high-income countries over the past twenty-five years that improvements to HDI come with disproportionately larger increases in Ecological Footprint, showing a movement away from sustainability. Some lower-income countries, however, have achieved higher levels of development without a corresponding increase in per capita demand on ecosystem resources.

Kuznets's Inverted U hypothesis has been tested in many studies. There is little evidence for a common inverted U-shaped pathway that countries follow as their income rises. For all countries, the goal should be high human development and a low ecological footprint per capita. However unlike Kuznets, data shows high human development countries have high ecological footprint scores.

References

1. Global Footprint Network (GFN) Sustainable Development. (http://www.footprintnetwork. org/our-work/sustainable-development/)
2. Shekhawat P (2013) UNDP uses a new human development eco footprint. http://www. policyinnovations.org/ideas/briefings/data/000258
3. World Wildlife Fund (WWF) Living planet report 2014. http://awsassets.panda.org/ downloads/lpr_living_planet_report_2014.pdf
4. Arrow K, Bolin B, Costanza R, Folke C, Holling CS, Janson B, Levin S, Maler K, Perrings C, Pimental D (1995) Economic growth, carrying capacity, and the environment. Ecol Econ 15:91–95
5. Dinda S (2004) Environmental Kuznets Curve hypothesis: a survey. Ecol Econ 49:431–455
6. Stern DI (2004) The rise and fall of the Environmental Kuznets Curve. World Dev 32 (8):1419–1439
7. Dasgupta S, Laplante B, Wang H, Wheeler D (2002) Confronting the Environmental Kuznets Curve. J Econ Perspect 16(1):147–168
8. Yandle B, Vijayaraghavan M, Bhattarai M (2002) The Environmental Kuznets Curve a primer. PERC Research Study 02-1
9. United Nations Development Program (UNDP) (1990) Human development report: overview. Oxford University Press, New York
10. Constanza R, Hart M, Poznerve S, Talberth J (2009) Beyond GDP—the need for new measures of progress, the Pardee Papers No. 4, January. Boston University
11. UNDP, Human Development Index (HDI). http://hdr.undp.org/en/content/human-development-index-hdi
12. UNDP, International Human Development Indicators. http://hdr.undp.org/en/countries
13. Moran DD, Wackernagel M, Kitzes JA, Goldfinger SH, Boutaud A (2008) Measuring sustainable development—nation by nation. Ecol Econ 64:470–474
14. GFN. Ecological Wealth of Nations. http://www.footprintnetwork.org/ecological_footprint_ nations/

An Examination of Sustainability and United Nations Sustainable Development Goals: Turkey Case

Arzuhan Burcu Gültekin[1]([✉]) ⓘ, Çiğdem Belgin Dikmen[2] ⓘ,
Ahmet Hilmi Erciyes[1] ⓘ, and Demet Örgü[1] ⓘ

[1] Department of Real Estate Development and Management, Faculty of Applied
Sciences, Ankara University, Ankara, Turkey
arzuhanburcu@yahoo.com, ahmethilmierciyes@gmail.com,
demetorgu@hotmail.com
[2] Faculty of Engineering and Architecture, Bozok University, Yozgat, Turkey
cbelgin.dikmen@gmail.com

Abstract. This study focuses on international summits which have defined the framework of sustainability, and which have played an important role in the emergence and development of the concepts, sustainability and sustainable development in the recent past. The paper concentrates on the effects of international summits between the United Nations Conference on the Human Environment (1972) and HABITAT III (2016) (Quito), which can be defined as milestones. Furthermore it explores the developments in Turkey with respect to the selected indicators for sustainable development goals and the process of achieving such goals that are identified in a global scale in the 2015 United Nations (UN) Sustainable Development Summit.

Keywords: Environment · Sustainability · Sustainable development
United Nations (UN) · Sustainable development goals

1 Introduction

Nature and natural resources have been considered as infinite assets since the very beginning of the human history. Increased consumption of natural resources with the emergence of the concepts of industrialization and colonialism has led to environmental problems. These environmental problems have proved that natural resources, which are very important for the existence of human race, are finite, and are under threat due to excessive use. Locally enforced measures aimed at environmental protection date back to very old times. Nevertheless, as environment is an entity which belongs to the humanity as a whole, it was comprehended that any measure aimed at environmental problems would be possible only with a joint movement. Economic development activities as one of the main reasons behind the environmental damage made it clear that we need to take a number of measures with respect to development. The requirement to further development without damaging the environment has led to the emergence of the concept of "sustainability".

The concept of sustainability is defined as "the ability to ensure the continuity for an extended period of time without any damage or with minimum damage to the environment" [1]. In this context, this concept which can also be interpreted as the ability of continuity refers to the concepts of development or growth rather than individual use. The emergence of the concept of sustainability owes itself to the raised awareness around the global environmental issues, and the pressure this situation places on the manufacturing and industrial actors. Ecologists believe that environmental protection is more important than any other goal. For economic actors, on the other hand, this demand of environmentalists in a world which competes for economic development translates into giving up fully on development which is accepted as a global policy [2]. This led to a conflict between the interests of economy and ecology. Sustainable development has emerged as a solution for this conflict.

Sustainable development is considered as the development of a country or a region which does not use natural resources at a level and which cannot be replaced in a way that does not harm the nature [3, 4]. The first definition of this concept was suggested in the Brundtland Report in 1987 which goes by "development that meets the needs of the present without compromising the ability of future generations to meet their own needs" [5, 6].

The notion of sustainable development has shaped in long years. The role of the efforts made by UN in the conceptualization of sustainable developments is especially important. In this context, many studies were conducted, reports were developed and conferences were organized [7]. The concept of sustainable development is a dynamic concept which changes with the developments achieved as a result of the efforts made. Summits organized by UN and organizations acting under the UN have played an effective role in the development of the concepts of sustainability and sustainable development in that it brings nations together.

In this study, the steps taken in sustainable development are explored within the context of summits, conferences and sessions organized between 1972 and 2016. The consequences of the decisions and the global sustainability development goals defined in these meetings are discussed for Turkey.

2 Historical and International Development of Sustainable Development

This chapter focuses on international summits which have defined the framework of sustainability and which have played an important role in the emergence and development of the concepts, sustainability and sustainable development, in the recent past and in international platforms.

2.1 1972 UN Conference on the Human Environment (Stockholm)

The awareness raised after the WWII regarding the increased environmental issues, and the threat they pose created the need for a movement against the environmental issues on an international platform. In this context, the first step was taken by the UN, and a conference was held in June, 1972 with the participation of 113 countries. The purpose of this conference was to develop a shared viewpoint in terms of the protection of human environment and its development, and to define the common principles. As a result of the conference, the "Report of the UN Conference on the Human Environment" which included 26 principles was accepted. This report stated that human beings have the basic right to live in an environment which offers freedom, equality and quality of life and that it has a serious responsibility to provide such an environment and to protect it for the future generations, and it emphasized that racists and discriminating policies must be abandoned. The report drew attention to the need to attend to the environmental issues surfacing in the nature and artificial environment which humans have interaction with, which are emerging in parallel with the increasing world population and which must be considered as a threat to the humanity both in regional and global scale. The world was warned about taking measures in order to decrease the number of environmental issues, having enforced local and global policies in order to leave a better environment for the future generations. Adopting the need to protect any element of the human environment for the next generation, the report also underlined one of the reasons behind the environmental issues as underdevelopment. It was decided that developing countries should receive assistance on international techniques in order for them to protect resources and the environment in their development process [8].

2.2   1976 HABITAT I Conference (Vancouver)

The importance of the human settlements and environment was understood as a result of the 1972 UN Conference on the Human Environment. Held in accordance with the UN recommendation to organize a conference on human settlements, the first UN Conference on Human Settlements, HABITAT I, which was held in Vancouver in 1976 included significant consequences for environment [9]. It is know that the period this conference was held has seen an acceleration in the migration from the suburban areas to urban areas, while still the two third of the world population was living in suburban areas. The rapid urbanization, the preference of people to live in urban areas instead of suburban areas and the understanding that most of the environmental issues are focused on cities show that the scale of environmental issues was not yet to be comprehended fully. In the conference, actors which may facilitate sustainable urbanization and the methods to be followed were discussed, and the important role local governments and local people play in urbanization were emphasized. A UN Center for Human Settlements was established in 1978 [10]. The declaration published at the end of the conference included 19 general principles and a national action plan consisting of 64 articles [11]. HABITAT conferences proved effective in defining goals for sustainable cities and communities.

2.3   1982 UN Conference on Environment Programme (Nairobi)

At the 10th anniversary of the UN Conference of Human Environment held in Stockholm, a UN Environment Programme (UNEP) was established in order to ensure coordination in actions taken for the solution of environmental issues. As part of this programme, the UN Conference of Environment Program was held in Nairobi in 1982 [12]. The purpose of the programme was to measure the practices applied after the decisions taken in Stockholm, to contribute to the process in line with national policies and demands and the need for intensifying the protection and development efforts in global, regional and local scales. A declaration of 10 articles was published at the end of the conference. UN Conference on Environment Programme stated that the decision taken in the Stockholm Summit were suitable and that they are still relevant, the awareness the conference raised was discussed, it was emphasized that the action plan was partly in place, and new needs emerged in the ten years period were defined. The declaration pointed out to the fact that however the actions which harm the environment are taken under the supervision of governments, international collaboration will be required if the consequences of these actions go over the borders, and the World Commission on Environment and Development was established after the conference. The Conference also played a role in the establishment of the World Charter for Nature in 1982 by the UN.

2.4   1987 Report of the World Commission on Environment and Development: Our Common Future (Brundtland Report)

Organized under the leadership of Gro Harlem Brundtland, former Norwegian President, in 1987, and published by the UN World Commission on Environment and Development, the "Our Future" report is considered the first official document to use the concept of Sustainable Development. Our Future Report defined sustainable development as the development that meets the needs of the present without compromising the ability of future generations to meet their own needs. The report addresses the environmental issues discussed in the previous declaration and the conferences, and the evolution of these issues. The characteristic of Brundtland Report which differentiates it from other conferences and their results was that it accepts the necessity of economic development and the possibility to serve for this necessity with environmentally-friendly methods [13]. Building on the assumption that economic growth is possible with an environmentally-friendly perspective, the report suggests that developing countries will play an important role in the solution of global environmental problems and prevention of poverty and that it is necessary to initiate a long-term growth age that will facilitate restructuring. Also including information about the demographic structure of the world, Brundtland Report assumes that the world population will be stabilized between 7.7 billion and 14.2 billion, and that the world population will largely be living in urban areas rather than suburban ones [14].

2.5   1992 UN Conference on Environment and Development (UNCED)— Rio Summit (Rio De Janeiro)

In order to be able to overcome the problems in practice which we have faced after the introduction of the concept of sustainable development addressed in Brundtland

Report, UN Conference on Environment and Development—Rio Summit (Earth Summit) was held with the participation of 172 countries under the leadership of the UN Rio Summit is considered to be an important step in the adoption of a number of principles aimed at the environmentally-friendly government methods used by the nations. Rio Summit differentiates from its predecessors as it addresses environmental issues and development concepts in conjunction. As a result of the Rio Summit, five documents serving as international conventions were created which can be used as a roadmap for issues about environment and development internationally. Among these documents were Rio Declaration on Environment and Development, Agenda 21, Forest Principles, Convention on Biological Diversity, Framework Convention on Climate Change [2].

Agenda 21, the product of the conference which offers an action plan for both environmental and development issues in 4 Chapters and 40 Articles as a methodology plan for sustainable development, is a guiding document about reforms to be done in national policies of countries in terms of trade, environment and economy. It is also known that 1992 Rio Summit and Agenda 21 were important steps in the way to the creation of the goals for 2030. In this context, it can be said that Rio Summit with the purpose of adoption of sustainable development principles in the national policies with respect to environmental sustainability contributed greatly to the specification of sustainable development goals about economy with report and international conventions developed as a result of the summit. As it was the case in 1982 Nairobi Summit, decisions taken in the Stockholm Conference were once again confirmed, and 27 principle decisions were published in the Rio Declaration after the conference. Having discussed sustainable development as a more specific concept, the declaration emphasized the fact that environmental preservation is a part of the development process and that it cannot be addressed separately. In the principles of the declaration, the goals for 2030 were adopted as a principle, and it was underlined that the process must be performed under an international cooperation [15].

2.6   1996 HABITAT II Conference (İstanbul)

HABITAT II Conference was held in the 20th anniversary of the first conference on this subject held in 1976, and aimed at creating sustainable living spaces and finding solutions to accommodation problems. Changing social, economic, politic and environmental conditions of the day required a change in the agenda of the first conference, and a new agenda was proposed in this respect. Decisions made on creating a sustainable global environment and provision of humane accommodation to those who live under substandard conditions while having a clean environment, and the agreed Istanbul Declaration addressed the current and potential consequences of urban population increase. It was noted after the conference that population movements focused on the urban areas in the last two decades, the cities are growing and turning into big hubs in terms of administration, and cities will become important actors economically, socially and administratively in the years to come. Therefore, recommendations such as increasing the authorities of urban administrations, and provision of residence areas and sustainable environment to central governments with the human-centred structuring in mind were made as an important act for sustainability [16, 17]. The declaration has the quality of a guiding document in order for the nations to find the required funds to actualize their settlement policies.

2.7 1997 Special Session of the UN General Assembly—Rio+5 Earth Summit (New York)

5 years after the Rio Summit held in 1992, Rio +5 Earth Summit was held in order to ensure that sustainable development is not only in the agenda but is in practice [5]. As part of the session, the effectiveness of the decisions made in the Rio Earth Summit was discussed, it was noted that the goaled development was not possible although important steps were taken after Rio Conference, and the importance of taking more substantial steps was emphasized [18]. It was agreed that Agenda 21, defined as the agenda of 21st century, will be adapted to the nations and that each nation will implement its own National Agenda 21 Action Plan. It was a historical moment when the fresh water and energy subjects were adopted as fundamental subjects with regards to clean and renewable energy goals.

2.8 1997 Kyoto Protocol (Kyoto)

The purpose of the Convention on Climate Change and Environment developed in 1992 Rio Earth Summit was to prevent greenhouse gas emissions, and to eliminate or take under control any factors which cause climate change and environmental problems. Decrease of greenhouse gas emissions due to human activities is significant for sustainable development. Held in order to find a solution to this problem, Kyoto Protocol is the only framework aimed at fighting with global warming and climate change. The countries which have signed the protocol are dedicated to reduce their GHG emissions or to increase their rights through carbon trade, if they are not able to reduce their GHG emissions. Nevertheless, any country which has signed the protocol needs to reduce their GHG emissions to the level estimated for 1990. Signed in 1997, the protocol became effective in 2005. The reason behind this delay was the need to reach at a level of 55% of the total GHG emissions in the world with the countries which confirmed the protocol, and that was only possible with the participation of Russian Federation 8 years later. Kyoto Protocol now includes 160 countries and more than 55% of the GHG emissions in the world [19]. With this protocol, industrial nations were forced to reduce their GHG emissions to a level not less than 5% of the level estimated for 1990. Turkey signed the protocol in 2009. Kyoto Protocol can be considered as an important step in prevention of environmental pollution with tax regulations implemented to those countries causing pollution, responsible production and consumption as one of the sustainable development goals and development of green systems, and exploring principles such as climate action.

2.9   2002 UN World Summit on Sustainable Development—Rio+10 Summit (Johannesburg)

Rio+10 Summit was held in order to assess the practices performed between the period following Rio and Rio+5 Summits. The main difference of this summit when compared to Rio and Rio+5 Summits was that it received extensive participation not only from nations but also from farmers, politicians, international organizations, NGOs, etc. as part of one of the most important principles of sustainable development, participation. One of the two documents drafted after Rio+10 Summit was the action plan, and the other was the political manifest which reflects the political will. Consisting of 10 chapters, the action plan brings forth updated goals related to sustainable development. Goals such as energy, health, agriculture, biodiversity and protection of water were presented with the action plan. The fact that nations have showed their dedication to halve the negative conditions related to clean water, hygiene and public health with this action plan was considered a tangible step [20]. When the action plan and declaration were reviewed, it can be said that all the sustainable development goals consisted the agenda of Rio+10 Summit. Among the subjects discussed widely in the political manifest was the importance of international cooperation as it was the case for other declarations published in previous international summits.

2.10   2012 UN Conference on Sustainable Development—Rio+20 Summit (Rio de Janeiro)

Held with the participation of world leaders in 2012, Rio+20 was important as it coincided with the 40th anniversary of 1972 UN Conference on the Human Environment (Stockholm), 20th anniversary of 1992 UN Conference on Environment and Development-Rio Summit (Rio De Janeiro), and 10th anniversary of 2002 UN World Summit on Sustainable Development-Rio+10 Summit (Johannesburg). Described as the 2nd Earth Summit, Rio+20 Summit aimed to assess the developments in the past 20 years, to evaluate the situation and to discuss new decisions. The assessment showed that however the economic development goals were met individually, the global sustainable development goal was not met and that indicators for the performance of the decisions regarding environment which were updated in the summits held since 1992 proved unsuccessful [21]. Rio+20 Summit involved the identification of the global goals and indicators which led to the development of sustainable development goals which were agreed on in the New York Summit held in 2015.

As a result of the Rio+20, a report, "Future We Want", was published including 283 articles and 53 pages. A very detailed work, the report included principle decisions on sustainable development under relevant categories. Millennium Development Goals decided in the UN meeting held in 2000, and action plans such as Agenda 21 were referred to under the title of Sustainable Development Goals in this report and the importance of defining common goals was emphasized. The period after 2015 was pointed at for common goals and the identification of goals based on taking action was underlined. The need for global, holistic and scientific information regarding sustainable development was also emphasized in Rio+20 Summit [22]. The fact that decisions taken in the previous summits and the application of principles defined did not meet the expectations and that environmental and economic indicators did not meet the expectations with the exception of regional or local actions contributed to an effort in Rio+20 Summit which offers realistic and reachable goals.

2.11   2015 UN Summit on Sustainable Development (New York): The 2030 Agenda for Sustainable Development

In order to define the universal goals of sustainable development in line with the decisions and goals defined in Rio+20 Summit, a Sustainable Development Summit was held in New York at the level of the Board of UN in 2015. Having gathered in order to create an agenda which was pointed at Rio+20, the nations defined a roadmap to achieve sustainable development fully, and they have adopted common goals in the 2015 UN Summit on Sustainable Development. "Transforming our world: the 2030 Agenda for Sustainable Development", a new agenda which was accepted by 193 countries, was created and the practices necessary to reach the goals were summarized with the expectation of actualizing 17 main goals and articles defined in order to ensure sustainable development globally by 2030, and efforts made during and before the summit regarding the goals.

2.12   2016 HABITAT III Conference (Kito)

HABITAT Conferences are important steps with regards to building sustainable cities and living spaces as one of the sustainable development goals for gradually growing urban areas. HABITAT III Conference was held in Kito, Ecuador with the participation of nations, governments, NGOs and private sector in order to ensure the cooperation on subjects such as global problems relevant to cities, habitation and sustainable urban development at an international level. The first UN global summit held after the 2030 Agenda for Sustainable Development and Sustainable Development Goals, HABITAT III Conference resulted in the publication of a draft report and Kito Action Plan named "New Urban Agenda". The estimation that the population of urban areas will be doubled by 2050 was developed with the New Urban Agenda. The Article 9 of the draft report emphasized that the New Urban Agenda will contribute to the 2030 goals, and that sustainable urban development and sustainable development can be achieved with the coordinated and holistic efforts at local, subnational, national, regional and global scales with the participation of the stakeholders of sustainable urban development and sustainable development [23].

3   A Review of UN Sustainable Development Goals at the Scale of Turkey

17 global sustainable development goals were defined and accepted by UN Commission on Sustainable Development to be actualized by 2030 in the 2015 UN Sustainable Development Summit as discussed in Article 3.11. 169 purposes and 230 indicators were defined as part of the goals in question for the scalability of the sustainable development. These indicators may be effective in defining local, national or regional goals as well as facilitating retrospective analyses. In order to review the reflections of sustainable development goals in Turkey, sustainable development indicators published in 2015 by Turkish Statistical Institute (TUIK) were used. The status of sustainable development in Turkey between 2000 and 2014 is discussed.

3.1   No Poverty

The first sustainable development goal of UN aims to end poverty in all its forms everywhere [24]. According to Fig. 1, it is seen that the defined conditions are met as of 2005, GDP per capita has assumed an upward trend and the population under the risk of poverty gradually reduced in years.

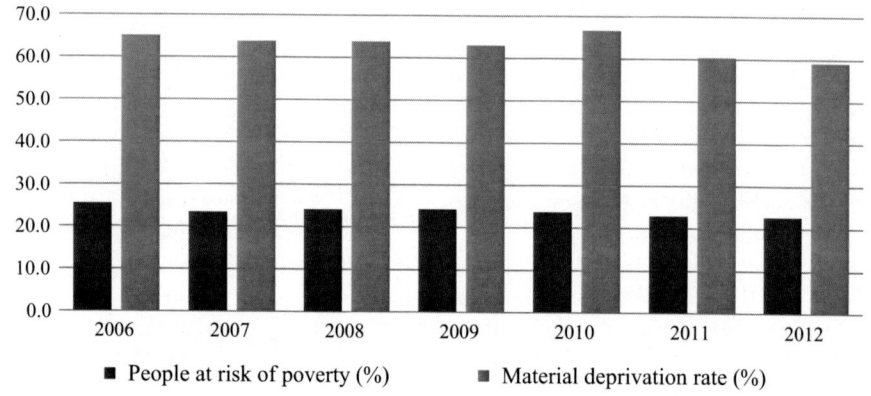

Fig. 1.   Risk of poverty and material deprivation rate (%) [25]

3.2   Zero Hunger

Goal 2 intends to end hunger, achieve food security and improved nutrition and promote sustainable agriculture by 2030 [24]. According to Fig. 2, it is seen that cultivated agricultural lands which are under the pressure of urbanization were opened to zoning activities in order to provide easier and increased revenues, thus agricultural land has been gradually reducing every year. Therefore, goals were met for other agricultural activities, however it was not the case for cultivated agricultural lands. This is an element which needs critical importance if it is to achieve sustainability.

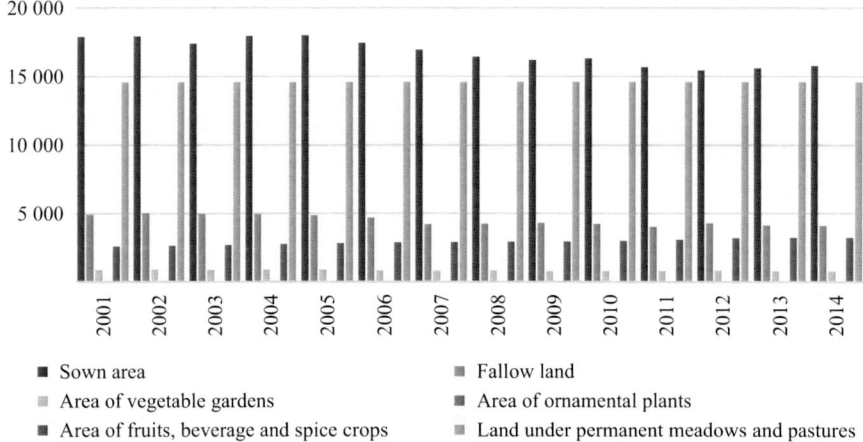

■ Sown area   ■ Fallow land
■ Area of vegetable gardens   ■ Area of ornamental plants
■ Area of fruits, beverage and spice crops ■ Land under permanent meadows and pastures

Fig. 2. Total statistical classification of products (1000 ha) [25]

3.3 Good Health and Well-Being

Goal 3 aims at ensuring healthy lives and promote well-being for all at all ages [24]. According to Fig. 3, looking into the age-gender relationship of Turkey's population, it's seen that average age has been increasing for both genders, and that average life expectancy is longer in women when compared to men. It is believed that life expectancy is increased depending on the advancements in medicine, increased health investments and improved environmental conditions as it is the case in the world.

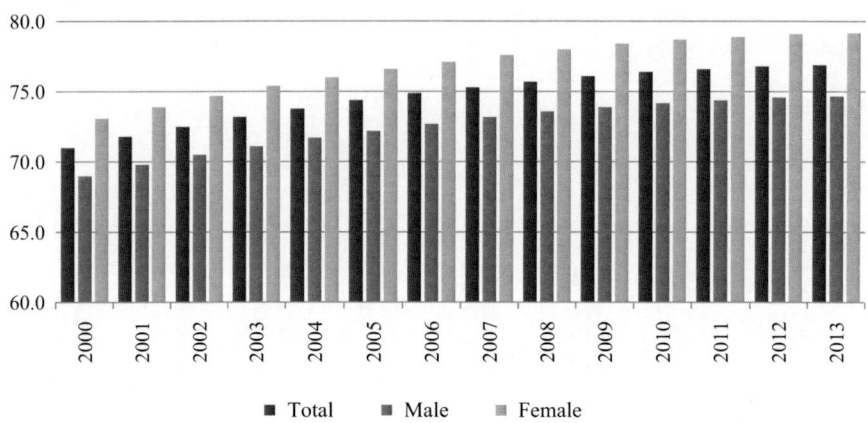

■ Total ■ Male ■ Female

Fig. 3. Life expectancy at birth by sex (years) [25]

3.4   Quality Education

Ensuring inclusive and equitable quality education and promoting lifelong learning opportunities for all are aimed in the context of Goal 4 [24]. According to Fig. 4, it is seen that the education level in Turkey is increased in parallel with the increased education period. Increased education period and education level for both genders can be considered as a positive development in terms of the gender equality.

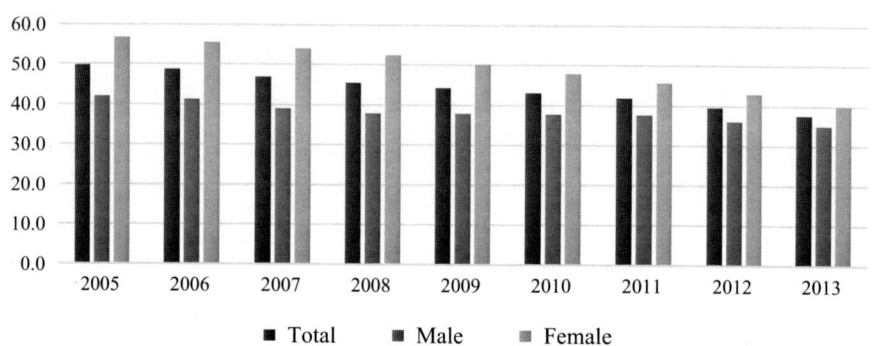

Fig. 4. Early leavers from education and training (%) [25]

3.5   Gender Equality

Goal 5 aims to achieve gender equality and empower all women and girls [24]. According to Fig. 5, it is seen that unemployment rate for women is decreased once was higher comparatively when the long-term unemployment rates are reviewed for gender. However, labour force participation rate of women is found to be lower when compared to that of men's.

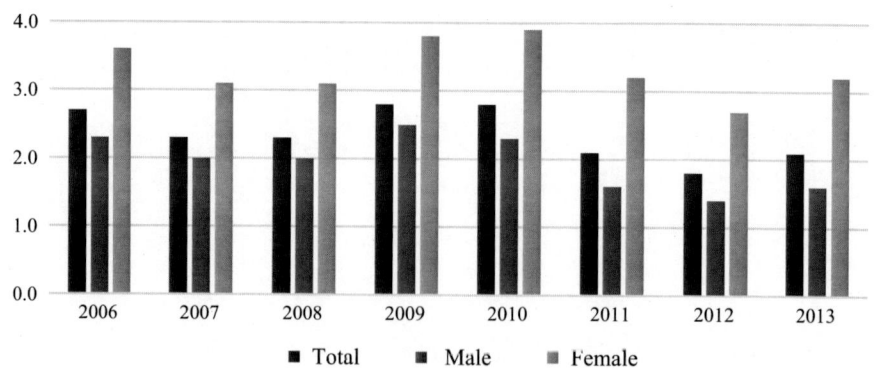

Fig. 5. Long-term unemployment rate by sex (%) [25]

3.6   Clean Water and Sanitation

Goal 6 intends to ensure availability and sustainable management of water and sanitation for all [24]. According to Fig. 6, it is seen that ground water as a fresh water resource available in Turkey is partly reduced, however, waste water and the number of biological water treatment facilities are increased depending on the urbanization. Therefore, it is important for the water preservation and urban health to consider increasing amount of waste water as a threat, to take measures to decrease water use in urban areas, and to ensure widespread use of water treatment technologies.

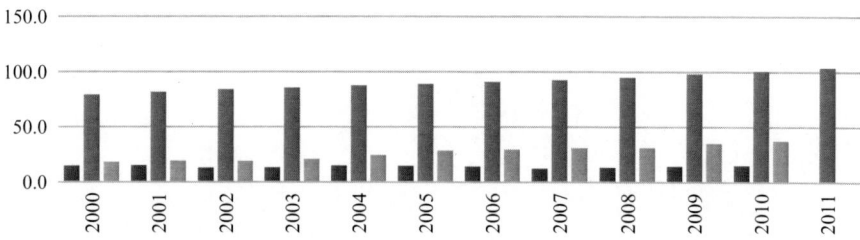

■ Fresh surface water
■ Fresh ground water
■ Population connected to urban wastewater treatment with at least secondary treatment

Fig. 6. Fresh water resources (%) [25]

3.7   Affordable and Clean Energy

Ensuring access to affordable, reliable, sustainable and modern energy for all is aimed in the context of Goal 7 [24]. According to Fig. 7, foreign-source dependency in Turkey has increased in years for all of the available energy resources. A slight increase in the renewable energy resources was seen, however, it can be said that the potential is not fully realized. Nevertheless, it was found that R&D activities related to solar and wind energy, i.e. renewable energy resources, were increased and the incentives provided for such projects were also increased.

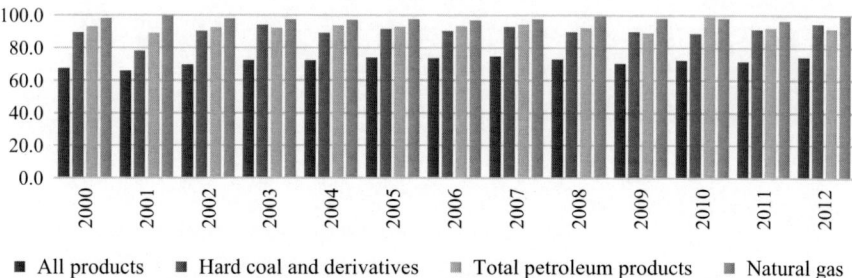

■ All products   ■ Hard coal and derivatives   ■ Total petroleum products   ■ Natural gas

Fig. 7. Energy dependency (%) [25]

3.8   Decent Work and Economic Growth

Goal 8 aims for promoting sustained, inclusive and sustainable economic growth, full and productive employment and decent work for all [24]. According to Fig. 8, it is seen that GDP has not been consistently increasing and negative values are observed in some years, and that sustainability was not achieved fully. It can be said that distribution of GDP shows irregularities due to the impact of regional and national economic crises.

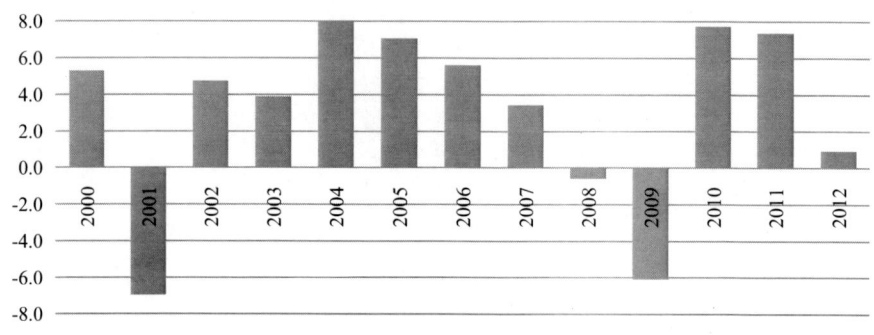

■ Growth rate of GDP per capita at 1998 prices

Fig. 8. Dispersion of regional GDP per inhabitant (%) [25]

3.9   Industry, Innovation and Infrastructure

In the context of Goal 9, building resilient infrastructure, promoting inclusive and sustainable industrialization and fostering innovation is intended [24]. According to Fig. 9, it is seen that the share of the industrial innovation and infrastructure budget allocated as part of R&D budget has increased in years.

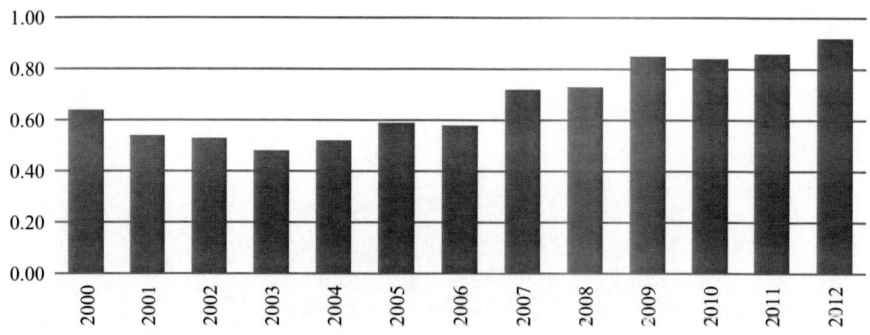

■ Share of Gross Domestic Expenditure on Research and Development (GERD) in GDP (%)

Fig. 9. Innovation, competitiveness and eco-efficiency (%) [25]

3.10 Reduced Inequalities

Goal 10 aims to reduce inequality within and among countries [24]. According to Fig. 10, it is seen that the inequality in income distribution in Turkey has assumed a downward trend in line with the goal. Further reduction of the inequalities in income distribution is important for the sustainable development, first step in the elimination of social inequalities.

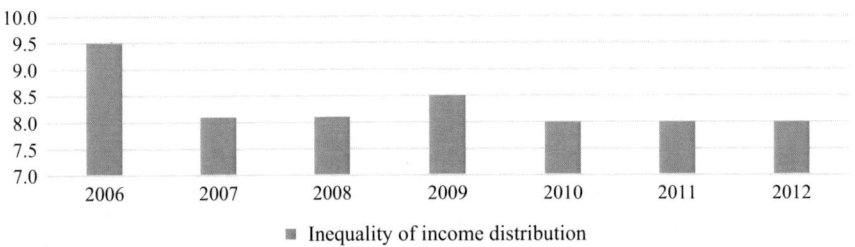

Fig. 10. Inequality of income distribution (%) [25]

3.11 Sustainable Cities and Communities

Goal 11 intends to make cities and human settlements inclusive, safe, resilient and sustainable [24]. According to Fig. 11, it can be seen that official development support and incentives for manufacturing industry have been increased in years. According to the data, it is found that manufacturing industry has been developing consistently, however, the incentives for social infrastructure service investments were increased exponentially.

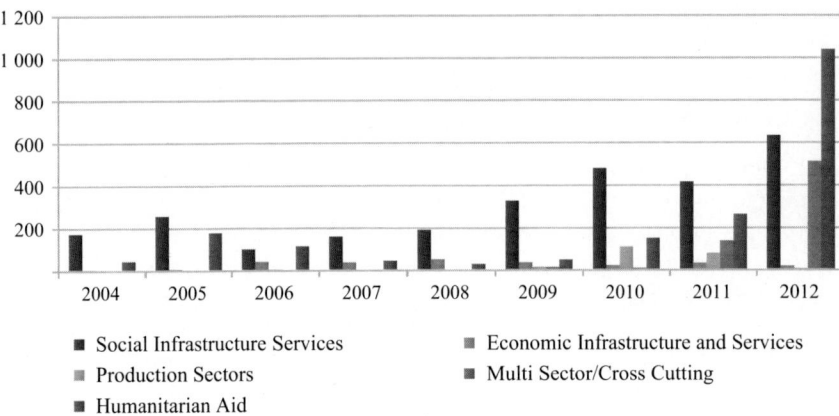

Fig. 11. Bilateral official development assistance by category (million $) [25]

3.12 Responsible Consumption and Production

Ensuring sustainable consumption and production patterns is aimed in the context of Goal 12 [24]. According to Fig. 12, a review of the resource efficiency and waste management in Turkey shows that the amount of waste has increased in urban areas in years, while the recycling and energy recovery rates are also increased. The fact that waste levels are stabilized in spite of the investments made in years is considered as a positive indicator. However, minimizing the amount of waste which harms the environment is a necessity of a responsible manufacturing and consumption approach.

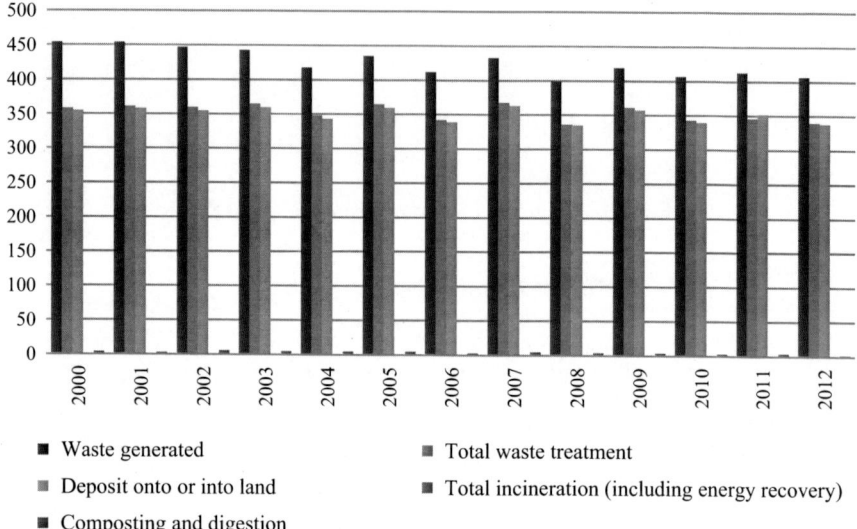

Fig. 12. Municipal waste generation and treatment, by type of treatment method (kg/capita-year) [25]

3.13 Climate Action

Goal 13 aims at taking urgent action to combat climate change and its impacts [24]. According to Fig. 13, a review of the GHG emissions per sector in Turkey shows that emissions are increased in transportation, industrial works and waste, while it was decreased in manufacturing sector. It is critical to reduce GHG emissions in order to prevent global warming and climate change.

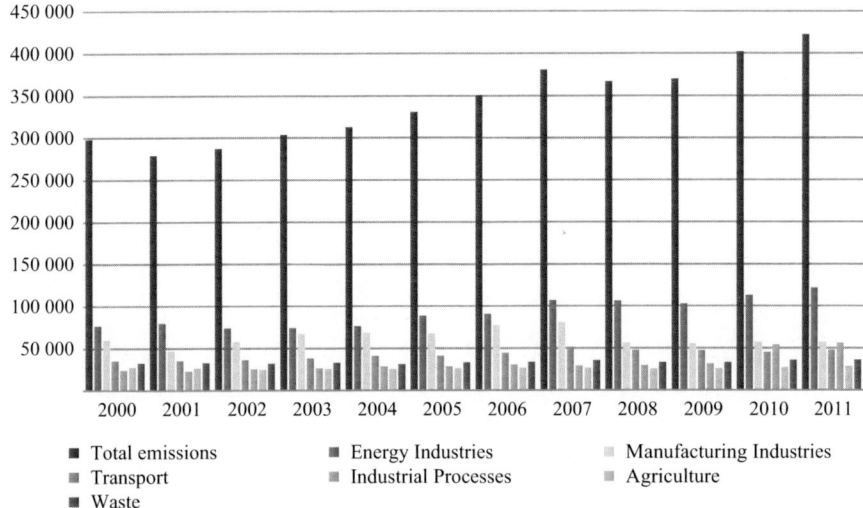

Fig. 13. Greenhouse gases emissions by sectors (Thousand tonnes CO_2 equivalent) [25]

3.14   Life Below Water

Conserving and sustainably using the oceans, seas and marine resources for sustainable development is intended in the context of Goal 14 [24]. According to Fig. 14, a closer look into the surface and ground water levels shows that ground waters were utilized more often in parallel with the fact that surface waters are reduced.

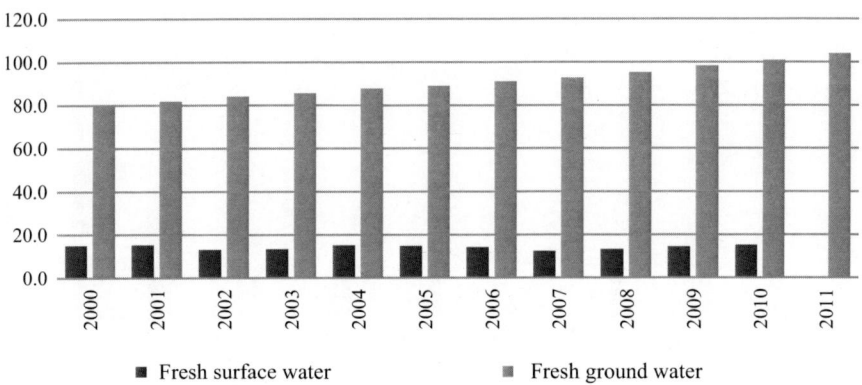

Fig. 14. Fresh water resources (%) [25]

3.15   Life on Land

Goal 15 aims to protect, restore and promote sustainable use of terrestrial ecosystems, sustainably manage forests, combat desertification, and halt and reverse land [24].

According to Fig. 15, it can be seen that forest areas are slightly increased in Turkey when the status of forest lands is reviewed. Data from the General Directorate of Forestry also confirms that the forestation activities are increased between 2002 and 2015 when compared to the period between 1946 and 2002.

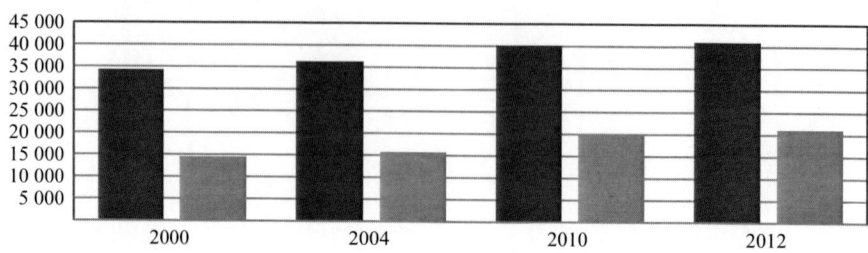

■ Net annual increment (bark) Thousand m³   ■ Annual fellings (bark) Thousand m³

Fig. 15. Land use (1000 m³) [25]

3.16   Peace, Justice and Strong Institutions

Goal 16 aims at promoting peaceful and inclusive societies for sustainable development, providing access to justice for all and build effective, accountable and inclusive institutions [24]. According to Fig. 16, a review of the sustainable development funding data shows that the funding provided to the private sector was decreased, while the total amount of funding was increased. Development subsidies included most commonly humanitarian aids and social infrastructure services.

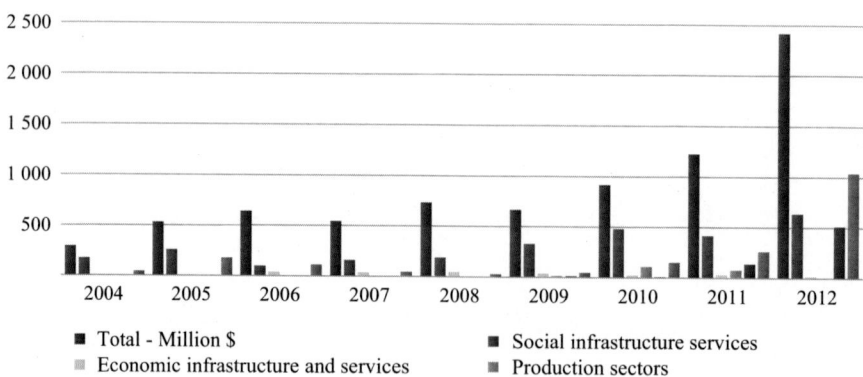

■ Total - Million $ ■ Social infrastructure services
▫ Economic infrastructure and services   ■ Production sectors

Fig. 16. Development assistance by type (Million $) [25]

3.17   Partnerships for the Goals

Goal 15 aims to protect, restore and promote sustainable use of terrestrial ecosystems, sustainably manage forests, combat desertification, and halt and reverse land [24]. According to Fig. 17, it can be seen that forest areas are slightly increased in Turkey when the status of forest lands is reviewed. Data from the General Directorate of Forestry also confirms that the forestation activities are increased between 2002 and 2015 when compared to the period between 1946 and 2002.

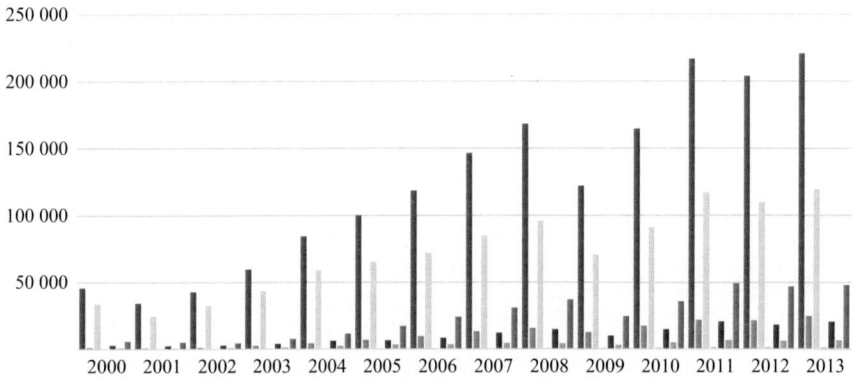

- Imports from developing countries by income group (Million $)
- China (including Hong Kong)
- Development assistance committee (DAC)
- Least developed countries
- Lower middle income countries
- Other low income countries
- Upper middle income countries

Fig. 17. Globalization of trade [25]

4   Results and Recommendations

Looking into the historical development of the concept of sustainability, it can be seen that awareness around this concept is raised as it is a living, changing and growing concept. Such awareness also led to the necessity of cooperation between past and future generations with a sense of responsibility in addition to individual obligations. First emerged from the possibility of disasters which may arise from environment and environmental issues, this concept has been developed and assumed a character which involves every aspect of life, not only the environment.

    One of the most important agenda items of the international process, sustainable development has found itself a place in almost every summit held in the past 50 years. International partnership has been a field which proved challenging with respect to consensus due to social, political, economic and cultural differences. This situation has reinforced the process in which sustainable development was conceptualized and became a common language. Mentioned for the first time in an international document in 1987, the roadmap for sustainable development was only drafted after 31 years as a result of regular meetings and summits. Existence of local, regional and global differences makes it harder to ensure the scalability of the concept. It is only possible to actualize the global goals set forth by the sustainable development movement with the realization of a globalization which will meet the expectations of every single person living on Earth.

In this study, data on the indicators used for the Turkey review belongs to the years before the goals were set. The subjects prepared in compliance with the Eurostat indicators correspond to those global goals also known as 2030 Agenda. This tells that in fact the goals were set long before the 2015 New York Summit, but it took a while until they were summarized [23]. Data shows that steps are taken in order to achieve sustainable development in Turkey even before the goals were fully defined. It is believed that the year 2030 which is pointed at for sustainable development goals will be a good time to make a sustainability test. In this process, international organizations, governments, local authorities and every individual assume separate and shared responsibilities. It is of utmost importance to create a sense of partnership in order to ensure the sustainability of resources for the next generations along with living in a healthy and suitable environment for human beings.

References

1. Cambridge Dictionary. http://dictionary.cambridge.org/dictionary/english/sustainable
2. Keleş R (2015) Urbanization policy (in Turkish: Kentleşme Politikası). İmge Booksotre Publications, Turkey
3. Macmillan Dictionary. http://www.macmillandictionary.com/dictionary/british/sustainable-development
4. Bartelmus P (1994) Environment, growth and development. Routledge Press, New York
5. Kara İ, Gültekin AB, Aliefendioğlu Y, Tanrıvermiş H (2016) An investigation of Turkey's real estate sector within the Scope of sustainable development and the human development index (HDI), smart metropoles-integrated solutions for sustainable and smart buildings & cities-SBE16 İstanbul. İMSAD, İstanbul, pp 420–433
6. WCED (World Commission on Environment and Development) (1987) Our common future. Oxford University Press, London
7. Bozlağan R, (1987) Historical background of sustainable development conception (in Turkish: Sürdürülebilir Gelişme Düşüncesinin Tarihsel Arka Planı), Sosyal Siyaset Konferansları Dergisi, No. 50, pp 101–1028
8. United Nations Environment Program. http://www.unep.org/documents.multilingual/default.asp?documentid=97&articleid=1503
9. Çamur D, Vaizoglu SA (2007) Important meetings and documents concerning environment. TAF Prev Med Bull 6(4):297–306 (Turkish)
10. Republic of Turkey Ministry of Foreign Affairs. http://www.mfa.gov.tr/birlesmis-milletler-insan-yerlesimleri-programi.tr.mfa
11. United Nations Human Settlements Programme. http://mirror.unhabitat.org/downloads/docs/The_Vancouver_Declaration.pdf
12. United Nations Documents Nairobi Declaration. http://www.un-documents.net/nair-dec.htm

13. United Nations Documents Our Common Future: Report of the World Commission on Environment and Development. http://www.un-documents.net/our-common-future.pdf
14. Michelle JE "Brundtland Report", Britannica Online (n.d.): Britannica Online, EBSCOhost
15. United Nations Documents, Rio Declaration on Environment and Development. http://www.unep.org/documents.multilingual/default.asp?documentid=78&articleid=1163
16. UN Habitat (1996) Istanbul declaration on human settlements, UN Doc. A/Conf. Vol. 165. https://documents-dds-ny.un.org/doc/UNDOC/GEN/N97/857/86/IMG/N9785786.pdf?OpenElement
17. United Nations Outcomes on Human Settlements. http://www.un.org/en/development/devagenda/habitat.shtml
18. United Nations Earth Summit+5. http://www.un.org/esa/earthsummit
19. Karakaya E, Özçağ M, (2004) Sustainable development and climate change: an analysis of application of economic instruments. In: 1st conference in fiscal policy and transition economies, University of Manas
20. Kavas K, Sezer S, (2002) After Johannesburg World summit on sustainable development (in Turkish: Johannesburg Dünya Sürdürülebilir Kalkınma Zirvesi'nin Ardından (in Turkish: Türk İdare Dergisi), 74(437), pp 1–25
21. Haines A (2012) From the earth summit to Rio+20: integration of health and sustainable development. The Lancet 379(9832):2189–2197
22. UN General Assembly (2012) The future we want. Resolution 66–288
23. Draft outcome document of the United Nations Conference on Housing and Sustainable Urban Development (Habitat III). https://www2.habitat3.org/bitcache/99d99fbd0824de50214e99f864459d8081a9be00?vid=591155&disposition=inline&op=view
24. Sustainable Development Knowledge Platform. https://sustainabledevelopment.un.org/sdgs
25. Turk Stat, Sustainable Development Indicators. http://www.tuik.gov.tr

Disseminating the Use of Sustainable Design Principles Through Architectural Competitions

Gizem Kuçak Toprak[(⊠)]

Faculty of Fine Arts Design and Architecture, Department of Architecture, Atilim University, Ankara, Turkey
gizem.toprak@atilim.edu.tr

Abstract. Sustainable architectural designs have become current issue, as in other disciplines, in architecture since 1970s because of these reasons, oil shock, global warming, to be consumed natural resources, get further away from the nature. Putting into practice and increase of awareness of sustainable architectural designs become a important topic 2000s in Turkey. It's hard to reach clear information about how much designs are carried into practice. Obtaining energy efficiency certificate for new buildings has become obligatory since 1 January 2011 and it's determined that 450,815 buildings have this certificate. But this certificate evaluates energy performance of buildings only. In fact, process of sustainable design cannot be evaluated by just energy performance analysis. Efficiently use of energy, water, building land and environment, relationship with environment and buildings, benefit/damage analysis in urban area, increase of user comfort in building come into prominence. 425 buildings obtained LEED certificate that make a comprehensive investigation. As it seen, the numbers, used above, show us, it's necessary that sustainable design principles must be disseminated and the methods, are been used, must be evaluated. This work tackles architectural competitions as a method for developing and disseminating sustainable design principles. Architectural competitions are not only didactic but also developer and independent design platform which all architects can join. The results are reached the large masses over draw up a contract, design process, evaluating the projects and discussion platform, colloquium. So that the establishment, designers, the jury and the other participants are involved in an interaction. Architectural competitions can used for development, evaluation, discussion and dissemination of sustainable design principles. The aim of this work is that evaluating how to tackle "sustainability" as a term in competitions, in 2010s and in accordance with the results, offering suggestions for disseminating the use of sustainable design principles with competitions also.

Keywords: Sustainable architecture · Architectural competitions

1 Introduction

The oil crisis of the 1970s and the emergence of the concept of global warming in the 1980s led humanity to reconsider its activities in every field [1]. As a result of this tendency, "sustainable architecture" became a major issue in the global agenda.

© Springer International Publishing AG, part of Springer Nature 2018
S. Fırat et al. (eds.), *Proceedings of 3rd International Sustainable Buildings Symposium (ISBS 2017)*, Lecture Notes in Civil Engineering 7,
https://doi.org/10.1007/978-3-319-64349-6_21

Today, sustainable architecture is interpreted in various ways and energy efficient building design, green buildings and ecological design are the most common definitions that come to mind. However, sustainable architecture is not only about energy or ecology. It has more extensive sub-meanings and it can even be considered a philosophical way of thinking.

Ayşin Sev defines sustainable architecture as "production of structures that prioritize the use of renewable energy sources; are environmentally conscious; use energy, water, materials and their locations effectively; and protect human health and comfort while taking future generations into consideration, under existing circumstances and at every stage of their existence," and lays out its principles in the diagram given below [1] (Table 1).

The strategies for implementing these principles are as follows (Table 2).

Table 1. A conceptual framework for sustainable design and construction, Principles [1]

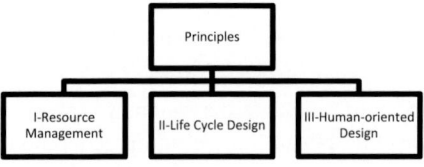

Table 2. A conceptual framework for sustainable design and construction, Strategies [1]

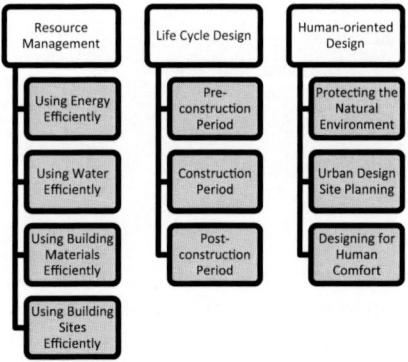

Jason F. Mclennan on the other hand, defines sustainable architecture in more general terms: "Sustainable Design is a design philosophy that seeks to maximize the quality of the built environment, while minimizing or eliminating negative impact to the natural environment" and relates it to day lighting, indoor air quality, passive solar heating, natural ventilation, energy efficiency, embodied energy, construction waste minimization, water conservation, commissioning, solid waste management, renewable energy, xeriscoping/natural landscaping and site preservation categories [2].

Paola Sassi points out that sustainable architecture has two main objectives: "First, sustainable buildings should metaphorically tread lightly on the Earth by minimizing the environmental impacts associated with their construction, their life in use and the end of their life. Sustainable buildings should have small ecological footprints. Second, buildings should make a positive and appropriate contribution to the social environment they inhabit by addressing peoples practical needs while enhancing their surrounding environment and their psychological and physical well being" [3].

As mentioned above, sustainable architectural design requires attention to detail, multidimensional thinking, correct assessment of data and a combination of various types of data in a single design, thus necessitating a comprehensive process. Teamwork is crucial for an adequate design process and these teams must involve collaboration between architecture and engineering professionals, consultants, clients, users, administrators, construction companies and building material manufacturing firms.

Kristel Del Mytlenaere lists three important questions that architects should ask themselves in this process, which requires multidimensional thinking [4]:

1. How architects propose to conciliate the human being with his natural environment?
2. How architects propose to articulate the different scales our humanity?
3. How architects propose to transmit to the future generations what they inherit from the past ones?

In a study conducted by RIBA this multidimensional process is divided into ten stages:

- commit the leadership
- benchmark practice performance
- demonstrate practice performance
- build on existing resources
- upskill with continuing professional development
- develop collaborative project methodologies
- consider the uses of software
- adopt a knowledge management framework
- follow the RIBA outline plan of work
- monitor sustainable projects [5].

As opposed to the comprehensive work in this field in Europe and the US, Turkey is lagging behind—particularly in terms of implementation—even though it is known that sustainable design has been on the agenda as "environmental design" in the 1970s, as "green design" in the 1980s, as "ecological design" in the 1990s and finally as "sustainable design" since mid-1990s [6].

Although this subject has been discussed conceptually since the 1970s, its appearance in practice happened much later. The subject became prevalent after 2009, when the KYOTO protocol was signed, and especially since 2011, when "Building Energy Performance" certificates became obligatory for new buildings. Nevertheless, the number of buildings that are built according to sustainable design principles in Turkey is still very small in comparison to the total building stock. The first place to look for data to evaluate this subject is the Turkish Statistical Institute (TUIK). According to TUIK, 1,348,211 buildings were given "occupancy permits" in Turkey since 2000 [7] and 450,815 buildings were given Energy Identity Certificates, which was made obligatory by the Ministry of Environment and Urbanism, since January 1, 2011 [9]. The Energy Identity Certificates evaluate buildings in terms of energy, water and building material use, but do not encompass the comprehensive scope of sustainable architecture. LEED on the other hand is a much broader certificate in this context and it has been received by 425 buildings [8].

These figures suggest that sustainable architectural design needs to be further promoted in Turkey. Many methods can be suggested for realizing this goal and this study aims to discuss the effectiveness of architectural design competitions for this purpose.

2 Disseminating an Idea Through Architectural Design Competitions

Architectural design competitions are, by nature, ways of obtaining projects by encouraging free thought. In these competitions ideas and designs are taken into consideration regardless of the architects' particulars (ideology, school, business contacts etc.).

> The concept of competition constitutes a distinct structure that involves elements of competitiveness, authentic production, comparison and rewarding. Perhaps the most influential feature of this method is the fact that it provides a basis for criticizing, testing and even challenging every aspect of our profession [9].

Architectural design competitions pave the way for disseminating opinions, movements and ideas. The features listed below are particularly confirming this aspect of competitions.

- Architects from every sector (public, private, academic etc.) can enter competitions and thus, dissemination of ideas among architects in every field is supported.
- Every submission is scrutinized by the jury. Jury reports that are published afterwards offer information on the submitted projects and their assessments. Therefore, competitions can inform us about the competency of the submitted ideas and encourage new ones.
- Designs that best comply with the rules of the competition are rewarded. The inclusion of sustainable design principles in competition rules—with the assumption that they are successfully implemented—and exhibiting successful projects will help these principles to spread.
- Competitors get a chance to observe and discuss their own designs in comparison with other submissions.
- From the announcement stage to the colloquium process, competitions reach a wide audience. The organizers, the members of the jury, the competitors and the colloquium participants are all involved in an interactive process.

Due to these characteristics, architectural competitions do constitute a ground for new ideas to emerge and spread. The Architects' Council of Europe confirms this and states that architectural competitions benefit all participants—the organizers, the public and architects—by offering alternative solutions to the client's building problems, allowing the public to witness the whole process from the beginning and forcing architects to face the challenge of comparing their strengths with others' to create appropriate and innovative solutions [10].

In this context, it is concluded that competitions have the potential to be a means to disseminate sustainable design principles.

3   Methodology

The study aims to identify the current status of architectural design competitions in Turkey in terms of sustainable design, and find out whether competing projects provide a means for spreading sustainable design principles. If these projects are not promoting these principles, then proposals will be offered to make this possible.

The study evaluates architectural design competitions organized in Turkey in terms of sustainable architecture principles. This evaluation focuses on the 2010s, which is the period that sustainable design principles started to reach wider audiences. Competition rules, jury reports, successful projects and project reports from this period are examined.

The evaluation tries to answer questions that are given below, the answers to which are expected to show at what level this subject is included in competitions, how they are handled by architects and juries, and whether competitions are effective in disseminating them.

- What does the competition expect from submissions in terms of sustainable design?
- How do the submissions deal with sustainable design criteria, and what types of competitions and projects include them?
- Do sustainable design criteria add value to the projects?
- How are competitions effected by the fact that the idea of sustainable design reached wider audiences in the 2010s?
- Is the philosophy of sustainable design an integral part of these projects?
- Are the newly developed ideas interacting with results of other competitions?
- Are the newly developed ideas repeating existing information or are they innovative?

4   Case Study

Similar to the development of sustainable architecture, development of architectural design competitions was also late in Turkey, compared to the rest of the world. The first competitions were held in the 1930s and the total number of competitions up to date is 867, 26% of them announced after the 2000s. Between 2010 and 2015 109 competitions were held and the rate is increasing.

The table below shows the types and numbers of competitions held between 2010 and 2015 (Table 3).

Table 3. Types and numbers of architectural design competitions held between 2010 and 2015

Type of project	Number of competitions
Prison	1
Religious building	4
Educational building	5
Call for ideas	30
Public building	29
Urban design	11
Residential	1
Tower	1
Culture and arts building	4
Tomb	2
Museum	2
Healthcare facility	3
Social facility	7
Sports	1
Commercial	1
Transportation	4

Below are some findings from the analysis of these competitions.

- Among these 109 competitions, 7 are directly about developing sustainable design ideas and one of those is an urban design competition (Çanakkale Municipality "Green" Urban Design for City Square and Surroundings Competition) while six of them (National Architectural Design Competition by Ytong: Roofs and Sustainability, Urban Dreams 5: EGO Hangars and Field Assessment National Student Project Competition, Sustainable Reception-Exhibition-Presentation Space National Student Project Competition, Urban Dreams 8: Saraçoğlu Neighborhood Assessment Project National Call for Ideas, 7 Climates 7 Regions National Architectural Design Competition, Houseboat Design Competition) are calls for ideas and are not application-oriented.
- Other competitions call for designs that comply with relevant legislation and therefore ask projects to comply with the Building Energy Performance Bylaw.
- Most competitions expect projects to take sustainable design criteria into consideration in one way or another. However, the subject is dealt with only a few superficial sentences in competition rules.
- The competition rules include concepts such as energy efficient building design, using renewable energy sources, environmentally sensitive design, design for all, sustainability of quality of life, multidimensional sustainability and sustainability of the architectural idea.

- Among these concepts, the submissions predominantly deal with energy efficient design. Double facade designs, using solar and wind energy, recycling water, green roofs, green shells, green wall designs, natural ventilation, orientation and natural temperature control using earth are some of the themes found in competing projects.
- Spatial sustainability, inheriting the design to future generations, sustainability of the building-environment relation and flexible space design are dealt with in project reports, despite not reflected in the designs themselves.
- Jury reports demand sustainable design principles to be considered at the implementation stage, when and if these projects are actually implemented.
- Sustainability are handled more effectively in call for ideas where it is the main theme of the competition.
- There is an interaction between competitions and especially the ways they handle energy efficient design are quite similar.

5 Conclusions and Proposals

As a result of the analysis, it is concluded that architectural design competitions can have a dissemination function for sustainable design criteria if improvements are made in aspects listed below.

- Competition rules only superficially indicate sustainable design expectations.
- Successful projects are mostly handling sustainability in terms of energy efficiency, mainly due to the fact that the Energy Performance Bylaw is one of the regulations that the projects are asked to comply with.
- While other aspects of sustainable design (spatial sustainability, social sustainability, economic sustainability etc.) are mentioned in project reports, these cannot be read in the designs.
- Sustainable design principles add value to architectural projects and this is indicated in jury reports.
- Calls for ideas deal with this subject in more detail and allow new ideas to be tested. These projects are focused more on improvements in this field rather than repeating existing methods.
- As mentioned earlier, implementing sustainable design principles requires comprehensive and systematic work by a large multidisciplinary team. Interviews with competitors show that the prescribed time for submissions are not long enough for both interpreting the given program and integrating sustainable design criteria in the design. This is mainly due to the fact that these two subjects are still considered separate and sustainable design is not adopted as a design philosophy.

These results show that architectural design competitions have not been effective tools in disseminating sustainable design criteria. "Sustainable design" needs to be included properly in competitions before it can be disseminated by them. For competitions to effectively disseminate these principles,

- Juries must include architects, and engineers if necessary, who are competent in this subject. The importance of the issue must be emphasized, the quality of the projects must be increased and successful projects should communicate these principles properly.

- Sustainable design expectations must be expressed in detail, not only in the prefaces of the competition rules but also in architectural programs, ensuring that the subject is brought to the attention of not only the competing architects but also everyone that is interested in the program.
- The prescribed submission periods must be revised in order to allow more in-depth study on the subject.
- The subject should be handled in all competitions as comprehensively as it is handled in calls for ideas.
- Application of sustainable design ideas should not be delayed to the implementation stage. This way competition colloquiums can host noteworthy discussions and these can be shared with everyone.
- Travelling exhibitions should be organized so that the audience is not limited to the city where the colloquium is held, and exhibitions and colloquiums should be open to the general public.

References

1. Sev A (2009) "Sustainable Architecture", Turkey, Yem Yayın 9, 31, 38
2. McLennan JF (2004) The philosophy of sustainable design: the future of architecture. Ecotone Publishing, Kansas City, p 4
3. Sassi P (2006) Strategies for sustainable architecture. Taylor & Francis Publishing, Boca Raton, p 13
4. De Mytleanere K. Towards sustainable architecture. In: Bodart M, Evrard A (eds) Architecture and sustainable development. Universitaire de Louvain, Louvain-la-Neuve, Belgium, p 21
5. Sulivan L (2012) The Riba, guide to sustainability in practice, pp 17–27
6. Durmuş Arsan Z (2008) Türkiye'de Sürdürülebilir Mimari, Mimarlık Dergisi, 340, March–April 2008
7. Türkiye İstatistik Kurumu (TUİK). https://biruni.tuik.gov.tr/yapiizin/giris.zul
8. http://www.usgbc.org/leed
9. Özbay H (1993) Yarışmalar Sahip Olduğumuz Tek Sağlıklı Kurumdur. Mimarlık Dergisi 251:s:34–s:35
10. The Architects' Council of Europe (2002) "Avrupa Mimarlık ve Yarın", Beyaz Kitap

The Healthcare Industry and Medical Devices in Environmental Sustainability: Medical Device Industry is Going Green

Meriç Yavuz Çolak[1]([⊠]), Levent Çolak[2], and Elif Gürel[3]

[1] Healthcare Management Department, Baskent University, Ankara, Turkey
meric@baskent.edu.tr
[2] Mechanical Engineering Department, Baskent University, Ankara, Turkey
lcolak@baskent.edu.tr
[3] Biomedical Engineering Department, Baskent University, Ankara, Turkey

Abstract. In the whole world there are several developments in order to make a more sustainable and qualified environment to handle the increase costs of the healthcare services industry. The changes in health sector also affect the medical device sector directly as the medical device sector is the most important key of health sector and that is why the concepts element such as environment, sustainability and green are very important in medical sector. The main purpose of the research is to point out the applications of medical device firms concerning environment and sustainability. The research has been planned and carried out as a cross sectional field work in defined type. The medical device firms running in the city center of Ankara are involved in research population. The research is carried out on 60 medical device firms which accept the research within a months of time. Compared with other sectors, it has been concluded that producing sustainable products in medical sector will be more effective, energy applications will be reducing the treatment costs in healthcare, some medical firms in Turkey are ready for energy efficiency applications and energy efficiency applications will reduce 50% energy costs and will protect 50% the environment according to the medical firm's opinions.

Keywords: First keyword · Second keyword · Third keyword

1 Introduction

The healthcare industry sector in Turkey is proportionally developing according to the supporting developments related to the healthcare sector and increasing purchasing power resulted from positive economic improvements and increasing accessibility. The changes in health sector also affect the medical device sector directly as the medical device sector is the most important key of health sector and that is why the concepts element such as environment, sustainability and green are very important in medical sector. Subsequent paragraphs, however, are indented.

Medical device sector is a sector that contain many different product groups and high level technologies and it may be different from countries to countries. Hospitals, laboratories, outpatient clinics, drugs and medical devices compose the health industry.

© Springer International Publishing AG, part of Springer Nature 2018
S. Fırat et al. (eds.), *Proceedings of 3rd International Sustainable Buildings Symposium (ISBS 2017)*, Lecture Notes in Civil Engineering 7,
https://doi.org/10.1007/978-3-319-64349-6_22

Generally drugs, cosmetics, living animal cells and human cell, organs out of transplant organs and most every instruments used in the hospitals are all within the scope of medical devices. Imaging devices, protheses, cardiac pacemakers, defibrillators, lighting devices and many devices and product groups can be classified in medical device sector.

In the whole world there are several developments in order to make a more sustainable and qualified environment to handle the increase costs of the healthcare services industry [3]. Although it is hard to difficult to say the total expense of R and D expenditures in the world, in global label, the total budget for R and D for the most biggest firmd that have a 60% or more part in market is 19.808 USD and this amount is greater than 10% of total income. This total is indicates that the significant amount of income in the sector is used to R and D activities and with compared other sectors, this amount is the indicator of this rate is much greater than the average amount [4].

1.1 Medical Device Sector in the World

Since 2010, world medical device market is reached to 250 million USD and more. If the medical device market ordered according to size, in 2010, the biggest markets in the world are; USA, Japan, Germany, France, England, Italy, China, Canada, Russian, Spain and Switzerland [3]. In Fig. 1, medical device sector market share according to countries were given.

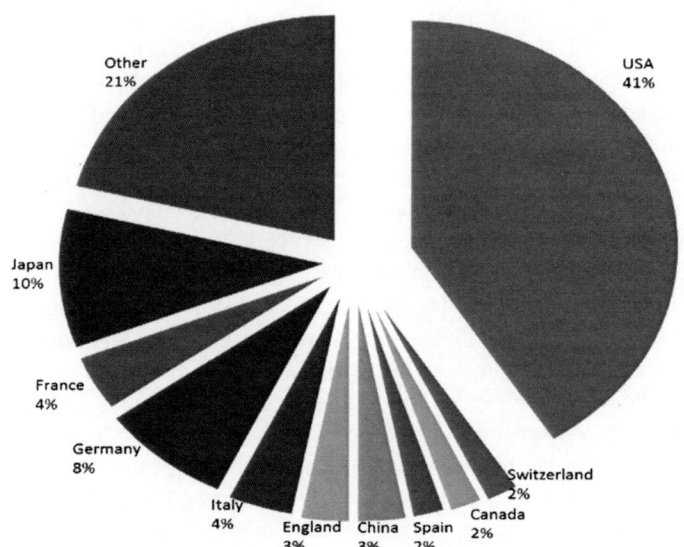

Fig. 1. Medical device sector market share according to countries—2010 (%)

In medical device sector especially USA, West European and Japan are coming to the forefront when the market sizes and researches and investments are considered [3].

If value added perspective in the world between the countries is considered, China is the country that the annual rate of increase is the maximum with the rate of 15.8%. Turkey with 6.0% annual rate of increase has value added parallel to the world average. Turkey is in the 19th order in the world sector with 1.9 million USD value added. Turkey is a country that progress consistently and important economic developments were existing in recent years. With economic improvements, health expenditures are increasing and parallel to this, the position of our country in health sector is growing with big acceleration compared with other countries.

1.2   Medical Device Sector in Turkey

Turkey medical device market has been growing gradually and is ranked as the largest in the region of the Middle East and Africa. According to BMI Espicom, the Turkish medical device market reached over USD 2.2 billion in 2012. Moreover, it is forecasted to grow at a CAGR of 8.5% through 2018, buoyed by strong imports. The expansion of healthcare facilities and rising health expenditure will be the main drivers of projected fastpacked growth. There are 6000 medical device company as small, medium and large scale. The number of suppliers are greater than 2500. 100 of 450 medium and large scale company are producer and supplier. According to 2012-year data's, the number of producer firm is 1548 [4].

The medical device sector has shown a stable growth rate between years 2005 and 2008. As a result of global crises after 2009, turkey medical device sector is reduced parallel with world medical device market. In 2010, the market again was recovered and at the end of 2010 the market size was reached to 1.9 billion USD [1].

By the end of year 2010, Turkey medical device market is the one in the 20 biggest medical device market in the world [3]. As Turkey's medical device sector grows, so does the need for high technology medical devices in parallel with the planned expansion in healthcare infrastructure. The number of MRI and CT devices per 1 million populations is 12.2 and 15.1, respectively. Although Turkey is above many of the upper income countries, it still lags behind the OECD average. Considering the projected growth potential, it is expected that Turkey will follow global trends and increase the number its MRI and CT devices to satisfy increasing need.

The Ministry of Health (MoH) sets targets and objectives to improve capacity, quality and distribution of health services as well as ensuring that both infrastructure and technology is sustainable and up to date. It also ensures accessibility, safety, efficacy and the effective use of medical devices. These areas are of the utmost importance to the Ministry. Since the production of medical devices is limited in Turkey, a large share of medical devices is imported. Therefore, the MoH aims to increase the export/import coverage ratio gradually from 12.7% in 2011 to 17% in 2017 and to 22% in 2023 [3]. Leading global medical device producers have chosen Turkey for their regional headquarters, R&D center(s) and/or for their production facilities. Globally renowned companies are among the major players of medical device industry of Turkey as supplier. Construction of new, city hospital complexes provide significant opportunities for medical device companies that are being commissioned to equip the [3].

1.3 Environmental Sustainability in Health Industry and in Medical Devices

The healthcare sector is a whole that include hospitals, outpatient clinics, laboratory, drugs and medical devices. The purpose of health services is to better the quality of life. The development of countries not only cover the economic indicators but also cover education, health, culture, social structure and technology indicators. In addition, sustainable development cover being healthy in longtime. Health expenditures are also development indicators therefore, healthy people and qualified human resources are very important for sustainable development.

The main equation of health sector is providing healthcare services in a qualified, sustainable and easily accessible. The balance between technology and finance is the main building block of sustainability. In Fig. 2, this balance is given.

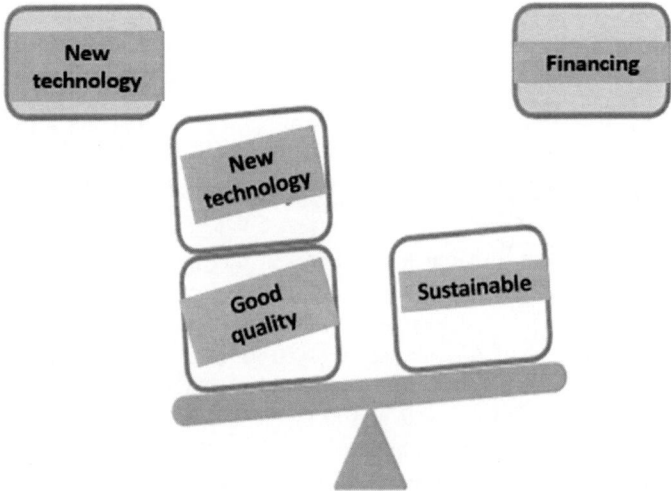

Fig. 2. Health sector in the context of financial and new technologies

Environmental sustainability in health sector is a connection between environmental operations and enhanced healthcare services. It can be easily said that environmental sustainability can decrease the operational cost in the health industry. In recent years, hospitals go green and although in health sector made major strides in sustainability, worldwide major strides will be needed to improve the health services.

Sustainable preventions designed to decrease the energy and water wastes provide direct financial return. For this reason, the importance of environmental sustainability in terms of finance is very much. Medical devices that comprise less chemicals, healthier foods, destruction of chemicals and toxic gases accepted as sustainable initiatives results in healthier results in terms of patients. If the complexity and diversity of health systems are considered, it is very difficult to construct sustainability into the clinical operations and to the hospitals. But the environmental friendly green hospital buildings and efficient usage of resources and using less chemicals in the products enable the sustainable healthcare services [5].

Medical device sector is the initiator sector of R and D an innovation activities and is the big actor of the innovation. The innovation of new product process is very long in medical device sector but the lifecycle of products is very short. Since 1990, the works of green firms on sustainability has been derived a profit to the firms and it has been seen that well conducted and strong firms have a strong environmental administrative systems and programs [2].

If price competitive products are a matter, consumers prefer sustainable products. But it is known that the percentage of consumers who prefer sustainable products to cost and more qualifications are very low. Successful firms and firms have a profit and have global purchasing and have significant voice on marketing have to take on a task on sustainability [2]. There is defined standards and directives to increase the value of medical devices designs in the world. According to European and international environmental standards, actions to be taken for environmental problems occurred during the design of products are standardized. Lean production and quality management systems in medical device and health sector decrease the production costs and support the environment performance.

2   Material and Method

The main purpose of the research is to point out the applications of medical device firms concerning environment and sustainability. The research has been planned and carried out as a cross sectional field work in defined type. The medical device firms running in the city center of Ankara are involved in research population. The research is carried out on 60 medical device firms which accept the research within a months of time. Field survey is carried out at R and D medical device firms and firms that sales medical devices and products in technicities of universities in Ankara, Turkey.

Data are collected by a questionnaire constructed by literature review. Questionnaire is consisting of two part. In the first part, questions about the properties of firms are exist and in the second part questions about the environmental sustainable applications of firms and questions about general responses to sustainability are exist. The name of the firm, foundation year of the firm and total number of employee, field of activity and the qualifications are questioned in the first part. In the second part, generally the environmental and sustainable applications of firms and their opinions about these subjects were questioned. Questionnaire was applied by face to face application. The average duration of questionnaire is approx. 20 min.

3   Findings

Findings of the study are given in three headings. Firstly, the general firm properties and then sustainable environmental properties of medical device firms are given.

3.1 Findings About the Properties of Medical Device Firms

Founding years of firms are changing between 1979 and 2016. Number of employees is at least 10 and large companies with 2000 employees are also exist. 51% of firms were established after the year 2000. The number of employees is under 25 in 64% of the firm. 83% of the firms have quality certificate. The most of the firms have ISO quality certificate (53.3%). 31.2% have CE certificate. Less stated certificates are FDA, GOHST-R, HSEQ. Most of the firm's field of activity are selling a product (88.3%) and 43.3% of firm's field of activity are product development.

3.2 Findings About Environment and Sustainability Properties of Medical Firms

71.6% of firms are indicated that they undertake environmental problems and 43.3% of these environmental issues are energy conservation issues. 59.3% of firms are defined their firm as partially environmentally friendly. 43.3% are applying managerial system considering the environmental risk factors (Fig. 3).

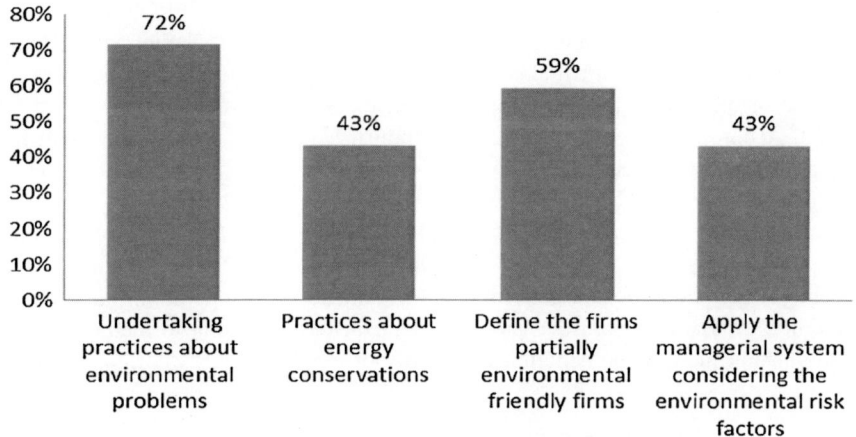

Fig. 3. Undertaking environmental issues of medical device firms

29.1% of 24 medical device R and D firms are stated that they are not applying eco design. But 16.6% of R and D firm are stated that they are targeted to apply in next years. Some of the firms not applying eco designs are stated that they have no ideas about these subject and some of them are stated that they have no financial resources and some of them stated that priority is sale in their firms.

63.3% of firms are defined their medical devices sold or designed as environmental friendly. 83.3% of firms stated that they are not using applications that reduce the greenhouse gases affect and energy efficient applications. In 16.7% of the firms, recycling of packing wastes, waste water discharge emission measurement and avoiding some toxic gases before spread to the air by exhausted filter line (Fig. 4).

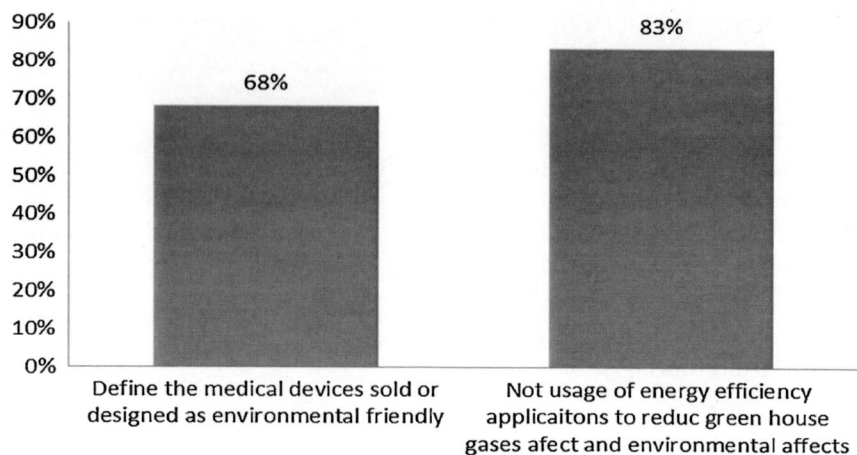

Fig. 4. Environmentally friendly medical device firms

Sustainability concept is very important for 86.7% of firms and 65% of them stated that sustainability would be effect According to results, sustainability and increasing the operational efficiency and reducing the cost is the most important factors in accomplishing the environmental sustainability of medical devices designs. Increasing the value of the firm is the other important factor according to them. their firm after 5 years. According to 65% of firm, green industry would be change the biomedical innovation studies in near future. In the questionnaire, the importance of some conditions accomplishing the environmental sustainable friendly medical devices are evaluated by giving them a score between 0 and 100 and then taking the of the scores. "0" is the less important factor and 100 is the most important factor. Results are given in Table 1. Sanctions of non-governmental organizations competition and individual preferences of designers are the less important factors according to medical firms in accomplishing the environmental sustainability of medical devices designs (Table 1).

Table 1. Opinions about the important factors in accomplishing the environmental sustainability of medical devices designs

Important factors in accomplishing the environmental sustainability of medical devices designs	Average
Satisfying the target of companies	60.0
Individual preferences of designers	51.5
Satisfying customer expectations	64.0
Sustainability	77.5
Increasing the operational efficiency and reducing the cost	77.2
Not commercial concern in environmental products due to the technologic developments	67.5
Increasing the value of firms	71.1
Sanctions of Governmental corporations	68.6
Legal sanctions	71.2
Competition	59.5
Sanctions of non-governmental organizations	59.0

In the same way barriers in accomplishing the environmental sustainability of medical devices designs are coded between 0 and 100 and averaged. Lack of education about these subjects and cost is the most important barrier according to medical firms. Priority of other designs and Risks (e.g. hygiene risk of recycle) are found as less important barriers in the study (Table 2). 48.3% of firms stated that lack of environmental sustainable products is the most important barrier in purchasing of these products in hospitals and financial factors are the most important barriers of green products usage in hospitals according to 68.3%. The most emergent subject in designing environmental sustainable products is legal sanction of governments to the hospitals and medical device producers about sustainable products. 60% of the firm stated this subject the second emergent factor is educating the designers of medical devices and service payer and increasing the preferences of disposable products instead of recycle products (Fig. 5).

Table 2. Opinions about the barriers in developing sustainable medical device design

Barriers	Average
Cost	72.3
Lack of education	77.6
Priority of demand and customer	65.9
Risks (e.g. hygiene risk of recycle)	52.5
Reliability of disposable products	57.8
Priority of other designs	52.0
Legislations	57.7

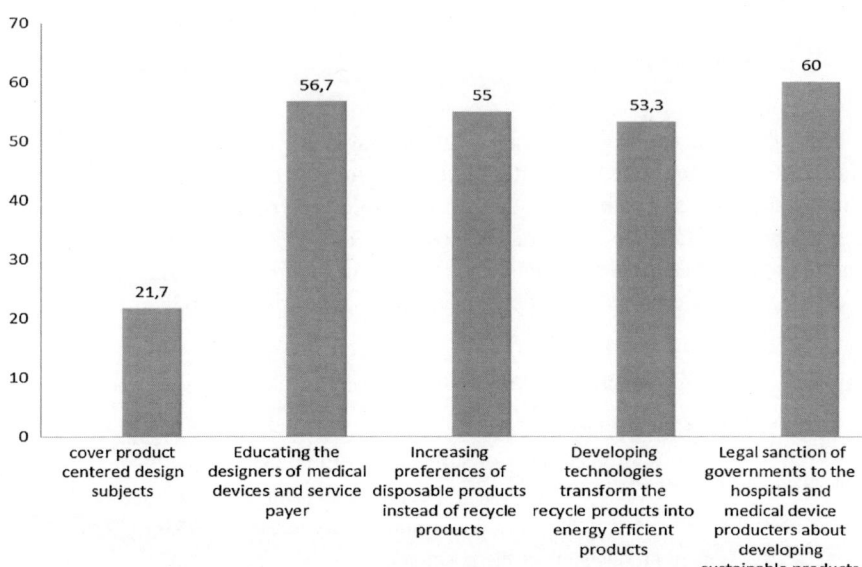

Fig. 5. Emergent subjects in designing environmental sustainable products

Opinions of firms about the importance of defined situations in the context of sustainability in health are scored between again 0–100 and then averaged. Cost is the most important situation and then productivity, energy efficiency, patient safety are the other important situations in the context of sustainability in health (Table 3). Sustainable solutions are more effective in medical device sector compared with other sectors according to 43.3% of firm. 58.3% of firm stated that, "green" concept has an important affect in purchasing diagnostic and treatment healthcare products decisions. 35.1% of firm stated that, the other important factor in purchasing medical products is the nonincluding of heavy metal in products. According to 29.8% of firm, energy efficiency of medical devices is important factor in purchasing these products.

Table 3. Opinions about the importance of defined situations in the context of sustainability in health

	Average
Long term care without giving any harm to the patient	68.1
Minimum cost medicine	69.7
Decreasing the harmful effect to environment	76.5
Usage of high technology devices	71.1
Productivity	78.7
Energy efficiency	76.8
Quality of life of employees	63.0
Recycle of medical devices used in diagnose and treatment	60.5
Patient safety	75.1
Improvement of health and life conditions	74.9
Cost	79.6
Design of medical devices	56.4
Less usage of materials	62.6
Legal applications	70.6

There is a direct utility of sustainability to humanity according to 86.7% of firm. About 72% of them stated that, sustainable products protect the hospital personnel. Opinions about green product usage in hospitals are very optimistic. 71.7% stated that green hospitals enrich the value of the hospitals and 65% stated that green products improve the health outputs and approximately half of the firm stated that green products effect the hospital choices of patients (Table 4). According to 65% of participants,

Table 4. General opinions about sustainability and green products

General evaluations	Agree %
There is a direct utility of sustainability to humanity	86.0
Sustainable products protect the hospital personnel	71.7
Green products improve the health outputs	65.0
Green products effect the hospital choices of patients	46.7
Green hospitals enrich the value of the hospitals	71.7

investments of hospitals to sustainable solutions is very important for hospital's success and also 85% stated that, in deciding purchasing products the energy considering the lifetime energy costs of medical devices and equipment's, energy strategy must be defined to accomplish the objective of energy conservation and efficiency in hospitals. Also 85% of them indicated that, in hospitals the purchasing units must be canalize to healthy products and devices in health sector. In addition, without legal sanctions, it is impossible to canalize to green products and eco designs and sustainable products in hospitals and also in medical device producer firms according to 80% of firm. But 66.7% are thinking that quality management system and lean production would be decrease the costs of product and would be increase the environmental performance.

4 Conclusions and Suggestions

The main purpose of the research is to point out the applications of medical device firms concerning environment and sustainability. Executives at all levels see an important business role for sustainability. According to executives, sustainability is becoming a more strategic and integral part of their businesses. Operational efficiency and reducing the cost is the most important factors for sustainable products. They believe that sustainability would be increase the value of the firm. Green industry in healthcare and especially in medical device sector would be the main actor in near future according to the medical device firms.

Most of the firms are undertaking environmental problems especially energy conservation and are managing the environmental risk factors. Although 65% see their firm as environmentally friendly but eco designs are not common in R and D medical device firms and 83.3% of firms stated that they are not using applications that reduce the greenhouse gases affect and energy efficient applications. Only 30% of R and D firms are applying eco designs. Lack of education on these subjects were given as primary reason for not using eco designs. Sanctions of non-governmental organizations and competition and individual preferences of designers are the less important factors according to medical firms in accomplishing the environmental sustainability of medical devices designs. Most of the firms stated that, there is a direct utility of sustainability to humanity and sustainable products protect the hospital personnel. Compared with other sectors, it has been concluded that producing sustainable products in medical sector will be more effective, energy applications will be reducing the treatment costs in healthcare, some medical firms in Turkey are ready for energy efficiency applications and energy efficiency applications will reduce 50% energy costs and will protect 50% the environment according to the medical firm's opinions.

"Green" concept has an important affect in purchasing diagnostic and treatment healthcare products decisions. According to their opinions, the purchasing units must be canalizing to healthy products and devices in health sector especially in hospitals. In addition, without legal sanctions, it is impossible to canalize to green products and eco designs and sustainable products in hospitals. Quality management system and lean production would be decrease the costs of product and would be increase the environmental performance according to the most of the opinion of firms. Healthcare costs in all industrial nations have increased and payers are starting to look at new ways to contain costs and at new funding models. The business model of medical device companies is also undergoing rapid changes. R&D costs have increased year on year, pressures on purchasing decisions and the efficacy and safety of devices and products are mounting. Change is therefore inevitable and already ongoing in healthcare systems and medical device companies alike. Even though medical devices are considered to be heterogeneous and classified in many other sectors such as chemicals, textiles and electronics, they have common features sufficient to be considered as a special product group and being an important part of the healthcare system, they are subject to common regulations. Medical devices sector also suffers from regulations that put cost on innovative activities, reimbursement policies that aim at cost containment, lower degrees of consumer support (in terms of user-producer relationship), high marketing costs due to the specific market they act in, in addition to the general obstacles such as scarce finance and human resources. Nonetheless, the ambiguity in entrance and allowance to reimbursement lists is also found to be a blocking factor on innovation.

Medical devices (MDs) are not only an innovative industry but also a key contributor to healthcare supply. Albeit unseen in daily routines, the medical devices are crucial to accurate diagnosis, treatment and even for prevention of diseases. Being able to prolong human life, medical technologies are also an item of expenditure with rising costs. Even though publicly it is normal to wish for the best healthcare possible, yet how to fund for these ever-increasing costs is a question hard to answer. Moreover, increasing healthcare expenditures are often related to innovation and R&D expenditures in medical devices and pharmacy. Mostly reimbursed by governments, health industry becomes an issue of importance that needs to be appropriately regulated.

References

1. Ankara Tıbbi Cihazlar Sektör Analizi (2013) Ankara: Kalkınma Ajansı (in Turkish)
2. Commonwealth of Massachusetts Executive Office of Environmental Affairs Office of Technical Assistance and Technology (2006) An environmental guide for the medical device industry in Massachusetts
3. Deloitte (2012) Türkiye Sağlık Sektörü Raporu. Uluslar Arası Yatırımcılar Derneği (in Turkish)
4. Dünyada ve Türkiye'de Tıbbi Cihaz Sektörü ve Strateji Önerisi (2013) Ankara: Türkiye Teknoloji Geliştirme Vakfı (in Turkish)
5. The Growing Importance of More Sustainable Products in the Global Health Care Industry (2012) Johnson & Johnson, New Brunswick

Assesment of Historical Kars Stone Bridge in Terms of Sustainability of Historical Heritage

Elif Tektas[✉], Zeynep Yesim İlerisoy, and Mehmer Emin Tuna

Faculty of Architecture, Department of Architecture, Gazi University,
Ankara, Turkey
{eliftektas,zyharmankaya,mtuna}@gazi.edu.tr

Abstract. Due to its geographical location, Anatolian lands which was home to many various civilizations in centuries, when examined in terms of Turkish history, includes many historical structures in the term from ancient ages until our day. Protecting historical structures is an inevitable input in terms of; continuity of functionality; development of the society and cities and improvement. Besides housing which is the most basic need, bridges connecting the two sides of lands in order to provide transportation and communication requirements, have an important place in terms of cultural heritage. While most bridges in Anatolia managed to stay alive over time, most of the took severed damages and even came to the point of destruction due to intense natural hazards like earthquakes, floods, wars and many other more external factors. Everyone is responsible for ensuring the protection, livelihood and sustainability of these structures, which are common to all people. The sustainability concept exploring the harmony/ balance between the preservation and use of historical and cultural assets was discussed in Taşköprü (Karahanoğlu Bridge), which connects the northern side to the south side in the center of Kars. The arched stone bridge belonging to the Ottoman period is a rare cultural asset of the city. Within the scope of the study, the repairs, the interventions and the current situation of the historical stone bridge were examined. In order to ensure the preservation of the building and the continuity of its function, the evaluation of the historical touch around the bridge, the continuity of the surrounding green areas, and the transfer of the bridge to the future are emphasized. As a result, this work will be a source for both the building itself and other buildings with similar characteristics, contributing to the preservation of the historical heritage and its transfer to future generations.

Keywords: Conservation · Historical buildings · Kars stone bridge
Sustainability · Sustainability kars stone bridge

© Springer International Publishing AG, part of Springer Nature 2018
S. Fırat et al. (eds.), *Proceedings of 3rd International Sustainable Buildings Symposium (ISBS 2017)*, Lecture Notes in Civil Engineering 7,
https://doi.org/10.1007/978-3-319-64349-6_23

1 Introduction

Space and society complement each other and form a whole. Since the earliest times, people have been seeking to create spaces where they can meet, meet their needs, and carry on their vital activities. Throughout history, there has been a number of factors that have affected the phenomenon of space creation in the process of change and transformation, and the transit of the nomadic settled life. Water entity, which has a great role in people's settled passages, has been associated with urban centers. In the developing societies, various water structures have been constructed to provide accessibility from water or to water.

In the Turkish art, bridges, hammams (Turkish bath), wells, aqueducts, cisterns and fountains can be considered as water architecture [1]. Bridges are among the oldest historical monuments of bridge civilization, which is one of the most important water structures in the field of architecture. The bridges that give way to all life and human production have become the symbols of cities with their splendor in many countries today. The country, which has a rich culture with its history, is also in the forefront with its stone bridges [2]. The Turks developed the bridge architecture against the geographical conditions, rivers and caravan roads of the regions where they dominated and dominated and provided great successes [1]. Bridges built from day to day; They vary according to their purpose, techniques and uses. But they were all made to be able to answer people's needs. Especially, the bridges that have been sung from the Ottoman times can be considered as an artwork. The important thing is to keep these things as meticulously as possible [2].

The concept of conservation is a multidimensional phenomenon with many components in it, and the most important feature that preservation covers is that the historical environments are kept alive with the values they carry. Historical structures, which are an important part of our historical heritage, need to be addressed with sustainable protection policies for their protection and survival in terms of their existence in the period and their social relations and design features. In order to achieve this, it is necessary to be aware of the changes and transformations that have occurred in historical circles and to develop appropriate legal and design methods. According to this philosophy, which is the foundation of sustainable environment and historical environment, The preservation and restoration of the structures, the maintenance of the continuity of its functions, and the development of the community and cities.

Historical buildings can be evaluated in terms of sustainability. The historical stone bridge hosted by Kars, which houses a rich historical texture as a gateway to the Caucasus region from Anatolia, was evaluated. Within the scope of the study, the repairs and the current situation of stone bridge throughout history are examined. The studies carried out to ensure the preservation and the continuity of the function are aimed at the study of the use of the bridge and the evaluation of the interaction with its immediate surroundings in the context of sustainability. For this purpose, a literature search was conducted in order to define what it is doing and to gather information about other historical buildings in the region. In the direction of the information obtained within the scope of the survey, a study tour of the study area was conducted. Conserve with consciousness how to transfer safely to future generations.

2 The Location and History of the Bridge

Since the earliest times of human history, Anatolia has witnessed many turning points due to its climate and geographical features. The province of Kars located in the north-east of this region has been used as a settlement since the Paleolithic Age. It is defined as the Civilized City of Kars East, which has long been home to a long history, incredible heroism and cultural treasures [3]. The geopolitical position that connects Anatolia and the Caucasus, and the fact that it has important trade routes and is very suitable for settlement and defense made it desired to acquire continuously throughout history [4]. It joined the Ottoman Empire in 1535 and became an important military superior to the Iranians and Russians until the Ottoman-Russian war of 1877–878. Between 1878 and 1918, it remained in Russian occupation for 40 years. Although it was taken back in 1918, it entered Turkish soil in 1920.

The Kars Stream passes through the center of the province of Kars located on the northeastern sides of the Eastern Anatolia region. One of the most important ecological values of the city, Kars Stream seperates, Kaleiçi and the skirts of the Ottoman-Seljuk settlement and the Russian-era Tahtdüzün's grid-planned pattern in the development of a natural border with functionality as its seen (Fig. 1). Taşköprü (Karahanoğlu Bridge),

Fig. 1. Satellite View of Kars

which is included in the scope of the study, is also located on the Kars Stream in the skirts of the castle, at the center.

In Kaleiçi District, stone bridge that is at the point which connects Mimar Oktay Ekinci Street and Erzurum Street (Figs. 2, 3). It was constructed with the order of Sultan Murat the Third by Lala Mustafa Pasha during the construction of the city at 1579 [5].

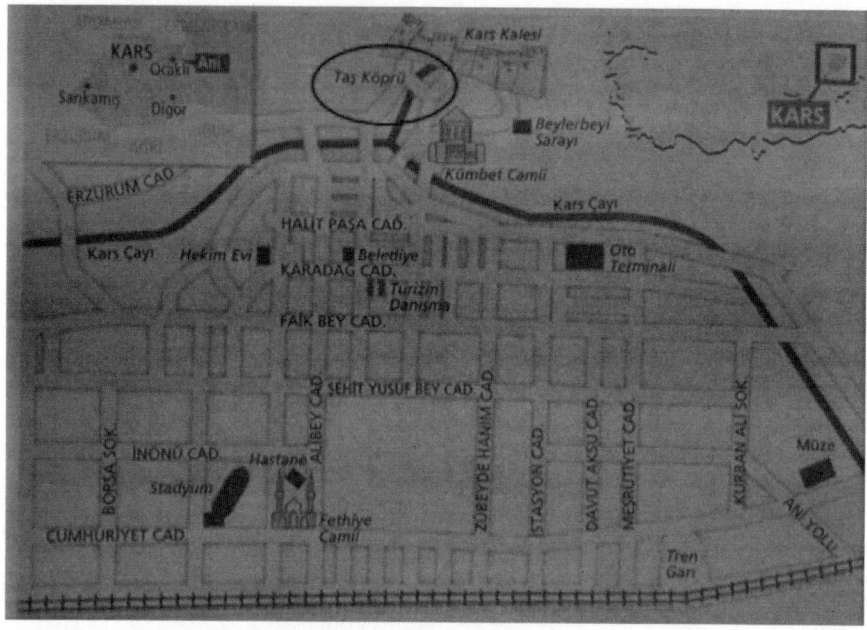

Fig. 2. General settlement plan of the bridge's environment [3]

Fig. 3. Kars Stonebridge at the year of 1875 [6]

The existence of functional values in bridge architecture and construction technology is of great importance. Due to both the facilitating and accelerating effect of the army's voyages and its role in the construction of the city, the permanence values of the bridges that hold high construction technologies in every period should be protected by the related institutions in the regions where they are located. Historical stone bridge which served for purposes such as military, commercial, education throughout the history and which continues to exist with this function today is also registered as a monumental work with the resolution numbered 330 of Erzurum Cultural and Natural Heritage Preservation Board on 12.09.1991.

3 Architectural Features of the Bridge

Like most architectural productions, bridge architecture reflects the characteristics of the geography in which it is located. It is shaped by the characteristics of the rivers on which the bridge is built [7]. Arch-bearing stone bridges are one of the most common applications for bridge typology with Ottomans in the fifteenth–eighteenth centuries. Though It is more common to construct a single-arched bridge with a wide span if there are limitations in terms of bridge construction preferences, otherwise arch-bearing stone bridges appear to be more common [2].

The historical stone bridge, which is among civil architecture works, is an inner city bridge. The width of the bridge examined ranges from 8.30 to 8.60 m. The bridge was 55 m in length, 9 m in height and 3 spans (Figs. 3, 4). When first made it was 4 spans but one span was covered under the ground [5]. The bridge is an arched structure with centers very close to each other and tilted in two directions from the center arch. There are also different angles in the slope. The arch forms are close to the circular form and there is a second arch line called archivolt on the arch stones. Arch stones average 60 cm and the archivolt arch stones are 25 cm wide (Fig. 5a).

Fig. 4. Kars Stonebridge

(a)   **(b)**

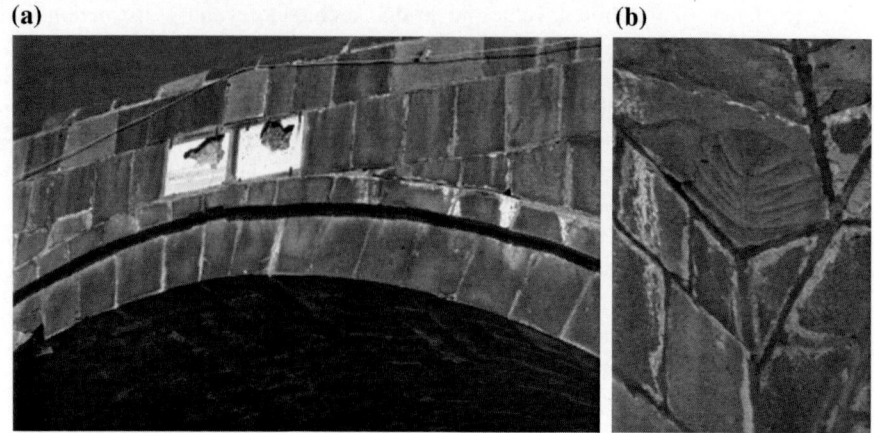

Fig. 5. a Arch and archivolt arch stones, **b** motif ornament on the wing walls

The diversity of the materials used in the bridges built by the Ottomans was limited. The stone, which is frequently used as material in this period and has high resistance to compression stresses, varies according to the geographical characteristics of the region used. The stone bridge from its time is also made of cut stone and all of the stones are basalt stone.

Since the bridge construction is carried out on the river, it is a troublesome work and it is necessary to make a preparation for the construction of the feet especially in the water. The feet should be specially supported when the feet forming the support points of the bridge openings are in the river bed.

In the middle legs of the three-span bridge, the spikes in the upstream direction (upstream) of the stream are articulated with circular-shaped spurs in the downstream (downstream) direction (Fig. 6). The task of both elements is to prevent the flow and destruction of the foot and foot foundations of the various heavy materials carried with it. The dimensions of these bridge elements differ from each other. On the two end sides, the footwalls and spurs on the feet, where the road connections are made, are made in half circle, without the circular shape being completed.

The railings on the side of the upstream and downstream sides were made with two rows of stones in the spandrel wall and a circular shape with a harp. The height of the railing is 80 cm and the railing stones are connected with lead iron clamps.

Except for a few examples of Anatolian stone bridges, there are not many ornamental elements, and there are rare constructions where vegetal and geometrical ornamentations are seen. The wing walls of the working bridge have vegetal and geometric motifs at the point where the arch ends are seen (Fig. 5b). However, sufficient information about the ornaments was not reached.

Fig. 6. Dikes and flood splitter of bridge

4 Concept of Sustainability in Historical Heritage

Historical buildings are an important part of cultural heritage, with daily life styles, aesthetic sensibilities, socio-economic and cultural characteristics of past periods. Historical heritage is an irreplaceable resource, which is an important key to connecting the daily living spaces of the past. For this reason, effective protection and sustainability must be ensured throughout.

The starting point of the concept of sustainability is focused on avoiding environmental problems that arise in parallel with economic and technological developments and protecting the ecosystem. The most widely used definition belongs to the Brundtland Commission established by the United Nations. In the so-called Brundtland Report, this concept is defined as "meeting the economic, environmental and social needs of the future generations without harming their living conditions" [8]. Daily and Ehrlich [9] describe the preservation of the assets needed by the social, economic and ecological systems as a process that expresses at least the maintenance of the required level.

Sustainability, a multidimensional concept, involves different dimensions and seeks to find a balance between conservation and use, not just conservation within every dimension. What is important in sustainability is not to spend the main capital in the hands of every generation, but rather to live the profit that the previous generation inherited from the inheritance. According to Beyhan [10], this is due to the fact that the concept of sustainability is long-term and human beings being entwined with both in nature-environment and in social life with future anxiety.

The historical heritage should be emphasized in the work to be done for sustainability, emphasizing as defined by Özlüer, *economic development, which is defined as the three main pillars of sustainable development, is the joint responsibility for social development and protection of the environment at local, national, regional and global levels* [11].

It is important that the people are conscious of the sustainability of the historical heritage. Nowadays, users who do not have adequate knowledge in the way of preservation and continuity of historical and cultural values need to be aware of these values.

It is crucial for sustainability that the structures that make up the inheritance should be transferred as genuinely as possible to the future with conscious and correct restoration works and living with the structures of the environment while having the continuing functionality.

4.1   Repairs of Kars Stone Bridge

Historical Bridges, which are of great importance in terms of technology history as well as cultural history, Project and implementation studies should be carried out in accordance with the provisions of national and international legislation in order to strengthen and survive, to preserve original identities and to transfer them to future generations [12]. Restoration works to be done on stone arch bridges, which have been proven for centuries to survive against the effects created by elements such as earthquakes, dams, variable water flows, heavy vehicle traffic; The original construction system of the bridge must be realized on the basis of the protection of the material properties and the minimum intervention principles.

Historical Kars Taşköprü which is considered in the study is used intensely in urban settlement as it is connected to the southern side of the city by the north side. The basalt of the material used increases the magnificence of the building one more time.

According to the inscription of the bridge and the records in hand, the historical stone bridge built during the reign of the city in 1579 was first destroyed during floods in the spring of 1715 and only the foundation of the foot remained. Four years later, was rebuilt in 1719 by Haji Aboubakir Karahanoglu from Kars [5].

During the Russian occupation at 1877–1915, the Russians dismantled or excavated the inscriptions on the constructions in order to erase the Turkish traces in Kars. The bridge has suffered destruction at this time, and the stone bridge which was dismantled has been rebuilt by the grandchildren of the people who reconstructed the bridge and placed on the middle arch [5].

Starting from the foundation for the repair of historical bridges, the basement and foot repairs are continuing as repair of belts, spandrel walls, cornices, balustrades, flooring and booklets respectively. The bridges must first be stabilized in order to allow a safe passage. When repairs are made, it is the basic principle to ensure that the historical structure is transferred to future generations by the method of consolidation without disturbing its original structure.

It is determined that one of the interventions made in the Republican Period of the historical Kars Stone Bridge belongs to 1991. In the repair work, the cement mortar applications were found all over the bridge and especially the foundation level. This structure, which has been neglected for a long time and started to deteriorate due to natural conditions, has been restored under the investment program of the Regional Directorate of Highways (2013) (Fig. 7).

Fig. 7. The Bridge during 2013 restoration applications

During the restoration work which began in 2013 and plants were overgrown on the bridge before restoration work (Fig. 8a). In the flood splitters and in the inner parts of the arches were found pieces that were partially broken or destroyed. The inscriptions of the bridge have undergone deformation, and some cement plaster was found in the spandrel walls of the bridge.

(a)   **(b)**

Fig. 8. a Plants seen from the bridge before restoration, **b** Cleaning of plants during restoration

In today's new repair works, all of the applications with cement mortar are removed and the original mortar of the bridge is reapplied.

During the restoration application, some findings related to the bridges and environment was found. These findings continued as a research excavation under the control of the 18th Highway. The excavation carried out on the side of the Mazlum Aga Hamam in the research excavation was made from 1.83 m radius basalt stone, starting from 3.60 m on the downstream side and 2.15 m on the upstream side (Fig. 9).

Fig. 9. Removal of the complete belt under the soil [5]

The entire belt found is not subject to any deterioration wear or tear while it is under the ground [5].

The roots of the plants found in the belly, on flood splitters and the heels, on the heels and on the spandrel walls were cleaned using chemical agents in limited area so as not to damage the bridge (Fig. 8b). Some examples of tufted stone stones were seen on bridge ceiling walls and on handrails, which were added to the structure during repairs of the structure (Fig. 10a).

Fig. 10. **a** Bridge stone railings, **b** Bridge stones replacement, **c** Stone floor [5]

The cement mortar repairs, which were found in the works of the bridge and especially the foundation level, were totally purified from the restoration works and were renovated in accordance with the mortar rates determined by the General Directorate of Highways.

The stones that were subjected to excessive surface loss or disintegrated were removed by decaying, and replaced with the same texture, color, and uniform cut basalt stone (Fig. 10b).

The greatest destruction was at the basic level. Over time, as the waters descended and rose, the stones that had been in the water for a long time came out to be in the water, and the stones were seen to be eroded. The existing cube stone pavement has been removed. Slab cover was changed to cut basalt stone (Fig. 10c). Finally, the Highways, along with the current epitaph, placed its translated version on the central arch.

4.2   Assessment of Vicinity of Kars Stone Bridge in Terms of Sustainability

The natural environment can be defined as the environment in which people continue their biological, social and economic activities. Considering the life times of the structures in this rapidly changing environment, they become an integral part of the ecosystem and are in fact similar to the human body. The necessity of making the most important key of making a healthy world and maintaining its continuity is healthy and in harmony with the structures around it during the usage period. The provision of this continuity is also related to the experts involved in the construction and the users in time.

The historical environment is one of the most important elements that reflects the original values of the cities and creates the identity of the city. Important developments in cities, historical environments that witness historical events, is a memory space that enhances the readability and perceptibility of cities. These circles, located somewhere between the past and the future, define a common language that provides familiar environments for people by defining strong place and contextual emotion. These features contribute to the cultural and urban continuity.

As one of the most concrete examples of the identity of the historical city is the bridges and the surroundings they create must be protected. It is very important that the values of today's cultures are taken into consideration with the participation of the city and transferred to future generations. In Özil (2001), it is stated that urban protection is more than just a custodian, that the preservation of the preserved items and participation in everyday life play an important role in social development [13].

From this point of view, historical bridges from traditional structures, both construction and use, are consumed in the process of using them, so protection must be done to ensure continuity. Because of their bridge scale and purpose, they are the dominant elements of historical urban architecture and thus constitute one of the focal points of the sustainable environment.

Historical Kars Stone Bridge, which is the subject of the study, functions actively and provides pedestrian and vehicle access. When it is taken to be evaluated with the environment in which the bridge is located, it is seen that it interacts with many historical buildings forming this environment.

When the vicinity of the bridge is examined, an enormous historical texture is formed with the views of İlbegioglu (Muradiye) and Mazlum Aga (Topcuoglu) Hammams, Shrines, Ottoman mansions, historical buildings and Kars castle with Ottoman architectural features on the downstream side. Ibeyoglu Hammam has a rectangular plan and was built in 1774 (Fig. 11a). It is not precisely known when and whom built the Mazlum Aga Hammams but it is known that it was built with Ottoman architectural technique in the early seventeenth century [14, 15] (Fig. 11b).

(a)   **(b)**   **(c)**

Fig. 11. a Ibegioglu Hammam, **b** Mazlum Aga Hammam, **c** Bridge and Castle walls

The Hammams made of cut stone and rubble stone show a similarity with the stone bridge and make a harmony with the historical texture. The bridges and hammams were destroyed by the Armenians in 1918 and later repaired. Despite the fact that the restoration work of the Ilbegioglu Hammam has been started from these hammams which have been neglected for many years, there have been no studies about Mazlum Aga Hammam.

Kars Fortress, the symbol of the city of Kars, was built by Saltuklu in the period of Seljuk State in 1153. Kars Fortress is one of the most important symbols of the city, where many victories were won during the wars against Mongolian, Georgian, Persian and Russian attacks for centuries. During the Ottoman period in 1579, Kars fortress and Kars City were reconstructed into a state center [14]. The stone bridge built during the same period of zoning studies is a connecting point connecting the two shafts to the trough (Fig. 11c). The castle walls of Kars have been destroyed today. On the side of Ilbegioglu hammam, there is a shrine on the Kars fortress (Fig. 12a).

(a)   **(b)**   **(c)**

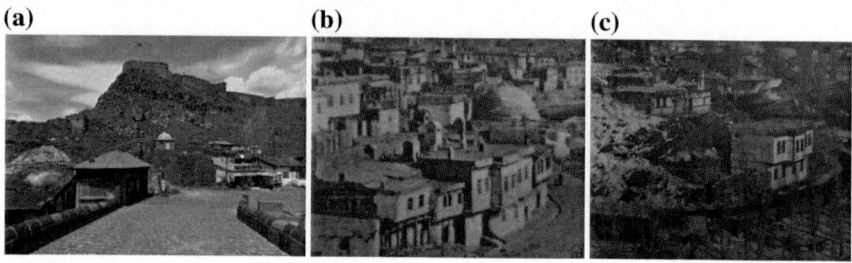

Fig. 12. **a** Shrine at near the Kars Castle, **b** 1920 Ottoman houses and the Red Church; **c** Ahmet Tevfik Paşa Mansion, which reached today [16]

When we look at the old photographs of Kars in 1870 years, there are Lacin Bey mansions on the side of Mazlum Aga Hammam [16], but today there are no remains of these mansions (Fig. 3). The photographs taken in 1870s on the skirts of the Kars Castle following the Ilbegioglu hammam show the Ottoman houses and the Red Church used until 1920s (Fig. 12b). Ahmet Tevfik Pasha Mansion, two storeys of a large part of the destroyed Ottoman houses, still has been survived (Fig. 12c). The Armenian Catholic Church (Frankish Church) was built on the side of the Kars Stream on the side of the Mazlum Aga Hammams in 1877. The church is in a state of obsolete and shows itself through its remains [16] (Fig. 13).

Fig. 13. Armenian Catholic Church at 1920 and its state in nowadays [16]

As we start with the historical buildings and residences that we have taken to examine the stone bridge near the center of the city, a rich and varied cultural texture is encountered. Although the city has hosted many historical buildings from different periods, most buildings have been destroyed or become unusable. Stone bridges, however, were able to arrive as solid as daylight and occupy an important position both as a function and as city tourism. The bridge is provided with terrestrial seating units built on the side of the Mazlum Aga Hammam.The presence of abundant green areas around the bridge is striking, but there is no parking arrangement in this area which has a unique texture in terms of historical heritage. There is also a lack of venues to exhibit this historic atmosphere, to provide space for visual and verbal presentations, and to visit visitors' needs (Fig. 14).

(a)   **(b)**

Fig. 14. Before restoration of bridge—after restoration of bridge 2013 **b** available sitting units

This silhouette, created by the rich historical buildings bearing the traces of the past, provides an inevitable opportunity to contribute to the city in terms of tourism, city development, traces of the ancestor to the next generation. The bridge will meet the transportation needs of its users. The rich historical texture is reinforced by the abundance of green spaces, but the historical texture that it forms together with the bridge and the historical structures around it is not given sufficient importance. Providing the sustainability of historic buildings in a life cycle with the environment is an important key to the formation of a good environment. It is important to analyze the existing texture, to determine the authentic and authentic values, and to make the spaces needed in accordance with the texture.

5   Recommendations and Conclusion

Given that the city is a living organism, each period is important and part of historical continuity. It is inevitable that the historical structures created by the environment will change in line with the current needs, and that change and transformation can only be achieved with conservation consciousness. Preservation of historical identity; The community needs to have an urban consciousness that respects reason and effort. Preservation of historical values and presentation of humanity as a common heritage have recently constituted an important civic duty. One of the most fundamental solutions within the scope of sustainability will start with the awareness of the users and the people of the environment. Trainings should be organized to contribute to the formation of a sustainable identity in the historical environment, public awareness about the cultural values of the city and support for the formation of civil society organizations should be provided.

This study gave information about the historical situation of the historical touch formed by traditional Ottoman hammams, the remains of the church, the Ottoman-Russian architecture on the skirts of the castle, the bridge between the historical buildings and the stone bridges of Kars with its history rich in heritage and evaluated it in terms of sustainability. The study was shaped by the collected data as a result of field studies, both in the relevant literature. The bridge should be transferred to future generations with the least intervention that can be done in the original scale for the effective use and sustainability of the historical heritage formed in the vicinity of the bridge. Within the scope of the study, repairs carried by the bridge have been mentioned. The bridge, which has undergone repairs at different times due to natural causes or human-caused demolitions, was restored to its original state at 2013. Achieving effective use with all the interventions is an important step in the context of sustainability.

Restoration Projects should be made in the historical environment to provide the continuity of the functioning of the bridge in the historical process together with the historical constructions around it, and to revitalize it in terms of tourism. First of all, there is a need for space solution suggestions that will meet the need for rest and recreation from the texture that the Stone Bridge has formed together with the historical buildings on the periphery. Already today, the use of a small buffet terrace and sitting units on the side of the Mazlumağa hammam is trying to meet this need. However, this solution is inadequate both in terms of positioning to historical texture and in terms of space possibilities. In order to address the needs, the old structure in the historical structure can be re-functionalized or new additional designs can be made. The first criterion to be taken into consideration in the framework of sustainability is that the necessity of additional designs to be proposed at this point should not disrupt the historic texture. The urban movement and circulatory system of the location where the additional features to be added must be analysed well. Integrity, transparency, pedestrian and vehicle transit organizations to be achieved with existing theme should be considered.

There are rich green areas and historical building remains around the Bridge. In the present case there has been no intervention in this region for many years. In order to ensure the sustainability of historic touch, first of all, historical buildings, restitution and restoration projects should be prepared for the historical building residences around it keep that historic touch. In addition, the green areas left to their own should be assessed and needed landscape arrangements should be made. The objective here is to support the sustainability of the historical touch with these proposals mentioned for the purpose of conveying the interaction of the city with the city safely in the future and the continuity of the green areas around it.

At the end of the study, the extent to which continuity has been sustained in the exchange balance has been assessed in the context of historical heritage. It is intended to contribute to the efforts to ensure that the preservation and legalization of historical buildings in modern cities are sustainable.

References

1. Yeşilbaş E (2007) Diyarbakır'da Su Mimarisi. Yüksek Lisans Tezi, Selçuk Üniversitesi Sosyal Bilimler Enstitüsü, Sanat Tarihi Bilim Dalı, Konya
2. Atak E (2008) Erken Osmanlı Köprüler. Çanakkale On Sekiz Mart Üniversitesi Sosyal Bilimler Enstitüsü, Sanat Tarihi Anabilim Dalı, Yüksek Lisans Tezi, Çanakkale
3. Ekinci O (2006) Kars Kitabı. Anahtar Kitaplar, İstanbul
4. Bingöl A (2011) Kars Ve Çevresinde Demir Çağı Yerleşmeleri. J Institute Soc Sci 8:20–40
5. Terzi SG (2013) Kars İli Tarihi Taş Köprüsü As-Built Raporu. T.C. Ulaştırma Bakanlığı, Karayolları Genel Müdürlüğü, Ankara
6. Akçayöz V, Öztürkkan Y (2010) Eski Yeni Fotoğraflarla Kars An Ilustrated Story. Kars Kültür ve Sanat Derneği, Kars
7. Özkök MK, Azsöz G, Erşan Ş (2015) Erken Ve Klasik Dönem (14.-17. Yy) Osmanlı Köprülerinin Tarihsel Gelişimlerinin Ve Yapım Tekniklerinin İncelenmesi: Edirne/ Uzunköprü Örneği, 3.Köprüler Viyadükler Sempozyumu, İnşaat Mühendisleri Odası, Bursa Şubesi, Mayıs
8. Cook J, Özkeresteci V (2001) Ekolojinin Mimarisi. Domus M Dergisi 10:53–57
9. Daily GC, Ehrlich PR (1996) Socioeconomik equity, sustainability and earth's carrying capacity. Ecol Appl 6(4):991–1001
10. Beyhan SG (2004) Kültürel Süreklilik ve Çağdaş Gereksinmeler Bağlamında Sürdürülebilir Turizm ve Kimlik Kavramsal Modeli: Pamukkale Modeli. İstanbul Teknik Üniversitesi Fen Bilimleri Enstitüsü, Doktora Tezi, İstanbul
11. Özlüer F (2007) Sürdürülebilir Kalkınmanın Ekonomik Politigi [Elektronik Hali]. Mimarlar Odası Ankara Subesi Bülten 51, Dosya 05, Sürdürülebilirlik: Kent Ve Mimarlık
12. Tarihi Köprüler, T.C. Ulaştırma Bakanlığı, Karayolları Genel Müdürlüğü, Köprüler Dairesi Başkanlığı, Tarihi Köprüler Şubesi, Müdürlüğü, Ankara 2009
13. Baysan O (2003) Sürdürülebilirlik Kavramı Ve Mimarlıkta Tasarıma Yansıması, Yüksek Lisans Tezi, İ.T.Ü. Fen Bilimleri Enstitüsü, İstanbul, s.11–18–19
14. Web; Doğanay, H., "Kars'ın turizm potansiyelinin değerlendirilmesi" http://www. karskulturturizm.gov.tr/TR,54935/karsin-turizm-potansiyelinin-degerlendirilmesi.html. Last Accessed 14 Dec 2016
15. Web; Kars Provincial Directorate of Culture and Tourism, "Hamamlar" http://www. karskulturturizm.gov.tr/TR,54875/hamamlar.html. Last accessed 14 Dec 2016
16. Akçayöz V, Öztürkkan Y (2010) Eski Yeni Fotoğraflarla Kars With Old And New Pictures. Kars Kültür ve Sanat Derneği, Kars
17. Büyükoksal T (2012) Kentsel Korumanın Sürdürülebilirliğinin Olabilirliliği Üzerine Bir Yaklasım Önerisi Eskişehir Odunpazarı Örneği, Uzmanlık Tezi. T.C. Kültür Ve Turizm Bakanlığı Kültür Varlıkları Ve Müzeler Genel Müdürlüğü, Ankara
18. Sey Y. SÜRDÜRÜLEBİLİR KALKINMA/ "TARİHİ KÜLTÜREL MİRASIN KORUN-MASI", VİZYON 2023 ÖNGÖRÜ PANELLERİ
19. Göksal T (2003) Mimaride Sürdürülebilirlik Teknoloji İlişkisi: Güneş Pili uygulamaları. Arredamento Mimarlık Dergisi, S.154, s.76–79
20. Özcan Z (2002) Cultural heritage and urban sustainability: a case study from Central Anatolia, the sustainable City II, Urban Regeneration and Sustainability, sayfa, Segovia-Spain, pp 377–386
21. Yenişehirlioğlu F (2002) Türkiyede Tarihi Kent Dokularının Korunması Ve Geleceğe Güvenle Taşınması Sempozyumu. T.C Kültür Bakalığı Kültür ve Tabiat Varlıklarını Koruma Genel Müdürlüğü, Kemer, Antalya
22. Tanyeli G (2002) Türkiye Köprüleri, Türkler (Cilt 12). Yeni Türkiye Yayınları, Ankara

Refugee Problems in Turkey and Its Evaluation in the Context of Sustainability

Sefik Tas[1](✉), Mursel Erdal[2], and Huseyin Yilmaz Aruntas[2]

[1] Disaster Emergency Directorate, Branch Manager, Van, Turkey
sefiktas65@gmail.com
[2] Technology Faculty, Civil Engineering Department, Gazi University, Ankara, Turkey
{merdal,aruntas}@gazi.edu.tr

Abstract. Today, a flow of refugee which has turned into the biggest problem of the happened towards various countries due to the disturbances emerged in the Middle East states headed particularly by Turkey. These refugees have failed to ensure the intercultural adaption in the countries where they have taken shelter. At the same time, the refugees are in position of asylum seekers and the physical conditions required in the countries where they are the guests should also be provided for them to be able to survive. In this study, it has been focused on immigration, immigration movements and their conclusions, refugee problems and the standardization of sustainable basic and social habitation spaces in the countries where the humans in state of asylum seekers have taken shelter and the situation of the Syrians living in Turkey is compared with the examples in Europe. The data obtained has shown that the Syrian refugees living in Turkey are in a better condition than the refugees living in European countries.

Keywords: Refugee problem · War · Migration · Camp · Sustainability

1 Introduction

Immigration is being defined as the movement of a person or a group of people to another place by crossing an international border or relocation within a state [1]. Post-relocation movements emerged population movements; Irrespective of time, structure and cause are being comprised to the extent of migration. Such displacement movements can take place voluntarily or due to obligatory causes. Today, millions of people evict the territories where they were born and grew up due to wars, natural disasters, political or economic reasons. Therefore, the refugees, asylum seekers, economic migrants, irregular migrants and the groups of people displaced from their places due to various means are also added to the concept of migration. While the migration used to be based on usually the economic or natural causes before the technology age, today, the wars being the greatest and most dangerous problem of world are the greatest cause procreating the need for migration.

© Springer International Publishing AG, part of Springer Nature 2018
S. Fırat et al. (eds.), *Proceedings of 3rd International Sustainable Buildings Symposium (ISBS 2017)*, Lecture Notes in Civil Engineering 7,
https://doi.org/10.1007/978-3-319-64349-6_24

While the migration movements deliver affirmative outcomes when it is managed well, well it accompanies a number of negative consequences when failed to be managed properly resulting even the human rights violations positive consequences when managed well, they firstly threaten the public order and security that leads up to human rights violations accompanied hereto.

An effective management system which would support an economic development in the country gone to and maintain the public safety and include the framework values such as protection of human rights of migrants. For this reason, in order to develop an effective migration management, it is worth for examining the causes of migration and its consequences.

The concept of migration; is closely associated with the political, economic, social and cultural life. In particular, the international migration simultaneously affects more than one state. While this type of migration provides the labor force contribution to the destination country, it is likely to be also leading to loss of qualified labor force in the source countries. Therefore the immigrants; don not only concern the destination countries but also the countries they have left behind and leave permanent traces by shaping up the interaction between these countries. Migration movements, On the other hand, it is expected that the immigration movements would continue also in future depending on many causes [2].

2 Refugee Problem

2.1 Arab Spring and Internal Disturbances in Syria

The events called Arab Spring and initiated first in Tunisia being one of the Northern Africa countries where a peddler named Muhammed Bouazizi burned himself in 2010 have spread to the countries like Libya, Egypt, Syria and Yemen. These events have led to the commencement of a civil war in Libya and international military intervention has been made to this country upon decision of the United Nations Security Council. The civil war in Libya resulted in killing of Gaddafi, the Head of State and establishment of two different governments in the country. Shortly after that, the Head of State, Mursi has been overthrown in consequence of a military coup made in Egypt and a new government was established. The events that started with the Arab Spring have jumped also to Syria and as a result of turmoil started; the country has been drifted along a civil war in 2011. Thousands of people affected by the civil war in Syria have been impelled to the borders of neighboring countries like Turkey, Jordan, Lebanon and Iraq in order to save their lives. Upon decision of the UN Security Council taken in 2014, military intervention began under the leadership of countries like USA, UK and France to Syria. Later on, Russia has joined the war alongside the current Syrian government and civil war still continues in the country.

2.2 Refugee Problem in Europe

As a result of the Arab Spring, particularly people who escaped from Libya began to migrate to Italy being the closest European country by sea. While Italy was struggling to cope with this immigration problem, some of the Syrians who migrated to Turkey following the civil war started within Syria in 2011 have turned towards the border of Bulgaria to go to Europe via Turkey. These people have tried to reach at Germany and France by following the route of Bulgaria, Serbia, Hungary, Croatia, Slovenia and Austria. As for another part of Syrians in Turkey, they have passed to the Greek islands in the Aegean Sea by unsafe marine vessels like rubber boats and boats. Unfortunately, thousands of Syrian children, women and men have died by being drowned during these transitions. The EU countries have been alarmed in front of this refugee flood and have drawn barbed wire to the borderlines and have caused the Syrian refugees to pile up at the borderlines by closing border gates. EU gathered under the leadership of Germany has fallen into a big challenge due to such refugee flow and has sat at the table with Turkey to eliminate this migration. As a result of long negotiations conducted, the readmission agreement for the refugees has been concluded in exchange of financial aids to be made between Turkey and the EU. On the other hand, EU countries headed by Germany have proposed selective and restrictive criteria such as qualified persons to limit the refugee flow [3].

2.3   Post-Arab Spring Asylum Seekers in Turkey

Due to its geographical structure and increasingly growing economy, Turkey is affected much more from the irregular migration movements of this nature. Adopting an open door policy from the past up to today, Turkey had hosted approximately 345,000 Turks forced to emigrate from Bulgaria due to the government's assimilation policy in 1989 and 467,489 asylum seekers from northern Iraq during the gulf war happened therein in 1991 [4]. Likewise, Turkey has similarly begun taking in also the Syrian refugees who escaped from the civil war and piled up at its border for the first time from Hatay-Yağladağ border gate on April 28th, 2011 and the needs of these asylum seekers were met. For this purpose, the asylum seekers were firstly placed into guesthouses belonging to public institutions and general-purpose tents. In priority, the accommodation and nutrition requirements of these asylum seekers have been met. Predicting that the number of asylum seekers to come to Turkey would exceed hundred thousands, Turkey has begun building the temporary housing centers in accordance with international standards in the provinces in the vicinity of Syrian borderline. Later on, the asylum seekers who came to Turkey were relocated at the housing centres completed within a few months. In line with the non-returnable principle specified in "Convention on the Legal Status of Refugees" signed by the Member States of United Nations in Geneva in 1951 and adopted by Turkey as well [5]. At present, fulfillment of accommodation, nutrition, health, education and the like requirements of the Syrians residing within Turkey are still being maintained.

3   Minimum Standards of Sphere Project

Sphere Project is a program jointly carried out by the Steering Committee for Humanitarian Response (SCHR), the International Council of Voluntary Agencies (ICVA) and VOICE. This initiative was initiated in 1997 by a group of non-governmental, humanitarian aid organization (NGO), Red Cross and Red Crescent [6].

The Humanitarian Aid Convention is concerned with meeting the needs of people exposed to disaster or skirmish and with the basic requirements needed for them to return to a respectable life. Minimum standards include the services of water supply and healthy environment, nutrition, food aid, shelter, location planning and healthcare to the humans.

3.1   Water Supply and Healthy Environment (Sanitation)

- Adequate water should be available for drinking, cooking and cleaning to all people at the housing camp. The water should be as close to a location where everyone living at the camp can easily reach at meet the minimum amount of water.
- Toilets should be designed and constructed so that all people, including people with disabilities, can use them safely under comfortable, healthy conditions.
- The toilets from the camp should be designed so as to be utilized by all humans including the handicaps, in comfortable, safe and healthy conditions.
- All people at the camp, the facility for living in a solid waste—free, cleaned environment, including medical wastes should be provided with the opportunity to live in an environment. When the disposal of solid waste is being made for hygiene, garbage should be collected daily and placed in garbage containers or buried in a special waste pit.
- The camps must have an environment that is free from the risks of water erosion, stagnant water, rain shower water, flood water, away from the waste waters of the houses and medical centres.
- For hygiene, remaining own traditions of the people and in a way not to create any risk for the health, necessary sources should be provided to the funerals of people [6].

3.2   Nutrition

- All people should always be able to find the food needed for active and healthy life.
- While the nutritional needs of all people are being met, required nutritional values must be met by taking into consideration their cultural lifestyle.
- In order to prevent the nutritional deficiencies of people exposed to the disaster, the nutrition should be provided in accordance with the values given in Table 1 [6].

Table 1. During initial step of the disaster event, the nutrition program has been adapted from (WHO (1997, draft) and WFP/UNHCR (December 1997))

Nutrient	Population average need
Energy	2100 kcal
Protein	10–12% of total energy
Fat	17% of total energy
Vitamin A	1666 IU
Thiamine (B1)	0.9 mg
Riboflavin (B2)	1.4 mg
Niacin (B3)	12.0 mg
Vitamin C	28.0 mg
Vitamin D	3.2–3.8 pg calciferol
Iron	22 mg
İodine	150 pg

3.3 Food Aid

All people are in need of consuming adequate and quality food to maintain a healthy life.

- Food items must be stored in safe and clean places, not under decomposition conditions.
- The method used with the food aid should be fair and in be conforming to the local conditions.
- People who received aid should be informed about their share and how it was decided.
- Distributed food items should not be contrary against the religious and cultural traditions of catastrophes [6].

3.4 Housing and Settlement Planning

It is obligatory that the honor of people under harsh conditions should not be hurt and continuation of a family and community lifestyle must be maintained.

People are moved and placed into a new society or temporary camps.

- In the designed settlement zones, a 45 m^2—space as per person should be provided, including infrastructure.
- The land should be adapted to build a sheltering center (drainage etc.).
- The material of shelter to be used should be of high heat—resistant in hot and dry climate conditions.
- As for the cold climates, the material for shelter and construction should ensure optimum insulation.

- The temperature at which those who live at the shelters would be comfortable is obtained with the insulated shelter, proper clothes, blankets, bed material, good heating and eating the energizing foods.
- The social facilities such as grocery stores, schools, places of worship, cemeteries, health centers, solid waste centers, water resources, community and nutrition centers, workshops, woodland and recreational areas should be provided.
- Quarantine camps should be established to prevent the spread of epidemics or places away from general settlements should be determined and prepared for those with such diseases to stay at [6].

3.5 Healthcare Services

The primary objective of the healthcare services rendered to the persons affected from the disaster is the diseases and deaths to be reduced.

- The health services should be arranged to allow fighting against the identified main causes leading to excessive death, illness and injuries.
- The health services should be implemented by the personnel who enjoy appropriate competence and experience and are well-managed and received necessary support.
- The diseases having the risk of creating epidemic should be investigated and controlled under international norms and standards.
- How the disaster affected the human health and what the needs are and what the health programs should be directed should be specified.
- Emergency health services should be developed making use of local capacities and skills [6].

4 Services Provided in Turkey to Syrian Asylum Seekers

4.1 General

It is seen that Turkey has accepted to meet the asylum seekers as a historical duty, which have run away from war in the best way. Within the framework of its responsibilities arising from the international law, Turkey provides temporary protection status of the Syrians and causes them to have access to the basic services and does not send them back and take them under protection [7]. With this understanding, Turkey has provided all kinds of facilities offered in their own country such as education, health, sportive activities and has gained the appreciation of all the countries across the globe.

Turkey has undergone tremendous financial burden due to Syrian asylum seekers who have been given temporary protection status and has given humanity lesson to the whole world. It is stated that currently more or less 3 million Syrians are being sheltered and the sum of funds spent up to now is about TL 36,000,000,000 [8]. As for the International support provided to Turkey for the Syrian refugees has remained at a level not worth to be mentioned when looked at the figure spent. As it would be recall, once only a few thousand Tunisians set foot to Italy on early days of the Arab Spring has led to the chattering of internally unlimited Schengen area, being the largest project of Europe [9]. At the same time, necessary physical conditions are required to be provided for the immigrants who are in the position of asylum seekers also in other countries where they became guests to maintain their lives as in Turkey.

In year 2013, biometric identity study has been initiated by the Turkish Government to allow the Syrians who took refuge in Turkey utilize and follow the social benefits like education and health and to follow those who involved in crime. With the efforts carried out until 12.02.2016, 2,790,767 Syrians whose fingerprints, identity and domicile information were taken have been registered. Thus, "temporary protection status" has been granted to the Syrians who were officially registered. The endeavors of taking incoming new Syrian guests are being kept on daily. Whenever the Syrians being in this status went out of the cities where they reside, they just get their records revised by the city where they moved to by applying with ID cards in order to utilize the health and education services maintained under the coordination of AFAD [10].

258,571 Syrian and 5993 Iraqi refugee's shelter at the AFAD—established and managed 24 housing centers. 2,484,997 Syrians have settled many cities, headed by Istanbul outside the housing centers in Turkey [10].

4.2 Water

Water supply in adequate amount for drinking, cooking, for personal and household cleaning is provided to the temporary sheltering centers with the water taken from the city network or drilling wells. Water is provided at the houses of the containers in the cities and at certain zones within the tent cities. WC and shower needs have been met in the containers of the container cities. And as for the tent cities, it is designed as a WC as per 20 persons and a shower as per 25 persons. The corpses are shrouded and buried in accordance with religious beliefs and cultures [10].

4.3 Nutrition

Markets where the Syrians living in the temporary shelter centers can meet their basic needs, especially nutrition have been established and an aid I.D. cards valid from these markets has been given to every family. According to the list regularly taken from the camp management software every month, money is being loaded to the card of every family in pro rata with the number of individuals in the family. The families can shop from these markets according to their own culture and tastes and cook their food at their own living spaces [10].

4.4 Food Aid

No further food aid is being made to the Syria's asylum seekers residing in the temporary sheltering center. For those living outside the shelter center, the demands are being entered to the system with the Electronic Help Distribution System (EYDAS) developed by AFAD and aid is being actualized in an effective and rapid manner. With this system, duplication of aids is being avoided and the donors to make aid are able to see where and who need what. Since the aids are being so-recorded into system, how much aid was made is being calculated [10].

4.5   Housing and Settlement Area Planning

According to the official records, there are 2,790,767 Syrians in Turkey. Totally 26 temporary shelters; 20 tent cities and 6 container cities have been established in the first stage, including infrastructure in 10 provinces and a total of 264,564 people was allowed to have accommodation at these housing centers. As from 2016, 3 each of the temporary shelters established to be tent have been converted into container towns. These conversion operations have been accomplished using Turkey's own resources. Sphere Project and Şanlıurfa-Akcakale Tent city and Kilis Container city were taken as references for ensuring the maximum standards while the tent and container cities were being established [10].

4.6   Healthcare Services

Practitioners and specialist physicians available at the hospitals within the temporary sheltering centers function. Syrian physicians are also assigned to the health centers according to their branches. There are also emergency response teams and a pharmacy to obtain the drugs. Health service is being provided by 84 doctors and 1174 health personnel from the temporary shelter centers in Turkey. In addition, Syrian asylum seekers were also provided services from the hospitals found across Turkey, and 184,390 deliveries and 797,450 operations have been realized [11].

4.7   Training

In temporary shelter centers, educational services are provided from kindergarten to high school. A total of 325,000 students studies at the temporary shelters and public schools in Turkey. In addition, the vocational training courses are given to all adults, especially children, women, elderly and disabled [10].

4.8   Financial Expenditures

Temporary shelter centers have playgrounds for children, interpreter services, sportive activities, socio-cultural activities and women and youth committees formed demo-cratically through elections. In particular, in order to ensure that the social integration is obtained, all social facilities and services aiming to develop awareness of working together and prevention of social conflict where the Syrian children are taken away from battle ground and their ages—necessitated environment is provided and the efforts of causing them to take part in the activities and possess a profession through courses and make it possible for them to get income are carried out [10].

5 Services Provided in Turkey to Syrian Asylum Seekers

According to official figures of AFAD and international organizations, 12 million persons out of 20 million persons living within Syria were obliged to migrate. While 8 million of them have left their homes and assets in the country and migrated to other regions, 4 million and 200 thousand Syrians have passed to neighboring countries from the regions where the conflicts were active. More than 50% of Syrians run away from the country took refuge in Turkey. With its open door policy, Turkey has not sent any Syrians back.

Among the countries where the Syrian asylum seekers most frequently go, Lebanon heads first. According to UN figures, there are 1,048,275 refugees in Lebanon. Since the Lebanese government has refused to establish any camps, there are small camps set up by international aid agencies in this country. More than 2 thousands of Syrian refugees stay at Zehle Camp in the city of El Bika. The facilities at the camp are very limited and there are big problems in subjects of heating, clean water, electricity and the like. Even though the conditions of camp are not very good, there is a tent serving as a health center and a school [12]. In another neighboring country, in Jordan, according to the UN figures, 655,217 Syrians live at 5 camps established for them across the country. 83,000 Syrians live at Zaatari established to be the largest camp in 2012 below the minimum settlement standards without adequate amount of food (Fig. 1). Many criminal offenses have been reported, including prostitution and drug trafficking in the public sector [13]. Too many crimes have been reported from the camp, including prostitution, drug trafficking as well [13].

Fig. 1. Zaatari refugee camp in Jordan

Again, according to UN figures, 247,339 Syrians have sheltered in Iraq and 89,039 hereof live in 17 each camps established. At the Domiz tent city, many deficiencies from dysentery, sanitation, health, food and up to education are being observed [13].

On the other hand, there are registered 115,204 Syrians in Egypt being an African country and 29,275 Syrians in other northern African countries. In these countries, there are settlement centers established for Syrian refugees. It has been reported that 512,909 Syrians have sheltered in other countries staying outside these countries [13].

6   Refugees Living in European Countries

After the civil war started in Syria in 2011, the flow of refugees also to the European countries has taken place. The Syrian refugee flow has caused a great panic in many countries, especially in EU countries headed by Germany, Austria, France and Hungary. As a result hereof, EU countries, especially headed by Germany have proposed selective and restrictive reasons such as qualified persons to restrict the refugee flow [3].

Greece heads first among EU countries where refugees passed through. To do so, the refugees see this country as a bridge for passing to Europe and they are trying to pass to the Greek islands closer to Turkey. According to the records, hundreds of Syrian refugees who tried to pass to the islands have died as a result of sinking of their boats. Even though, Idomeni camp built on the border of Macedonia in Greece had been planned for 2500 persons, 12,000 refugees shelter at this camp (Fig. 2). 54.9% of these people are Syrians, 24.4% are Afghans and 11% are the Iraqi refugees. At the camp, difficulty of water and sanitation is being suffered and is likened to a Nazi camp [14].

Fig. 2. Idomeni refugee camp in Greece

Another EU country where refugees go is Italy. The reason why refugees choose Italy is because of its proximity to the Mediterranean Sea and Africa. Italy's refugee camp on Lampedusa Island is seen as an important point for African and Libyan refugees who want to pass to Europe through the Mediterranean Sea (Fig. 3). Thousands of refugees are taking refuge to this island every year, but many of the refugees are losing their lives in the Mediterranean Sea while trying to reach the island. It is believed that since October 2015, at least 360 refugees have lost their lives on the way to Lampedusa. On the other side, it is stated that the conditions of the refugee camp in the island are very bad. The inhuman treatment at Lampedusa refugee camp has also shocked Italian Prime Minister Enrico Letta. Letta said at a press conference he organized; "They say that the images are extraordinarily striking, that the matter will be investigated promptly and criminal sanctions will be imposed on those responsible." The UN High Commissioner for Refugees warned Italy regarding improvement of conditions from that center [15].

Fig. 3. Lampedusa refugee camp in Italy

Another country where refugees are sheltered to reach Europe is France, being both permanent member of UN Security Council and an EU state. As a cause for refugees to prefer taking shelter at France can be pointed out existence of its coastline to both Mediterranean Sea and Atlantic Ocean. The number of refugees desiring to go to England from Junge camp found in the port city of France, Calais on the coastline of English Channel has risen up two fold and reached at 6000 persons in October 2015

(Fig. 4). It has been observed that the refugees who live in this camp stayed in the hut-like shelters they have built with their own means. These shelters seem to be multi sized tents deprived from all kinds of social facilities [15].

Fig. 4. Calais Junge refugee camp in France

7   Comparison of Situations of Refugees in Europe and Turkey

It is seen that the borders of all the refugees and other refugees who have escaped from the civil war in Syria in general all over the world, the borders of the borders are closed, the wires are drawn and the walls are covered. It was reported at the end of 2016 that over 100,000 refugees await at the border of EU States [16]. According to UN High Commissioner for Refugees António Guterres; "This is the largest refugee population displaced by a single conflict in one generation". A population that needs the support of the world, but instead lives in very bad conditions and drifts to poverty, has emerged. When we look at Macedonia and Greece, where we can think of Europe as an entry, it is obvious that those countries are working as a filter [16]. Today, 54% of the Syrian refugees in Turkey are under the age of 18, while the remaining 46% is composed of women and the elderly, with more than half of the Europeans coming from young people and men [17]. European countries accept refugees who will be able to withstand the hardships and who can afford everything for them, and the refugee camp images obtained confirm this view. Refugees in Europe, like all sports clubs, have to go to the language and/or vocational courses after they have passed all the difficulties and have stayed in the 1st and 2nd level camps. Home, work and social security are given to the successful refugees in these camps. However, when these refugees risk their lives in some way and go to Europe, if they have a qualification and their bodies are resistant, they are taken to European countries. For example, a four-lingual qualification is accepted as an engineer refugee, and a distinction is made between "good refugee" and "non-refugee". It is expected that refugees will be able to show their sacrifice in order to stay in Europe [17].

On the other side, while 43,429 Syrians passing to Europe through Aegean Sea and Mediterranean Sea have been rescued by Turkish security units due to the sinking of their boats. 670,685 immigrants containing the citizens of many countries including Africa, Afghanistan and Libya have migrated to Europe through the Mediterranean Sea. Refugees not admitted by official application have endangered their lives in order to illegally enter these countries and thousands of people have lost their lives in the Mediterranean waters. Even though the shame photos of baby Aylan baby and other refugees whose lifeless bodies have hit the coast were bumped into their faces, the intensively of United Nations in this respect continues [16].

In the countries sharing an adjacent border with Turkey and Syria, open door policy is being applied without discrimination of religion, language, race and profession [17]. Only in Turkey, 2,743,497 Syrian asylum seekers are being hosted by presenting a sustainable life. Turkey spent totally $12,000,000,000 for Syrian asylum seekers. This figure exceeds the Ministry of Interior budget of 2016.

This figure exceeds the 2016—Ministry of Interior budget. As for the contribution of international organizations to this expenditure is only $418,000,000, or TL 1,200,000,000. As for the place of annual average expenditure TL 53,000,000,000 made by Turkey for the Syrians within the economy, it can be understood better in consideration of budgetary figures [18].

The number of Syrian refugees who took shelter in the neighboring countries of Turkey and Syria is approximately 5,500,000 persons. The number of refugees admitted to European countries is given in Table 2. Only the number of official applications made to European countries due to the civil war in Syria has reached up to 1,177,914. However, the total number of refugees admitted to European countries from various countries is around 327,000 [19].

Member States of the EU have agreed to provide €3,000,000,000 aid to Turkey in exchange of prevention of Syrian refugees in Turkey from coming to European countries. However, Turkey states that a very low rate of this money has been sent and that the aid made is inadequate.

Table 2. Number of refugees admitted to European countries

European states	Number of Refugees
Germany	140,910
Sweden	32,215
Italy	29,615
France	20,630
Holland	16,450
Austria	14,045
Switzerland	14,000
England	13,905
Belgium	10,475
Denmark	9920
Norway	6250
Bulgaria	5595
Greece	4030
Finland	1680
Southern Cyprus Greek Section	1585
Malta	1250
Spain	1020
Poland	640
Hungary	505
Ireland	485
Romania	480
Czech Republic	460
Portugal	195
Luxembourg	185
Lithuania	85
Estonia	80
Slovakia	65
Island	50
Slovenia	45
Serbia	40
Latvia	50
Liechtenstein	5

8 Conclusion

The requirements of accommodation and nutrition of the Syrian asylum seekers who have entered Turkey as a result of the civil war in Syria have been furnished. Approximately 3 million Syrian asylum seekers have been given temporary protection status in the direction of the non-return policy adopted by Turkey thus their needs for housing, nutrition, health and education have been met. The contribution made by

European states to the financial expenditures made by Turkey for the refugees has remained at a quite a low percentage.

Although approximately 1,180,000 official asylum requests happened to the European countries following the ongoing war experienced in Syria, the number of refugees accepted is slightly more than one fourths of the applications made. There are 2,790,767 Syrian refugees registered in Turkey, whereas in European countries, this number is only 512,990 refugees.

As a result, being a refugee for people is not a choice. This can be understood better if we can replace ourselves with the refugees. The refugees had a profession in the country they lived in, houses, shops, or cars. On one day, some of these people were bombed at work and their homes, their families disappeared or some of them lost their lives and were obliged to run away from the country leaving their assets behind. After all these things, they had to live in tents with a little help from the country. Every human being needs to earn money, live and receive good education for their children as a means of living humanely. Refugees are looking for a home environment where they can study, have children and have a regular life.

References

1. Cicekli B (2013) Explanatory migration and asylum glossary (print location not specified) pp 39–40
2. Ministry of Interior Affairs İmmigration Management General Directorate (2015) Immigration design periodical pp 6–7
3. www.gercekhayat.com.tr/roportaj/avrupa-hiristiyan-multeci-seviyor/
4. Çolak F (2013) Bulgaristan Türklerinin Türkiye'ye Göç Hareketi/Tarih Okulu 197:113–117 (in Turkish)
5. http://www.goc.gov.tr/files/files/multecilerin Hukuki Statüsüne İlişkin Sözlesme. pdf
6. AFAD (2016) Sphere project: minimum standards in disaster intervention and humanitarian aid agreement, Ankara
7. http://www.goc.gov.tr/icerik6/2015-turkiye-goc-raporuyayinlandi_350_359_10058_içerik
8. http://www.sozcu.com.tr/2016/ekonomi/suriyelilere-36-milyar-tl-harcadik-1342152/Austos 2016
9. http://www.sde.org.tr/tr/newsdetail/-dogudan-gelen-goc-baskisi/3065
10. AFAD (2016) Our Syrian Siblings in the Sibling Territories, Ankara
11. Balcilar M (2016) Study of health situation of Syrian refugees in Turkey
12. http://www.aljazeera.com.tr/haber/4-ulke-4-milyon-siginmaci
13. https://en.wikipedia.org/wiki/Zaatari_refugee_camp (Police disperse rioting Syrians at Zaatari camp). The Jordan Times. 2012-09-24. Retrieved 01 Feb 2013
14. http://www.independent.co.uk/news/world/europe/idomeni-refugee-dachau-nazi-concentration-camp-greek-minister-a6938826.html
15. www.hurriyet.com.tr/lampedusa-dan-sok-goruntuler-25395474
16. http://bianet.org/bianet/insan-haklari/167862-dunyaya-siginamayan-multeciler
17. Bosphorus University European Studies Center Student Forum Bulletin (2016) President of International Refugee Rigths Association Lawyer Ugur Yildirim Interview 5: 36–37
18. http://www.hurriyet.com.tr/turkiye-2011den-bu-yana-2-1-milyon-suriyeli-multeciye-8-milyar-dolar-harcadi-40007235
19. http://www.haber3.com/hangi-ulkede-kac-multeci-var-3643929h.htm

Understanding Sustainability Through Mindfulness: A Systematic Review

Ecem Tezel$^{(\boxtimes)}$ and Heyecan Giritli

Faculty of Architecture, Istanbul Technical University, Istanbul, Turkey
{tezele, giritli}@itu.edu.tr

Abstract. The construction industry is central to how the future of humanity will be shaped and can be sustained. Sustainable development and construction is the responsibility not only of researchers, but also all stakeholders involved in the creation and use of the built environment. Despite increasing global awareness on environmental issues, creating a sustainable built environment seems not to be easy. This may be attributed to the gap between the possession of environmental knowledge and environmental awareness, and displaying pro-environmental behavior. It is epitomized by the attitude, "I know I should, but I don't."; exemplary, even replacing shorter automobile trips with walking or cycling may cause reduction on environmental impact, driving is the overall tendency of individuals. The characteristic of such an attitude, mindfulness, is ascribed to people who have a relatively greater awareness of the surroundings and of other people. Although mindfulness research has seen exponential growth in the last two decades, a limited number of studies has been adapted towards environmental purposes. This study aims to assess the link between mindfulness and sustainable behavior. In the study, a meta-analytical approach will be adopted, which assumes the conceptualization of mindfulness as a form of behavioral regulation, and the relationship between mindfulness and pro-environmental behavior.

Keywords: Mindfulness sustainability · Pro-environmental behavior Meta-analysis

1 Introduction

Construction industry's negative impacts on the environment have being mentioned more than 30 years [1]. In order to limit the harm on nature that construction activities cause, sustainable building approach raised which aims to generate more efficient living areas with consuming and damaging less [2]. On the other hand, users do not always act in buildings as designers projected [3], thus the gap between predicted and actual sustainability performance become inevitable.

Despite the critical role of human behavior in ensuring a sustainable future, there exist individual differences regarding environmental concern. Even though many people are aware of all environmental issues such as air and water pollution, climate change, deforestation and loss of biodiversity, some others still engage in environmentally damaging behaviors. Consequently, increasing attention has been paid to

© Springer International Publishing AG, part of Springer Nature 2018
S. Fırat et al. (eds.), *Proceedings of 3rd International Sustainable Buildings Symposium (ISBS 2017)*, Lecture Notes in Civil Engineering 7,
https://doi.org/10.1007/978-3-319-64349-6_25

understanding the impact of human behavior on climate change and environmental decline. It may be claimed that people must change their unsustainable behaviors to protect the environment and sustainability [4]. However, this simple statement disguises the complexity of the constituents which make up individuals' behavior and particularly, an individual's sustainable behavior. In an attempt to explain sustainable behavior, many studies have focused on analyzing the factors that have been found to have influence on sustainable behavior. Along with these studies, several researchers have tended to focus on the effect of psychological factors on human behavior patterns in general and sustainable behavior in particular.

The science of psychology can play an important role in studying human-environment interactions and understanding environmental problems which are rooted in human behavior. These problems can thus be managed by behavioral changes so as to promote environmentally responsible behavior. The psychology of environmental behavior can make a substantial contribution to this change. There has been a growing body of research investigating the psychological underpinnings of behavior that minimize environmental issues [5]. Mindfulness has been viewed as a psychological construct for overcoming behavioral barriers to empower sustainability [6]. Research has also demonstrated the value of mindfulness in behavior change interventions. Eco-psychologists have suggested that mindful awareness of people's interdependence with nature may help us to behave more sustainably [7]. In the context of sustainable behavior, it is reasonable to expect that, mindful individuals are more likely to act more sustainably. This paper discusses the effect of mindfulness on sustainable behavior and search for a path how sustainable architecture can be used as a behavioral regulation tool.

2   Mindfulness and Sustainable Behavior

There has been theoretical and empirical research illustrating the impact of human behavior on nature and indicating the behavioral change as a solution against environmental damage [8, 9]. Several studies have contributed to our understanding of what motivates changes in behavior to minimize environmental issues.

2.1   Mindfulness

In recent years, there has been a growing interest on the construct of mindfulness issues. Its roots go back thousands of years to Buddhist spiritual tradition. The word "mindfulness" originally comes from the Pali word "sati", which means having awareness, attention and remembering [10]. A general definition of mindfulness is a way of paying attention to the present moment, on purpose and without judgment [11, 12]. Several other researchers provided similar definitions of the term mindfulness. Baer [13], for example, defines mindfulness as "the nonjudgmental observation of the ongoing stream of internal and external stimuli as they arise". A common point for these various definitions is that mindfulness is an intentional exclusively on the attentional aspects of mindfulness (e.g., [14]), others follow the model of Bishop et al.

[15], which proposed that mindfulness consist of two components: attention and awareness of experience.

It may be helpful to review questionnaires that have been developed to measure mindfulness. Measurement of the mindfulness construct varies ranging from knowledge based questionnaires [16] to validated questionnaires, such as; Kentucky Inventory of Mindfulness Scale (KIMS), the Cognitive and Affective Mindfulness Scale (CAMS) of Feldman et al. [17], the Freiburg Mindfulness Inventory (FMI) of Buchheld et al. [18] and the Mindfulness Attention Awareness Scale (MAAS) of Brown and Ryan [14]. A detailed review of these different mindfulness inventories to date is beyond the scope of this paper and the reader is referred to Baer et al. [19] and Bergomi et al. [20].

Although a number of self-report questionnaires have been developed to measure mindfulness, the lack of a consensually agreed-upon definition and nature of mindfulness particularly in the cross-cultural setting remain among scholars [21]. This may be attributed to the view that mindfulness appears to be usually difficult to define and conceptualize [14].

2.2 Sustainable Behavior

Although pro-environmental behavior is, in practical terms, synonymous with sustainable behavior, the former has been used to emphasize efforts to protect the natural environment, while the latter specifies actions aimed to protect both the natural and the human (social) environment [22].

Pro-environmental behavior focus on individual's tendency to act in environmentally responsible way and named through different names such as, ecological behavior or environment-friendly behavior [23]. According to Kollmuss and Agyeman [24], pro-environmental behavior is the term used to define a conscious behavior related with natural and built environment that aims to limit adverse effect of human actions. Similarly, Steg and Vlek [8] also defined PEBs through emphasizing their minimized damage on nature and even their contribution to nature.

Pro-environmental behaviors are not only investigated by academic researches but also the governments to motive people to minimize human activities' negative impact on nature. For instance, governmental agencies of Japan, United Kingdom and United States were provided lists of PEBs and total 200 behaviors summarized currently by Kurisu [23].

Many researchers investigated the influential factors of sustainable/pro-environmental behaviors and claimed different behavior models including the relationship of those factors on specific behavior such as; recycling behavior, energy-saving behavior, etc. [23]. Related with those models, questionnaire surveys have also been developed in order to determine the connection with attitude and behavioral tendency. The New Environmental/Ecological Paradigm (NEP) Scale of Dunlap and Van Liere [25], Ecocentric, Anthropocentric and Environmental Apathy Scale of Thompson and Barton [26], Ecological World View Scale of Blaikie [27], Environmental Perception Scale (ENV) of Bogner and Wiseman [28] are some of the measurement methods explained in detail in Kurisu's book [23].

2.3   Relationship Between Mindfulness and Sustainable Behavior

To what extent mindfulness can be correlated to sustainable behavior? It is believed that, mindfulness can make a difference for fostering sustainable behavior. A review of the empirical literature concludes that one's level of mindfulness has an independent and significant relationship with sustainable behavior [29–31]. A few number of studies report that higher level of mindfulness may be associated with lower carbon footprints. Mindful individuals are more apt to pay attention and intentionally process information about environmental impact. Amel et al. [7] found supporting evidence that "acting with mindful awareness was significantly positively correlated with self-reported sustainable behavior". In addition, Brown and Kasser [32] present additional evidence for the correlation between the level of mindfulness and ecological concern. They found that dispositional mindfulness, along with an intrinsic value orientation, are related to more ecologically responsible behavior. Ericson et al. [33] condemn that mindfulness may promote sustainable behavior. These authors suggest that promoting the practice of mindfulness in schools and organizations may be a key to a more sustainable society.

In 2014, according to their study about hotel guests' behavior, Barber and Deale [34] announced that the level of mindfulness lead individuals make choices which are beneficial for both people and nature. Barbaro and Pickett [6] conducted two studies, reporting that "mindfulness is significantly associated with pro-environmental behaviors" and that mindful traits of observing, non-reactivity and connectedness are particularly important. Jacob et al. [35] reported results of a survey of 821 people, finding significant associations among mindfulness, ecologically sustainable behavior and subjective well-being.

3   Interplay Between Sustainable Buildings and Mindfulness-Based Behavior

Despite increasing focus on mindfulness and sustainable behavior issues, their connection with sustainable buildings is still ambiguous. For example, current literature about mindfulness in journal publications is shown in Fig. 1 [36]. According to American Mindfulness Research Association (AMRA) database, there is a residual trend on mindfulness research since mid-1980s, and an immediate growth of the number of studies about mindfulness, especially for last decade [36].

Considering the increasing research trend about mindfulness, the relationship between mindfulness and sustainable behavior has been mentioned in just few researches seen in Table 1.

However, since it is mentioned at the very beginning of the paper, construction of sustainable buildings is favorable but underpowered target according to environmental deterioration. As it is stated by Wu et al., "We believe that intentionality and mindfulness in the design process of sustainable spaces will bring about intentionality and mindfulness in the thoughts and actions of people using the space" [37]. So, buildings, especially sustainable buildings, should be considered through their occupants' behavioral tendencies and as mediums to shape these tendencies in more sustainable

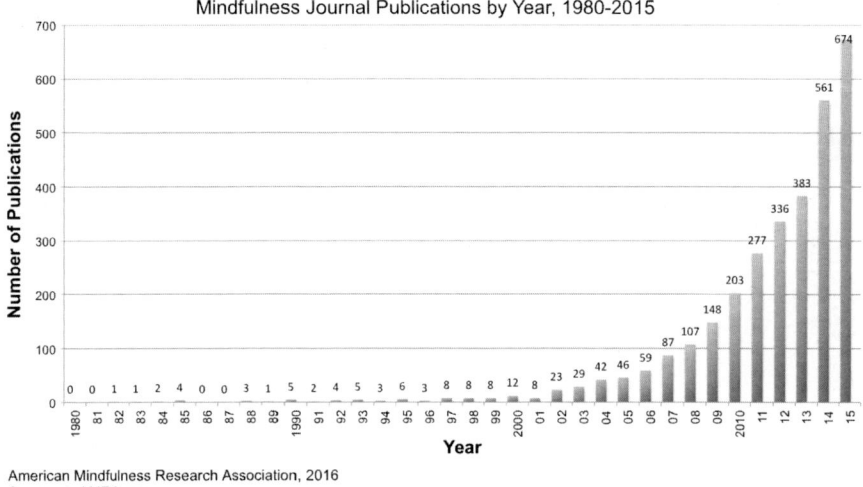

American Mindfulness Research Association, 2016
Source: goAMRA.org

Fig. 1. Mindfulness journal publications by year, 1980-2015 [36]

Table 1. An overview of studies about mindfulness concept into sustainable behavior

Year	Author(s)	Scope of study
1997	Jacob and Brinkerhoff	Effect of mindfulness on sustainable behavior and personal well-being
1999	Brinkerhoff and Jacob	
1999	Jacob and Brinkerhoff	
2005	Brown and Kasser	Ecological concern
2009	Amel et al.	Sustainable behavior
2009	Jacob et al.	
2014	Barber and Deale	Tendency to environmental benefit
2014	Ericson et al.	Well-being and sustainable behavior
2016	Barbaro and Pickett	Mindfulness and pro-environmental behavior

way to ensure the environmentally oriented efforts of construction industry will serve the purpose.

4 Conclusion

User behaviors have non-ignorable role on buildings' actual performance. According to Lockton et al. [3] "The design of products, systems and environments can be used to influence user behavior". However, despite the great effort of creating sustainable built environment especially in last two decades, an important performance gap still exists, due to the environmental concern and behavioral tendencies of buildings' occupants.

Notwithstanding limited studies, mindfulness, as a psychological construct, has become increasingly important in human behavior, especially in sustainable behavior. Application of mindfulness to environmentally responsible behavior can provide sustainability researchers with novel roads of research. This paper aimed to conceptualize mindfulness as a form of behavioral regulation that needs to be investigated in future research within the context of sustainable construction. In other respect, since people are connected with buildings in some way, improvement of mindfulness about sustainability may be achieved by considering architecture as an agent [1]. Thus, construction professionals should focus on designing sustainable buildings by considering them not only as service or products with minimized environmental harm but also as guides for environmentally oriented behavior for users.

References

1. Chansomsak S, Vale B (2008) Can architecture really educate people for sustainability. World congress SB08, pp. 2262–2269
2. Barnett DL, Browning WD (1995) A primer on sustainable building. ISBN: 1-881071-05-7. Rocky Mountain Institute, USA
3. Lockton D, Harrison D, Stanton NA (2010) Design for behavior change. Appl Sci Psychol Res 67(69):123–141
4. Schultz PW, Zelezny LC (2000) Promoting environmentalism. J Soc Issues 56:443–457
5. Bamberg S, Möser G (2007) Twenty years after Hines, Hungerford and Tomera: a new meta-analysis of psycho-social determinants of pro-environmental behavior. J Environ Psychol 27:14–25
6. Barbaro N, Pickett SM (2016) Mindfully green: examining the effect of connectedness to nature on the relationship between mindfulness and engagement in pro-environmental behavior. Personal Individ Differ 93:137–142
7. Amel EL, Manning CM, Scott BA (2009) Mindfulness and sustainable behavior: pondering attention and awareness means for increasing green behavior. Ecopsychology 1:14–25
8. Steg L, Vlek C (2009) Encouraging pro-environmental behavior: an integrative review and research agenda. J Environ Psychol 29:309–317
9. Turaga RMR, Howarth RB, Borsuk ME (2010) Pro-environmental behavior. Ann N Y Acad Sci 1185:211–224
10. Bodhi B (2000) The connected discourses of the Buddha. ISBN: 978-0-861-71973-0. Wisdom Publications, USA
11. Kabat-Zinn J (1990) Full catastrophe living: using the wisdom of your mind to face stress, pain and illness. Dell Publishing, New York
12. Kabat-Zinn J (2005) Coming to our senses: healing ourselves and the world through mindfulness. ISBN: 0-7499-2588-4. Piatkus Books Ltd., UK
13. Baer RA (2003) Mindfulness training as a clinical intervention: a conceptual and empirical review. Clin Psychol Sci Pract 10:125–143
14. Brown KW, Ryan RM (2003) The benefits of being present: mindfulness and its role in psychological well-being. J Pers Soc Psychol 84:822–848
15. Bishop SR, Lau M, Shapiro S, Carlson L, Anderson ND, Carmody J, Segal ZV, Abbey S, Speca M, Velting D, Devins G (2004) Midfulness: a proposed operational definition. Clin Psychol Sci Pract 11:230–241

16. Moscardo G, Pearce PL (1986) Visitor centres and environmental interpretation: an exploration of the relationship among visitor enjoyment, understanding and mindfulness. J Environ Psychol 6:89–108
17. Feldman GC, Hayes AM, Kumar SM, Greeson JM (2004) Development, factor structure and initial validation of the cognitive and Affective mindfulness scale. Unpublished manuscript
18. Buchheld N, Grossman P, Walach H (2001)Measuring mindfulness in insight meditation (Vipassana) and meditation-based psychotherapy. The development of the Freiburg mindfulness inventory (FMI). J Meditation Med Res 1:11–34
19. Baer RA, Smith GT, Hopkins J, Krietemeyer J, Toney L (2006) Using self-report assessment methods to explore facets of mindfulness. Assessment 13:27–45
20. Bergomi C, Tschache W, Kupper Z (2013) The assessment of mindfulness with self-report measures: existing scales and open issues. Mindfulness 4:191–202
21. Brown KW, Ryan RM, Creswell JD (2007) Mindfulness: theoretical foundations and evidence for its salutary effects. Psychol Inq 18:211–237
22. Tapia-Fonllem C, Corral-Verdugo V, Fraijo-Sing B, Duron-Ramos MF (2013) Assessing sustainable behavior and its correlates: a measure of pro-ecological, frugal, altruistic and equitable actions. Sustainability 5:711–723
23. Kurisu K (2015) Pro-environmental behaviors. ISBN: 978-4-431-55832-3. Springer, Japan
24. Kollmuss A, Agyeman J (2002) Mind the gap: Why do people act environmentally and what are the barriers to pro-environmental behavior. Environ Educ Res 8:239–260
25. Dunlap RE, Van Liere KD (1978) The new environmental paradigm. J Environ Educ 9:10–19
26. Thompson SCG, Barton MA (1994) Ecocentric and anthropocentric attitudes toward the environment. J Environ Psychol 14:149–157
27. Blaikie NWH (1992) The nature and origins of ecological world views: an Australian study. Soc Sci Quart 73:144–165
28. Bogner FX, Wiseman M (1999) Toward measuring adolescent environmental perception. European Psychol 4:139–151
29. Brinkerhoff MB, Jacob JC (1999) Mindfulness and Quasi-religious meaning systems: an empirical exploration within the context of ecological sustainability and deep ecology. J Sci Study Relig 38:524–542
30. Jacob JC, Brinkerhoff MB (1997) Values, performance and subjective well-being in the sustainability movement: an elaboration of multiple discrepancies theory. Soc Indic Res 42:171–204
31. Jacob JC, Brinkerhoff MB (1999) Mindfulness and subjective well-being in the sustainability movement: a further elaboration of multiple discrepancies theory. Soc Indic Res 46:341–368
32. Brown KW, Kasser T (2005) Are psychological and ecological well-being campatible? the role of values, mindfulness and lifestyle. Soc Indic Res 74:349–368
33. Ericson T, Kjonstad BG, Barstad A (2014) Mindfulness and sustainability. Ecol Econ 104:73–79
34. Barber NA, Deale C (2014) Tapping mindfulness to shape hotel guests' sustainable behavior. Cornell Hosp Q 55:100–114
35. Jacob JC, Jovic E, Binkerhoff MB (2009) Personal and planetary well-being: mindfulness mediation, pro-environmental behavior and personal quality of life in a survey from the social justice and ecological sustainability movement. Soc Indic Res 93:275–294
36. American Mindfulness Research Association (AMRA) (2016) AMRA database-mindfulness journal publications by year, 1980–2015. https://goamra.org/resources
37. Wu DW-L, DiGiacomo A, Kingstone A (2013) A sustainable building promotes pro-environmental behavior: an observational study on food disposa. PLoS ONE 8:1–4

Estidama Pearl Building Rating System of Abu Dhabi and Al Sa'fat of Dubai: Comparison and Analysis

Omair Awadh$^{(\boxtimes)}$ 🆔

British University in Dubai, Dubai, United Arab Emirates
omairawadh@gmail.com

Abstract. In the built environment industry, sustainable building assessment systems provide a framework and tool to follow in order to embed sustainable development measures. While green building assessment systems are mostly focused on the environmental design and systems performance. To help developers, designers and construction stakeholders in defining the projects' sustainability objectives and design indicators, two UAE-based sustainability regulations have been objectively assessed and major differences have been identified. A comparative analysis of (1) Estidama Pearl Building Rating System (PBRS) of Abu Dhabi and (2) Al Sa'fat of Dubai has been conducted. The analysis has been structured in accordance to an international standard; ISO/AWI 21929 Sustainability Indicators, and to another similar system; SBAT of South Africa. As construction is constantly increasing in the UAE and other developing countries, climate change impacts of buildings and adaptability measures need to be considered more frequently through such systems. In addition, the social aspect of sustainable development could be further supported by new building techniques suitable to the region. The local green building assessment systems should look at the life cycle analysis and operational phase assessment as they directly support the economic aspect of the stainability along with their environmental-social positive impact. It is expected that the results presented in this research can contribute to a better understanding in the field of sustainability regulations and environmental designs.

Keywords: Sustainability · Green building · Estidama · Al Sa'fat

1 Introduction

The first generation of green building assessment systems, such as LEED and BREEAM, were created in developed countries. Subsequently, developing countries saw the need and expressed interest in creating their own systems. The development of these systems has been mainly driven by the three pillars of sustainability, being environmental; social; and economics, and customized in line with the region's needs. This has resulted in a difference in quality between systems emanating from developed and developing countries [1]. On the other hand, some developing countries saw no need to follow such systems as they are in transition phase with no requirement of maintaining the present [2]. According to Castro et al. [3], due to the social and economic glitches in developing countries, the sustainability of developments and

© Springer International Publishing AG, part of Springer Nature 2018
S. Fırat et al. (eds.), *Proceedings of 3rd International Sustainable
Buildings Symposium (ISBS 2017),* Lecture Notes in Civil Engineering 7,
https://doi.org/10.1007/978-3-319-64349-6_26

buildings is extremely important. The sustainable buildings' aim in developing countries is to maximise beneficial social and economic impact while minimising negative environmental impacts [4].

In the Middle East, the United Arab Emirates (UAE) recognized the need to adopt sustainability practices in the construction industry early on and on that basis further actions have been taken. Acknowledging the sustainable development needs in the UAE, green building codes and regulations have been mandated for new constructions in Abu Dhabi and Dubai. The Estidama Pearl Rating Systems (PRSs) have been developed for villas, buildings and communities as the regulatory sustainability code in Abu Dhabi. The first version of PRSs was released in April 2010 in line with Abu Dhabi's 2030 vision [5]. Pearl 1 is the minimum rating for privately-funded projects and Pearl 2 for government-funded projects. Pearl 3–5 rating is targeted voluntarily. Another local system; the Dubai Green Building Regulations and Specifications (DGBR), has been developed as part of the Dubai 2015 Strategic Plan. This code was mandated in 2011 for all government projects at the piloting stage, then for private projects in March, 2014. DGBR is a set of mandatory requirements for buildings with no different ratings [6].

In September 2016, the new Dubai Green Buildings Evaluation System; Al Sa'fat, was introduced to the construction industry stakeholders and professionals. An updated Al Sa'fat version v1.1 has since been released at the end of 2016, in line with Dubai Plan 2021 [7]. This system has four different ratings for building, with Bronze Sa'fa being the minimum rating requirement. Silver, Gold and Platinum ratings are optional ratings that are expected to be encouraged to target through governmental incentives.

As sustainable developments are mainly driven by codes and regulations in the UAE and the Middle East [8], this study compares the Al Sa'fat of Dubai and Estidama Pearl Building Rating System (PBRS) of Abu Dhabi regulations. The objective of this comparison is to assess and analyze sustainability practices in the UAE's built environment. The differences and strengths of each system are the main focus of the study, while further possible improvements have also been highlighted.

2 Research Methodology

The selected sustainable development regulations within the UAE, Estidama PBRS and Al Sa'fat, will be assessed and analyzed against the frame of the following;

- ISO 21929-1: 2011 Sustainable construction works core indicators and the sustainable building assessment tool as an international standardization body [9].
- The Sustainable Building Assessment Tool (SBAT) of South Africa as a developing country system that sensibly addresses the three pillars of sustainability; environmental, social and economic. The SBAT describes 15 sets of objectives that should be aimed for in buildings' development. It suggests that the extent to which these objectives are achieved in buildings provide a simple, yet reasonably effective, measure of the level of support for sustainable development [4].

A quantitative and qualitative analysis of the selected systems has been conducted and objectively presented. The selection of the frame works to which the two systems are compared against is based on state-of-the-art review, how they best fit the study objectives, and the area of the author's own experience. SBAT represents another developing country case of similar environment to the UAE while the International Organization for Standardization (ISO) indicators represent an international frame.

According to Berardi [10] the sustainability assessment systems differ in structure and they do not preferably overlap. The assessment is structured according to the two frame works for better understanding and parallel analysis. This should simplify the process and highlight the main differences. Critical analysis has been adopted throughout the discussion as a way to identify potential for improvements in buildings' sustainability standards and to propose a roadmap for the overall green building systems' approach, with the mention of future areas of research.

3   Discussion and Results

Castro et al. [3] differentiate between Green buildings and Sustainable buildings based on the aspects they consider and address. The list in "Fig. 1" demonstrates how sustainable building design considers several additional aspects as compared to green

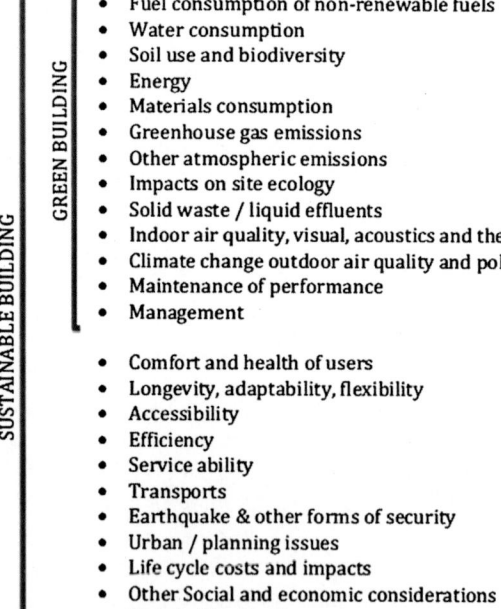

Fig. 1. Aspects considered in green building and sustainable building design (*Source* Castro et al. [3])

building. Green buildings tend to focus on the environmental aspects of sustainability, while Sustainable buildings also cover social and economic and additional environmental related aspects [3].

Fenner and Ryce [11] define the Total Quality Assessment (TQA) sustainability system as a system that evaluates ecological, economic and social aspects. Estidama Pearl Rating System can be considered a TQA point-based system, whereby projects are awarded points for different credits that are grouped under a number of categories. Prerequisite credits need to be achieved, with no associated points, in order to accommodate the minimum system requirements. In addition, project specific matrices should be followed for additional credit points to achieve different certification levels. According to Elgendy [12] Estidama is mostly developed using LEED and BREEAM elements whilst adapting the system to the unique local needs and environment.

Estidama Pearl Rating System (PRS) was established by the Abu Dhabi Urban Planning Council (UPC) on a five ratings basis that can be obtained based on achieving prerequisites and optional credit points. The building ratings under the PRS are; 1 Pearl (20 prerequisites), 2 Pearl (20 prerequisites + 60 points), 3 Pearl (20 prerequisites + 85 points), 4 Pearl (20 prerequisites + 115 points), and 5 Pearl (20 prerequisites + 140 points). Eight credit categories are available in the Pearl Building Rating System (PBRS v1.0) with a maximum of 180 points [5]. Figure 2 illustrates the credit categories and their points' weighting.

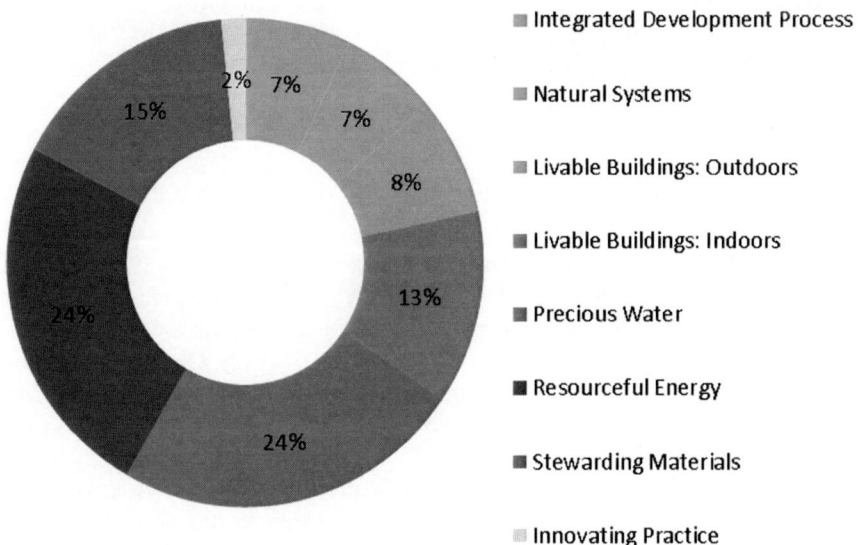

Fig. 2. Estidama PBRS V1.0 available credit points weighting per category

Al Sa'fat can also be categorized as a TQA system that has 4 different ratings with a set of requirements for each rating. This provides less flexibility than the point-based system, and makes it most stringent as per Shareef and Altan [8]. The latest v1.1 of Al Sa'fat, developed by the Dubai Municipality (DM), consists of 5 categories and has four ratings for buildings only. Bronze Sa'fa requires for 42 prerequisites. Silver, Gold and Platinum Sa'fa rating require 81, 98, and 102 prerequisites, respectively. Figure 3 illustrates Al Sa'fat categories and their prerequisites' weighting.

To simplify the categories' weighting comparison, similar categories have been color-coded in Figs. 2 and 3. The intent of the three categories of PBRS; Integrated Development, Natural Systems and Livable Building Outdoors, is the same as that of Al Sa'fat's Ecology and Planning category intent, and thus have been given the same color code.

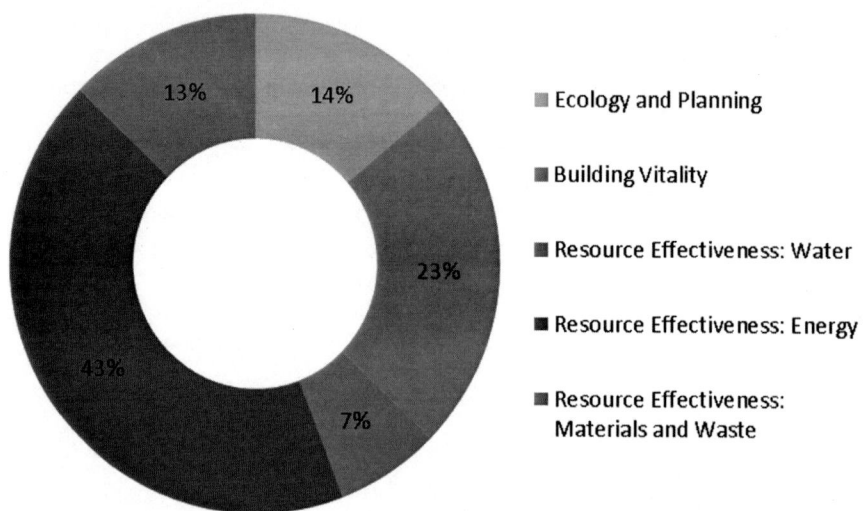

Fig. 3. Al Sa'fat V1.1 available credit points weighting per category

Based on the simplified Figs. 1 and 2, it is clear that Al Sa'fat is an energy-focused system with a 43% weighting for energy efficiency requirements, while PBRS equally prioritizes and balances between energy and water categories with a weighting of 24% each. Water category weighting is the least in Al Sa'fat, 7% only. On the other hand, the Building Vitality (which is equivalent to PBRS's category Livable Building: Indoors) category is given a higher weighting in Al Sa'fat than in PBRS. This category is mostly related to the indoor environmental quality and occupants comfort and wellbeing.

According to ISO/TR 21932: 2013, a projects' six phases of decision-making process are: strategic planning; project definition; design; construction and handover; operation and maintenance; and end-of-life strategy [9]. Both PBRS and Al Sa'fat consider only the first 4 phases of most developments. The Operation and Maintenance and End-of-life phases are currently not being considered. It should be noted that an Estidama system for the operation phase has been developed but has not yet been mandated to date. Similarly, Al Sa'fat includes a re-certification requirement post-construction completion for building operations. Therefore, more attention should be given to the operational phase practices. It is preferable for new constructions to consider building life cycle analysis in the early stages of design to examine the building systems suitability throughout the development's life, including the operation and end-of-life phases.

As for existing buildings' sustainability in the UAE, no regulations have been enforced yet. It is worth noting that the 'Technical Guidelines for Retrofitting Existing Buildings' has been developed by the Emirates Breen Building Council in 2015 [13] however it has not been enforced. Existing buildings' monitoring, auditing and benchmarking are the first steps to identify ways of improvements.

Sustainability indicators of construction works according to ISO TS 2011 mandate are classified under 14 categories and 25 indicators. This part of ISO 21929 describes and gives guidelines for the development of sustainability indicators related to buildings and defines the aspects of buildings to consider when developing sustainability indicators system. The PBRS and Al Sa'fat comparison, in line with ISO sustainability indicators, is presented in Table 1 hereafter. A point-based comparison has been followed, where '1' reflects positive indicator consideration and '0' means non-considered indicator.

7 of the 14 categories are fully addressed in both systems. Out of 7, 5 of those are environmental related, one is environmental-social related and the seventh is economic related. Site selection and site accessibility indicators are considered in PBRS only while planning for maintainability and building accessibility indicators are considered in Al Sa'fat only. Access to services and Aesthetic quality categories are better covered in PBRS than in Al Sa'fat. Safety and Adaptability categories are poorly considered in both, where Resistance, Fire Safety and Adaptability for Climate Change are not covered. However, Stability and Adaptability for changed use purpose are considered in PBRS.

In summary, most of the environmental indicators are considered in both systems with no direct consideration given to climate change adaptability. On the other hand, both assessed systems consider the social indicators the least with clear shortfall in the safety category. As for the economical aspect, only one indicator is present within ISO indicators and it is considered in both Al Sa'fat and PBRS.

Taking another developing country's case, South Africa's Sustainable Building Assessment Tool (SBAT) considers the three pillars of sustainability equally. 5 objectives have been assigned under each one of these pillars. SBAT consists of 9 structured stages of development: Briefing; Site Analysis; Target Setting; Design; Design development; Construction; Handover; Operation; and Reuse/refurbish/recycle.

Gibberd [4] suggests that SBAT objectives provide a reasonably effective and simple measure of the development sustainability. Table 2 demonstrates that PBRS credits cover all of SBAT's 15 objectives. Two objectives have not been taken into consideration within Al Sa'fat. One of these is a social objective (Access to facilities) and the other one is an economic objective (Adaptability and flexibility).

In general, PBRS and Al Sa'fat follow the environmental objectives of similar local sustainable buildings' assessment systems of similar developing country, like SBAT of South Africa. However, not all environmental indicators of ISO have been considered by those. On the other hand, Al Sa'fat considers the economic indicators of ISO and follows 4 of SBAT's 5 economic objectives. SBAT social objectives are fully followed by PBRS while only 6 out of 9 social indicators of ISO are considered. In terms of Al Sa'fat social performance, 3 of 9 ISO indicators are considered and 4 of 5 SBAT

Table 1. PBRS and Al Sa'fat assessment based on ISO 21929 sustainability indicators

ISO 21929-1: 2011 core indicators		PBRS	Al Sa'fat
Category/indicator			
a	*Access to services*		
1	Public transportation	1	0
2	Personal modes of transportation	1	1
3	Green and open spaces	1	1
4	User relevant basic services	1	0
b	*Aesthetic quality*		
5	Integration with the surrounding	1	0
6	Impact of building in site	1	1
7	Local concerns	1	1
c	*Land*		
8	Site selection	1	0
d	*Accessibility*		
9	Building site	1	0
10	Building	0	1
e	*Harmful emissions*		
11	Potential impact on climate	1	1
12	Potential impact on the depletion of stratospheric ozone layer	1	1
f	*Non-renewable resources*		
13	Use of resources	1	1
g	*Fresh water*		
14	Use/consumption	1	1
h	*Waste*		
15	Production	1	1
i	*Indoor environmental*		
16	Indoor conditions	1	1
17	Indoor air quality	1	1
j	*Safety*		
18	Stability	1	0
19	Resistance	0	0
20	Fire safety	0	0
k	*Serviceability*		
21	Planning/measurement	1	1
l	*Adaptability*		
22	Adaptability for changed use purpose	1	0
23	Adaptability for climate change	0	0
m	*Costs*		
24	Planning/measurement	1	1
n	*Maintainability*		
25	Planning/assessment	0	1
	Total	20	15

Table 2. PBRS and Al Sa'fat assessment in line with SBAT objectives

SBAT objectives		PBRS	Al Sa'fat
a	*Social*		
1	Occupant comfort	1	1
2	Inclusive environments	1	1
3	Access to facilities	1	0
4	Participation and control	1	1
5	Education, health and safety	1	1
b	*Economic*		
6	Local economy	1	1
7	Efficiency of use	1	1
8	Adaptability and flexibility	1	0
9	Ongoing costs	1	1
10	Capital costs	1	1
c	*Environmental*		
11	Water	1	1
12	Energy	1	1
13	Waste	1	1
14	Site	1	1
15	Materials and components	1	1
	Total	15	13

objectives are followed. It can be concluded that PBRS and Al Sa'fat objectives are very similar to other developing countries' sustainable building assessment tools, while they do not necessarily consider all sustainability core indicators of ISO 21929, 2011.

According to Sinha et al. [14], green buildings are directly related development's sustainability and support the sustainable development practices, but they are not the same. Castro et al. [3] also identified the difference between Green buildings and Sustainable buildings design aspects, presented in Fig. 1. Sustainable buildings' assessment tools or systems should be considering the same parameters as these design aspects.

The discussion presented demonstrates that sustainable building assessment systems in developing countries are environmentally focused and need to sensibly consider the social aspects. It is worth mentioning that Estidama PRS has considered a fourth pillar for Abu Dhabi's sustainable development, the Cultural pillar, along with the three typical pillars of sustainability, which should contribute to the overarching social pillar. Future research in developing countries needs to assess the social performance of current sustainable developments, and consider improvements through sustainability regulations/systems. This is also reinforced by Castor et al. [3] suggestion to prioritize social and economic support in the developing countries. In order to encourage positive change in the economic and social system through buildings, new knowledge and techniques need to be developed specifically to the region.

The ongoing international standardization work related to sustainability in construction will enable more transparent assessments of the sustainability criteria of new and existing construction works and the results more comparable than what is possible with today's methods [15]. According to Brophy [16] it is crucial that these systems are simple to make them useful as design tools throughout the construction development. The assessment tools should contribute towards the balancing of sustainability dimensions (environmental, social, and economical), with the enhancement of practicality and resiliency. These systems should be able to consider constant technological development and multi-level application.

4   Conclusion

The two studied regulations of Abu Dhabi and Dubai; Estidama PBRS and Al Sa'fat, differ in structure and they do not perfectly overlap, making direct comparisons a challenge. When comparing the categories' weighting, PBRS is more balanced in terms of credits categorization. Energy and water categories of PBRS are given the same weighting, both contributing to 48% of the overall 180 points. The energy category weighting alone is 43% of 102 regulations of Al Sa'fat, while water category weighting is 7% only. On the other hand, indoor environmental quality and occupants' comfort & wellbeing are given higher weighting in Al Sa'fat than in PBRS.

Following two different frame works; (1) the international standardization body ISO 21929-1: 2011 sustainable construction core indicators and (2) the South African local sustainable building assessment tool (SBAT), the study suggests categorizing PBRS and Al Sa'fat as sustainability assessment systems as they target aspects beyond green buildings' criteria. Note that green buildings' design mostly focuses on the environmental aspects of sustainability, yet they are of direct support to sustainable development.

It was noted that no direct measure of the climate change impact and future adaptability is the main shortage found in both systems' environmental indicators. It is critical to account for the social design parameters and improve current characteristics of PBRS and Al Sa'fat. As for the economic criteria, it is recommended to focus on life cycle analysis for new projects and to enforce the implementation of the operational regulations for both new and existing projects.

References

1. Cole RJ (2005) Building environmental assessment methods: redefining intentions. In: The 2005 world sustainable building conference, Tokyo, pp 1934–1939
2. Voinov A, Farley J (2006) Reconciling sustainability, systems theory and discounting. Ecol Econ 63:104–113
3. Castro MF, Mateus R, Braganca L (2014) A critical analysis of building sustainability assessment methods for healthcare buildings. Environ Dev Sustain. doi:10.1007/s10668-014-9611-0

4. Gibberd J (2002) The sustainable building assessment tool assessing how buildings can support sustainability in developing countries. Built Environ Prof Convention, Johannesburg, South Africa
5. Abu Dhabi urban planning council (2010) Pearl building rating system. Des Constr. Version 1.0, April
6. Dubai municipality (2011) Green building regulations and specifications. www.dm.gov.ae. Accessed 10 Dec 2016
7. Dubai municipality (2016) Al Sa'fat—Dubai green buildings evaluation system. www.dm.gov.ae. Accessed 10 Dec 2016
8. Shareef SL, Altan H (2016) Building sustainability rating systems in the Middle East. Eng Sustain. doi:10.1680/jensu.16.00035
9. International Organization of Standardization (ISO) (2011) TS: ISO/TS 21929-1: 2011, sustainability in building construction—Sustainability indicators—part 1: framework for the development of indicators for buildings 2011 edition. ISO, Geneva, pp 1–24
10. Berardi U (2011) Sustainability assessment in the construction sector: rating systems and rated buildings. Sustain Dev. doi:10.1002/sd.532
11. Fenner RA, Ryce T (2007) A comparative analysis of two building rating systems. Part 1: evaluation. Eng Sustain 161:55–63. doi:10.1680/ensu.2008.161.1.55
12. Elgendy K (2010) Comparing Estidama's pearls rating system to LEED and BREEAM. www.carboun.com/sustainable-urbanism/. Accessed 15 Dec 2016
13. Emirates Green Building Council (2015) EmiratesGBC technical guidelines for retrofitting existing buildings. www.emiratesgbc.org/academy/emiratesgbc-technical-guidelines/. Accessed 22 Dec 2016
14. Sinha A, Gupta R, Kutnar A (2013) Sustainable development and green buildings. Drv Ind 64(1):45–53
15. Krigsvoll G, Fumo M, Morbiducci R (2010) National and international standardization (International Organization for Standardization and European Committee for Standardization) relevant for sustainability in construction. Sustainability. doi:10.3390/su2123777
16. Brophy V (2014) Building environmental assessment—a useful tool in the future delivery of holistic sustainability? In: The 2014 world sustainable building conference, Barcelona: Paper 119

An Investigation into University Students' Perceptions of Sustainability

Begüm Sertyeşilışık[1](✉), Heyecan Giritli[1], Ecem Tezel[1],
and Egemen Sertyeşilışık[2]

[1] Faculty of Architecture, Istanbul Technical University, Istanbul, Turkey
{bsertyesilisik,giritli,tezele}@itu.edu.tr
[2] Istanbul, Turkey
egemens@alumni.bilkent.edu.tr

Abstract. Sustainability efforts are consequences of concerns about future generations. These efforts need to be supported by all countries for permanent success. For this reason, universities, as multidisciplinary and multicultural platforms, are important for establishing international and multidisciplinary collaboration and move towards sustainability. This paper aims to analyze general tendencies, behaviors and consciousness about sustainability studies among universities' students. With this aim, after providing informative classes about sustainability to a class of international and multidisciplinary students studying in different universities in various countries, students were asked to fill in the questionnaire and to participate in the brainstorming session about sustainability education and sustainable behavior. Findings revealed that the awareness about sustainability should be improved through education and that universities play an important role in the fight against climate change especially with the help of courses on sustainability, sustainability practices and their green campuses.

Keywords: Sustainability · Sustainability tendency · Green campus

1 Introduction

The world's habitat is being deteriorated at an accelerated rate especially due to climate change fostered by the greenhouse gas (GHG) emissions. The European Union has set the following sustainable growth targets to fight against climate change [1]: reducing GHG emissions by 20% by 2020; increasing the share of renewables to 20%; increasing energy efficiency by 20%. These targets can be achieved through international and interdisciplinary collaboration as well as through sustainability conscious people acting as change agents for sustainability and having 'pro-environmental behavior'. The term "pro-environmental behavior" is one of the items considered in creation of effective resistance for this serious problem [2]. Investigating environmental effects of humankind in individual scale is one of the contemporary study areas of researchers (e.g. [2, 3]). Gatersleben et al. [4] indicated that "people are not always aware of the environmental impact of their behavior". However, due to direct and

© Springer International Publishing AG, part of Springer Nature 2018
S. Fırat et al. (eds.), *Proceedings of 3rd International Sustainable Buildings Symposium (ISBS 2017)*, Lecture Notes in Civil Engineering 7,
https://doi.org/10.1007/978-3-319-64349-6_27

indirect effect of individuals' consumption habits on environment, [5] and direct effect of occupants on buildings' occupancy and maintenance process [6] the actions taken against global warming is still insufficient. Therefore, a deep focus on human behavior should be implemented in sustainability studies [7] and current gap of insufficient action for preventing environmental harm may be limited by education.

According to Erten [8], regulations or technology won't be enough to solve environmental problems solely, but education and behavioral changes lead human-kind's immediate response against negatively changing conditions of nature. As sustainability certificates of buildings is not a symbol of buildings' sustainable occupancy [9], green campuses may be achieved with green buildings and certifications, and environmentally oriented education which enhance pro-environmental behaviors.

Universities are sources of future professionals in different industries. They need to enhance their students' sustainability consciousness level and transform their behavior into pro-environmental one. These students equipped with sustainability requirements from technical, social and economic aspects can transform their professional fields into sustainable ones. Furthermore, they can contribute to the interdisciplinary and inter-national action needed to fight against climate change. This paper analyses general tendencies, behaviors and consciousness about sustainability studies and behaviors among universities' students.

2 Universities' Roles in Sustainable Development

Green universities play an important role in regional sustainable development and in transforming societies into ecologically sound ones [10]. The United Nations 'Decade of Education for Sustainable Development' underlined the importance of integrating sustainable development values into learning and pro-environmental behavior [11, 12]. People's pro-environmental behavior and their consumption patterns are critical factors influencing success for transition towards sustainability and reducing carbon emissions [12]. Their behaviors can be influenced by education and education facilities. For this reason, the green curricula is important for Education for Sustainable Development [10]. Cotton et al. [12] emphasized the importance of knowledge in transforming the behavior and stated that "... a lack of knowledge may make it more difficult for them to select the most appropriate behavior" [12]. In the 4th UNESCO Chair Conference on Higher Education for Sustainable Development (HESD) in 2011, three roundtables on universities' roles in the sustainable development revealed the importance of inte-grating the campuses into the HESD curriculum and recommended the HESD courses to be elective to accelerate their opening procedure [13]. Students' pro-environmental behavior can be shaped not only through the courses but also through the green campuses. As there is relationship between campus sustainability and education [14], sustainability across the campus and the curriculum are inter-related [12].

Many universities have transformed their campuses into green ones [15, 16]. For example, Harvard University's sustainability plan [17] shows university's commitment to sustainability and provides roadmap for building and operating a more sustainable campus community [18]. Similarly, University of Copenhagen's strategy entitled as "Green Campus 2020: A Strategy for Resource Efficiency and Sustainability" establishes university's 2020 targets for CO_2 reduction, energy consumption reduction, and in the fields of resources, chemicals, organization and behavior and campus as a living lab [19]. Furthermore, McGill University in Montreal inform their students about the sustainable food choices and provide them information on the environmental and resource use impacts of the most common meals available in the cafeterias [20]. Universities can get benefit from their sustainable campuses [16, 21]. For example, University of Copenhagen saved DKK 35 million a year and reduced its CO_2 footprint due to its green efforts, and reduced energy consumption [22].

National and international networks have been established to create synergy among sustainable universities and to foster sustainability performance. These networks enable the universities to learn from the practices. Examples for these networks include:

- The China Green University Network (CGUN) to enhance collaboration among different campuses, innovation, and to promote energy saving ideas [23, 24].
- The International Sustainable Campus Network (ISCN) "to support leading colleges, universities, and corporate campuses in the exchange of information, ideas, and best practices for achieving sustainable campus operations and integrating sustainability in research and teaching" [25].
- International Alliance of Research Universities (IARU)'s the campus sustainability initiative "to promote collaboration between member institutions and develop best practices strategies in environmental management" [26].
- Nordic Sustainable Campus Network (NSCN) of the Nordic higher education institutions "to strengthen Nordic collaboration, find new partners to collaborate with, and to have international visibility" [27].

3 Research Methods

This paper aims to analyze general tendencies, behaviors and consciousness about sustainability studies and behaviors among universities' students. After providing informative classes about sustainability to a class of international and multidisciplinary students studying in different European universities, a brainstorming session has been carried and a questionnaire has been applied. In the questionnaire, students evaluated according to their consciousness levels on (1) CO_2 emission, (2) electric consumption, (3) water consumption and (4) waste generation and recycling issues. The first part of the questionnaire was adopted from a book related to pro-environmental behaviors and their measurement [28]. These questions constituted of a 28-item list of pro-environmental behaviors proposed by Defra, UK. According to researchers, "pro-environmental behaviors were especially related to attitudinal variables" [4]. In this part, students were asked to self-report their general attitude about sustainability by indicating each item's importance level for them on a five-point Likert scale where 1 means "strongly disagree" and 5 means "strongly agree". In the second part of the questionnaire students were asked to self-report their daily life routines in their campuses by indicating "Yes" in terms of "I do" and "No" in terms of "I don't". 14 questions have been included in the second part of the questionnaire. Furthermore, the brainstorming session has been conducted about sustainability education and sustainable behavior.

4 Results and Discussion

4.1 Questionnaire

The sample consisted of 53 students 18 foreign students from different countries of Europe and 35 Turkish students respectively. The questionnaire has been distributed to the sample as hardcopy. Students were given 20 min to complete the questionnaire. The questionnaire was applied as a part of class study, thus there was no invalid answer, and all papers have been used for evaluation.

28 questions in the first part of the questionnaire aimed at identifying general consciousness level of students about sustainability into four categories; CO_2 emission, electric consumption, water consumption and waste generation and recycling. Students ranked each item according to the 5-point Likert scale. The results are shown in Fig. 1.

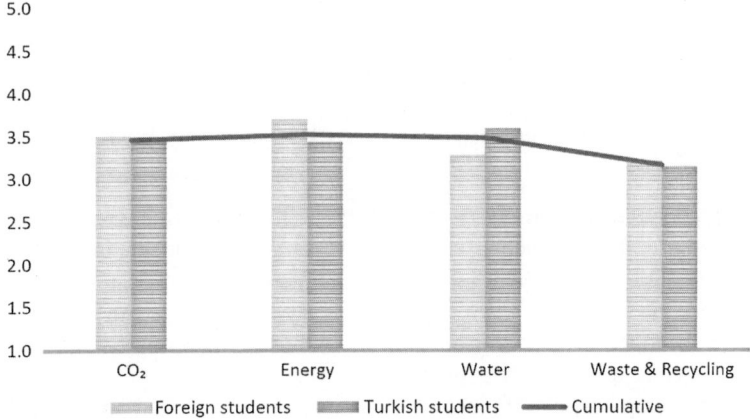

Fig. 1. Sustainability consciousness level of student groups

As it is seen Fig. 1, there are little differences in each item between European and Turkish students. Consciousness level on CO_2 emission (3.50), electric consumption (3.70) and waste generation and recycling (3.21) of foreign students are slightly higher than Turkish ones (3.45; 3.43 and 3.14) whereas Turkish students have higher consciousness level (3.59) on water consumption compared to others (3.28). Nevertheless, overall consciousness levels of all students on the four items indicate a relatively moderate distribution and revealed the importance and need for sustainability education at the undergraduate level.

14 questions in the second part of the questionnaire were designed in order to understand general behavioral pattern of students in their campuses. Students marked each item as "Yes" if they do the defined act and "No" if they don't, considering their routine behaviors. The descriptive results are shown in Table 1.

Considering Table 1, it can be said that, there exist much similarities between European and Turkish students' behaviors. There are two prominent behaviors namely; use of electronic resources for study and less energy mode of appliances. Turkish

Table 1. Daily life behavior of students (in percentage)

Behavior	Foreign		Turkish	
	Yes	No	Yes	No
Public transport/walk-bike	100	–	88.6	11.4
In-campus shuttle	16.7	83.8	28.6	71.4
Electronic source for study	33.3	61.1	82.9	14.3
Double-side use of paper	94.4	5.6	85.7	14.3
Turn off electrical appliance	88.9	5.6	85.7	14.3
Report broken lighting ap	44.4	55.6	28.6	71.4
Report broken plumbing fix	66.7	33.3	34.3	65.7
Less energy mode of app	38.9	55.6	82.9	17.1
Less water mode of plumb	38.9	44.4	54.3	45.7
Recycling bin	88.9	11.1	65.7	31.4
Multilayer wearing	61.1	38.9	68.6	28.6
Native food/drink	88.9	11.1	71.4	28.6
Sustainability education	100	–	94.3	5.7
Education in ITU	77.8	16.7	82.9	14.3

F foreign students, *T* Turkish students

students highly prefer electronic resources for their studies, whereas foreign students tend to study in traditional methods. In addition, sample of students in Turkey tends to pay more attention to use lighting appliances in less energy consuming way, while others don't. This difference between students may be connected to the situation that some of the European students in the sample indicated that they do not know about "less energy mode of electrical appliances".

The relationship between students' conscious level of sustainability issues and their behaviors has been analyzed through Pearson correlation analysis. Some statistically significant results were observed as shown in Table 2.

Table 2. Correlation analysis between conscious level and behavior of students

Behavior	Conscious			
	CO_2 emis.	Elect. con.	Water con.	Waste Rec.
In-campus shuttle	−0.301*			
Electronic source	−0.358**			−0.339*
Turning off elec. app		0.545**		
Recycling bins				−0.362**

*Correlation is significant at the 0.05 level (2-tailed)
**Correlation is significant at the 0.01 level (2-tailed)

Students tend not to prefer travelling with shuttles in campus according to their high conscious level of CO_2 emission. This negative correlation might be interpreted in two folds. Since the distances from one place to other is not too much inside a campus, students tend to prefer walking these distances. Furthermore, shuttle services are not available in some of campuses students come from.

Students tend to prefer print-out resources for their study or researches despite their high consciousness on electricity consumption and waste generation topics. Since this was also an unexpected result, students indicated that they study better if they have hard-copied resources. Furthermore, they added that it is much easier to find any subject from print-out sheets rather than electronic resources since it needs turning on computer, waiting it for getting ready to use, network connection, etc.

Results revealed that the higher consciousness level on electricity consumption generates a tendency on turning the appliances and lights off when leaving the space, they use in their campus. Despite their high consciousness on waste generation and recycling items, students tend not to pay attention to throw their recyclable wastes into fragmented bins, due to lack of available bins for recyclable wastes in their campuses.

Finally, students were asked to evaluate the importance of education about sustainability in their universities and whether or not the classes they had in ITU improved their knowledge about sustainability issues. Respondents' answers have been demonstrated in Table 3. According to the table, majority of the students (96.23%) think that sustainability education should be given as a part of university education. Besides, most of them (81.13%) think that the classes about sustainability in ITU have contributed to their point of view about sustainability topics.

Table 3. Response of students on sustainability education questions

	European students			Turkish students'		
	Yes	No	N/A	Yes	No	N/A
Importance of education about sustainability	18	0	0	33	2	0
Contributions of sustainability classes in ITU	14	3	1	29	5	1

4.2 Brainstorming

Following the questionnaire session, students have been invited to contribute to the study through brainstorming session. During this phase, a collaborative medium has been provided that ensured each student had the chance to reveal his/her opinion on the education of sustainability.

European students in the sample were asked to compare the practices in their own universities with the classes they have taken in ITU. They said that in their undergraduate programs, none of them have any compulsory or selective course related to sustainability. Just few of them mentioned that some courses included sustainability issue as a chapter in overall course flow and that they were not asked any question about sustainability part of the course in examinations.

Turkish students in the sample replied that there are some specific courses on "sustainability" and some courses covering sustainability in one or more chapters in overall course flow. Differing from the current application in Europe, the Turkish students in the sample indicated that examinations of those classes in ITU mostly have at least one question about sustainability topic that lead students to study and focus on the issue more deeply.

The brainstorming session revealed the following recommendations for enhancing sustainability performance of the universities and for enhancing sustainability consciousness of the students:

- Using piezoelectric technology in crowded areas (e.g. university and department entrances; library entrance; sports halls; catering area entrance etc.) and informing the students about the availability of the technology in these areas
- Using impressive banners (which are framed and constantly changing) on sustainability
- Using interesting posters published abroad to motivate research
- Providing articles on food waste in the catering center
- Establishing students' clubs to provide short lectures and speeches on sustainability within 15–20 min during lunch breaks
- Providing sustainability related posters on the elevators
- Establishing sustainability projects competitions and giving awards
- Enabling disposal of waste
- Allocating broader space for paper waste in the Architecture Department
- Informing students from different departments that waste from their departments can be used by the students of the architecture department for the design purposes
- Giving lessons on the sustainability subject
- Giving sustainability courses to the students coming from different departments
- Enabling the companies to introduce and to provide environmentally sensitive products to the students in their professional field
- Providing design courses in the area of sustainability.
- Giving importance to students' sustainable design proposals and ensuring that their design is implemented in the campus

5 Conclusions

This paper has analysed general tendencies, behaviours and consciousness about sustainability studies and behaviours among universities' students. The findings can be summarized as it follows:

- Consciousness level of the sample on CO_2, energy, water, and waste and recycling aspects ranged between 3 and 3.5 in the Likert scale revealing moderate and homogenous consciousness level in these aspects. Consciousness level of students needs to be further enhanced especially in the field of waste and recycling.
- Pro-enviromental behaviors of the European and Turkish students in the sample showed similar patterns. They differed mainly in the usage of electronic resources for study and of energy mode of appliances. Correlation analysis between conscious level and behavior of students in the sample revealed that students tend to prefer

print-out resources for their study and that they tend not to pay attention to throw their recyclable wastes into fragmented bins despite their high consciousness.

- The importance of courses on sustainability has been emphasized. Majority of the students agreed that sustainability education should be given as a part of university education and that the sustainability course provided them has contributed to their knowledge of sustainability.

- Sustainability performance of the universities and sustainability consciousness of the students are recommended to be enhanced with the help of courses and sustainability practices in the campus including application of new technologies in the campuses, information boards and articles on sustainability, student clubs, project competitions, waste disposals, interdisciplinary collaboration among departments.

As the awareness about sustainability should be improved through education, universities play an important role in the fight against climate change especially with the help of courses on sustainability, sustainability practices and their green campuses.

As today's university students are the future's professionals, their consciousness about sustainability and their pro-environmental behaviors are important. Student exchange programs play an important role due to the need for international collaboration in the fight against climate change. Further research is recommended to be carried out on how to enhance collaboration among universities, industries, and politics globally in transforming the society into sustainable one.

References

1. The EU website: Sustainable Growth-for a Resource Efficient, Greener and More Competitive Economy. http://ec.europa.eu/europe2020/europe-2020-in-a-nutshell/priorities/sustainable-growth/index_en.htm
2. Turaga RMR, Howarth RB, Borsuk ME (2010) Pro-environmental behavior. Ann NY Acad Sci 1185:211–224
3. Brown ZB, Dowlatabadi H, Cole RJ (2009) Feedback and adaptive behavior in green buildings. Intell Build Int 1:296–315
4. Gatersleben B, Steg L, Vlek C (2002) Measurement and determinants of environmentally significant consumer behavior. Environ Behav 34:335–362
5. Fransson N, Garling T (1999) Environmental concern: conceptual definitions, measurement methods and research findings. J Environ Psychol 19:369–382
6. Hei C, Yu LSS (2014) Determinants of residential building occupants' behavior in sustainable living: a questionnaire survey in Hong Kong. In: ECSEE—the European conference on sustainability, energy and the environment, UK
7. Steg L, Vlek C (2009) Encouraging pro-environmental behavior: an integrative review and research agenda. J Environ Psychol 29:309–317
8. Erten S (2003) 5. sınıf öğrencilerinde çöplerin azaltılması bilincinin kazandırılmasına yönelik bir öğretim modeli. Hacet Üni Eğitim Fak Derg 25:94–103
9. Abdalla G, Maas G, Huyghe J, Oostra M (2011) Criticism on environmental assessment tools. Int Proc Chem Biol Environ Eng 6:443–446

10. Wang Y, Shi H, Sun M, Huisingh D, Hansson L, Wang R (2013) Moving towards an ecologically sound society? Starting from green universities and environmental higher education. J Clean Prod 61:1–5
11. UNESCO (2005) UN decade of education for sustainable development 2005–2014. UNESCO education for sustainable development (ED/PEQ/ESD). http://unesdoc.unesco.org/images/0014/001416/141629e.pdf
12. Cotton D, Shiel C, Paço A (2016) Energy saving on campus: a comparison of students' attitudes and reported behaviors in the UK and Portugal. J Clean Prod 129:586–595
13. Müller-Christ G, Sterling S, van Dam-Mieras R, Adomßent M, Fischer D, Rieckmann M (2014) The role of campus, curriculum, and community in higher education for sustainable development: a conference report. J Clean Prod 62:134–137
14. Jones P, Selby D, Sterling S (2010) Sustainability education: perspectives and practice across higher education. Earthscan Ltd., London. ISBN 978-1-84407-877-6
15. Isiaka A, Siong HC (2008) Developing sustainable index for university campus. In: EASTS international symposium on sustainable transportation incorporating Malaysian universities transport research forum conference 2008 (MUTRFC08). Malaysia
16. Abd-Razak MZ, Mustafa NKF, Che-Ani AI, Abdullah NAG, Mohd-Nor MFI (2011) Campus sustainability: student's perception on campus physical development planning in Malaysia. Procedia Eng 20:230–237
17. Harvard University Sustainability Plan (2015–2020). https://green.harvard.edu/sites/green.harvard.edu/files/Harvard%20Sustainability%20Plan-Web.pdf
18. Waddell J. Resident dean of Freshman, Elm Yard. https://green.harvard.edu/campaign/our-plan
19. Green Campus 2020: A strategy for resource efficiency and sustainability. http://greencampus.ku.dk/strategy2020/english_version_pixi_GC2020_webversion.pdf
20. Cheni DM, Tucker B, Badami MG, Ramankutty N, Rhemtulla JM (2016) A multi-dimensional metric for facilitating sustainable food choices in campus cafeterias. J Clean Prod 135:1351–1362
21. Alfieri T, Damon D, dan Smith Z (2009) From living building to living campuses. Plan High Educ 38:51–59
22. Schaltz T (2016) Green efforts save DKK 35 million a year at the University of Copenhagen Denmark. https://nordicsustainablecampusnetwork.wordpress.com/
23. China Green University Network. http://www.cgun.org/en/content.aspx?info_lb=107&flag=1
24. Li X, Tan H, Rackes A (2015) Carbon footprint analysis of student behavior for a sustainable university campus in China. J Clean Prod 106:97–108
25. International Sustainable Campus Network. http://www.international-sustainable-campus-network.org/
26. International Alliance of Research Universities. http://www.iaruni.org/sustainability/10y-green-campus/
27. Nordic Sustainable Campus Network. https://nordicsustainablecampusnetwork.wordpress.com/
28. Kurisu K (2015) Pro-environmental behaviors. Springer, Japan. ISBN 978-4-431-55832-3

Properties of Mortars Produced with PKF Press Filter Waste

İlker Bekir Topçu[(✉)] and Taylan Sofuoğlu

Department of Civil Engineering, Faculty of Engineering and Architecture,
Eskişehir Osmangazi University, Eskişehir, Turkey
ilkerbt@ogu.edu.tr, taylansofuoglu@hotmail.com

Abstract. The limestone is used in different stages of the sugar industry. In a phase of the sugar manufacturing process, limestone, approximately 36 h after loading the top of a limestone quarry, which is cylindrically shaped with lime coke is taken as the bottom of a limestone quarry. Lime obtained by quenching hot water, is filtered and used in the production slurry treatment. At given temperature, the raw juice obtained from sugar beet lime is mixed in the apparatus. Some of the muddy slurry of mud here is precipitated in the decanter. The precipitated sludge is filtered by means of filter presses portions. This land including 70–80% dry material is called "PKF press filter waste soil". The effect of the waste material referred as PKF press filter waste soil that is used rather than cement in mortar mixture by weight of cement in different ratios have been investigated on mortar properties. In the scope of this aim, the results obtained from the experiments carried out with reference mortar produced in laboratories and compressive strength values are compared with mortar samples containing waste material. Cement weight of waste, PKF waste used in the mixture of 5, 10, 15, 20% sand, water and CEM I 42.5, CEM II 32.5 type of cement were obtained from mortar mixture and was poured into prismatic molds with in size $4 \times 4 \times 16$ cm and 2 in size $4 \times 4 \times 8$ cm samples were prepared. Samples are over 7 and 28-day compressive strength tests were performed. 5% by weight of the cement that is used PKF waste in mortar samples 7-day average compressive strength is determined as 29.25 MPa for CEM I 42.5. These strength values; in the framework of the ongoing research process of mortars, have been shown to have the appropriate standards.

Keywords: Industrial waste · PKF filter press waste · Pozzolanic materials Mortar · Compressive strength

1 Introduction

Many studies on the components of concrete have been carried out with the help of the improvements of concrete technologies nowadays. In accordance with these studies, concrete components, which are used in concrete production, can be different compared to the traditional usage in order to be more economical and provide more quality according to where it is used. For example, another material having pozzolanic property can be added providing that the binder ratio in concrete is fixed.

© Springer International Publishing AG, part of Springer Nature 2018
S. Fırat et al. (eds.), *Proceedings of 3rd International Sustainable Buildings Symposium (ISBS 2017)*, Lecture Notes in Civil Engineering 7,
https://doi.org/10.1007/978-3-319-64349-6_28

In recent years, industrial wastes with the result of development of industrial technology in the world have been quite a lot and when nature damages were analyzed, this made these wastes to be examined. In industry, it is quite important for production waste to be eliminated in a safe, reliable and economical way [1]. The subject of reuse of waste materials has gained much attention recently. Effective utilization of these conditioned materials for reuse needs some new investigations [2].

Concrete has been the most commonly used material for many years in the construction industry [3] and it consists of cement, aggregates, water and admixtures. A lot of energy and raw material are needed to manufacture cement, which is a concrete component. In addition, the large amount of CO_2 released into the atmosphere in the cement production leads to environmental problems [4]. The contribution of cement production worldwide to the greenhouse gas emission is estimated to be about 1.5 million tons annually or about 8% [5]. For this reason, when it is evaluated within the framework of economy, technology and ecology, ecofriendly binder materials play an important role in the construction industry. Not only a good workability, an excellent mechanical properties and durability of the concrete, but also its environmental impact and economic profitability must be taken into consideration [6]. Use developing technologies let waste materials arisen from industrial processes in the mixture of concrete rather than cement by reforming them [7]. From the perspective of environment, the use of binding materials instead of cement in the mixture of concrete has two main advantages:

- Reducing the energy during the process, the fall of the value of CO_2 emission,
- Waste materials can be used actively in recycling rather than be stored and disposed [8].

The benefit of this solid and ongoing recycling waste can considered as both the reduction of environmental harmful effects legally and the positive contribution to the economy [1]. Due to pozzolanic materials partially used in the concrete mixture, the damages in the inner and outer layers of concrete, caused by the heat resulting from exothermic reactions observed in the process of hydration, can be also reduced [9].

The limestone provided from the market contains at least 90% of $CaCO_3$. In a phase of sugar manufacturing process, limestone is taken as lime from the bottom of limestone quarry approximately 36 h after being loaded with lime coke into limestone quarry, which is cylindrically shaped, from above. Lime obtained is quenched with hot water, filtered and used in the production of slurry treatment. Hydraulic lime is obtained in whitewash consistency. Raw juice, which is obtained from sugar beet at a certain temperature, is mixed with lime in the apparatus. It is precipitated with colloidal substances in raw juice. Excess lime added into juice is removed from the juice with CO_2 in the saturation tanks. Muddy part of muddy juice received from here is precipitated in the decanters. The precipitated sludge part is filtered by means of filter presses. The soil containing almost 70–80% dry matter, which is removed from filter presses, is called "PKF Press Filter Waste Soil".

In this study, the effect of mortar properties on compressive strength has been investigated by using the waste material referred as waste PKF press filter waste soil as a replacement of cement by weight at different ratios and compressive strength values evaluated which is obtained by the use of PKF press filter waste soil which is nonhazardous waste in the mortar mixture.

2 Materials and Methods

2.1 Methods

In the scope of this study, the effect of the mortar properties on compressive strength was investigated by using PKF press filter waste soil, which is a nonhazardous waste of sugar factory in the mortar mixture at certain ratios. Experimental study firstly started with obtaining a control mortar according to specified standards. In the study, mortar mixtures obtained by using PKF press filter waste soil by weight cement at certain ratios together with standard sand, water, CEM I 42.5R, CEM II 32.5R types of cement and the compressive strength behavior examined by stabilizing values such as water-binder ratio, standard sand apart from PKF press filter waste soil which used as a replacement of cement.

After the ratio of PKF press filter waste soil to be used rather than cement was determined, fresh concrete test was done. Compressive strength tests were performed on the 7 and 28-day samples relatively. The strength values were determined by doing compressive strength experiments on two pieces of samples with size of 4 × 4 × nearly 8 cm. In this experimental study, the mixture was prepared by changing PKF waste at certain ratios in the mortar mixture. The values of PKF waste soil to be used in the mixture was determined after being evaluated. Three pieces of samples must be produced in order to determine the 7 and 28-day strength for each ratio and to make a statistical analysis. According to determined mixture ratios, required number of samples have been produced and have been subjected to all experiments.

In this experimental study (7 and 28-day compressive strength) compressive strength behavior of cement evaluated with experimental analysis and data upon the mortar samples obtained from the mixture of two pieces of different cement and PKF press filter waste soil at certain ratios. The mortar mixtures containing two different types of cement and PKF press filter waste soil obtained within the framework of ratios determined by weight of cement. According to the 7 and 28-day periods, the samples removed from molds and the compressive strength behaviors of hardened molds assigned with a variety of experiments. The compressive strength experiments were done on two pieces of 4 × 4 × nearly 8 cm sized samples, which are obtained from hardened mortar samples. Thus, the compressive strength values of samples having different mixtures ratios were determined.

2.2 Materials

In the study, for preparing control mortar and mortars which containing PKF press filter waste soil; at certain ratios two types of cement (CEM I 42.5R, CEM II 32.5R), standard sand, water and PKF press filter waste soil were used.

In other mortar mixtures, Type CEM II 32.5R cement were used. Table 1 shows the chemical properties of cement, the physical and mechanical properties of cement are shown in Table 2. The chemical properties of PKF press filter waste soil used as a replacement of cement in mortar mixtures at certain rates are also shown in Table 3.

Table 1. Chemical properties of cements

Components	Percentage of components	
	CEM I 42.5R	CEM II 32.5R
Total SiO_2	20.11	20.52
Insoluble residue	0.57	0.63
Al_2O_3	4.90	4.85
Fe_2O_3	2.83	2.87
CaO	64.37	64.45
MgO	1.49	1.52
SO_3	2.98	2.98
K_2O	0.72	0.75
Na_2O	–	–
LOI	2.40	2.42
Free CaO	0.67	0.69
Chloride	0.02	0.02
C_3S	61.41	61.43
C_2S	11.33	11.33
C_3A	8.20	8.20

Table 2. Physical and mechanical properties of cements

	CEM I 42.5R	CEM II 32.5R
Strength day (MPa)		
7-day	–	–
28-day	≥ 42.5	≥ 32.5
Specific gravity	3.14	3.14
Setting time		
Initial	150 min	150 min
Final	180 min	180 min
Volume stability (Le Chatelier) total (mm)	0.5	0.5
Fineness		
Specific surface (Blaine) cm^2/gr	3490	3490
200 μm percentage of remain on the sieve (%)	0.1	0.1
90 μm percentage of remain on the sieve (%)	1	1

Table 3. Chemical composition of PKF waste soil

Component	Mass (%)
MgO	2.0242
Al_2O_3	1.2064
SiO_2	3.7480
P_2O_5	0.1948
SO_3	0.2142
K_2O	0.0910
CaO	56.3714
TiO_2	0.0783
Cr_2O_3	0.0392
Fe_2O_3	1.0255
BaO	0.2331
LOI	34.7740

2.2.1 Binders

Type CEM I 42.5R cement has been firstly used in the mortar mixtures. The chemical properties of used cement types are in Table 1 and the physical and mechanical properties are shown in the Table 2.

2.2.2 Aggregate

Standard sand has been used in the study and its loose density and specific gravity are indicated in Table 4.

Table 4. Loose unit and specific gravity of standard sand

	Loose unit weight (kg/m^3)	Specific gravity
Standard sand	1350	2550

2.3 Mortar Mixtures

As CEM I 42.5R cement used in mortar mixtures is indicated in Table 5, at ambient temperature, by stabilizing water-binder ratio and standard sand amount, PKF press filter waste soil has been used as a replacement of cement in proportions of 5, 10, 15 and 20% respectively.

As to other mortar mixtures, CEM II 32.5R cement was used, which is shown in Table 5. At ambient temperature, by stabilizing water-binder ratio and standard sand amount, PKF press filter waste soil was used as a replacement of cement in proportions of 5, 10, 15 and 20% respectively. As it is indicated in Table 5, in the experimental study, After PKF press filter waste soil used as a replacement of cement in a determined ratio by stabilizing 450 g binder content had been mixed with primarily 1350 g standard experiment sand and 225 g water in the mortar mixer sufficiently in a required consistency, it was poured into three pieces of mortar molds. Mortar mixtures poured into their molds were left to wait for their setting. Two pieces of 4 × 4 × 8 cm sized

Table 5. Mortars prepared with two types cement and PKF

Sample code	Cement weight (g)	PKF press filter soil weight (g)	Standard sand weight (g)	Water weight (g)
Control42.5	450.0	0.00	1350	225
42.5PKF5	427.5	22.5	1350	225
42.5PKF10	405.0	45.0	1350	225
42.5PKF15	382.5	67.5	1350	225
42.5PKF20	360.0	90.0	1350	225
Control32.5	450.0	0.00	1350	225
32.5PKF5	427.5	22.5	1350	225
32.5PKF10	405.0	45.0	1350	225
32.5PKF15	382.5	67.5	1350	225
32.5PKF20	360.0	90.0	1350	225

samples gained from mortar mixtures poured into a mold with the size of $4 \times 4 \times 16$ cm were removed from mold 7 days later and they were subject to compressive strength tests. Other mortar mixtures were left to wait in the curing pool by being removed from their molds. After 28 days, four pieces of $4 \times 4 \times 8$ cm sized mortar mixtures gained from two pieces of $4 \times 4 \times 16$ cm sized samples were broken and their 28-days compressive strength were determined.

3   Results and Discussion

Mortar mixtures, which had been prepared with PKF press filter waste soil as a replacement of cement at certain ratios, CEM I 42.5R and CEM II 32.5R types of cement, were subject to compressive strength tests and the compressive strength values of mortar mixtures having different PKF press filter waste soil ratios were obtained.

3.1   Mortars Prepared with Type CEM I 42.5R

3.1.1   Mortar Mixtures Containing PKF Waste Soil

When the 7 and 28-day compressive strengths of the mortar mixtures containing 5% PKF press filter waste soil and 95% cement were examined, the values given in Table 6 were determined. The 7-day average compressive strength value is 29.25 MPa and the 28-day compressive strength value is 40.76 MPa. When Table 6 and Fig. 1 were evaluated, it was seen that there was a similarity between the compressive strength values gained from mortars containing 5% PKF press filter waste soil and the 28-day compressive strength value of the control mortar mixture given in Table 2. However, it was also seen that the mortar samples containing 5% PKF press filter waste strength value could not reach the 28-day control mortar's strength.

When the 7 and 28-day compressive strengths of the mortar mixtures containing 10% PKF press filter waste soil and 90% cement were examined, the values given in Table 6 were determined. The 7-day compressive strength value is 27.97 MPa and the

Table 6. Compressive strength of waste PKF, MPa

Sample code	7-day	28-day
42.5PKF5	29.25	40.76
42.5PKF10	27.97	39.51
42.5PKF15	26.88	37.91
42.5PKF20	26.09	35.97
32.5PKF5	16.38	30.86
32.5PKF10	14.22	29.60
32.5PKF15	13.25	27.45
32.5PKF20	12.47	25.77

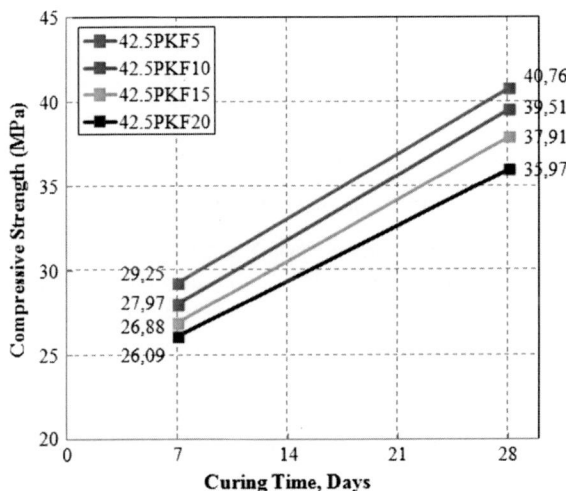

Fig. 1. Strength change CEM I 42.5 with PKF with days

28-day compressive strength value is 39.51 MPa. When Table 6 and Fig. 1 were evaluated, it was seen that there was a similarity between the compressive strength values gained from mortars containing 10% PKF press filter waste soil and the 28-day compressive strength value of the control mortar mixture given in Table 2. However, it was also seen that the mortar mixtures containing 10% PKF press filter waste soil strength value could not reach the 28-day control mortar's strength value. Besides, it was observed that the mortar samples containing 10% PKF press filter waste soil strength values were under the mortar samples containing 5% PKF press filter waste soil strength values.

When the 7 and 28-day compressive strengths of the mortar mixtures containing 15% PKF press filter waste soil and 85% cement were examined, the values given in Table 6 were determined. The 7-day average compressive strength value is 26.88 MPa and the 28-day compressive strength value is 37.91 MPa. When Table 6 and Fig. 1 were evaluated, it was seen that the compressive strength values gained from mortars containing 15% PKF press filter waste soil started receding from the 28-day compressive strength value of the control mortar mixture given in Table 2 and it could not reach the 28-day strength value of the control mortar. However, it was also seen that the mortar samples containing 15% PKF press filter waste soil strength values were under the mortar samples containing 5 and 10% PKF press filter waste soil strength values.

When the 7 and 28-day compressive strengths of the mortar mixtures containing 20% PKF press filter waste soil and 80% cement were examined, the values given in Table 6 were determined. The 7-day average compressive strength value is 26.09 MPa and the 28-day compressive strength value is 35.97 MPa. When Table 6 and Fig. 1 were evaluated, it was seen that the compressive strength values gained from mortars containing 20% PKF press filter waste soil started receding from the 28-day compressive strength value of the control mortar mixture given in Table 2 and it could not reach the 28-day compressive strength value of the control mortar. However, it was also seen that the mortar samples containing 20% PKF press filter waste soil strength values were under mortar samples containing 5, 10 and 15% PKF press filter waste soil strength values.

3.2   Mortars Prepared with Type CEM II 32.5R

3.2.1   Mortar Mixtures Containing PKF Waste Soil

When the 7 and 28-day compressive strengths of the mortar mixtures containing 5% PKF press filter waste soil and 95% cement were examined, the values given in Table 6 were determined. The 7-day average compressive strength value is 16.38 MPa and the 28-day compressive strength value is 30.86 MPa. When Table 6 and Fig. 2 were evaluated, it was seen that there was a similarity between the compressive strength values gained from mortars containing 5% PKF press filter waste soil and the 28-day

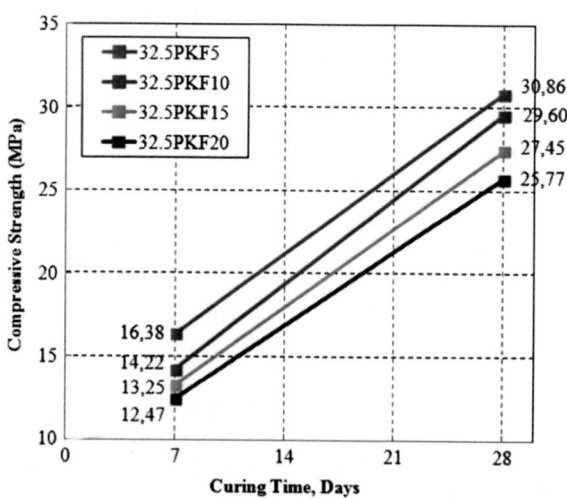

Fig. 2. Strength change CEM II 32.5 with PKF with days

compressive strength value of the control mortar mixture given in Table 2. However, it was also seen that the mortar samples containing 5% PKF press waste strength value could not reach the 28-day control mortar's strength value.

When the 7 and 28-day compressive strengths of the mortar mixtures containing 10% PKF press filter waste soil and 90% cement were examined, the values given in Table 6 were determined. The 7-day average compressive strength value is 14.22 MPa and the 28-day compressive strength value is 29.60 MPa. When Table 6 and Fig. 1 were evaluated, it was seen that there was a similarity between the compressive strength values gained from mortars containing 10% PKF press filter waste soil and the 28-day compressive strength value of the control mortar mixture given in Table 2. However, it was also seen that the mortar samples containing 10% PKF press filter waste soil strength value could not reach the 28-day control mortar's strength value. Besides, it was observed that the mortar samples containing 10% PKF press filter waste soil strength values were under the mortar samples containing 5% PKF press filter waste soil strength values.

When the 7 and 28-day compressive strengths of the mortar mixtures containing 15% PKF press filter waste soil and 85% cement were examined, the values given in Table 6 were determined. The 7-day average compressive strength value is 14.22 MPa and the 28-day compressive strength value is 29.60 MPa. When Table 6 and Fig. 2 were evaluated, it was seen that the compressive strength values gained from mortars containing 15% PKF press filter waste soil started receding from the 28-day compressive strength value of the control mortar mixture given in Table 2 and it could not reach the 28-day strength value of the control mortar. However, it was also seen that the mortar samples containing 15% PKF press filter waste soil strength values were under the mortar samples containing 5 and10% PKF press filter waste soil strength values.

When the 7 and 28-day compressive strengths of the mortar mixtures containing 20% PKF press filter waste soil and 80% cement were examined, the values given in Table 6 were determined. The 7-day average compressive strength value is 12.47 MPa and the 28-day compressive strength value is 25.77 MPa. When Table 6 and Fig. 2 were evaluated, it was seen that the compressive strength values gained from mortars containing 20% PKF press filter waste soil started receding from the 28-day compressive strength value of the control mortar mixture given in Table 2 and it could not reach the 28-day compressive strength value of the control mortar. However, it was also seen that the mortar samples containing 20% PKF press filter waste soil strength values were under the mortar samples containing 5, 10, 15% PKF press filter waste soil strength values.

When Fig. 3 was evaluated, it was determined that the mixtures containing type CEM I 42.5R of cement with 5% PKF press filter waste soil strength met control mortars strength in proportion of 96%. When PKF press filter waste soil was used as a replacement of cement at 10% in the mortar mixture, it was seen that the average compressive strength of samples met up to 93% value of compressive strength of the control mortar. When PKF press filter waste soil was used as a replacement of cement at 15% by cement weight, the strength of samples met up to 89% value of compressive strength of the control mortar containing CEM I 42.5R cement.

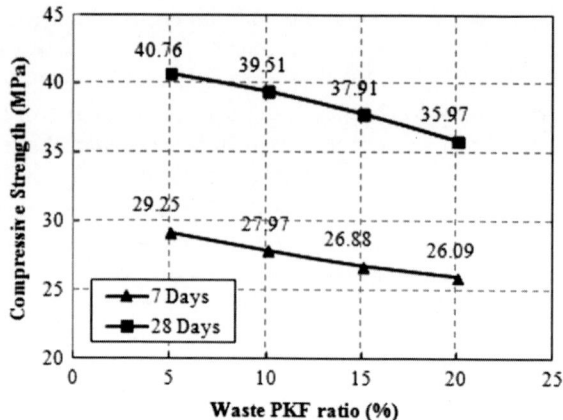

Fig. 3. Strength change CEM I 42.5R with PKF ratio

When the ratio of PKF press filter waste soil was replaced with 20% as a replacement of cement in the mortar mixture, it was seen that the average compressive strength of samples reached up to 85% of compressive strength of the control mortar containing CEM I 42.5R cement. Because of experimental studies, it was concluded that increasing PKF press filter waste soil at certain ratios by using it as a replacement of cement, the 7 and 28-day strength values of samples decreased. Consequently, it has been inferred that the samples gained from the mortar mixtures containing PKF waste less than 5% can be more advantageous in obtaining intended strength.

When Fig. 4 was evaluated, it was determined that the mixtures containing type CEM II 32.5R of cement with 5% PKF press filter waste soil strength met control mortars strength in proportion of 95%. When PKF press filter waste soil was used as a

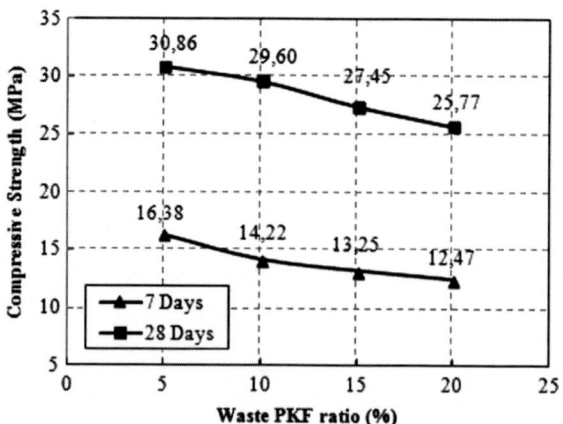

Fig. 4. Strength change CEM II 32.5R with PKF ratio

replacement of cement at 10% in the mortar mixture, it was seen that the average compressive strength of samples met up to 91% value of compressive strength of the control mortar.

When PKF press filter waste soil was used as a replacement of cement at 15% by cement weight, the average compressive strength of samples met up to 85% of compressive strength of the control mortar containing type CEM II 32.5R cement. When PKF press filter waste soil was used as a replacement of cement at 20% in the mortar mixture, it was seen that the average compressive strength of samples reached up to 80% of compressive strength of the control mortar containing type CEM II 32.5R cement.

As a result of experimental studies, it was concluded that increasing PKF press filter waste soil at certain ratios by using it as a replacement of cement, made the 7 and 28-day compressive strengths of samples decreased. Consequently, it has been inferred that the samples gained from the mortar mixtures containing PKF press filter waste soil less than 5% can be more advantageous in meeting intended compressive strength.

When Fig. 1 was evaluated, the compressive strength values of mortar samples containing PKF press filter waste soil as a replacement of cement in proportions of 5, 10, 15 and 20% and type CEM I 42.5R cement under the 7 and 28-day curing conditions were seen. It was determined that as the ratio of waste increased in samples containing PKF press filter waste soil, the strength values decreased. It was also seen that at the end of the 28-day curing period, samples containing PKF press filter waste soil with type CEM I 42.5R cement meet at most 96% and at least 85% of compressive strength of control mortar samples.

When Fig. 2 was evaluated, the compressive strength values of mortar samples containing PKF press filter waste soil as a replacement of cement in proportions of 5, 10, 15 and 20% and type CEM II 32.5R cement under the 7 and 28-day curing conditions were seen. It was determined that as the rate of waste increased in samples containing PKF press filter waste soil, the compressive strength values decreased. It was also seen at the end of the 28-day curing period samples containing PKF press filter waste soil with type CEM II 32.5R cement meet at most 95% and at least 80% of compressive strength of control mortar samples.

4 Conclusions

The 7 and 28-day compressive strength values of the mortar samples gained by use of PKF press filter waste soil at different ratios and type CEM I 42.5R and CEM II 32.5R were determined under curing conditions. Accordingly;

- It was determined that the 28-day compressive strength values of the mortar containing 5% PKF press filter waste soil and also prepared with type CEM I 42.5R cement were quite close (96%) to the control mortar.
- It was seen that the 7 and 28-day compressive strength values of the mortar mixtures containing PKF press filter waste soil in proportions of 10, 15 and 20% and type CEM I 42.5R cement started receding from the compressive strength value of the control mortar because of the increase of waste in the mortar samples.

- It was determined that the 28-day compressive strength values of the mortar mixtures containing 5% PKF press filter waste soil and type CEM II 32.5R cement were quite close (95%) to the control mortar. It was seen that the 7 and 28-day compressive strength values of the mortar mixtures containing PKF press filter waste soil in proportions of 10, 15 and 20% and type CEM II 32.5R cement started receding from the compressive strength value of the control mortar because of the increase of waste in the mortar samples.

In a study on high-performance/self-compacting concrete, it was seen that as the cement kiln dust used in the mixture increased, the compressive strength values, which were gained from the samples containing cement kiln dust used as a replacement of cement at certain ratios by weight cement, decreased. This decrease in the compressive strength was because of the increase in porosity [10].

In addition, in a study done with limestone, the increase of porosity during hydration was because of running out of calcium carbonate [11]. In another study, the mortar compressive strength values of concrete samples, gained with lime and ferrochrome ash used instead of cement at certain ratios by cement weight, were evaluated by being conserved between the 28 and 91-day in curing conditions. Because of this evaluation, that the hydration process continued after 28-day and the increase was observed in the strength of samples [12].

In this study, the compressive strength of samples can also be evaluated under curing conditions for the ages over 28 by producing more sample. In a study done with PKF press filter waste soil, the low of the compressive strength of samples gained by PKF press filter waste soil containing different chemical wastes causes the more P_2O_5 compound, forming α-C_2S by converting C_3S compound into C_2S compound and occurring $C_2S.0.5C_3P_2$ phase. This phase affects the compressive strength of samples in a negative way [13]. In another study on the compressive strength of samples prepared with sugarcane bagasse ash, it was shown that required strength could be gained more than 28-day due to the pozzolanic reaction period between silica and CH compounds [14]. Thus as the ratio of PKF press filter waste soil in the mortar mixtures increases, the pozzolanic reaction slows down and hydration reaction does not occur usually as a result of porosity. These affect the strength of samples in a negative way at the least.

As a result of experimental study, it has been stated that as the use of waste increases in the mortar mixtures gained by mixing PKF press filter waste soil at different ratios with two types of cement, the 7 and 28-day compressive strength values decreases according to the compressive strength values of the control mortar. Since the compressive strength values of samples containing PKF press filter waste soil is close to the compressive strength values of the control mortar, this waste will be able to use in different fields of construction industry. In accordance with these ongoing studies, it is considered that this waste material can be used as a raw material of cement in ceramic and construction chemicals sectors. Thus;

- It has been seen that as the ratio of PKF press filter waste soil used in mixtures increases, the duration of cement hydration takes long.
- It has been determined that as long as it is used as a cement raw material, the required energy will decrease due to the high of the waste heat value.

- It has been known that since it is a nonhazardous waste, it does not have a chemical damage to the nature on recycling and transporting.
- It has been seen that since it is used as raw material in cement production, there are not any changes in the value of CO_2 emission released into the nature as a waste gas.
- It has been considered that as long as it is used in ceramic and construction chemicals sectors, it will be an easily provided, low-cost and valuable waste.

Acknowledgements. Eskişehir Osmangazi University Scientific Research Fund (ESOGU BAP) supported this work under the Project Code 2016-1172. The authors wish to express their gratitude to the ESOGU for its financial assistance.

References

1. Pamukçu S, Topçu İ-B, Lynn J-D, Jablonski C-E (1991) Reuse of solidified steel industry sludge waste for transportation facilities, transportation research record no. 1310, materials, construction and maintenance, construction innovations, p 93
2. Topçu İ-B (1992) Effect of environmental factors on the behavior of solidified steel process residue, environmental geotechnology. In: Proceedings of the mediterranean conference on environmental geotechnology, Çeşme Turkey, p 467
3. Vejmelkova E, Konakova D, Kulovana T, Keppert M, Zumar J, Rovnanikova P, Kersner Z, Sedlmajer M, Cerny R (2015) Engineering properties of concrete containing natural zeolite as supplementary cementitious material: strength, toughness, durability, and hygrothermal performance. Cem Concr Compos 55:259
4. Teixeira E-R, Mateus R, Camoes A-F, Bragança L, Branco F-G (2016) Comparative environmental life-cycle analysis of concretes using biomass and coal fly ashes as partial cement replacement material. J Clean Prod 112:2221
5. Mehta P-K (2010) Sustainable cements and concrete for the climate change, era—a review. In: Proceedings of the second international conference on sustainable construction materials and technologies, Universita Politecnica dele Marche, Italy, pp 1–10
6. Kabay N, Tüfekçi M-M, Kızılkanat A-B, Oktay D (2015) Properties of concrete with pumice powder and fly ash as cement replacement materials. Constr Build Mater 85:1
7. Ballan J, Paone P (2014) Supplementary cementitious materials, concepts for the treatment of raw materials. IEEE Ind Appl Mag 20:65
8. Juenger M-C-G, Siddique R (2015) Recent advances in understanding the role of supplementary cementitious materials in concrete. Cem Concr Res 2015(78):71
9. Shabab M-E, Shahzada K, Gençtürk B, Ashraf M, Fahad M (2015) Synergistic effect of fly ash and bentonite as partial replacement of cement in mass concrete. KSCE J Civil Eng 20:1987–1995
10. Khalid N-B, Al-Jumaily I, Atea A-M (2016) Characterization of sustainable high performance/self compacting concrete produced using CKD as a cement replacement material. Constr Build Mater 2016(103):127
11. Pliya P, Cree D (2015) Limestone derived eggshell powder as a replacement in Portland cement. Constr Build Mater 95:8

12. Prasanna K-A, Sanjaya K-P (2015) Effect of lime and ferrochrome ash (FA) as partial replacement of cement on strength, ultrasonic pulse velocity and permeability of concrete. Constr Build Mater 94:451–452
13. Li H, Xu W, Yang X, Wu J (2014) Preparation of Portland cement with sugar filter mud as lime-based raw material. J Clean Prod 66:109
14. Arenas-Piedrahita J-C, Montes-Garcia P, Mendoza-Rangel J-M, Lopez Calvo H-Z, Valdez-Tamez P-L, Martinez-Reyes J (2016) Mechanical and durability properties of mortars prepared with untreated sugarcane bagasse ash and untreated fly ash. Constr Build Mater 105:74

Evaluation of Occupational Safety Culture in Construction Sector in the Context of Sustainability

Mursel Erdal$^{(\boxtimes)}$, Nihat Sinan Isik, and Seyhan Fırat

Civil Engineering Department, Technology Faculty, Gazi University, Ankara, Turkey
{merdal, nihatsinan, sfirat}@gazi.edu.tr

Abstract. Construction sector is one of the sectors where occupational-work accidents are highest. A way to minimize accident ratio in this sector is to manage the subject of occupational health and safety beyond the legal necessities. In this study, necessary precautions that company management has to take are explained for the development of safety culture in construction sector and they are evaluated.

Keywords: Occupational safety culture · Sustainability · Construction sector
Occupational health and safety

1 Introduction

Turkey is a developing country and therefore its requirements are increasing. As in the case of other developing countries, construction sector is the main sector.

All infrastructure and life sustaining structures like dams, energy producing plants, roads, hospitals, airports, city spaces, and factories are built by construction activities. It can be stated that the share of construction sector is about 30% in gross national product if all construction sector and related sectors taken into account. Construction sector is the demand creating sector for sub sectors which are more than 200. Construction sector is the main locomotive sector of economy because no other sector than construction sector can create this much direct or indirect effects [1].

One of the most important problems of construction sector is occupational work accidents. Every year about 1.2 million people lost their lives doe to occupational work accidents around the world. Occupational work accidents continue to be one of the main problems in Turkey [2].

In Turkey, a total of 241,547 number of accidents occurred and 1252 people have lost their lives due to these accidents in 2015 according to statistics of Social Security Institution (Table 1) [2].

A lot of legal and institutional arrangements have been made to present occupational work hazard from past to today. However it can be understood from statistics that these arrangements are not successful enough. This situation implies that it is not enough to take occupational accidents as a technical issue alone. It is inevitable that

© Springer International Publishing AG, part of Springer Nature 2018
S. Fırat et al. (eds.), *Proceedings of 3rd International Sustainable Buildings Symposium (ISBS 2017)*, Lecture Notes in Civil Engineering 7,
https://doi.org/10.1007/978-3-319-64349-6_29

human factor should be taken into account as much as technical issues in working life. It is stated that 80–90% of occupational accidents are occurred due to unsafe behaviour of working personal. For this reason, in addition to legal necessities, development of safety culture should be considered as an important factor for the prevention of accidents [3].

Table 1. Occupational accidents and death tolls in Turkey in 2015 [2]

Activity code	Groups of activity	Work accident number			Number of death tolls		
					Number of work accident		
		Male	Female	Total	Male	Female	Total
41	Building construction	14,980	85	15,065	239	0	239
42	Construction of structures other than buildings	7861	42	7903	124	0	124
43	Special construction activities	10,301	92	10,393	110	0	110
Total of construction activities		33,142	219	33,361	473	0	473
Total of all sectors		206,922	34,625	241,547	1219	33	1252

2   Theoretical Foundations

2.1   Occupational Safety Culture

Concept of occupational safety culture was first used in a report which was prepared after the nuclear accident occurred in Chernobyl in 1986. This report indicates organizational errors and infringement of working personal play important role [4]. Either in Chernobyl accident or in other important accidents "safety culture" arise as a key factor to explain the role of human factor for providing security especially where risk is high [5–8].

There are a lot of definitions related with safety culture done by various scientists and researchers. However this concept was defined as a product of behavioural, competence, attitude, value patterns of individuals and groups who decide to institution's efficiency of health and safety programs and their persistence in practices, according to International Atomic Energy Agency in 1991 [9].

According to Cooper [10] safety culture is a product of interaction between individual, work and organization. Uttal [11] defines safety culture as "The interaction between control systems which produce behavioural norm and organization". According to these definitions safety culture has two dimensions. These are values, attitudes believes related to safety; controls, applications structures which are used to improve safety of organization [12].

Aytaç [13] in her study handled occupational safety as a cultural subject and she remarked the importance of the development of safety culture for the development of safety culture for the prevention of occupational accidents.

Although many companies organize trainings to increase occupational safety and to assign responsible personal they could not reach intended occupational safety standards. Authorities have attributed the reason of not reaching intended-safety levels to lack of safety culture [14].

There are a lot of factors affecting the safety culture. One of these factors is the social culture in which companies are living in. Interest of workers to occupational safety and their sensitivity are affected from the properties of social culture. Creation of organizational culture can be easy or difficult depending on the social cultural properties [15].

If it is thought that occupational accidents create important losses for both country and organizations, one of the most important subject is to develop a safety culture. However there are a lot of factors which should be firstly actualized before developing safety culture in organizations [16].

Basis of creating a safety culture in organizations is the perspective of management to safety culture. According to the article number 5 of "Regulations related with the methods of training of working personal related with occupational health and safety" which was entered into force in 05.13.2013 at 28,648 numbered Official Gazette, management is supposed to take necessary precautions to generate healthy and safe working environment [17].

2.2 Requirements of Management to Improve Occupational Safety Culture

Construction companies in the sector should adopt an understanding in which organizational standards are continuously developed in their applications. Informing employees, sharing information about the changes in legislations using an effective communications system, objectives of occupational health and safety should become a company culture.

Occupational health and safety system should be supported by giving information, control and improvement efforts. Structure of occupational health and safety comities, employee representatives, assignment of occupational health and safety experts and workplace medical doctor, risk analysis, fire drill, emergency action plan issues should be managed according to legal regulations. Aims of occupational health and safety and performance results should be controlled periodically and these should be evaluated in occupational health and safety board's collective agreements signed in union should cover occupational health and safety subjects because of the importance of personal health and safety. Protective measures, reporting of illnesses and accidents, occupational safety rules to be followed should be organized within the scope of collective agreements. Employer should organize training activities to increase the level of knowledge and consciousness.

Employer should aim zero accident and should aim its sustainability. The most important tool to achieve this goal is systematic site controls. A methodology should be developed for foreseeing employees to take precautions when there a danger in working environment during site control. Improvement methods should be developed against the inconsistencies defected at the end of site controls and they should be applied as soon as possible. Number of control teams should be determined according to the size of construction site. In this context, it will be important to monitor and record personal occupational safety performance.

Tools and activities such as verbal instructions, training and warming signs must be used, near miss events should be reported, reporting culture should be wide spread, and it must not be forgotten that the creation of safety culture is a long strategy requiring a constant effort and dedication, for developing safe behaviours. It is necessary to clearly explain to all employees the necessity of recording not only accidents but also near miss events. It would be possible to develop a personal profile that shows and sustains exemplary behaviour in environmental and occupational health issue that achieve targets that exceed these set by values that respect environment and environmentally friendly standards that comply with legal legislations.

Employer should be evaluate new practices arising from technological developments in the context of occupational health and safety culture. He has to perform necessary studies in order to remove the risks and he has to adopt corrective and preventive actions against risky situations by evaluating the results. Employer must make analyses for the accidents from occurring according to the results of the accident analysis.

Management should provide a safe, secure and motivating working environment. It has to be proven that it has been acted with the aim of being fair and equitable work place where the best standards are applied in occupational health and safety, where employee rights are protected and supervised, skills are developed.

Management should arrange end of year meetings, should consider the expectations and thoughts of the employees at these meetings. Management should constantly monitor and improve occupational health and safety application by making necessary improvements in line with management organizational needs, technological developments, changing regulations and employees' expectations.

Management has to carry out a series of in house training and education programs to improve talents and skills of employees. These programs should be organized according to defined procedures and policies and employee needs. The program should improve occupational health and safety, improve workforce productivity and should be sustainable. New employees are required to adapt quickly to the organization and be able to get the information they need in a fast and accurate manner with the help of effective guidance process.

A risk management unit should be established to identify and manage the potential risks that company may face. This unit should identify risks related to occupational health and safety as well as investment and business based risks under the context of risks management. Risk analysis should be performed in the context of occupational health and safety, risky subjects should be identified and this process should be managed.

3   Results

Construction companies must ensure that both the employees and the subcontractors contribute to this process by making security cultures as an indispensable part of their working life. Security culture should be made sustainable by integrating employees into decision making, implementation and business process. Apart from all the efforts that the management has taken to create a security culture in the company and the goals it has taken, it is also very important for the employees to take contribution and responsibility. All personal working in the company will be more interested and involved in health and safety studies in the workplace when they are aware of their responsibilities for preventing accidents and injuries.

References

1. Türkiye İnşaat Sanayicileri İşveren Sendikası (2016) İnşaat Sektörü Raporu, Ankara, Turkey (in Turkish)
2. http://www.sgk.gov.tr/wps/portal/sgk/tr/kurumsal/istatistik/sgk_istatistik_yilliklari
3. Aytaç S (2011) İş Kazalarını Önlemede Güvenlik Kültürünün Önemi (1. Bölüm), Türkmetal Dergisi, 147: 30-33 (in Turkish)
4. The Post-Accident Review Meeting on the Chernobyl Accident (1986) Safety series no. 75-INSAG-1
5. Cox S, Flin R (1998) Safety culture: philosopher's stone or man of straw? Work Stress 12(3):189–201
6. Dursun S (2012) İş Güvenliği Kültürü – Kavram, Modeller, Uygulama, Beta Yayınları, İstanbul, Turkey (in Turkish)
7. Dursun S (2011) Güvenlik Kültürünün Güvenlik Performansı Üzerine Etkisine Yönelik Bir Uygulama. Ph.D. thesis, Uludağ University, Institute of Social Science, Bursa, Turkey (in Turkish)
8. Demirbilek T (2005) İş Güvenliği Kültürü, Legal Yayıncılık, İzmir, Turkey (in Turkish)
9. The Safety Culture (1991) Safety series no. 75-INSAG-4
10. Cooper MD (2000) Towards a model of safety culture. Saf Sci 36:111–136
11. Uttal B (1983) The corporate culture vultures. Fortune 108(8):66–72
12. Reason J (1998) Achieving a safety culture: theory and practice. Work Stress 12(3):293–306
13. Aytaç S (2011) İş Kazalarını Önlemede Güvenlik Kültürünün Önemi (2. Bölüm). Türkm Derg 148:36–39 (in Turkish)
14. http://docplayer.biz.tr/7991499-Peryon-towers-watson-is-guvenligi-ve-isci-sagligi-arastirmasi-2012.html
15. T.C. Cumhurbaşkanlığı Devlet Denetleme Kurulu (2008) Tersanecilik Sektörü ile İş Sağlığı ve Güvenliği Açısından Tuzla Tersaneler Bölgesinin İncelenmesi ve Değerlendirilmesi, Ankara, pp 356–358 (in Turkish)
16. Akalp G, Yamankaradeniz N (2013) The importance and the role of the management to create the safety culture on the companies. J Soc Secur 3(2):96–109
17. http://www.resmigazete.gov.tr/eskiler/2013/05/20130515-1.htm

Part III
Strengthening and Rehabilitation of Structures

The Investigation of Alkali Silica Reaction of Mortars Containing Borogypsum

Kursat Yildiz[(⊠)]

Technology Faculty, Civil Engineering Department, Gazi University, Ankara, Turkey
kursaty@gazi.edu.tr

Abstract. In this study, various properties and the alkali silica reaction (ASR) of mortars containing borogypsum were investigated. In the first stage a complete physical, chemical, mineralogical, molecular characterization of the borogypsum and cement were performed. In the second stage, the alkali silica reaction, mechanical and porosity properties of mortars, replacement 0, 1, 2, 3 and 4% borogypsum by weight were determined. According to experimental result, from a mineralogical perspective, the borogypsum was observed to have a sharp crystal structure and possess a mineralogical composition of calcite, colemanite and celestine. Looking from a molecular perspective, the similar OH, CaO, B–OH and H–OH bonds were displayed in the close wavenumbers. As the results of 14 days measurements were analyzed, the average maximum length extension was detected to be in the control mortar bars and at the detrimental zone from the ASR perspective. The average minimum length extension at the borogypsum with a additive rate of 3 and 4%, was observed to be at the control zone from ASR perspective. The lowest strength loss in mortars exposed to ASR effect occured in those with substitution level of 3% borogypsum. When the flexural strength of mortars exposed to ASR effect was examined, the increase in strength was observed for borogypsum with all substitution levels (1–4%). The maximum recovery concerning the minimum volume of mercury intruding the sample and vacancy rate was observed for the substitution level 3% borogypsum.

Keywords: Porogypsum · Alkali silica reaction
Mineralogical-mechanical property · Porosimetry

1 Introduction

Its use as aggregate in materials may in certain circumstances lead to alkali silica reaction (ASR) related problems. ASR is one of the most studied deleterious degradation mechanisms of concrete, which is particularly harmful. Once detected in a concrete structure, ASR is very difficult to stop. It is nowadays possible to use mineral additions like natural pozzolans or sub-products with pozzolanic reactivity to inhibit ASR in new concrete [1].

Alkali silica reaction (ASR) is one of the degradation process and refers to chemical reaction between the reactive silica phases in the aggregates and the alkali hydroxides

© Springer International Publishing AG, part of Springer Nature 2018
S. Fırat et al. (eds.), *Proceedings of 3rd International Sustainable
Buildings Symposium (ISBS 2017)*, Lecture Notes in Civil Engineering 7,
https://doi.org/10.1007/978-3-319-64349-6_30

in the pore solution of cement-based materials. The reaction results in the formation of alkali–silica gels and/or alkali–calcium–silica gels. These gels which absorb water and swell can lead to the formation of micro cracks [2–10].

Adjustments and improvements to the present concrete making methods are essential in order to address environmental and economic issues. This has encouraged researchers in the area of concrete engineering and technology to investigate and identify supplementary by-product materials that can be used as substitutes for constituent materials in concrete production [11]. Borogypsum is formed by reacting colemanite with sulphuric acid in the production of boric acid and is obtained by filtering the reaction mixture on the filter presses. The material mainly consists of gypsum, B_2O_3 and some impurities and causes various environmental and storage problems. B_2O_3 is dissolved by rain water and mixed with soil. High amount of boron content, which has toxicological effect, leads economic losses and also causes environmental pollution [12–14].

Boron is one of the most important ores in Turkey which possesses 72% of the world boron reserves. The most important boron minerals in Turkey are colemanite, ulexite and tinkal. The amount of Borogypsum waste is approximately 550,000 tons per year. Therefore, its positive effects on concrete properties borogypsum have recently been drawn attention by some researchers [15].

There are a number of studies in the literature which investigated the use of borogypsum as a cement additive [15–18].

Some studies focused on the use of borogypsum in mortar production as a cement replacement [19–22]. On the other hand, some researchers studied leaching kinetics of borogypsum by leaching with water [20–25]. However, only a few studies investigated the effect of borogypsum on the properties of concrete specimens [26, 27]. It is believed that has an effect borogypsum on the strength and consistency properties of concrete warrants further research.

There are a number of studies in the literature which investigated the use of pore size distributions which were measured using mercury intrusion porosimetry (MIP). Mercury intrusion porosimetry (MIP) is a widely used technique for characterizing the distribution of pore sizes in cement-based materials. It is a simple and quick indirect technique, but it has limitations when applied to materials that have irregular pore geometry [28–31].

The primary objective of this study is to investigate the usability of borogypsum in cementbased mortars a mineral additive and its effect on XRD analysis, FT-IR analysis, expansion test, compressive and flexural strength and mercury intrusion porosimetry (MIP) tests.

2   Materials and Methods

In this study, standard Portland Cement CEM I (PC 42.5 N/mm^2) with a specific gravity of 3.17 g/cm^3 was used. Initial and final setting times of the cement were 160 and 210 min, respectively. The blaine specific surface area was 3625 cm^2/g. Borogypsum used in the experiments was supplied by Etibor Bandırma Borax and Boric Acid Plants in Turkey. The specific gravity of Borogypsum was 2.25 g/cm^3.

Blaine specific surface area was 19170 cm^2/g. The remaining of the borogypsum on 50 and 20% sieves were 5 and 1.3 μm respectively. Elemental analyses of samples were carried out by X-ray fluorescence (XRF) spectrometry technique. Chemical, physical, XRD, FT-IR and Porosimetry analyses were conducted for the samples used in the experiments. Chemical analyses of cement and borogypsum are performed on ARL 8680 S X-ray diffraction. Surface areas are determined as Blaine values by Toni Technic 6565 Blaine and specific weights are determined by Quantachrome MVP-3. The mineralogical properties are determined by Rikagu miniflex XRD device using Cu K$_\alpha$ ($\lambda = 1.54$ Å) radiation. FT-IR analyses are conducted using Bruker Vertex 70 in the wave number range of 400–4000 cm^{-1}. Pore size distributions were measured using mercury intrusion porosimetry (MIP) [32].

The chemical compositions of cement and borogypsum are presented in Table 1. The grain distribution of borogypsum is shown in (Fig. 1).

Table 1. Chemical composition of cement and borogypsum

Materials	Cement	Borogypsum
Chemical composition: wt%		
SiO$_2$	20.01	4.55
Al$_2$O$_3$	5.15	1.30
Fe$_2$O$_3$	3.51	0.35
CaO	62.97	27.72
MgO	2.25	1.48
SO$_3$	2.95	37.75
Na$_2$O	0.47	–
K$_2$O	0.87	0.77
B$_2$O$_3$	–	4.30
Loss on ignition	2.17	21.78
CaO free	–	–

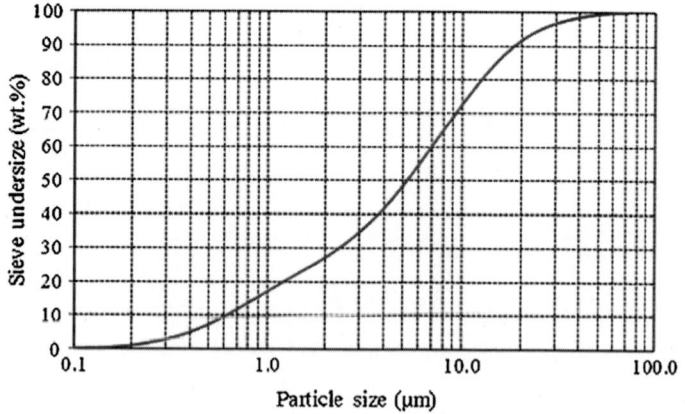

Fig. 1. The grain distribution of borogypsum

Dry and clean natural crushed stone aggregates were used with a maximum size of 4 mm. The absorption value and relative density at saturated (A-RD-SSD) condition of aggregate were $WA_{24} = 1.2\%$ and $\rho_a = 2.71$, $\rho_{rd} = 2.65$, $\rho_{ssd} = 2.66$ respectively. The methylene blue value of aggregate was $M_B = 0.75$ g/kg. Other characteristics of the aggregates are provided in Table 2. The compositions of each aggregate group are illustrated in Table 3. City water of the province of Ankara was used.

Table 2. The characteristics of the aggregates

Characteristics	Standards	Results
Very fine material	TS EN 933-1	%9
Bulk density (mg/m^3)	TS EN 1097-3	1.65
Chlorides (%)	TS EN 1744-1	0.0042
Acid-soluble sulfate (%)	TS EN 1744-1	AS0.2 $\leq$ 0.2
Total Sulfur (%)	TS EN 1744-1	0.009
Organic matter	TS EN 1744-1:15.1	–
Fulvo acid content	TS EN 1744-1:15.2	–
Lightweight organic pollutants	TS EN 1744-1:14.2	MLPC = 0.0%

Table 3. Aggregate gradation

Sieve size	The remaining material between of the two sieve (%)
4.75 mm (no: 4)–2.36 mm (no: 8)	10
2.36 mm (no: 8)–1.18 mm (no: 16)	25
1.18 mm (no: 16)–600 μm (no: 30)	25
600 μm (no: 30)–300 μm (no: 50)	25
300 μm (no: 50)–150 μm (no: 100)	15

In this study, the cement was used for the preparation of reference samples. Borogypsum was blended in this cement at rates of 0, 1, 2, 3 and 4%. Therefore, a total of five different cements were used and those were codified as (Boron Gypsum-BG) BG0, BG1, BJ2, BG3 and BG4. The compositions of each mortar sample (25 × 25 × 285 mm^3) mixture are illustrated in Table 4. Mortar bar were prepared according to the

Table 4. Mortar samples mixture ratio for ASR (g)

Mixture	Borogypsum	Cement	Aggregate	Water	W/C
BG0	0	440	990	206.8	0.47
BG1	4.4	435.6	990	206.8	
BG2	8.8	431.2	990	206.8	
BG3	13.2	426.8	990	206.8	
BG4	17.6	422.4	990	206.8	

ASTM C 1260-07 standard [33]. They were demoulded 24 h after casting, and after were immersed in water at 23 °C and stored in a climatic chamber at 80 °C for 24 h. After this period, the initial length of each prism was measured (zero point or initial point), after which the prisms were immersed in a NaOH 1 M solution at 80 °C for 14 days. During this period of time, regular length measurements were taken at various ages to check the progress of the expansion. According to the test limits a mix that has an expansion at 2, 7 and 14 days below 0.10% are considered non-reactive (harmless zone) and are considered reactive (harmful zone) if that expansion is higher than 0.20%.

The prepared mortar samples were poured into the $(40 \times 40 \times 160 \ mm^3)$ prismatic formworks for flexural and compressive strength tests. The compositions of each mortar sample mixture are illustrated in Table 5 [34]. These mortar specimens were shaken for one minute on a shaking table then settled into the formworks. These specimens were kept in a laboratory environment for 24 h. At the end of this duration, the specimens were taken out of the formworks and kept in the half ASR cure other half water curing pool. The specimens were taken from the water and ASR pool after 28 day and were tested for flexural and compressive strengths in accordance with [35].

Table 5. Mortar samples mixture ratio for flexural and compressive strength tests

Mixture	Borogypsum	Cement	Aggregate	Water	W/C
BG0	0	660	1485	310.2	0.47
BG1	6.6	653.4	1485	310.2	
BG2	13.2	646.8	1485	310.2	
BG3	19.8	640.2	1485	310.2	
BG4	26.4	633.6	1485	310.2	

3 Experimental Results and Discussion

3.1 Chemical Analysis

According to the chemical analysis results, cement consists of CaO and SiO_2 with higher proportion and Al_2O_3, Fe_2O_3 and SO_3 compounds with lower proportion. The main component of borogypsum is SO_3 and CaO the ratio of SiO_2/Al_2O_3 (S/A) is 3.50 in weight. The total of $SiO_2 + Al_2O_3 + Fe_2O_3$ is 6.20 in weight. To be accepted as a mineral pozzolan S/A > 4 and $SiO_2 + Al_2O_3 + Fe_2O_3 > 70$ should be. Therefore, the use of borogypsum is not true as a pozzolan [36].

3.2 Mineralogical Analysis

XRD analyses are conducted to determine the mineralogical structure of cement and zeolite (Figs. 2 and 3). According to the XRD patterns, the main components of cement

CEM I (PC 42.5 N/mm^2) are [C$_3$S-Alite (3CaOSiO$_3$), C$_2$S-Belit (2CaOSiO$_3$), Cc-Calcite (CaCO$_3$), CH-Portlandite (Ca(OH)$_2$) and brownmillerite (Ca$_2$(Al,Fe$_2^{3+}$)O$_5$]. The XRD results indicate that the structure of cement is regular (crystal) (Fig. 2) Borogypsum consists of [Calcite –(CaCO$_3$), Colemanite-(Ca$_2$B$_6$O$_{11}$·5H$_2$O), Celestine-(SrSO$_4$)]. The XRD results indicate that the structure of borogypsum is sharp (crystal) (Fig. 3) [37].

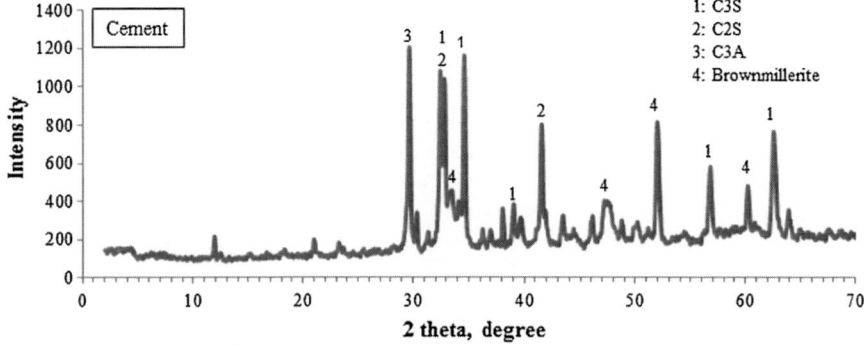

Fig. 2. The mineralogical structure of cement

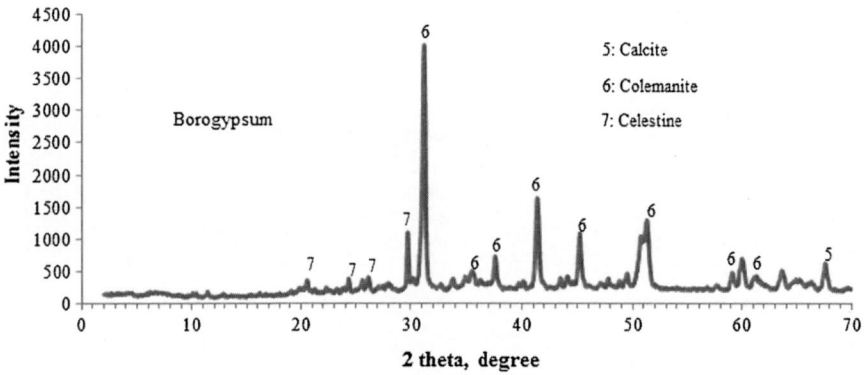

Fig. 3. The mineralogical structure of borogypsum

3.3   FT-IR Analysis

FT-IR analyses are available to characterize molecule groups in a particle. In the FT-IR studies are related to cement and pozzolan, the infrared spectrum to be regarded as in mainly 4 wide band regions. They are consisting of peaks corresponding to the deviations in Si–Al, S, C and OH bonds [38]. Surface structures of the molecules are determined from the FT-IR results indicated from the analyses and shown in (Figs. 4 and 5) in the manner of a schematic.

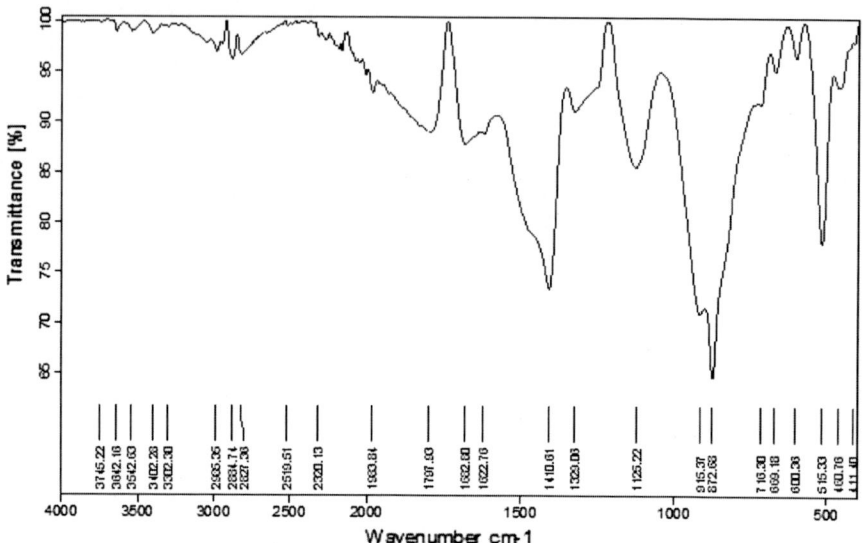

Fig. 4. FT-IR spectrum of cement

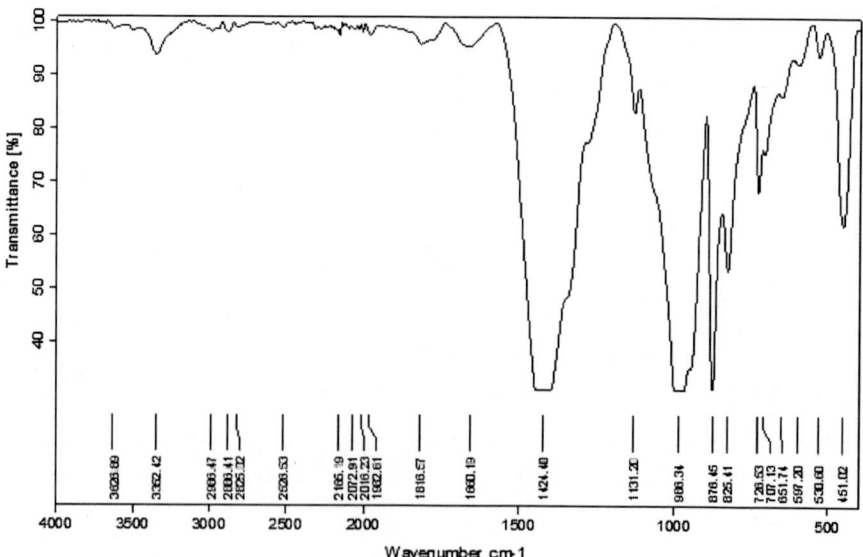

Fig. 5. FT-IR spectrum of borogypsum

In FT-IR spectroscopy, vibration of the atoms forming solid cages and molecular vibrations are observed in 400–1600 and 1600–4000 cm^{-1} region, respectively. Vibration peaks are displayed at 461, 516, 601, 669, 873, 915, 1125, 1410, 3402 and 3642 cm^{-1} wave numbers from FT-IR analysis of cement (Fig. 4). Si–O bonds present with Al–O give vibration peaks of 461 and 516 cm^{-1}. Si–O bonds in cage structures are in the form of a vibration peak at 872 cm^{-1} wave number. S–O bonds which show the plaster in cement is seen at 601, 1125 and 1623 cm^{-1}. Vibration peak of water ions and molecules in the structure is at 3302 and 3642 cm^{-1} wave number [39, 40].

FT-IR spectra can obtain advantageous information on borogypsum mineral structure. The spectrum of borogypsum over the 500–4000 cm^{-1} spectral range is displayed in (Fig. 5). Vibration peaks are displayed at 451, 597, 727, 825, 876, 986, 1424, 1660, 1816, 3352 and 3627 cm^{-1} wave numbers from FT-IR analysis of borogypsum. The absorptions at 1660, 1816, 3626 and 3352 cm^{-1} are due to intermolecular and weakly H bonded OH because of water of crystallizations [41]. CaO vibrations are observed in 451 cm^{-1} region. The spectrum of borogypsum is dominated by a strongest vibration band at 986 cm^{-1}. This peak is attributed to the bending mode in the shape of trigonal symmetric vibration. Symmetric vibration peak at 876 cm^{-1} wave number tetrahedral the presence of boron in the structure, a strong peak at 1424 cm^{-1} wave number in the trigonal structure indicates the presence of boron. The vibration peaks at 727 and 825 cm^{-1} may be assigned to the out-of-plane B-OH bending modes [42].

3.4   Expansion Results

CEM I and borogypsum were determined of equivalent alkaline value as (1.04) and (0.50) respectively. These values were calculated with equation of state ($Na_2O + 0.658K_2O$). Figure 6 shows the evolution of the average expansion of (BG1, BG2, BG3 and BG4) mortar and (BG0) mortar.

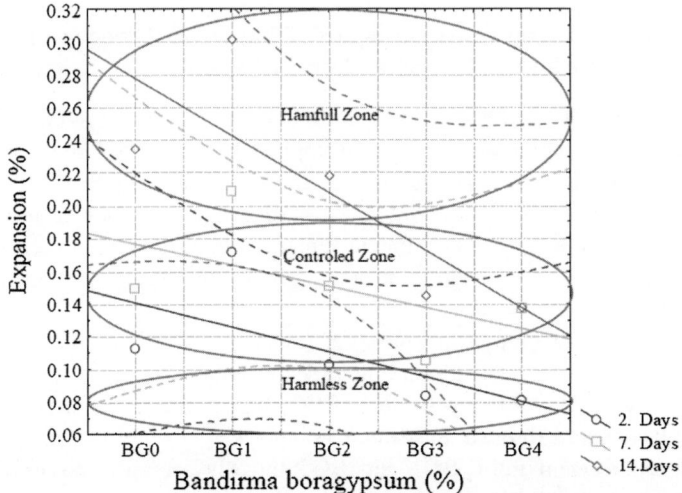

Fig. 6. The expansion of borogypsum

When the longitudinal elongation values depending on time were evaluated in 2, 7 and 14 days of periods;

At the end of 2 days measurements, while the control mortar bar with no boro-gypsum replacement took part in BG0 controlled zone (harmful-harmless zone) BG1 mortar bar with 1% borogypsum replacement (although staying in the controlled zone) exhibited maximum value in the 2 days group. This situation, depending on the increase of borogypsum replacement, stayed at limit values in BG2, stayed at the harmless zone in BG3 and BG4 mortar bars from the point of longitudinal elongation. When generally taking a look at at the 2 days measurement group, it is thought that the usage of 2, 3 and 4% replacement ratios will provide positive results in the first years of service period of cement mortars from the point of ASR.

At the end of 7 days measurements, while the control mortar bar with no boro-gypsum replacement took part in BG0 controlled zone (harmful-harmless zone) BG1 mortar bar with 1% borogypsum replacement stayed in the harmful zone and exhibited a maximum longitudinal elongation value in the 7 days group. Depending on the increase of borogypsum replacement ratio BG2 and BG4 took part in the controlled zone. BG3 exhibited optimal value in the 7 days measurement group and took a 0.105% value which is very close to the harmless zone. When taking a look at the 7 days measurement group, as in the 2 days measurement group, it is thought that the usage of 2, 3 and 4% replacement ratios will provide positive results from the point of ASR at the middle ages of the service life of cement mortars and the usage of 3% borogypsum will exhibit optimal result.

At the end of 14 days measurements, it was observed that the control mortar bar BG0 with no borogypsum replacement took part in the harmful zone from the point of ASR. Depending on the increase of borogypsum replacement ratio, the harmful zone could not be overcome at the BG1, BG2 and BG3 mortar samples. During the service life of cement mortars, it was observed from the point of ASR that 4% borogypsum replacement could be used for all of the three age group.

3.5 Flexural and Compressive Strengths of Bandirma Borogypsum Replacement Mortars Under the Effect of Alkali–Silica Reaction

For the purpose of determining the mechanical strength loss in the mortars due to ASR effect, flexural and compressive tests were applied to the control species kept in the water curing pool for 28 days and to the mortars subjected to ASR effect. The experimental data was evaluated at the 95% confidence band among each other in each group containing three species in different replacement ratios and was observed to be statistically reasonable. The scatter plots belonging to this statistical analysis are given in Figs. 7 and 8.

From the point of Compressive Strength;

When the mortars with Bandırma borogypsum replacement (not subjected to ASR effect) were evaluated among each other, BG0 control mortar sample with no boro-gypsum replacement exhibited maximum compressive strength. Depending on the increase of replacement ratio, BG1 and BG3 mortars samples and BG2 and BG4 samples exhibited similar behavior from the point of compressive strength. When these

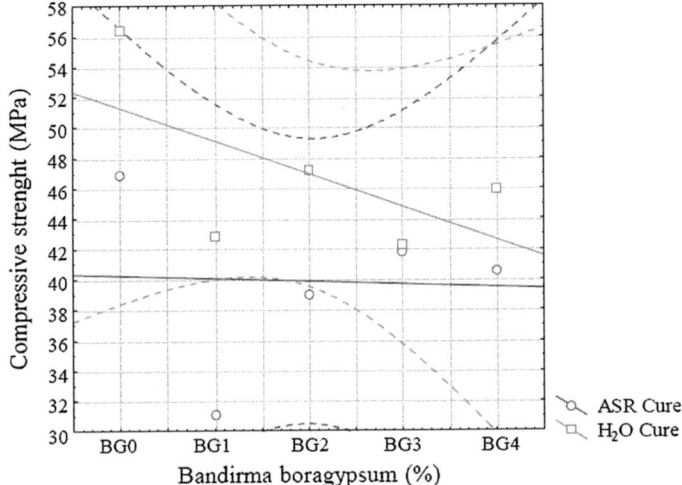

Fig. 7. The compressive strength of borogypsum

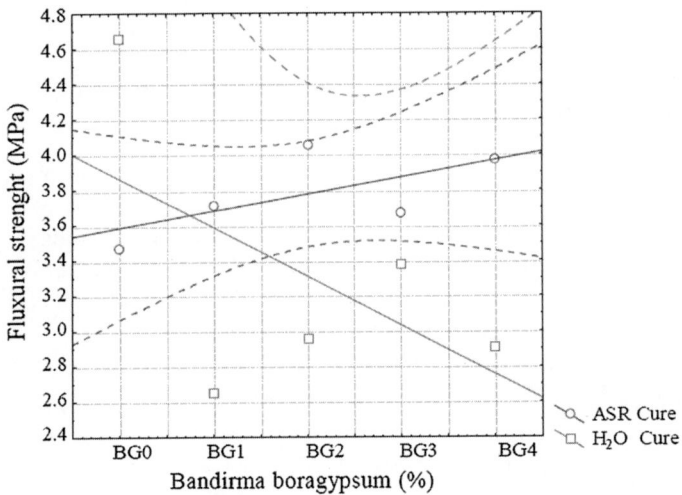

Fig. 8. The flexural strength of borogypsum

groups were considered, BG2 and BG4 mortars groups showed higher compressive strength compared to BG1 and BG3 mortars groups. When the samples with no ASR effect were evaluated, it was observed that increase of borogypsum replacement caused compressive strength loss.

When the mortars with borogypsum replacement (subjected to ASR effect) were examined among each other, control mortar sample exhibited maximum compressive strength, whereas BG1 mortar sample showed minimum compressive strength. While the

BG2 and BG4 samples were exhibiting values close to each other, maximum compressive strength in the group with replacement was observed in the BG3 mortar sample.

When the samples with ASR effect and with no ASR effect were compared with each other, compressive strength loss was observed in all of the samples under ASR effect. This loss occurred at an acceptable level in BG3 mortar group.

From the point of Flexural Strength;

When the mortar samples having no ASR effect were examined, a flexural loss was observed in all of the replacement ratios compared to control group BJ0 mortar samples at different levels. This loss occurred in BG1, BG2, BG4 and BG3 mortar groups from the highest to the lowest respectively.

When the mortar samples (subjected to ASR effect) were examined, no flexural loss was observed in all of the replacement ratios compared to BG0 mortar samples. On the contrary an increase in the flexural strength was seen in BG2, BG4, BG1 and BG3 mortar groups from the highest to the lowest respectively.

When the samples which were subjected and not subjected to ASR effect were compared with each other a serious flexural loss was observed in the control mortar group with no replacement, whereas in the other, replacement ratios at different levels a flexural strength increase was observed. Like in the compressive strength, the closest flexural strength was observed in the BG3 replacement ratio.

3.6 Porosity Volumes of Mortars with Bandırma Borogypsum Replacement

The relationship between the porosity diameter belonging to Bandırma borogypsum and mercury volume (intruding cumulatively) is given in in the graphs of Fig. 9. From the graphic, it is seen that the cumulative mercury volume intruding to samples is observed in the control group BG0 mortar sample in the highest amount. Whereas in all

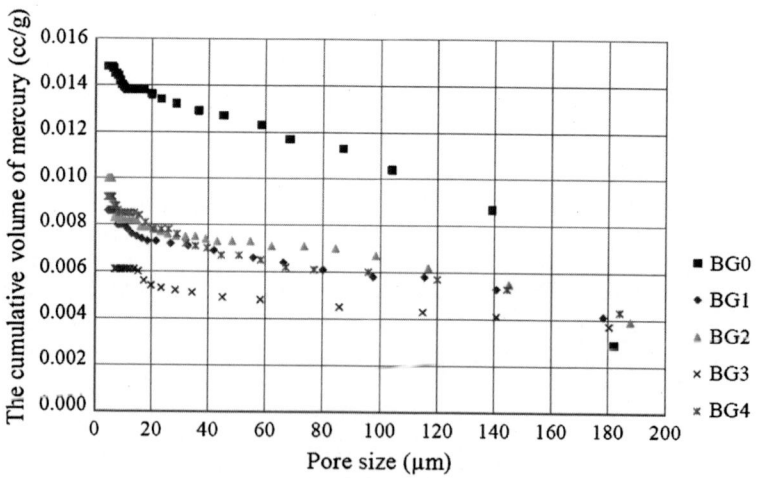

Fig. 9. Pore size distribution of mortars

of the replacement ratios less mercury intrusion was realized compared to control group. This can be accepted as an indication of Bandırma borogypsum mortar samples reducing the porosity volume in various portions. Because mortar samples exhibited different porosities at various borogypsum ratios. This occurred in the BG2, BG4, BG1 and BG3 mortar groups from the highest to the lowest respectively. BG2, BG4 and BG1 mortar samples exhibited very close values from the point of cumulative intruding mercury volume. Whereas BG3 provided a maximum improvement from the aspect of porosity ratio with minimum mercury intrusion.

4   Conclusions and Suggestions

Longitudinal elongation, compressive strength and flexural strength tests were carried on for the determination of alkali silica reaction behavior of borogypsum in mortars. Besides, mineralogical and molecular analyses of the used materials were made. Furthermore, the effect of borogypsum containing mortars on the porosity ratio was analyzed.

From the point of mineralogical and molecular structure;

- When the XRD analysis result of borogypsum was examined, it was observed that it was formed by [Calcite–($CaCO_3$), Colemanite–($Ca_2B_6O_{11}·5H_2O$), Celestine ($SrSO_4$)]. It was a sharp crystal from the point of general structural view.
- From the molecular point, it exhibited –OH, B–O and B–OH bonds.

From the aspect of ASR;

- When a general look was taken at the 2 days measurement group it was observed that the 2, 3 and 4% usage of replacement ratios could provide positive results in the first years of cement mortar service life from the point of ASR.
- When the 7 days measurement group was considered, as in the 2 days measurement group, the 2, 3 and 4% usage of replacement ratios could provide positive results in the middle ages of cement mortar service life from the point of ASR and 3% usage of borogypsum could exhibit optimal results.
- During the service life of cement mortars, from the point of ASR, it was observed that 4% borogypsum replacement could be used in all of the three age groups.

The optimal replacement amount for ASR must be determined not only according to the length elongation but also be made by considering the mechanical strengths. From the point of mechanical properties;

- When the samples that were not subjected to the effect of ASR were evaluated, it was observed that the increase in the borogypsum replacement caused compressive strength loss.
- When the samples that were subjected and not subjected to ASR effect were compared with each other, a compressive strength loss was observed in all of the samples subjected to ASR effect. This loss was at an acceptable level in the BG3 mortar group.

- When the samples that were subjected and not subjected to ASR effect were compared with each other a serious flexural loss was observed in the control mortar group with no replacement whereas in all of the other replacement ratios an increase in the flexural strength was observed at various levels. As in the compressive strength the closest flexural strength was determined with the BG3 replacement ratio.

From the point of porosity volume;

- The cumulative mercury volume intruding to samples was observed in the control group BJ0 mortar sample in the highest amount.
- Mortar samples exhibited different porosities at various borogypsum ratios. This situation occurred in the BG2, BG4, BG1 and BG3 mortar groups from the highest to the lowest respectively.
- Whereas in the BG3 replacement ratio, maximum improvement was obtained from the point of porosity ratio with minimum mercury intrusion.

Acknowledgements. This study was supported by Gazi University Scientific Researches Project Department with 07/2012-03 project code number. I thank to my university and to the valuable members for their supports.

References

1. Serpa D, Santos Silva A, de Brito J, Pontes J, Soares D (2013) ASR of mortars containing glass. Constr Build Mater 47:489–495
2. Neville AM (1995) Properties of concrete, 4th edn. Longman Group, Harlow
3. Carles-Gibergues A, Hornain H (2008) La durabilité des bétons face aux réactions de gonflement endogène. In: Ollivier JP, Vichot A (eds) La durabilité des bétons: bases scientifiques pour la formulation de bétons durables dans leur environnement. Presses de l'Ecole Nationale des Pons et Chaussées, Paris, pp 487–611 (in French)
4. Fournier B, Bérubé MA, Thomas MDA, Smaoui N, Folliard KJ (2004) Evaluation and management of concrete structures affected by alkali–silica reaction—a review. MTL 2004–2011 (OP), Natural Resources Canada, Ottawa
5. Godart B, Le Roux A (1995) Alcali-réaction dans le béton: mécanisme, pathologie et prévention. Techniques de l'Ingénieur, traité de Construction, C 2 252 (in French)
6. Marzouk H, Langdon S (2003) The effect of alkali-aggregate reactivity on the mechanical properties of high and normal strength concrete. Cem Concr Compos 25:549–556
7. Poyet S, Sellier A, Capra B, Foray G, Torrenti JM, Cognon H et al (2007) Chemical modelling of alkali silica reaction: influence of the reactive aggregate size distribution. Mater Struct 40:229–239
8. Garcia-Diaz E, Riche J, Bulteel D, Vernet C (2006) Mechanism of damage for the alkalisilica reaction. Cem Concr Res 36:395–400
9. BenHaha M, Gallucci E, Guidoum A, Scrivener KL (2007) Relation of expansion due to alkali silica reaction to the degree of reaction measured by SEM image analysis. Cem Concr Res 37:1206–1214
10. Giaccio G, Zerbino R, Ponce JM, Batic OR (2008) Mechanical behaviour of concretes damaged by alkali–silica reaction. Cem Concr Res 38:993–1004

11. Roma SN, Ngo T, Mendis P, Mahmud HB (2011) High-strength rice husk ash concrete incorporing quarry dust as a partial substituten for sand. Constr Build Mater 25:3123–3130
12. Okay O, Guclu H, Soner E (1985) Boron pollution in the Simav river, Turkey and various methods of boron removal. Water Res 19:857–862
13. Goldberg S (1997) Reactions of boron with soil. Plant Soil 193:35–48
14. Howe PD (1998) A review of boron effects in the environment. Biol Trace Elem Res 66:153–166
15. Boncukoglu R, Yılmaz TM, Kocakerim MM, Tosunoglu V (2002) Utilization of borogypsum as set retarder in Portland cement production. Cem Concr Res 32:471–475
16. Elbeyli İF, Derun EM, Gülen J, Pişkin S (2003) Thermal analysis of borogypsum and its effects on the physical properties of Portland cement. Cem Concr Res 33:1729–1735
17. Erdoğan Y, Genç H, Demirbaş A (1992) Utilization of borogysum for cement. Cem Concr Res 22:841–844
18. Erdoğan Y, Demirbaş A, Genç H (1994) Partly-refined chemical by-product gypsum as cement additives. Cem Concr Res 24:601–604
19. Kavas T, Olgun A, Erdoğan Y (2005) Setting and hardening of borogypsum-Portland cement clinker-fly ash blends. Studies on effects of molasses on properties of mortar containing borogypsum. Cem Concr Res 35:711–718
20. Demir D, Keleş G (2006) Radiation transmission of concrete including boron waste for 59.54 and 80.99 keV gamma rays. Nucl Instrum Methods Phys Res B 245:501–504
21. Demirbaş A, Karslıoğlu S (1995) The effect of boric acid sludges containing borogypsum on properties of cement. Cem Concr Res 25:1381–1384
22. Demirbaş A (1996) Optimizing the physical and technological properties of cement additives in concrete mixtures. Cem Concr Res 26:1737–1744
23. Demirbaş A (1999) Recyciling of lithium from borogypsum by leaching with water and leaching kinetics. Resour Conserv Recycl 25:125–131
24. Demirbaş A, Yüksek H, Çakmak İ, Küçük MM, Cengiz M, Alkan M (2000) Recovery of boric acid from boronic wastes by leaching with water, carbon dioxide- or sulfur dioxide-saturated water and leaching kinetics. Resour Conserv Recycl 28:135–146
25. Alp I, Deveci H, Süngün YH, Yazici EY, Savaş M, Demirci S (2009) Leachable characteristics of arsenical borogypsum wastes and their potential use in cement production. Environ Sci Technol 43:6939–6943
26. Tümen Y (2008) Effect of borogypsum on some properties of concrete. Masters thesis, Department of Civil Engineering, Mustafa Kemal University (in Turkish)
27. Sevim UK, Tümen Y (2013) Strength and fresh properties of borogypsum concrete. Constr Build Mater 48:342–347
28. Abell AB, Willis KL, Lange DA (1999) Mercury intrusion porosimetry and image analysis of cement-based materials. J Colloid Interface Sci 211:39–44
29. Khatib JM, Mangat PS (2003) Porosity of cement paste cured At 45 °C as a function of location relative to casting position. Cem Concr Res 25:97–108
30. Caré S (2008) Effect of temperature on porosity and on chloride diffusion in cement pastes. Constr Build Mater 22(7):1560–1573
31. Durmuş G, Arslan M (2010) Physical properties of cement mortars exposed to high temperature in various cooling conditions. J Fac Eng Arch Gazi Univ 25(3):541–548
32. ASTM D (1984) 4404: Standard test method for determination of pore volume and pore volume distribution of soil and rock by mercury intrusion porosimetry. ASTM International, Aug 31
33. ASTM C 1260–07 (2010) Standard test method for potential alkali reactivity of aggregates (Mortar-Bar method), American Society for Testing and Materials, West Conshohocken, Pennsylvania

34. Demir I (2010) The mechanical properties of alkali–silica reactive mortars containing same amounts of silica fume and fly ash. J Fac Eng Arch Gazi Univ 25(4):749–758
35. TS EN 196-1 (2002) Methods of testing cement—part 1: determination of strength. Turkish Standards, Ankara (in Turkey)
36. ASTM C 618 (2002) Standard specification for coal fly ash and raw or calcined natural pozzolan for use as a mineral admixture in Portland cement concrete. Annual book of ASTM standards, Pennsylvania
37. Bayca SU (2013) Microwave radiation leaching of colemanite in sulfuric acid solutions. Sep Purif Technol 105:24–32
38. Koçak Y, Taşcı E, Kaya U (2013) The effect of using natural zeolite on the properties and hydration characteristics of blended cements. Constr Build Mater 47:720–727
39. Gomes CEM, Ferreira OP, Fernandes MR (2005) Influence of vinyl acetate–versatic vinylester copolymer on the microstructural characteristics of cement pastes. Mater Res 8 (1):51–56
40. Govin A, Peschard A, Guyonnet R (2006) Modification of cement hydration at early ages by natural and heated wood. Cem Concr Compos 28(1):12–20
41. Cvetkovic J, Petrusevski VM, Soptrajanov B (1997) Reinvestigation of the water bending region in the spectra of gypsum-like compounds an FTIR study. J Mol Struct 408(409):463–466
42. Frost RL, Xi Y, Scholz R, Belotti FM, Filho MC (2013) Infrared and Raman spectroscopic characterization of the borate mineral colemanite—$CaB_3O_4(OH)_3 \cdot H_2O$—implications for the molecular structure. J Mol Struct 1037:23–28

Structural Stability Analysis of Large-Scale Masonry Historic City Walls

Rüya Kılıç Demircan[1]([⊠]) and Ali İhsan Ünay[2]

[1] Civil Engineering Department, Technology Faculty, Gazi University, Ankara, Turkey
ruyakilic86@gmail.com
[2] Department of Architecture, Faculty of Architecture, Gazi University, Ankara, Turkey
unay@gazi.edu.tr

Abstract. Owing to the developments in computer aided structural analysis methods in the last two decades, it has been possible to realize structural analysis of historic masonry structures in a very detailed and precise way. Besides the structural analysis package programs, special software have also been developed based on the characteristics of the case in question. While concentrating on the advanced methods of mathematics and mechanics, it might be possible to overlook the main point, which is the relationship between structural stability and the natural events. In this study, structural stability issues and the consequent structural damages on the fortification walls of Sinop Castle under the effects of sea waves have been explored. Sinop Castle is known to have been built by the immigrants who settled in the region at eighth century BC. The fortification walls that were destroyed by the attacks of Cimmerians in seventh century BC were fixed during the reign of King Mithridat IV and they were expanded to the present day limits. Having undergone constant maintenance during the periods of Romans and Byzantines, the castle was invaded the Seljouks between 1214 and 1261. During this period, an inner castle was built to fortify the defense of the repaired castle. The fortification walls of the Sinop Castle completely surround the narrowest pass of the peninsula. The northern walls are 1800 m, northern walls are 400 m, eastern walls are 500 m and the western walls are 273 m long. The walls reach a height of 30 m with a thickness that varies between 3 and 8 m. The northern walls have sustained excessive deterioration due to the sea waves and only the Kumkapi and Lonca gates have survived. In this study, stability analysis has been performed by using the finite element method on a partial analytical model of the northern walls of the Sinop Castle. The erosion caused by the waves at the bottom of the walls have been taken into consideration in a gradual manner to identify the origin of the damage, which has been caused by the excessive stress due to balance degradation in the massive walls; and consequently, appropriate strengthening methods have been proposed.

Keywords: Structural analysis · City wall · Large-scale masonry structures
Finite element methods

© Springer International Publishing AG, part of Springer Nature 2018
S. Fırat et al. (eds.), *Proceedings of 3rd International Sustainable Buildings Symposium (ISBS 2017)*, Lecture Notes in Civil Engineering 7,
https://doi.org/10.1007/978-3-319-64349-6_31

1 Introduction

Protecting historic structures and monuments is a responsibility that should be acknowledged without monetary concerns and regardless of the country, culture or religion that they represent. The primary concern in structural protection of historic structures should be preventing collapse. Scientific and technological developments on this topic have rapidly been improved over the last 25 years. Especially the possibility of realizing the structural analysis methods more accurately and the striking development of repair techniques and materials have inspired the researchers and professionals in protecting historic structures and monuments.

Structural analysis of historic structures usually differs from the analysis performed for the design of modern structural systems. Structural analysis of historic buildings are performed to determine the stress and internal forces in the structural members due to the loads caused by the self-weight of the structure, wind, earthquake, foundation settlements and occupational use. This kind of calculations and comparisons are called elastic approach. However, the structural members of historic masonry structures do not display linear elastic material behavior. There are analysis methods that are based on nonlinear material properties; however, they take considerable amount of time. Besides, any small error made in the assumptions related to material properties might lead to more serious mistakes and errors during the iterations.

Based on the principles of mechanics it is easy to understand the structural behavior of a building that has a regular structural layout. However, understanding structural behavior of historic structures that have their own spatial geometry is not an easy task. It becomes especially difficult in structures that is composed of domes, vaults, pendentives and arches, of which the shear force diagrams, bending moment diagrams and axial force distributions cannot easily be determined by using approximate methods.

The finite element analysis is a numerical method used to solve problems that can be formulated with partial differential equations. It is appropriate for structures with complex geometries and large masses under various loads and with various boundary conditions. In this study, a finite element model of Sinop Castle, which has a massive geometry has been prepared and analyzed under various loads and environmental factors.

2 Sinop City Walls

2.1 The City of Sinop and Its History

The city of Sinop is settled at the north of Turkey, on a 1.5 km long tombolo at the junction of the mainland and Boztepe Peninsula, which is a ridge as a result of tectonic movements. The narrowest part of the tombolo is 300 m and it is 15–20 m above the sea level [1]. Figure 1 shows the city of Sinop at the narrowest part of the coastline that connects the island to the mainland. On the east and west of the peninsula are the Interior Harbor (Sinop Harbor) and the Exterior Harbor (Ak Harbor), respectively. The archaeological ruins are an evidence to prove that the region was inhabited in the very

Fig. 1. Settlement of the city of Sinop [4]

old times. This can be attributed to the fact that Sinop was the most sheltered natural harbor in the Black Sea region [2, 3].

Based on the archaeological evidence, inhabitance in the region dates back 5000 years to the Chalcolithic Age. The town of Sinop was first settled by a Miletus colony in eighth century BC and became one of the most important cities of the Black Sea in the Archaic Age [2]. Between 546–547 and 340 BC, it was dominated by the Persians, who built various monuments and hosted important scholars and philosophers in the region. Right after the Persians, the town was taken by the Pontus Empire, and had its peak in history until 30 BC. Then it was conquered and dominated until 395 AD by the Roman Empire, who built aqueducts to sustain the primary needs of the community [3]. After the Romans, the region was dominated by the Byzantines for almost eight centuries and was conquered by the Seljuk Empire in the early thirteenth century. After having been governed by several principalities, between thirteenth and fifteenth centuries, it was annexed by the Ottoman Empire in 1461. During all those years throughout its history, Sinop had retained its identity as a cultural and trade center and became the most important naval base in the region [2].

2.2   The History of the Sinop Castle

The date of construction of Sinop Castle is unknown. Demir [5] suggests that the castle was built after the colonization period. According to Demir [5] the castle walls protected the city from being conquered even though it was surrounded in 220 BC. The walls were fortified during the Pontus Empire and new additions—along with various maintenance and repairs, were made during the time of the Romans, Byzantines, Seljuks, Principalities, and Ottomans. The Seljuk Empire Izzeddin Keykavus I added a keep by means of new walls and bastions at the western part of the city. In order to obtain better control over the harbor, eleven bastions were added along the north–south axis. The bastions on the southern side are 22 m high while the 3 m thick castle walls have 18 m height. Today, the northern castle walls are 880 m, southern walls are 400 m, eastern walls are 500 m, and western walls are 273 m long and the castle has one main gate for each one of these walls. The northern keep and the southern keep, respectively, cover 16,875 and 9500 m^2 [2].

With effect from 1568, the keep served as the dungeon. The castle was kept well until the Constitutional Period [6]. Having witnessed various battles between the Ottomans and Russians, the castle sustained the most severe damage during the raid in 1853 (Fig. 2). Figure 3 displays the continuity of the castle walls during the World War 1.

Fig. 2. Ottoman–Russian battle 1853, after the raid (taken from Cengizhan Ersoy archive)

Fig. 3. 1915–1920 World War 1, picture taken by the German planes

However, during the first years of the Turkish Republic, the western walls of the castle and the western and eastern walls of the keep were demolished to create the main access to the city. The majority of the keep is still visible today, while a significant part of the castle walls is gone.

3 Structural Damages Observed in Sinop City Walls

3.1   Eastern City Walls and Bastions of the Castle

The major part on the northern side of the 500 m long eastern city walls were demolished and replaced by buildings. The bastions, which are over these walls, were made of rubble and ashlar stone with mortar as the binding material. The stairs that lead to the two-story bastion consist of 35 steps. The first floor has two vaulted walkways at the eastern and western sides. These walkways have arched openings. The western walkway leads to the second floor and also involves arched openings. There is a clock tower over the western walkway. A 20 m part of the city walls on the southern side of the bastion was demolished to build a road in Fig. 4 [7].

Fig. 4. Clock tower of the Sinop Castle eastern bastions (from Mustafa Cambaz Archive)

3.2   Southern City Walls and Bastions of the Castle

The southern city walls of the castle were entirely destroyed. There is a sound looking bastion at the southern end of the castle walls that run in the north–south axis as can be seen in the Figs. 5 and 6.

Fig. 5. Southern walls and bastions of the Sinop Castle from 1935 to 1940 (from Zeynel Zeki Özcanoğlu archive)

Fig. 6. Southern walls and bastion of the Sinop Castle in 2014 (from Çetin Tek archive)

3.3   Western and Northern City Walls and Bastions of the Castle

Almost all of these 270 m long city walls have been destroyed. The western city walls, which are the main entrance to the city were demolished in order to provide road access.

The northern city walls, which are approximately 880 m long, run between the northwestern and northeastern bastions along the coastline. As it can be seen in Fig. 7, parts of these walls collapsed in the sea because of strong waves while some other parts leaned towards the sea [6].

Fig. 7. Northern city walls (www.sinopthk.com)

3.4   The Keep, Northern City Walls, Eastern City Walls and Bastions

The keep is at the western part of the city, the Seljuks, right after they conquered Sinop, built two separate inner walls that constitute the keep. One of these inner walls run on the north–south axis between the northern and southern city walls while the other is at the north. These additions were made when the castle was enlarged to include the western part of the city. The keep, which hosts several Seljuk inscriptions, consist of two parts. The northern and southern parts cover areas of 16,875 and 65,000 m^2, respectively. The southern part, which hosts the old prison, is lower in elevation than the northern part and is used as a museum today.

At the northwestern corner of the keep, a bastion, which was constructed of rubble and mortar inside and covered with ashlar stone at the surface, takes place. This bastion, once started to collapse, was repaired close to its original characteristics as it can be seen in Fig. 8.

Fig. 8. The keep-northern city walls, 1930 (from Zeynel Zeki Özcanoğlu archive)

The eastern walls of the keep, on the other hand, run as two city walls parallel to each other between the ends of the northern walls and southern walls by the coast (see Fig. 9).

Fig. 9. The keep-eastern city walls (www.sinopthk.com)

4   Structural Analysis of Sinop City Walls

4.1   Finite Element Analysis and Main Considerations for Calculation

Finite element model of Sinop City Walls has been prepared according to modeling rules of SAP2000. All the geometrical dimensions and other related measures were obtained from the building survey conducted. The parameters for calculations and the model are given below:

- Majority of the structure is suitable for the definition of 3D solid elements.
- As shown in Fig. 10, the finite element model consists of 11,280 joints and 7248 solid elements.
- Material properties of the structure have been obtained from the literature on the similar type of structures and suggested values of the masonry material properties of the current earthquake code.
- Material properties are shown in Table 1.
- For the determination of the elasticity modulus and unit weight of the material, masonry units and the mortar have been assumed as one single material.
- On the model, spectrum has been applied in two fundamental directions as EQx and EQy.

For the easiness in the evaluation of the results, two different load combinations have been made as G + EQx and G + EQy.

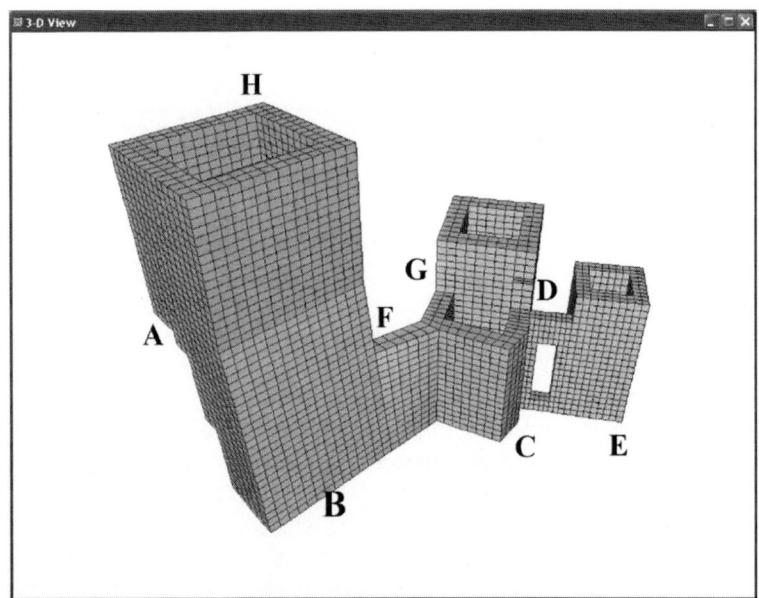

Fig. 10. Finite element model of Sinop Castle with remarkable point for the evaluation of the analysis results

Table 1. Material properties of the masonry

Structural elements	Modulus of elasticity (kN/m^2)	Specific weight (kN/m^3)	Mass (t/m^3)
Stone walls (including the mortar)	450,000 (450 MPa)	30	3.06

4.2   The Results of Finite Element Analysis

Structural analysis of Sinop City Walls was carried out with SAP2000 using the load combinations provided above. Since the evaluation of displacement, forces and stresses at every single joints and load carrying elements is not efficient, interpretation of results is based on the graphs of stress distribution and deformed shape considering the unfavorable case. Base shear and vertical reactions are provided for load combinations in Table 2.

Total weight of the structure is seen to be 44,154 kN while total base shear is 6205 and 4193 kN in x direction (north–south) and in y direction (west–east), respectively. According to these results, base shear force that the structure is subjected is 14% of the total weight in x-direction and 1% of the total weight in y-direction.

In the evaluation analysis results among the stresses calculated for solid elements, stresses that are vertical to the local axis of each element (in SAP 2000, S33) is used. The stresses S33 for the load combinations G + EQ$_x$ and G + EQ$_y$ are evaluated walls.

Table 2. Base shear and vertical reactions

Loading type	Analysis type	Base shear in x-dir. (kN)	Base shear in y-dir. (kN)	Vertical reactions (kN)
G	Linear static			44,154
EQ$_x$	Linear spectrum	6205	1493	457
EQ$_y$	Linear spectrum	1493	6798	967
G + E$_x$	Combination	6205	1492	44,612
G + E$_y$	Combination	1493	6798	45,122

Maximum stress distribution on the structural elements for G + EQ$_x$ and G + EQ$_y$ load combinations is provided in Figs. 11 and 12 respectively. Similarly, Table 3 points out the maximum tensile and compressive strength values for remarkable points in the structure.

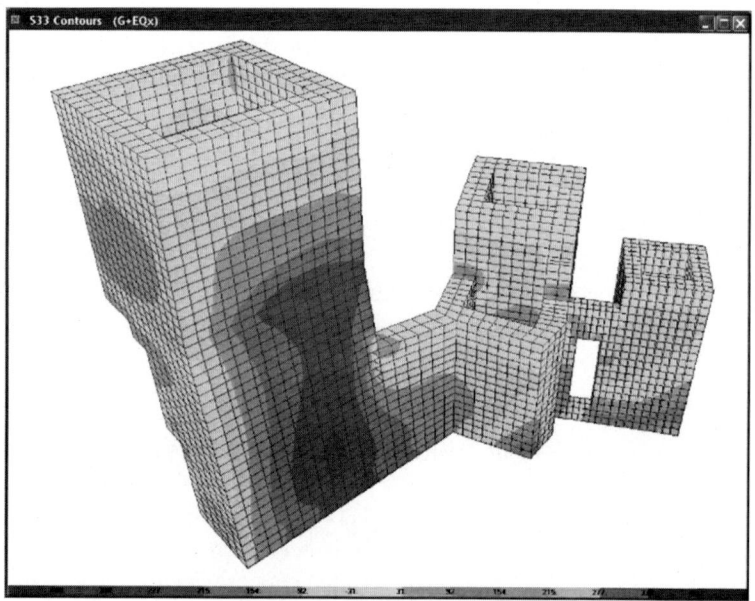

Fig. 11. Vertical stress distribution for G + EQ$_x$ load combination

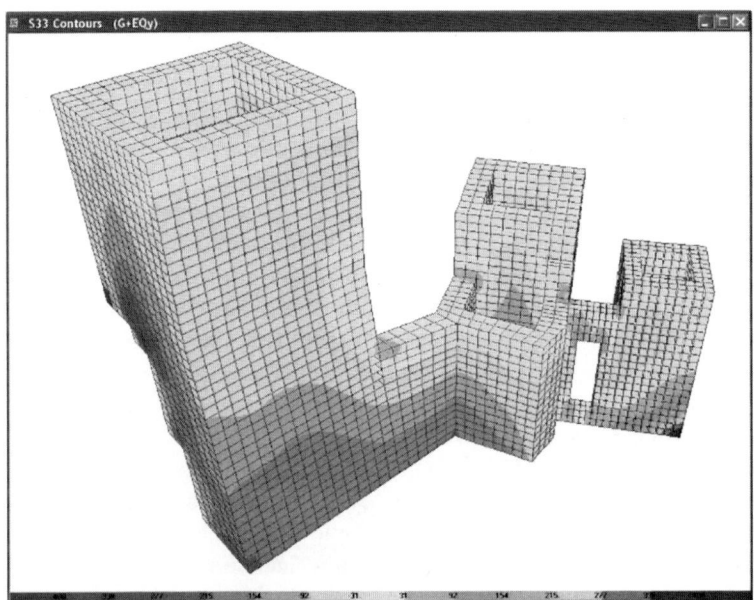

Fig. 12. Vertical stress distribution for G + EQ$_y$ load combination

Table 3. Maximum stresses on remarkable points in the structure

Location	G + EQ$_x$ (kN/m^2)	G + EQ$_y$ (kN/m^2)
A	92	400
B	−338	−182
C	63	28
D	134	54
E	272	338
F	−215	−29
G	118	78
H	27	31

5   Conclusions

The aim of this study is not creating the most realistic and accurate simulation of the actual structural behavior and structural capacity of Sinop City Walls but to acquire an overall understanding of the interval in which this structure could behave during prospective environmental disturbances. A detailed repair and restoration history of the structure must be known to assess the local variations of the material properties. As a result, to use approximate material values and analytical methods to understand the seismic behavior of the structure as accurately as possible with the available means. This study can be considered as the foundation or a starting point for a possible future restoration or strengthening works.

Following observations are made on the behavior of the structural elements of the building:

The structural system of Sinop City Walls has high rigidity. Largest displacements (Δx) in the x direction is observed as $\Delta x = 12$ mm; under combined gravity and spectral loading (G + EQy) in y direction, largest displacements (Δy) in the y direction is observed as $\Delta y = 14$ mm.

The high rigidity level of Sinop City Walls is verified by the values obtained for the periods of the seismic modes. The value of the period for the first mode is equal to $T1 = 0.44$ s.

The finite element analysis of Sinop City Walls has demonstrated that Turkish Earthquake Code's allowable compressive or shear stress levels for masonry materials were not exceeded anywhere under the applied load cases.

Allowable tensile stress levels were only locally exceeded near the corners of the openings and lower corners of the walls. Such local stresses are expected and allowable due to the mesh structure of the finite elements and used support conditions.

It should be kept in mind that the material properties used in the finite element model were not actual values obtained from the testing of the material samples from the actual structure but values taken from scientific literature. As a result of this, it is possible for certain partial damages to happen due to the non-homogenous behavior of the actual structural elements or material deterioration. However, since the obtained maximum stress levels are well within safety intervals, it is very unlikely that Sinop City Walls will be subjected to major damage during several possible environmental disturbances.

The analyses reveal that the structure is not under a serious stability threat, however, it should be remembered that in the study the effects of current damages, probable material deteriorations, and the probable foundation failures are not taken into account. Moreover, the material properties used are the common values and might greatly change in the actual case. Within this context, the structure seems to need some precautions to prevent any further damage.

References

1. Akkan E (1975) Sinop Yarımadasının Jeomorfolojisi. Ankara University, DTCF Publishing, No: 261, Ankara
2. Yılmaz C (2009) TARİHİ SİNOP KALESİ CEZAEVİ, Doğu Coğrafya Dergisi. e-dergi. atauni.edu.tr (in Turkish)
3. Üstün Demirkaya F, Tuluk Ö (2012) İskender 'Eflatun'un "Kurbağa"Sı Sinope'den Sinop'a: Kaynaklara Göre Sinop Kentinin Fiziksel Gelişimi'. Metu Jfa 29(1):45–68
4. www.sinopthk.com
5. Demir Y (2001) Antik Anadolu'da Bir Kozmopolitik Şehir Sinope. Basılmamış yayın, İstanbul (in Turkish)
6. Ulus İ (2014) Açıklamalı Sinop Kitabeleri. Sinop Belediyesi Kültür Yayınları, İstanbul (in Turkish)
7. Gökoğlu A (1952) Paphlagonia. Doğrusöz Matbaası, Kastamonu (in Turkish)

The Reliability of the ACI 209R-92 Method in Predicting Column Shortening in High-Rise Concrete Buildings

Alaa Habrah[1]([⊠]) [iD] and Abid I. Abu-Tair[2]

[1] Development, Research and Urban Planning Consultancy, Fujairah, UAE
alaa84-eng@hotmail.com
[2] The British University in Dubai, Dubai, UAE
abid.abu-tair@buid.ac.ae

Abstract. The early prediction of column shortening during the design of high-rise concrete buildings is of high importance to allow for shortening compensation measures to be proposed and taken in the construction. However, a column shortening that consists of elastic, creep and shrinkage shortening can be very complex to predict influenced by many uncertainties, such as inelasticity and non-homogeneity of the concrete structures and highly affected by the chosen method for time-dependent effects. In this paper, the ACI 209R-92 method (ACI Committee 209 in Prediction of creep, shrinkage, and temperature effects in concrete structures (ACI 209R-92). American Concrete Institute, Farmington Hills [1]) for predicting creep and shrinkage effects is evaluated by comparing its results to actual site readings taken during the construction of a high-rise building presented as a case study in this research. For that purpose, an Excel (Microsoft Corporation in Excel (computer program), https://www.office.com/, [2]) spreadsheet including all the factors of this method is developed to provide a simple interface to predict the elastic and time-dependent column shortening and settlement. Furthermore, the presented case study building was analyzed using a finite element software, Etabs (Computers and Structures in Etabs 2015 ultimate, version 15.0.0 (computer program), http://www.csiamerica.com/, [3]) where column shortening and settlement were predicted based on the CEB-FIB 90 method (Comite Euro-International Du Beton in CEB-FIB model code, Thomas Telford, London, [4]) for time-dependent effects. By comparing the two methods' results to the actual readings, it was found that both methods overestimated column settlement in all floors. Whilst the average overestimation of the developed Excel sheet based on the ACI 209R-92 model was 630%, Etabs analysis based on the CEB-FIB 90 model had more accurate results with average overestimation of 258%.

Keywords: Column shortening · Creep · Shrinkage · Modulus of elasticity

1 Introduction

A concrete column under sequential construction loading exhibits two types of shortening, elastic due to immediate loading and inelastic as a result of creep and shrinkage effects. Prediction of elastic effects includes less variables than those of

© Springer International Publishing AG, part of Springer Nature 2018
S. Fırat et al. (eds.), *Proceedings of 3rd International Sustainable*
Buildings Symposium (ISBS 2017), Lecture Notes in Civil Engineering 7,
https://doi.org/10.1007/978-3-319-64349-6_32

time-dependent (creep and shrinkage) since it's related directly to the applied load and modules of elasticity which can be directly predicted by experimental tests on specimens taken from the casting concrete. On the other hand, predicting creep and shrinkage effects is highly affected by the prediction model.

1.1 Research Significance

The accumulation of elastic and time-dependent effects due to sequential loading during the construction of tall buildings leads to considerable columns shortening which affects the overall serviceability of the building and may destruct the non-structural members such as services ducts, finishes and curtain walls which are installed prior to end of construction.

In tall buildings, the difference in the applied loads and the axial stiffness between vertical members is significant leading to considerable differential shortening between those members, more precisely between columns and shear walls. Ignoring this deferential shortening in tall buildings compromises in addition to the non-structural members, the structural elements due to the induced additional forces in the connecting horizontal members.

This study provides a simple and easy-to-use interface using Excel sheet to predict the elastic, time-dependent and total column shortening. The prediction depends on information available during the design for practicing engineers, so the effects of this phenomena can be accounted for in the design and proper measures and compensations can be specified.

1.2 Research Challenge

Although the prediction of column shortening behavior is sophisticated and accompanied with much uncertainties, a good prediction can be based on estimated information during the design stage. Creep and shrinkage, elastic modulus, construction method and environmental conditions are all affecting the accuracy of columns shortening prediction.

A sequential analysis based on the construction method and schedule rather than the total loading conventional analysis is essential to predict this behavior more accurately and realistically. For this purpose, the proposed Excel sheet is developed to predict the elastic, creep and shrinkage shortening of any concrete column or wall during a particular construction stage (after casting a certain floor) using the ACI 209R-92 method for time-dependent effects prediction.

It's essential to this research, in addition to provide column shortening prediction individually, to present the results of settlement in the same way that were provided in the presented case study building records to compare and assess those values. That is where the settlement is surveyed each five floors at chosen levels during the construction. Thus, the column shortening is incorporated with complex effects of the previous and the following floors which were considered in the developed excel sheet.

1.3   Research Objectives

This research intends to obtain the following targets:

1. To assess the reliability of the ACI 209R-92 method in predicting column shortening in high-rise concrete buildings.
2. To predict both elastic and time-dependent column shortening in high-rise buildings
   and provide a spreadsheet for that purpose.
3. To provide a simple procedure for engineers to predict the columns shortening at an
   early stage of the design.

1.4   Case Study

In order to evaluate the theoretical results of the considered ACI 209R-92 method, a
case study of 360 m high vertically asymmetric building was considered to compare
the site readings of columns shortening taken during the tower construction with the
theoretical results. The building is a 71 floors office tower located in Dubai, UAE. The
typical plan shapes into two intersecting ellipses which, after floor 52, only one ellipse
continues to floor 64. A reinforced concrete central core wall with peripheral frame
connected by spine beams at each floor form the main structural systems of the tower.
The floor is hollow core slab with 320 mm thickness and 80 mm topping slab.
A typical structural plan is shown in Fig. 1.

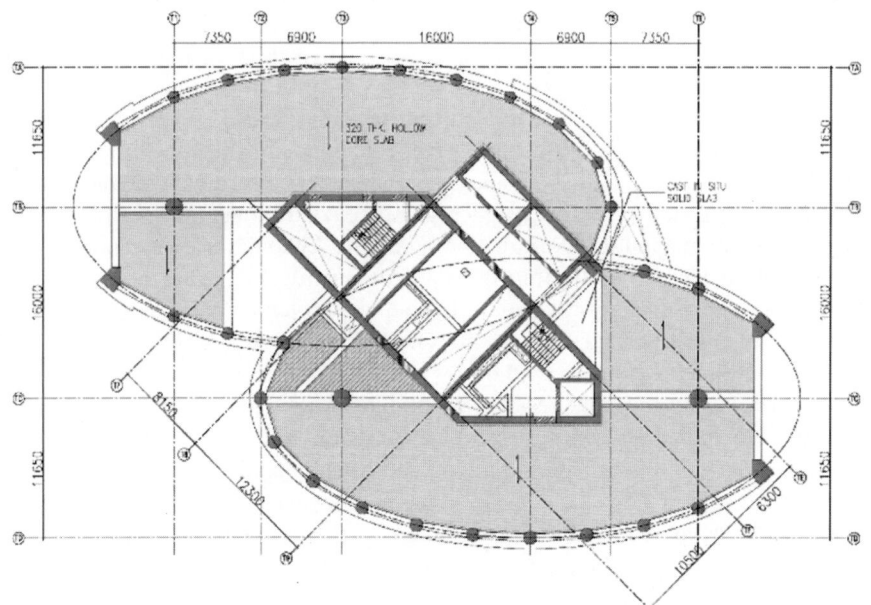

Fig. 1. Typical structural plan of the tower

A monitoring system was specified to allow for site readings of the column movement during the construction work. The monitored points were set at each floor for lateral movement and at different floors at peripheral columns and the core wall to monitor the vertical movement. The survey was done by laser device with reference to a benchmark located outside the building. Reading results was done after each 5 floors construction.

2 Axial Column Shortening

In reinforced concrete buildings, the effects of columns shortening increase in high-rise buildings and hence, need special attention during the design and construction. Columns shortening is a cumulative procedure of elastic and inelastic strains which are produced during the construction and extend over the life time of the building. The prediction of columns shortening is a complex task including the elastic and time-dependent strains with all the influencing factors. The deformation of axially loaded vertical members due to time dependent effects takes place from the moment the member is cast for shrinkage effects and the moment it is loaded for creep effects. It was reported [5–7] that creep and shrinkage effects have the most proportion of the total effect (represented as column strain) occurs during the first few months of the member age. As well, creep effects due to loads applied at later ages have much less effects than those applied at early age. As a consequence, the consideration of columns shortening in tall buildings during the construction stage is of higher concern compared to those caused by sustained loads applied at later stages (imposed loads). Fintel and Khan [8] firstly originated a model to quantify the axial shortening of a concrete column. The model was proposed in [6] as the PCA method for prediction and compensation of columns shortening in tall buildings. The method is based on numerous tests and formulations to calculate the elastic and inelastic post-slab installation deformations caused by the sequence of tall buildings construction. In the proposed procedure, it was suggested that columns shortening can be accurately predicted during the preliminary design while the designer has some control over the specifications and enough information about the construction sequence. Although that procedure suggests a compensation work by adjusting the slab form vertically, it has no limits to the columns shortening or slabs tilt due to differential shortening at which the slab cambering is cost effective and convenient. The method presented by Ghosh [9] is a refinement of Fintel and Khan method [8] which gives according to Jayasinghe and Jayasena [10] almost the same results. Swamy and Arumugasaamy [7] researched the complexity of reinforced concrete columns behavior in buildings by observing the movement of an eight-story building during construction and six years later in service. The study was based on separating the three components of the total strain, elastic, shrinkage and creep strains. The elastic strain was computed by knowing the stress history and the elastic modulus from laboratory control tests. The shrinkage and creep strains were predicted using tests on unreinforced concrete specimens which were under the same environmental conditions of the building. The good agreement between measured and predicted values reflected the accurate and precise procedure of the latter study utilizing controlled experiments. In addition to column shortening, a very recent study [11] investigated the effects of creep on the whole behavior of RC columns including geometric nonlinearity, cracking and aging.

3   Procedure of Excel Sheet Calculation

The procedure for creep and shrinkage strain prediction presented in this paper is a direct application of the ACI 209R-92 model. The ACI formulas were used to develop spreadsheets in order to predict the final settlement of a considered column in any floor caused by casting floors above as following:

1. An Excel sheet was developed to estimate the total axial shortening (elastic, creep and shrinkage) of any column caused by adding one single floor and get the value S_{ih}. Where S_{ih}: shortening of a column in floor (i) caused by the instantaneous and time-dependent effects of adding one single floor (h).
2. The spreadsheet was prepared to sum up all the column shortenings of that particular column caused by adding a group of floors to get the value TS_{kj}, the total shortening of a column in floor (k) below floor (n) caused by adding number of floors (j − n + 1).

$$TS_{kj} = \sum_{h=n}^{j} S_{ih} \tag{1}$$

3. The spreadsheet was replicated to calculate the value TS_{kj} for each column below the considered column caused by adding the same group of floors.
4. Another Excel sheet was prepared to sum all the column shortening values TS_{kj} to get the final value Δ_{nj}, the total displacement of a column in floor (n) after casting floor (j).

$$\Delta_{nj} = \sum_{k=1}^{n} TS_{kj} \tag{2}$$

5. Steps 3 and 4 were repeated for each required group of floors to get the final settlement of that particular column after each five floors construction above that column.
6. Displacement charts were constructed to represent the column displacement for each stage similar to the actual site reading charts taken during the construction of the case study building.

An example of the developed Excel sheet is shown in Fig. 2.

Shortening Prediction of Column in Floor (1) by Adding Floors (4 to 8)					
Input Data				**Output Data**	
Floors Data	Considered Column Floor	1		Elastic Shortening Δe (mm)	0.1044
	Floor Height (m)	3.2			
	Casting Floors from	4		Creep Shortening Δc (mm)	0.03
	to	8			
	Rate of Construction (days/floor)	3		Shrinkage Shortening Δsh (mm)	0.058
Elastic Data	$f'c$ (mpa)	70		Total Shortening TS_{kj} (mm)	**0.192**
	Column Cross Section (m²)	1.54			
	Load Increment (kN/floor)	320			
Creep & Shrinkage	Column volume/surface (mm)	350			
	Percentage of fine agg. (%)	30			
	Slump (mm)	100			
	Relative Humidity (%)	80			

Input Notes

Considered column floor : the floor at which the column deformation is required			
Casting floors: refers to the slab loads causing the considered column shortening			
Column in floor (*i*) suports slab (*i+1*)			
Column in basement (*i*) input as -*i*			
Column in ground floor input as 0			

Fig. 2. Developed excel sheet for column shortening prediction

4   Case Study Results

The developed spreadsheet in Sect. 3 is used to estimate the columns and walls shortening in each floor and then accumulate them to predict the settlement of the considered column in the same survey results style for comparison between actual and theoretical values.

4.1   Shortening of Monitored Column Below 4th Floor

The previous procedure in Sect. 3 is used to calculate the settlement of the monitored column and wall in 3rd, 10th, 21st, 29th, 4th, 50th and 60th floor. Those columns are the ones which were monitored during the tower construction. Below are the steps for the monitored column in 3rd floor and similarly used for other columns floors:

- The added floors are divided into groups, 3rd to 8th, 3rd to 13th, 3rd to 18th, …, 3rd to 63rd. Each group is used to calculate the shortening of the 3rd floor column after casting this group. The resulted values are similar to those of monitoring points survey used in the case study.
- The developed Excel sheet was replicated to estimate the shortening of each column below 4th floor for each of the twelve groups of floors, 3rd to 8th, 3rd to 13th, 3rd to 18th, …, 3rd to 63rd.

The total shortening of columns below 4th floor for the first floors group (3rd to 8th) are plotted as a chart in Fig. 3 for each shortening component.

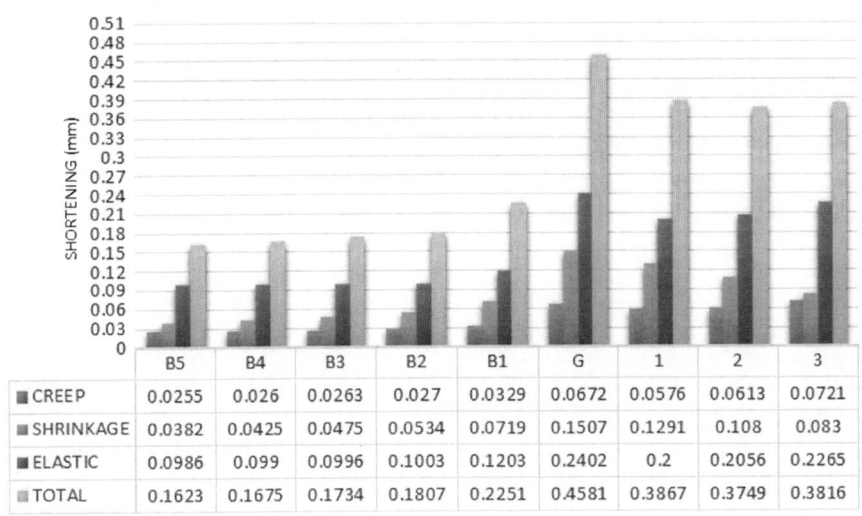

	B5	B4	B3	B2	B1	G	1	2	3
CREEP	0.0255	0.026	0.0263	0.027	0.0329	0.0672	0.0576	0.0613	0.0721
SHRINKAGE	0.0382	0.0425	0.0475	0.0534	0.0719	0.1507	0.1291	0.108	0.083
ELASTIC	0.0986	0.099	0.0996	0.1003	0.1203	0.2402	0.2	0.2056	0.2265
TOTAL	0.1623	0.1675	0.1734	0.1807	0.2251	0.4581	0.3867	0.3749	0.3816

FLOOR

Fig. 3. Shortening of monitored columns below 4th floor due to adding floors 4th–8th

To eliminate the influence of column length on the shortening and for a better understanding of this behavior, the shortening charts are transformed into strain charts by dividing each shortening component by the column length (Fig. 4).

The strain charts show a more regular pattern than shortening charts when elastic and creep strains keep increasing from B5 to 3rd floor. The shrinkage strain has a peak value in the 1st floor decreasing then for both fresher and older columns. The total strain is thus in line with the dominant pattern by increasing from B5 to 3rd floor except for one decrease in the 2nd floor for both column and wall strains. By examining the elastic strain, it can be

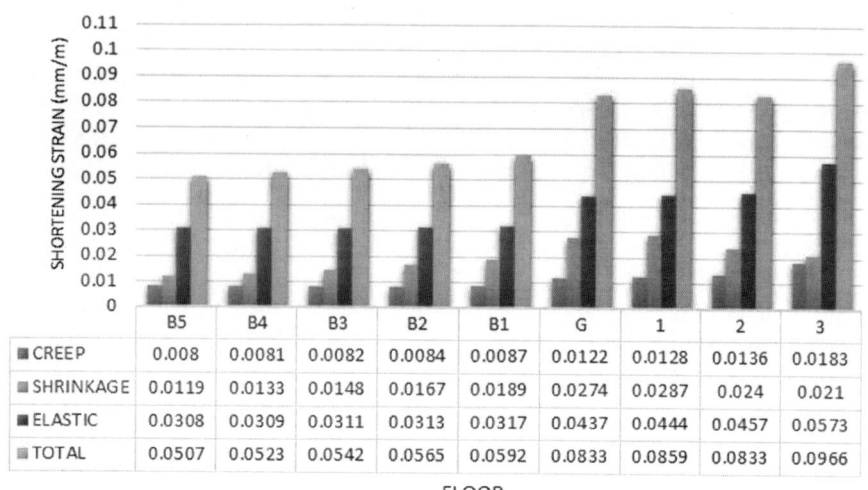

	B5	B4	B3	B2	B1	G	1	2	3
CREEP	0.008	0.0081	0.0082	0.0084	0.0087	0.0122	0.0128	0.0136	0.0183
SHRINKAGE	0.0119	0.0133	0.0148	0.0167	0.0189	0.0274	0.0287	0.024	0.021
ELASTIC	0.0308	0.0309	0.0311	0.0313	0.0317	0.0437	0.0444	0.0457	0.0573
TOTAL	0.0507	0.0523	0.0542	0.0565	0.0592	0.0833	0.0859	0.0833	0.0966

FLOOR

Fig. 4. Strain of monitored columns below 4th floor due to adding floors 4th–8th

noticed that the increase in elastic strain is much less than of creep and shrinkage. For further examination of the results, the differences in strain between floors are expressed as a percentage of column strain in floor B5 as shown in Table 1.

Table 1. Difference between columns/walls strain below 4th floor due to adding floors 4th to 8th

Floor	B5	B4	B3	B2	B1	G	1	2	3
Creep per. (%)									
Col.	100	101	102	105	108	152	160	170	228
Wall	100	102	104	105	109	112	118	126	184
Shrinkage per. (%)									
Col.	100	111	124	140	158	230	241	201	176
Wall	100	131	146	164	186	213	223	187	189
Elastic (%)									
Col.	100	100	101	102	103	141	144	148	186
Wall	100	100	101	102	103	104	106	109	149
Total (%)									
Col.	100	103	107	111	116	164	169	164	190
Wall	100	107	111	116	121	129	133	128	163

By examining the results in Table 1, it's possible to see the leap in all column strains in ground floor (108–152% for creep, 158–230% for shrinkage and 103–141% for elastic). However, the elastic strain revealed a very slight increase until the jump in ground floor.

4.2 Settlement of Monitored Column in 3rd Floor

For each group of floors, the shortenings of all the columns below 4th floor were accumulated to get the final settlement of the monitored column in 3rd floor after casting that considered group of floors. The resulted settlements of each added group of floors are used to construct settlement charts similar to the actual results style plotted on the same chart for comparison (Fig. 5).

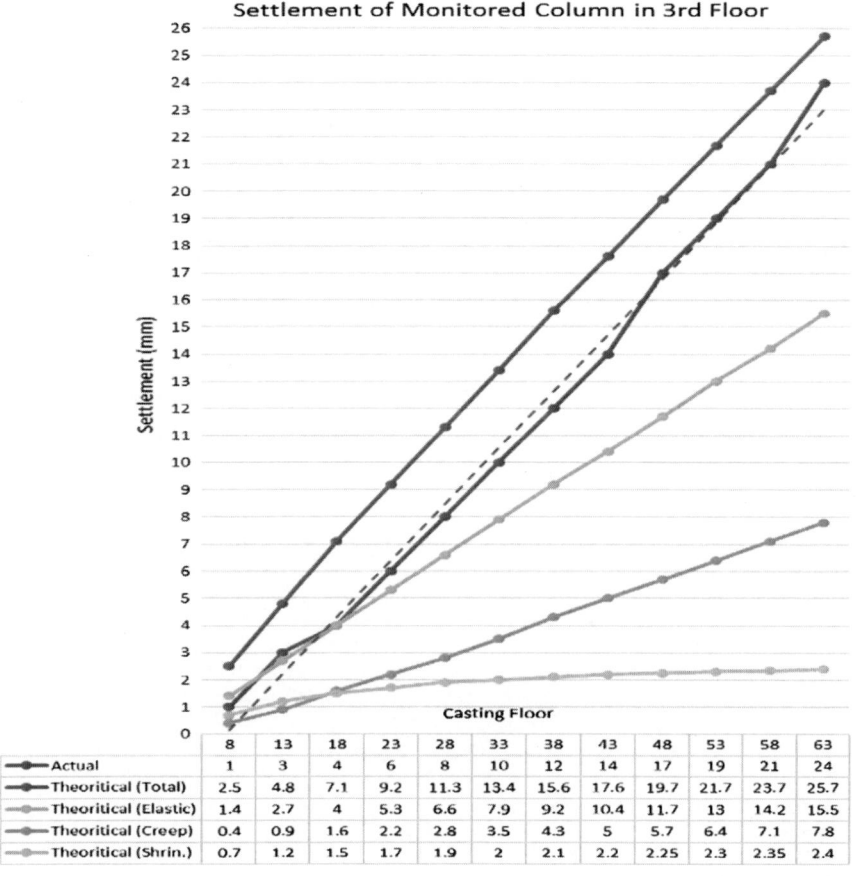

	8	13	18	23	28	33	38	43	48	53	58	63
Actual	1	3	4	6	8	10	12	14	17	19	21	24
Theoritical (Total)	2.5	4.8	7.1	9.2	11.3	13.4	15.6	17.6	19.7	21.7	23.7	25.7
Theoritical (Elastic)	1.4	2.7	4	5.3	6.6	7.9	9.2	10.4	11.7	13	14.2	15.5
Theoritical (Creep)	0.4	0.9	1.6	2.2	2.8	3.5	4.3	5	5.7	6.4	7.1	7.8
Theoritical (Shrin.)	0.7	1.2	1.5	1.7	1.9	2	2.1	2.2	2.25	2.3	2.35	2.4

Fig. 5. Actual and theoretical settlement of monitored column in 3rd floor

By studying all the charts individually, it can be observed that elastic and creep charts tend to have linear relation between settlement and adding floors. In contrast, shrinkage chart starts linear after adding the first few floors, then the effect of shrinkage starts to disappear in the horizontal segment of the chart. It's obvious that elastic settlement is highly overestimated before casting the 28th floor since it exceeds the actual settlement which includes the three settlement components. The elastic settlement is then deviated decreasingly from the actual settlement chart till the end of construction. By comparing the total theoretical settlement chart and the actual chart, it can be seen that in all construction stages the settlement is overestimated, however a consistent relation is maintained by the parallelism between the theoretical chart and the approximated linear line of the actual chart.

4.3   Settlement of Monitored Columns in All Monitored Floors

The procedure presented in Sects. 4.1 and 4.2 to predict the total settlement of monitored column in 3rd floor is repeated for the same monitored column in the other considered floors, 10th, 21st, 29th, 41st, 50th and 60th. The building was analyzed using Etabs software as a 3D finite element model using the CEB-FIB 90 model to predict the time-dependent settlement and the results were compared with the previous ACI model results and the actual site readings as well. The same inputs for the developed excel sheet are used in the Etabs model. The monitored column settlement charts are plotted for both ACI and CEB-FIP model with the actual settlement chart (Figs. 6, 7, 8, 9, 10 and 11).

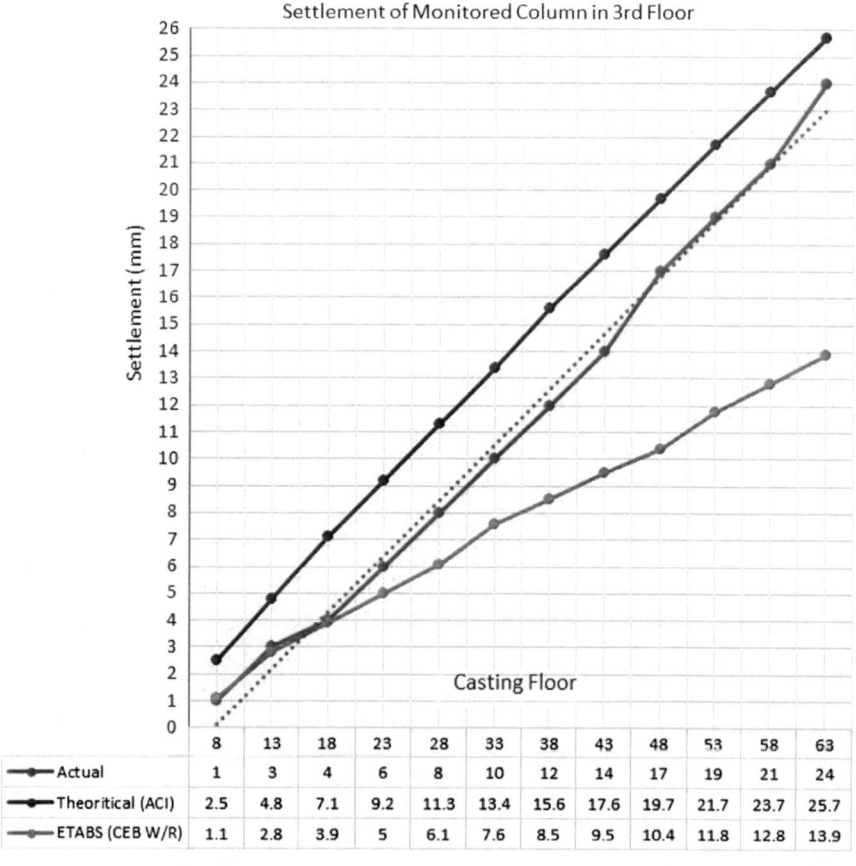

	8	13	18	23	28	33	38	43	48	53	58	63
Actual	1	3	4	6	8	10	12	14	17	19	21	24
Theoritical (ACI)	2.5	4.8	7.1	9.2	11.3	13.4	15.6	17.6	19.7	21.7	23.7	25.7
ETABS (CEB W/R)	1.1	2.8	3.9	5	6.1	7.6	8.5	9.5	10.4	11.8	12.8	13.9

Fig. 6. Actual, theoretical and ETABS settlement of monitored column in 3rd floor

In the 3rd floor column settlement, the CEB-FIP model results extrapolated from the Etabs model analysis show identical results to the actual ones in the first casted 20 floors. After that, the results start to deviate continuously underestimating the settlement until the end of construction where the Etabs settlement becomes around 57% of the actual one. Unlike the ACI model chart, the CEB-FIP chart shows inconsistent relation with the actual chart where parallelism is maintained between the ACI and actual charts but not between CEB-FIP and actual charts. However, overestimation by the ACI model is maintained during all construction stages.

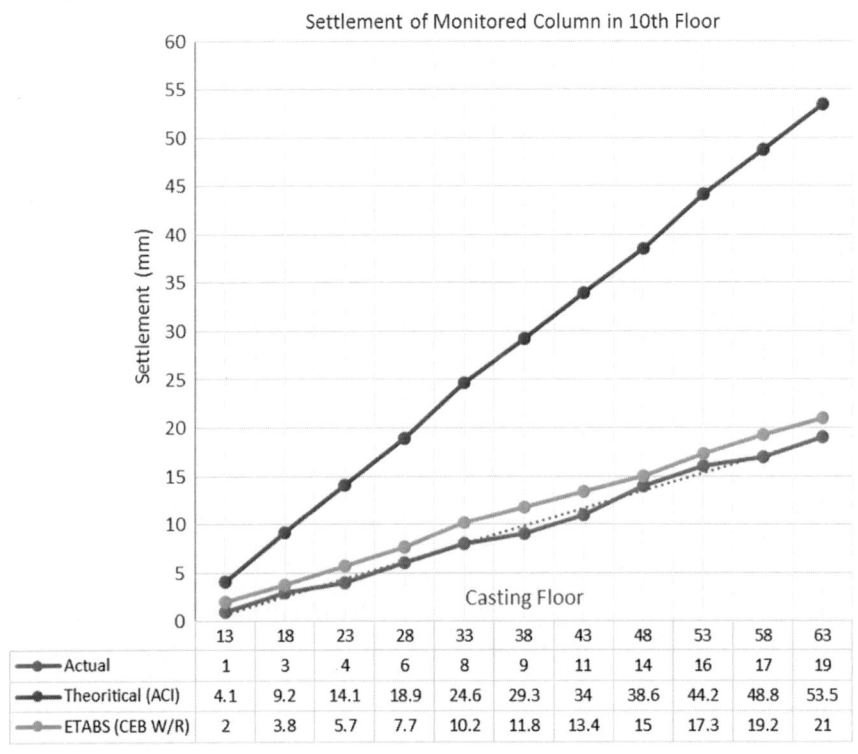

Fig. 7. Actual, theoretical and ETABS settlement of monitored column in 10th floor

By studying the 10th floor column settlement, it can be noticed that both models (CEB and ACI) overestimate the settlement in all construction stages. On the contrary to 3rd floor, the ACI chart gradually varies from the actual chart until the end of construction where the theoretical settlement reaches 280% of the actual one. In the contrast, the Etabs chart now is, unlike the 3rd floor, parallel to the actual chart with average overestimation of 120%. The continuous linear chart of the Etabs model (CEB-FIP method) represents the consistency with the actual results despite the overestimation of the total settlement.

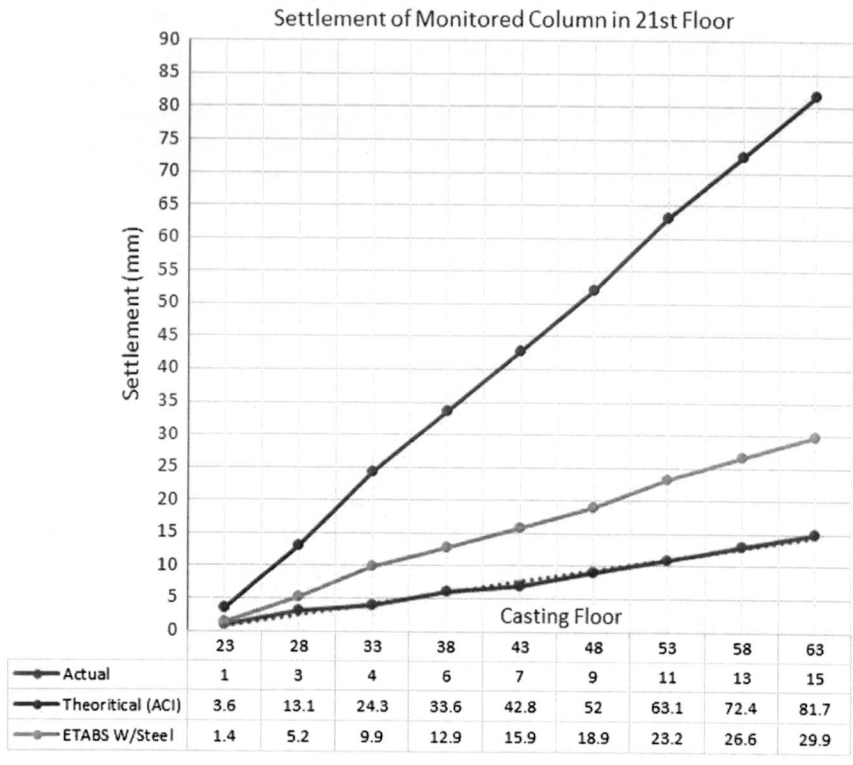

Fig. 8. Actual, theoretical and ETABS settlement of monitored column in 21st floor

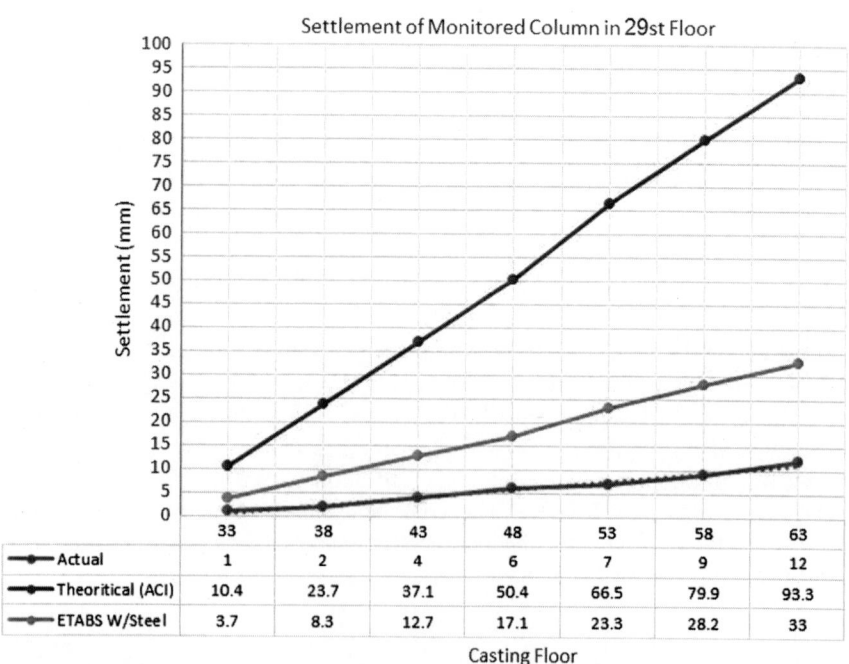

Fig. 9. Actual, Theoretical and ETABS settlement of monitored column in 29th floor

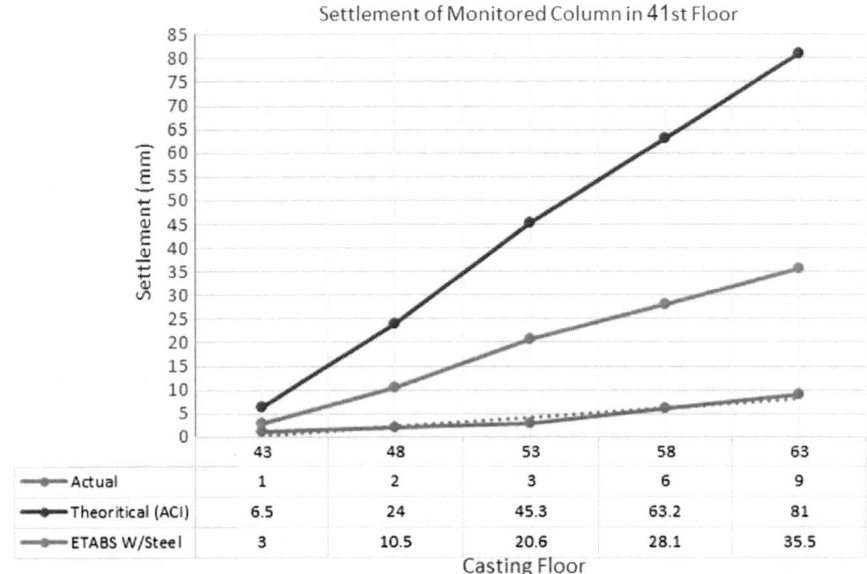

Fig. 10. Actual, theoretical and ETABS settlement of monitored column in 41st floor

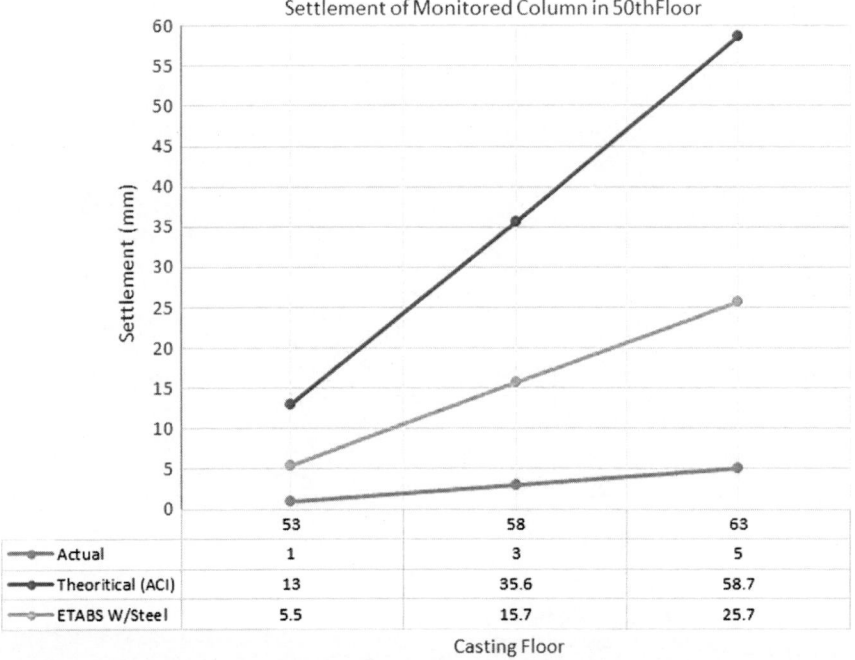

Fig. 11. Actual, theoretical and ETABS settlement of monitored column in 50th floor

By examining the charts for the other floors (21st, 29th, 41st and 59th) it can be noticed that both charts (ACI and CEB-FIP) start to deviate significantly from the actual chart reaching 1174 and 546% for ACI and CEB-FIP charts respectively for the total settlement of the 50th floor column after casting the 63rd floor. Another notice from the previous charts is that both models charts start close to the actual charts, that is with the first additional floors and then the different inclined slope represents the significant difference between the actual measurements and the predicted values.

4.4   Final Settlement of Monitored Column

The final settlement of the monitored column after completion of construction work (after casting floor 63) are derived from the previous charts and plotted on charts for theoretical, actual and Etabs results (Fig. 12).

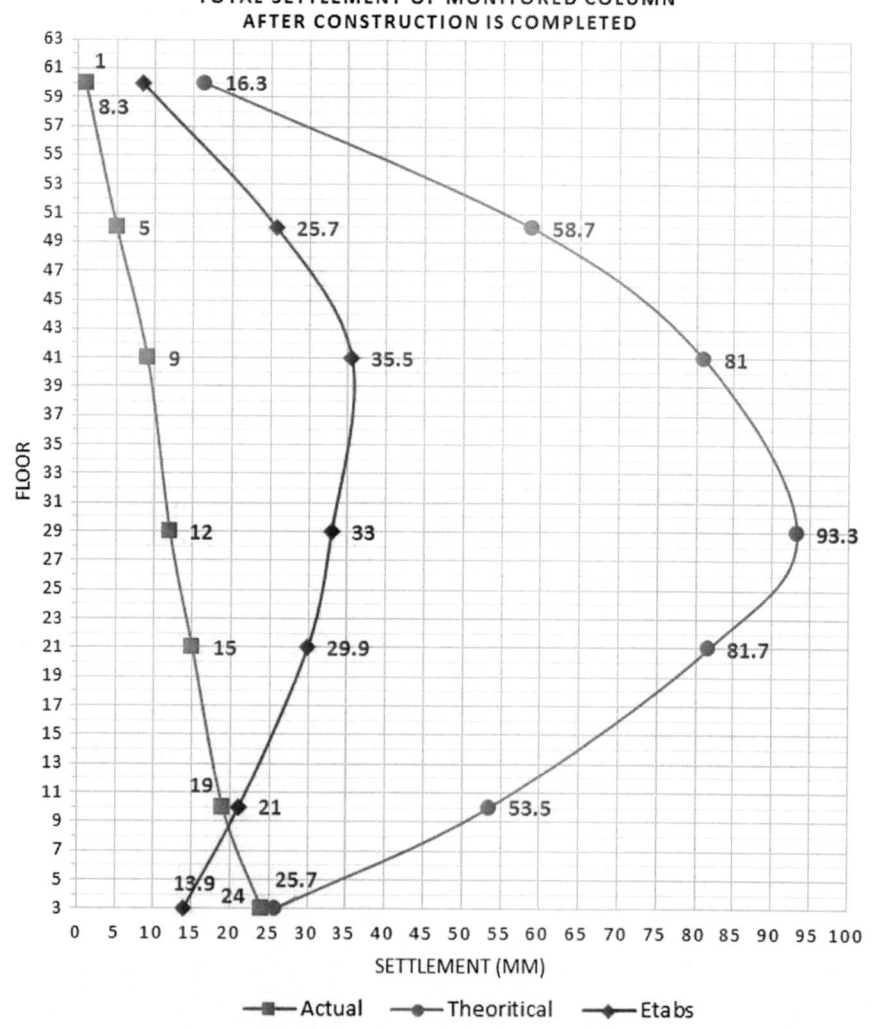

Fig. 12. Total settlement of monitored column after completion of construction

In the total settlement actual chart, the settlement has almost a linear relation with floors decreasing continuously from 24 mm in the 3rd floor to only 1 mm in the 60th floor. Both theoretical and Etabs charts show a totally different relation between floors and total settlement where the settlement increases when going up in floors until it reaches a maximum value near the building mid height and then decreases until the last floor. Where the actual and theoretical (ACI) settlement of 3rd floor are too close (24 and 25.7 mm), the other floors settlement differ significantly except for one intersection which reveal identical settlement between the two intersecting charts (actual and Etabs in the 9th floor as 20 mm). In general, the theoretical settlements are the highest, the actual settlements are the least where Etabs settlements are between those values.

5   Conclusions

Prediction of column shortening is a complex task associated with many uncertain factors that may not be available during the design stage. Also, prediction of time-dependent effects is an important part of column shortening estimation and is highly affected by the chosen method of prediction. The ACI 209R-92 method for predicting creep and shrinkage effects was evaluated by comparing its results to actual site readings of the case study presented in this research. That comparison led to the following conclusions:

- Theoretical elastic settlement of the monitored column in the 3rd floor was in the range of 60–65% of the total theoretical settlement.
- Theoretical total creep settlement keeps increasing with construction progress but with less effect for loads applied at later stages than those applied earlier.
- The effect of shrinkage on the total theoretical settlement of the 3rd floor column almost disappeared after the 33rd floor (about 90 days column age), only 0.4 mm additional settlement is caused by shrinkage for the remaining 90 days of the construction time.
- The proposed approach of this research based on developed Excel sheet program of the ACI 209R-92 method was able to predict the 3rd floor column settlement with reasonable accuracy compared to site measurements.
- The overestimation of settlement by the proposed theoretical approach increased steadily after the 3rd floor reaching 1174% of the actual settlement in the 50th floor.
- The average overestimation of the developed Excel sheet based on the ACI 209R-92 model was 630%.
- In any monitored floor, the proposed theoretical approach was able to predict the settlement in good accuracy only for the first added floors. The more added floors, the more deviation from the actual results takes place.
- In general, the proposed approach was unable to provide reliable prediction of the column settlement after the 3rd floor.
- Etabs analysis using CEB-FIP method was able to predict the settlement better than the theoretical approach of the ACI model.
- The average overestimation of Etabs analysis based on the CEB-FIB 90 model was 258%.

- Due to the many variables and uncertainties of the two proposed methods (theoretical and Etabs), it was difficult to compare the ACI and CEB-FIP models on which is better in predicting the time-dependent effects.
- The developed simple excel sheet program associated with the proposed method of this research is found suitable for predicting shortening of individual columns and the total settlement of low-rise buildings up to ten stories.
- By representing the compensation technique, staged construction analysis (sequential loading) done by Etabs provided a realistic scenario on how the building was constructed, and thus the obtained results had more reliability than those of normal analysis.
- A reasonable accuracy of predicting settlement of high-rise building can be achieved with a three-dimensional computer analysis where the restraining effects of all the structural members and the steel–concrete interaction are considered.
- The accuracy of column shortening and settlement prediction can be greatly enhanced by firm knowledge and information about the materials, environmental conditions and construction method that are involved in the building construction.

Design Recommendations

The design stage should extend during the construction of the building to ensure that design results are well met. The following steps are general outlines on how to overcome the uncertainties in this type of buildings:

- After the strength design and member sizing are finalized, the column shortening verification should initiate.
- A prediction method of creep and shrinkage should be chosen to account for time-dependent effects.
- A three dimensional software model to be developed and all the parameters and concrete properties required by the chosen method to be inputted into the model.
- Based on shortening results, a compensation method to be specified to be adopted in the construction.
- During the construction, tests are necessary on concrete specimens taken from the concrete patch used in the construction to determine all relevant concrete properties, such as elasticity modulus, concrete specified strength and creep and shrinkage strain.
- The design model inputs are then to be modified to account for the results of those tests. A modification to the compensation method to be done if needed.
- A monitoring survey system for settlement is a very effective and helpful way to assess the design results.

References

1. ACI Committee 209 (1992) Prediction of creep, shrinkage, and temperature effects in concrete structures (ACI 209R-92). American Concrete Institute, Farmington Hills
2. Microsoft Corporation (2013) Excel (computer program). https://www.office.com/

3. Computers and Structures (2015) Etabs 2015 ultimate, version 15.0.0 (computer program). http://www.csiamerica.com/
4. Comite Euro-International Du Beton (1993) CEB-FIB model code. Thomas Telford, London
5. Glanville W, Thomas F (1933) Creep of concrete under load. Struct Eng 11(2):54–68
6. Fintel M, Ghosh S, Iyengar H (1987) Column shortening in tall structures-prediction and compensation. Portland Cement Association, Skokie
7. Swamy R, Arumugasaamy P (1978) Structural behaviour of reinforced concrete columns in service. Struct Eng 56(11):319–329
8. Fintel M, Khan F (1969) Effects of column creep and shrinkage in tall structures-prediction of inelastic column shortening. ACI J Proc 66(12):957–967
9. Ghosh S (1996) Estimation and accommodation of column length changes in tall buildings. In: Rangan B, Warner R (eds) Large concrete buildings. Longman, Harlow
10. Jayasinghe M, Jayasena W (2005) Effect of relative humidity on absolute and differential shortening of columns and walls in multistory reinforced concrete buildings. Pract Period Struct Des Constr 10(2):88–97
11. Hamed E, Lai C (2016) Geometrically and materially nonlinear creep behaviour of reinforced concrete columns. Structures 5:1–12

A Review on the Effect of Environmental Conditions on Concrete Evaporation and Bleeding

İlker Bekir Topçu[1]([⊠]) and Burak Işıkdağ[2]

[1] Faculty of Engineering and Architecture, Department of Civil Engineering, Eskişehir Osmangazi University, Eskişehir, Turkey
ilkerbt@ogu.edu.tr
[2] Porsuk Vocational School, Department of Construction, Anadolu University, Eskişehir, Turkey
bisikdag@anadolu.edu.tr

Abstract. The formation of plastic shrinkage crack increases on the fresh concrete placed in hot and windy conditions. The cracks affect the durability of the concrete in a negative way. This type of cracking is mostly seen on slabs, pavements, beams and other flat concrete surfaces. Many factors influence evaporation, which is the main source of cracks. Evaporation is the function of climatic variables like relative humidity, temperature and speed of wind. In this study, experimental results of plastic shrinkage obtained from the literature are examined and compared to predict the effect of environmental conditions on evaporation and bleeding of concrete.

Keywords: Concrete · Evaporation · Bleeding · Humidity · Temperature

1 Introduction

Main factors that accelerate the evaporation on the surface of fresh concrete are climatic ones; such as, low relative humidity, high temperature, high wind speed or sunbeams. Evaporation occurs when thermal energy like sunbeams affects liquid or pressure on the surface of liquids is less than the pressure inside the liquid. Evaporation takes place as active water molecules move away from liquid. Evaporation is accelerated if the wind that provides ongoing motion of water molecules is available. On the other hand, if surface water temperature decreases, evaporation slows down. It is necessary to know what the degree of water temperature on the surface is in order to set the difference of pressure between water surface and air. Temperature of concrete is taken as the temperature of bleeded water [1]. The formula that Dalton drew the main features of evaporation in 1802 is shown with the Eq. (1).

$$E = (e_o - e_a)f(u) \tag{1}$$

In this equation, E, $(e_o - e_a)$, and f(u) are respectively defined with evaporation speed, the difference of pressure, and function of wind (Dalton, 1802) [2]. In 1950–1952,

© Springer International Publishing AG, part of Springer Nature 2018
S. Fırat et al. (eds.), *Proceedings of 3rd International Sustainable Buildings Symposium (ISBS 2017)*, Lecture Notes in Civil Engineering 7,
https://doi.org/10.1007/978-3-319-64349-6_33

broad researches were carried out around Hefner Lake. As a result of these researches, the Eq. (2) was introduced by Kohler.

$$E = (e_o - e_a)^{0.88}(0.37 + 0.0041u) \qquad (2)$$

E, evaporation speed; u, wind speed; and $(e_o - e_a)$, the difference of pressure is respectively defined as in/day, miles/day, and in/Hg (Kohler, 1955) [3]. Menzel obtained Eq. (3) by using Kohler formula.

$$E = 0.44\,(e_o - e_a)(0.253 + 0.096v) \qquad (3)$$

In this equation, E represents weight of water that evaporates from 1 ft^2 surface in 1 h as $(lb/ft^2/h)$; e_o indicates vapor pressure on surface that evaporation occurs as psi; e_a represents vapor pressure of the weather as psi; and v represents wind speed that is measured 50 cm above from the evaporation surface as mile/h. Number (3) equation is presented in an article that was published by Menzel in Portland Cement Association (PCA) in 1954 (Menzel, 1954) [4]. This is the evaporation equation that is known as Menzel's equation. It is necessary to know vapor pressures in order to detect evaporation speed from these formulas. In 1960, Bloem points out temperature that is measured in the concrete and relative humidity of the air as variables, so he finds a solution for this problem. This solution takes place in an article for the name of National Ready Mixed Concrete Association (NRMCA) in a monograph form (Bloem, 1960) [5].

Uno creates formula (4) by using variables of concrete and air temperature without counting vapour pressures by putting Menzel's equation into consideration.

$$E = 5\left([T_c + 18]^{2.5} - r.\,[T_a + 18]^{2.5}\right)(v + 4)10^{-6} \qquad (4)$$

E, evaporation speed, with $kg/m^2/h$; T_c, concrete temperature, with °C; T_a, weather temperature, with °C; r, rate of relative humidity, with %; and v, wind speed, with km/h are represented in this equation [1, 6]. Evaporation speeds calculated using these formulas give different values from experiment results.

2   Experimental Results

As it is understood in the formulas, environmental factors that play a role in evaporation are air temperature, relative humidity, wind and water surface (concrete) temperature. Evaporation is the phenomenon that liquid becomes vapor or gas. In other words, evaporation is the penetration of thermal energy to the inside of the water, and the situation that pressure on the surface of liquid is less than the pressure that the water has. As a result of this, active water molecules move away from liquid by evaporating. If the wind that changes its place continuously and water molecules is available, evaporation gains momentum. The loss of energy on the surface of evaporated water means decrease in the temperature of water on the surface. In these conditions, evaporation slows down. The measurement of air temperature variable is certain and easy. The only condition is to be measured away from direct sunbeams and to reduce direct solar radiation. Relative humidity is the proportion of actual humidity that is available in the unit volume of the air (kg/m^3) to the amount of saturated air humidity, and it is assigned by percentage. When relative humidity accesses to 100 out of 100, evaporation ceases if other factors such as wind do not change saturated air with unsaturated air. The temperature of bleeded water (or the temperature of the surface) is necessary to be measured in order to set the difference of vapor pressure between water surface and air.

It is considered that the temperature of the bleed water is same with the temperature of the concrete. Therefore, what is obtained at the end is the temperature of concrete [1–3]. While Berhane [4] reaches to the maximum evaporation speed in a hot and dry environment half an hour later from its casting, it reaches to this speed in a hot and humid environment after 4–7 h later. Although values that are measured at the beginning of experiment are the same with the results of Menzel equation, values that are obtained in the latter phase of the equation are less than values got in the equation. Hasanain et al. [5] conduct some experimental studies on evaporation of concrete specimens that are produced for a building in clear weather, under the shade and on the conditions that concrete surface is closed with polyethylene plate. The graphic method shows almost the equal results in evaporation speed that is measured in the specimens that are kept in the clear weather in the beginning hours with experimental results. However, it does not indicate the right predications in evaporation speed 3–5 h latter after its casting. Özkul and Öztekin [7] compare the evaporation speeds that they have obtained from the experiential results at the end of their study on plastic shrinkage with the values that are drawn according to Menzel's correlation in graphic method. They also stated that graphic method gives lower values.

Saman et al. [8] state that concrete mixtures which have low resistance and contain high level of water show the maximum evaporation speed, but loss of water and evaporation speed is higher in the concretes that have high resistance because of low water content.

Furthermore, that the amount and speed of evaporation increase as the rate of mineral admixture increases in the specimens of concrete that have mineral additives is stated at the end of another experiment. It is also pointed out that evaporation speed in concrete specimens kept under conditions which 95% relative humidity is available is far less than in concrete specimens kept under the conditions which have 25% relative humidity [9].

The aggregates in fresh concrete decompose in cement paste and particles inside the cement paste are dissolved in the water. This distribution is up to a certain extent depends on attractive and repulsion force between these particulars. This dissolved situation occurs while concrete is being mixed. The particulars are under the effect of a downward force. This gravitational force causes the particulars to precipitate by decreasing the distance between the particulars in verticular axis. Because of precipitation, volume of the hardened concrete is less than the volume that the concrete gains right after being put the concrete to its place. Nonetheless, this change in volume cannot be restricted only to gravitational force. There is also the effect of internal forces such as change in temperature and hydrostatics tensile excessive that affects all of the volume. Aggregate and concrete particles exert pressure to the mixture of water.

Throughout this process, concrete particulars hold too much water in the concrete, but they throw out the extra water by flocculating. In the course of observation, bleeded water seems as if it rose on the surface or boil over occurred on the surface of concrete. Yet, any kind of rising in the level does not take place actually [10]. Powers claims that bleeding speed is low not only because it is dependent on the type of concrete, but also because distribution of aggregate is random [11]. Steinour makes some experiments on various concrete with the same water levels and high water level. According to his observations, if the speed of chemical reaction is high at the beginning, the amount and speed of the perspiration is low in various materials that contain same amount of concrete [12]. Öztekin and Özkul [7] record, in their studies, increased concrete dosage lessens the amount of bleeding by increasing water bond of concrete. The effects of superplasticizers on the properties of new concrete are examined in a study conducted by Akman. At the end of the experiments, it is stated that bleeding increases with the participation of the admixtures. This increase occurs by the help of retarder admixtures in a more obvious way [13]. Özkul et al. [14] point out in their studies that bleeding is correlated with the plug timeframe, and the longer the plug timeframe is, the more the amount of the bleeding is. Tamimi [15] stated that the mixture of concrete that is prepared in the respect of two-tier phase method shows much less the amount of bleeding the mixture of concrete prepared in respect the of common method.

3 Evaluation of Experimental Results

In the results of evaporation experiments conducted in four different environmental conditions, it is extracted that low rate of relative humidity and the amount of evaporation in the specimens that are kept under the windy conditions. Maximum amount and speed of evaporation observed in the specimens that are kept under the 50% relative humidity conditions are 28 °C and 15 km/h wind speed. On the other hand, minimum amount and speed of evaporation observed in the specimens that are kept under the 80% relative humidity conditions are 28 °C. The amount of evaporation that is obtained at the end of 5 h triples because the speed of the wind rises to 15 km/h from 0 under the permanent ambient temperature, and doubles as the rate of relative humidity decreases from 80 to 50%. When the amounts of evaporation are examined, the specimens that have the maximum amount of evaporation at the end of the 5 h are the ones that have C 25/30 quality. It is also observed under four different environment conditions that the concrete specimens in C 25/30 quality that are produced by using CEM II/B-M 32.5 and CEM I 42.5 concrete have much more the amount of evaporation than the specimens of concrete in C 35/45 quality. Why less amount of evaporation occurs in the specimens in C 35/45 quality can be related to the idea that bleeding water on their surfaces is less. In the experiments that are conducted with two different types of concrete, it is noticed that the type of concrete does not have a direct effect on the amount and speed of evaporation.

Evaporation speeds of concretes of C 25/30 and C 35/45 under 28 °C and 50% humidity are given in Figs. 1 and 2. Experimental evaporation speeds are bigger than the ones that are obtained in Menzel and Uno's correlations. At the end of the experiment, maximum evaporation speeds that are measured in the C 25/30 and C

35/45 concrete specimens kept on the level of 28 °C and 50% relative humidity; and produced with both types of the concrete according to Menzel and Uno's correlations are obtained after 60 min. While the experimental values for C 25/30 concrete change between change between 0.350 and 0.125, the ones for C 35/45 concrete change between 0.325 and 0.125.

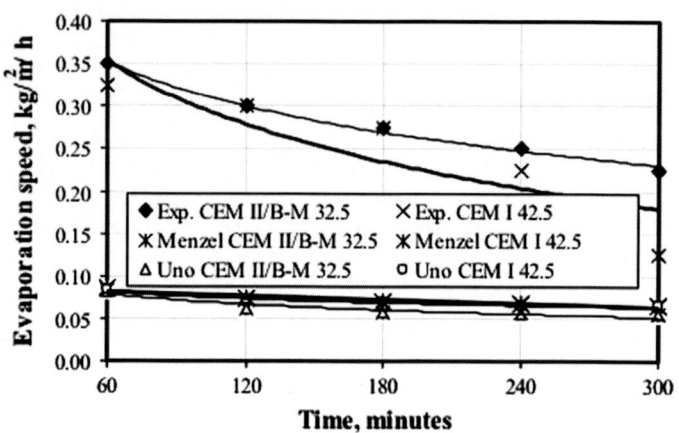

Fig. 1. Evaporation speeds of concretes of C 25/30 under 28 °C and 50% humidity

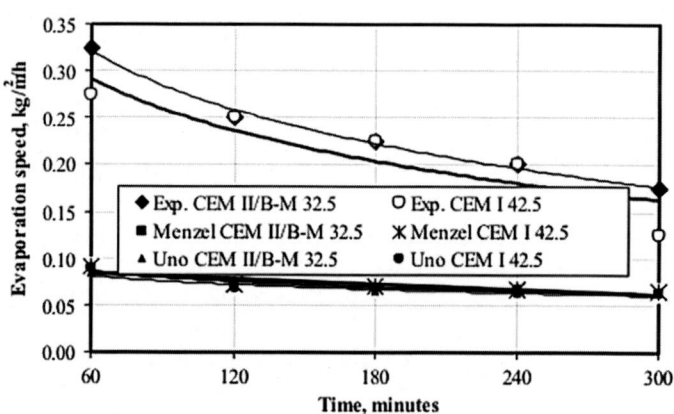

Fig. 2. Evaporation speeds of concretes of C 35/45 under 28 °C and 50% humidity

In Figs. 3 and 4, evaporation speed of C 25/30 and C 35/45 concrete specimens under 28 °C, 15 km/h and 50% humidity are given. Experimental evaporation speeds that are bigger than the evaporation speeds got in Menzel and Uno correlations are obtained. The evaporation speeds for C 25/30 concrete specimens that are kept under 28 °C temperature, 15 km/h speed of wind and 50% the rate of relative humidity change between the values 0.900 and 0.525. In addition to this, the values for C 35/45 concrete specimens change between 0.875 and 0.475. It gives very close results with evaporation speeds obtained in Menzel and Uno's correlations.

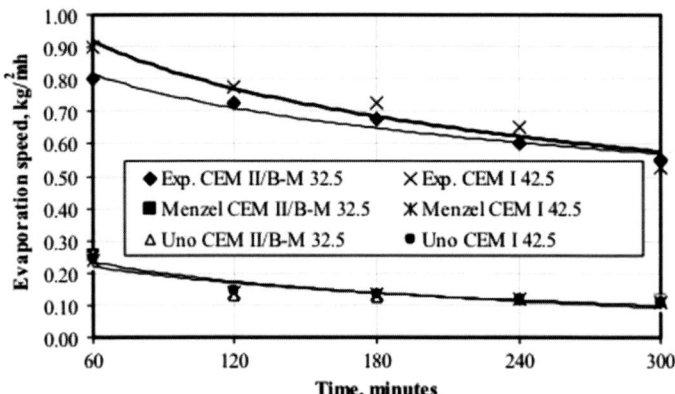

Fig. 3. Evaporation speeds of concretes of C 25/30 under 28 °C, 15 km/h and 50% humidity

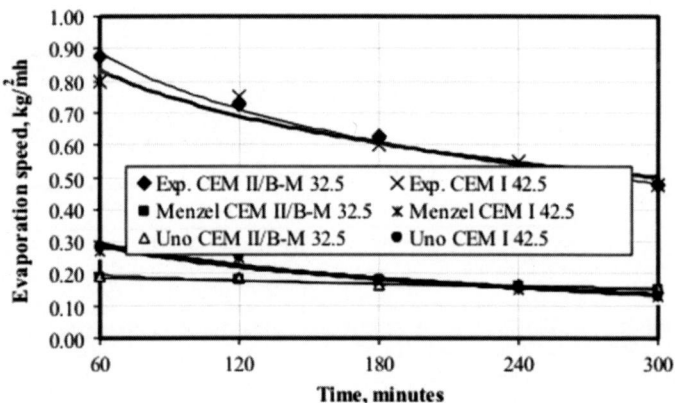

Fig. 4. Evaporation speeds of C 35/45 under 28 °C, 15 km/h and 50% humidity

In Figs. 5 and 6, evaporation speed of C 25/30 and C 35/45 concrete specimens under 28 °C and 80% humidity are given. Experimental evaporation speeds that are bigger than the evaporation speeds got in Menzel and Uno's correlations are obtained. The evaporation speeds for C 25/30 concrete specimens that are kept under 28 °C temperature and 80% the rate of relative humidity changes between the values 0.200 and 0.125. In addition to this, the values for C 35/45 concrete specimens change between 0.175 and 0.075. It gives very close results with evaporation speeds obtained in Menzel and Uno's correlations.

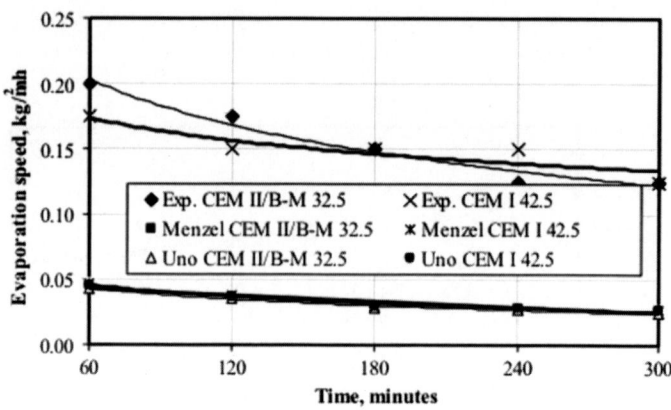

Fig. 5. Evaporation speeds of C 25/30 under 28 °C and 80% humidity

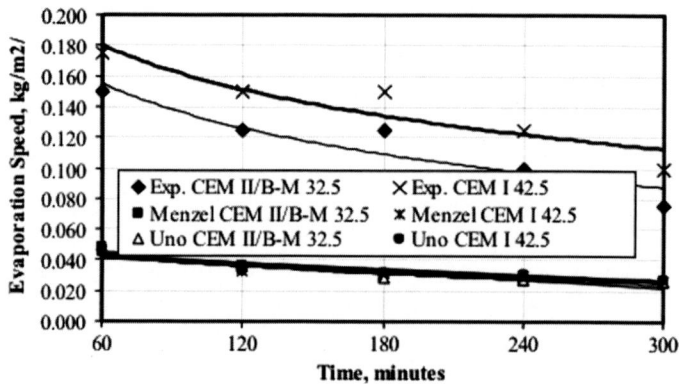

Fig. 6. Evaporation speeds of C 35/45 under 28 °C and 80% humidity

In Figs. 7 and 8, evaporation speed of C 25/30 and C 35/45 concrete specimens under 28 °C, 15 km/h wind, and 80% humidity are given. Experimental evaporation speeds that are bigger than the evaporation speeds got in Menzel and Uno's equations are obtained. The evaporation speeds for C 25/30 concrete specimens that are kept under 28 °C temperature, 15 km/h wind speed, and 80% the rate of relative humidity changes between the values 0.475 and 0.275. In addition to this, the values for C 35/45 concrete specimens change between 0.175 and 0.075. It gives very close results with evaporation speeds obtained in Menzel and Uno's correlations.

Relations that can change according to environmental conditions are observed in different cement concrete specimens that are in the same concrete category. If evaporation experiment results of concrete specimens in C 25/30 quality under the conditions that same rate of temperature and relative humidity exist are analyzed, while C 25/30 concrete made of CEM II/B-M 32.5 cement indicates bigger amount of evaporation under the conditions that there is no wind C 25/30 concrete made of CEM I 42.5 cement shows bigger amount of

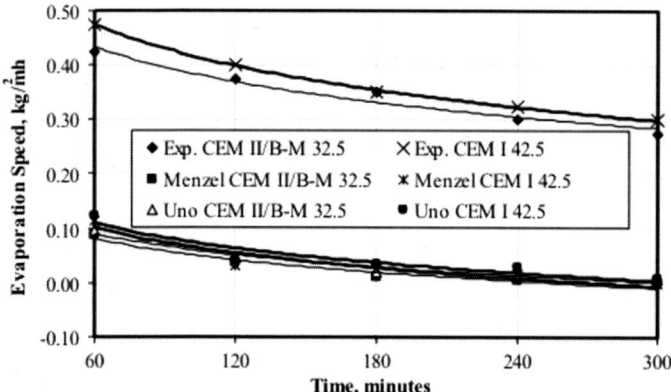

Fig. 7. Evaporation speeds of C 25/30 under 28 °C, 15 km/h wind, and 80% humidity

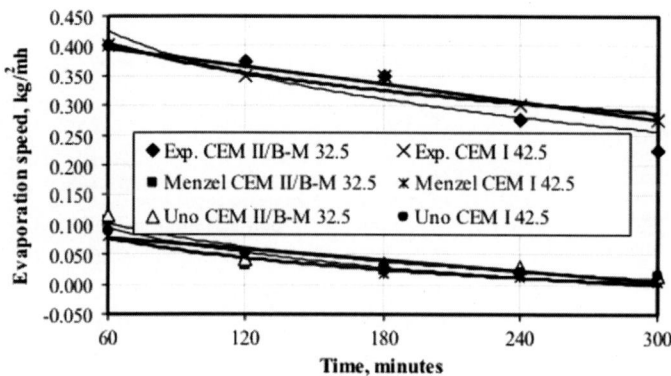

Fig. 8. Evaporation speeds of C 35/45 under 28 °C, 15 km/h wind, and 80% humidity

evaporation when 15 km/h wind speed is available. On the other hand, if evaporation experiment results of concrete specimens in C 35/45 quality are analyzed under the same wind speed and temperature conditions, bigger amount of evaporation is observed in C 35/45 concrete made of CEM II/B-M 32.5 cement under the condition that 50% relative humidity is available; however, under the conditions that high relative exists, much more amount of evaporation speed is observed in C 35/45 concrete made of CEM I 42.5 cement at the end of 5 h. The amount and speed of evaporation that are measured at the end of the evaporation experiments state that evaporation shows much bigger values in the existence of low rate of relative humidity and wind speed. Menzel and Uno's equations give lower values than evaporation speeds calculated by measuring in the experiments in four different environmental conditions including stable ambient temperature, variable wind speed, and the rate of relative humidity. ACI nomograph that is prepared in the respect of Menzel's equation used in the prediction of evaporation speed on the surface of new concrete does not provide the realistic values in these conditions that evaporation experiments are conducted. Furthermore, coefficients of correlation are supplied in the Table 1.

Table 1. Variation of evaporation speeds with and time

Cement type	Environment conditions	Concrete type	Equations (y = a.lnx + b)		a	b	R^2
CEM II/B-M 32.5	28 °C temperature 50% humidity	C 25/30	Experimental		−0.075	0.352	0.992
		C 35/45			−0.090	0.321	0.991
		C 25/30	Menzel		−0.017	0.080	0.871
		C 35/45			−0.013	0.082	0.883
		C 25/30	Uno		−0.017	0.079	0.873
		C 35/45			−0.013	0.081	0.884
	28 °C temperature 50% humidity 15 km/sa Wind	C 25/30	Experimental		−0.153	0.816	0.959
		C 35/45			−0.252	0.886	0.993
		C 25/30	Menzel		−0.083	0.229	0.791
		C 35/45			−0.023	0.191	0.900
		C 25/30	Uno		−0.087	0.234	0.792
		C 35/45			−0.025	0.196	0.900
	28 °C temperature 80% humidity	C 25/30	Experimental		−0.051	0.203	0.968
		C 35/45			−0.042	0.155	0.862
		C 25/30	Menzel		−0.013	0.045	0.986
		C 35/45			−0.015	0.047	0.970
		C 25/30	Uno		−0.012	0.044	0.986
		C 35/45			−0.013	0.046	0.970
	28 °C temperature 80% humidity 15 km/sa wind	C 25/30	Experimental		−0.092	0.433	0.961
		C 35/45			−0.104	0.425	0.824
		C 25/30	Menzel		−0.056	0.081	0.968
		C 35/45			−0.060	0.095	0.887
		C 25/30	Uno		−0.055	0.090	0.970
		C 35/45			−0.059	0.103	0.889
CEM I 42.5	28 °C temperature 50% humidity	C 25/30	Experimental		−0.108	0.353	0.747
		C 35/45			−0.081	0.292	0.794
		C 25/30	Menzel		−0.012	0.084	0.948
		C 35/45			−0.015	0.088	0.891
		C 25/30	Uno		−0.011	0.082	0.949
		C 35/45			−0.015	0.086	0.893
	28 °C temperature 50% humidity 15 km/sa wind	C 25/30	Experimental		−0.213	0.919	0.933
		C 35/45			−0.207	0.833	0.927
		C 25/30	Menzel		−0.076	0.221	0.885
		C 35/45			−0.094	0.286	0.931
		C 25/30	Uno		−0.080	0.227	0.887
		C 35/45			−0.098	0.294	0.930
	28 °C temperature 80% humidity	C 25/30	Experimental		−0.025	0.174	0.802
		C 35/45			−0.042	0.180	0.862
		C 25/30	Menzel		−0.013	0.046	0.975

(continued)

Table 1. (*continued*)

Cement type	Environment conditions	Concrete type	Equations (y = a.lnx + b)			
				a	b	R^2
		C 35/45		−0.011	0.044	0.924
		C 25/30	Uno	−0.012	0.045	0.975
		C 35/45		−0.010	0.043	0.925
	28 °C temperature 80% humidity 15 km/sa wind	C 25/30	Experimental	−0.109	0.474	0.999
		C 35/45		−0.073	0.405	0.911
		C 25/30	Menzel	−0.067	0.101	0.895
		C 35/45		−0.048	0.077	0.981
		C 25/30	Uno	−0.066	0.110	0.896
		C 35/45		−0.047	0.086	0.982

The results of evaporation experiments that are carried out in four different environmental conditions are analyzed in terms of the relation between evaporation speed and time that are obtained in Menzel and Uno's equations such as y=a.lnx+b. R^2 values that are obtained at the end of the regression analysis are in general terms found reliable.

Bleeding experiments are carried out in three different groups composed of type of concrete, the ratio of water–concrete, and dosage. In the results of the bleeding experiment that is made by choosing four different type of concrete, maximum amount of bleeding emerging at the end of the first hour is seen in the specimen of C 16/20 concrete made of CEM I 42.5 cement. However, minimum amount of bleeding is observed in the specimen of C 35/45 concrete made of CEM II/B-M 32.5 cement. Maximum and minimum amount of bleeding in the concrete specimens that are produced with both types of cement are respectively seen in the specimen of C 16/20 and C 35/45 concretes. The amount of bleeding increases as water and cement ratio increases at the end of the bleeding experiments that are carried out by changing water and cement ratio. Yet, an hour's amount of bleeding stays at the same values at the end of the bleeding experiment whose cement dosage is analyzed. The results of bleeding experiment are shown in Figs. 9, 10 and 11.

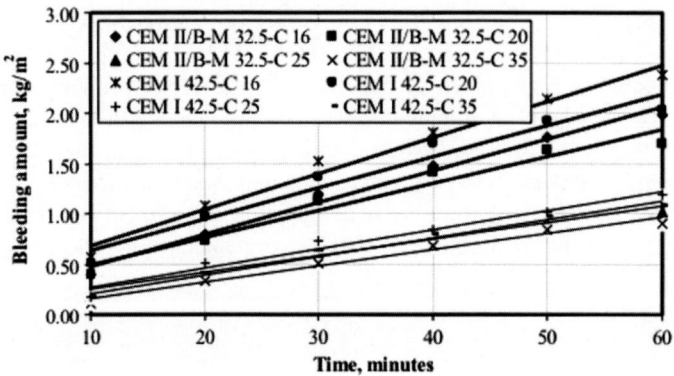

Fig. 9. The variation of amount of bleeding with time in different types of concrete

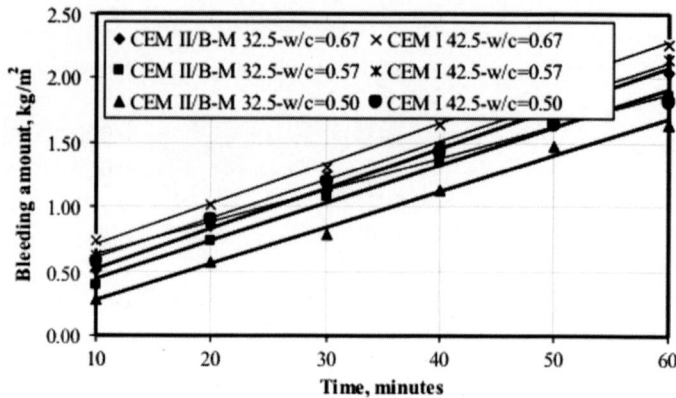

Fig. 10. The variation of amount of bleeding with time in different water/cement ratios

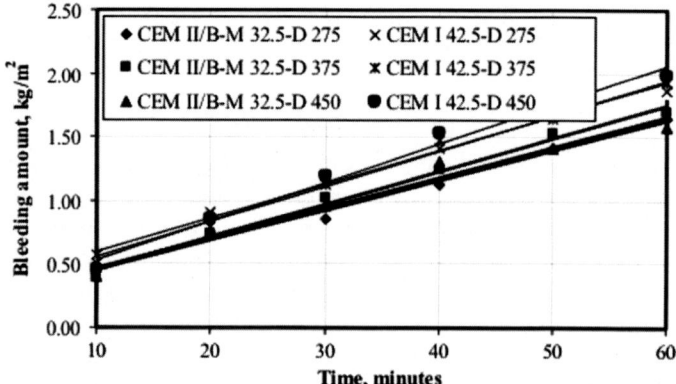

Fig. 11. The variation of amount of bleeding with time in different cement dosages

The direct proportion between the amount of perspiration and time in four different types of concrete composed of C 16/20, C 20/25, C 25/30 and C 35/45 is analyzed, and their correlation coefficients are shown in the Table 2. It is seen that correlation coefficients of all direct relations are high. Direct proportion between three different types of water and cement ratios including 0.67, 0.57 and 0.50 is analyzed, and their correlation coefficients are given in the Table 3. The direct proportion between the amount of bleeding and time in three different types of cement dosages as 275, 375 and 450 kg/m^3 is analyzed, and correlation coefficients are given in the Table 4. CEM I 42.5 cement indicate much more amount of bleeding than CEM II/B-M 32.5 cement in all conditions at the end of the results of bleeding experiment carried out in three groups by applying different concrete types, water/cement ratio, and cement dosage.

Table 2. The equations amount of bleeding with time in different types of concrete

Concrete types	Cement type	Equations $(y = a.x + b)$		
		a	b	R^2
C 16/20	CEM II/B-M 32.5	0.317	0.154	0.988
	CEM I 42.5	0.359	0.328	0.981
C 20/25	CEM II/B-M 32.5	0.272	0.219	0.957
	CEM I 42.5	0.311	0.328	0.961
C 25/30	CEM II/B-M 32.5	0.163	0.098	0.954
	CEM I 42.5	0.192	0.072	0.969
C 35/45	CEM II/B-M 32.5	0.162	0.0001	0.978
	CEM I 42.5	0.184	0.015	0.991

Table 3. The equations amount of bleeding with time in different water/cement ratios

Water–cement ratio	Cement type	Equations $(y = a.x + b)$		
		a	b	R^2
0.67	CEM II/B-M 32.5	0.310	0.215	0.997
	CEM I 42.5	0.315	0.397	0.996
0.57	CEM II/B-M 32.5	0.296	0.144	0.995
	CEM I 42.5	0.302	0.309	0.997
0.50	CEM II/B-M 32.5	0.281	−0.004	0.994
	CEM I 42.5	0.247	0.388	0.987

Table 4. The equations amount of bleeding with time in different cement dosages

Cement dosage	Cement type	Equations $(y = a.x + b)$		
		a	b	R^2
275	CEM II/B-M 32.5	0.236	0.212	0.991
	CEM I 42.5	0.269	0.324	0.986
375	CEM II/B-M 32.5	0.260	0.193	0.991
	CEM I 42.5	0.277	0.287	0.993
450	CEM II/B-M 32.5	0.238	0.234	0.974
	CEM I 42.5	0.306	0.223	0.987

4   Discussion

In hot airs, the crack of plastic shrinkage sticks to the concrete pouring in arid climates. The crack occurs in the concrete that is set in an open place. The amount of water that rises by the help of bleeding initially in the flat elements, beams, foundations and on the surface of the concrete occurs in the other climatic conditions that much more amount of evaporation is available. Plastic shrinkage crack always sets an anxiety in the concrete technology. It is very important to know the possibility that to what extent the shrinkage crack can occur in terms of the problems that will be encountered in the future. High level of temperature and low relative humidity accelerate the plastic shrinkage of concrete [16]. According to ACI, it is suggested to take precautions when evaporation speed comes closer to the value 1.0 kg/m^2/h (0.2 lb/ft^2/h). While Australian Codes note the value that has the potential plastic shrinkage occurs as 1.0 kg/m^2/h and the value that necessary precautations are taken as 0.5 kg/m^2/h, Canadian Codes points 0.75 kg/m^2/h as the critical value. When the experiment results in the Table 1 are examined, a kind of danger is not visible according to ACI report.

5 Conclusions

At the end of the evaporation and bleeding experiments that are carried out on the concrete specimens prepared by using two different types of cement, high correlation coefficients are obtained. Therefore, it is observed that the amount of bleeding throughout 1 h and the amount of evaporation throughout 5 h on the new concrete provide nearly realistic results. As a result of this, the suggestions and the conclusions below are submitted.

In the bleeding experiments made by the help of different quality of concrete, water/cement ratio, and cement dosages, the concrete specimens prepared with CEM I 42.5 cement show much more amount of evaporation than the concrete specimens prepared with CEM II/B-M 32.5 cement at the end of 1 h. That the concrete specimens that are prepared with CEM I 42.5 cement show much more amount of evaporation than the concrete specimens prepared with CEM II/B-M 32.5 cement can be connected to smaller surface area of CEM I 42.5 cement. One hour's bleeding amount of new concrete decreases since the water/cement ratio decreases as being dependent on cement dosage in concrete combination. Bleeding decreases since growing in the cement dosage increases water bonding of concrete because of cement particles' water bonding features.

In evaporation experiments that are carried out in four different conditions of environment and in two different concrete quality as C 25/30 and C 35/45, more amount of evaporation is measured in the C 25/30 concrete specimens than in the concrete specimens that are prepared with CEM I 42.5 and CEM II/B-M 32.5 cement at the end of 5 h. This situation occurs because more amount of bleeding occurs in the specimens in C 25/30 concrete category compared to the specimens in C 35/45 concrete category. The effects of the environmental conditions to the amount and speed of evaporation are more important than the concrete combination. The amounts and speeds measured at the end of the evaporation experiments point out that evaporation shows bigger values in the existence of low relative humidity and wind speed. Menzel and Uno's equations that are used in the prediction of evaporation speed occurring on the surface of new concrete cannot provide realistic values in evaporation speed. Evaporation speeds found in an experimental way are bigger than the estimated evaporation speeds that are obtained from the equations. It is necessary to work on providing the validity of these equations by increasing the number of the experiment groups. If they are not valid, equations should be reviewed.

References

1. Uno PJ (1998) Plastic shrinkage and evaporation formulas. ACI Mater J 95:365–375
2. Topçu İB (2005) Raising of water and evaporation from freshly placed concrete surface. In: ICCE 12, twelve international conferences on composites or nano engineering, Tenerife
3. Topçu İB, Elgün VB (2004) Influence of concrete properties on bleeding and evaporation. Cem Concr Res 34:275–281
4. Berhane Z (1984) Evaporation of water from fresh mortar and concrete at different environmental conditions. ACI J Proc 81:560–565
5. Hasanain GS, Khalaf TA, Mahmood K (1989) Water evaporation from freshly placed concrete surfaces in hot weather. Cem Concr Res 19:465–475
6. Elgün VB (2003) Influence of concrete properties on bleeding and evaporation. M.Sc. Thesis, Osmangazi University, Science Institute, Eskişehir
7. Öztekin E, Özkul MH (1994) The effect of concrete composition and environmental conditions on plastic shrinkage. In: Proceedings of the national concrete congress, İstanbul, vol 3, pp 163–74
8. Saman TA, Mirza WH, Wafa FF (1996) Plastic shrinkage cracking of normal and high strength concrete: a comparative study. ACI Mater J 93:36–40
9. Almusallam AA, Maslehuddin M, Abdulwaris M, Dakhil FH, Al-Amoudi OSB (1999) Plastic shrinkage cracking of blended cement concretes in hot environments. Mag Concr Res 51:241–246
10. Powers TC (1968) The properties of fresh concrete. Wiley, London, pp 533–603
11. Powers TC (1939) The bleeding of Portland cement paste, mortar and concrete. Portland Cem Assoc Bull 2:465–480
12. Steinour HH (1945) Further studies of the bleeding of Portland cement paste. Portland Cem Assoc Bull 4:1–87
13. Akman MS (1996) The troubles of superplasticizers on fresh concrete workability. In: 4th National concrete congress, İstanbul, pp 55–71
14. Özkul MH, Başkoca A, Artırma S (1997) Influence of prolonged agitation on water movement related properties of water reducer and retarder admixtured concretes. Cem Concr Res 27:721–732
15. Tamimi AK (1994) The effects of a new mixing technique on the properties of the cement paste-aggregate interface. Cem Concr Res 24:1299–1304
16. Almusallam AA (2001) Effect of environmental conditions on the properties of fresh and hardened concrete. Cem Concr Compos 23:353–361

Triplet Shear Tests on Masonry Units with and Without Seismic Textile Reinforcement

Berna Istegun and Erkan Celebi[✉]

Department of Civil Engineering, Sakarya University, Sakarya, Turkey
{bernaistegun, ecelebi}@sakarya.edu.tr

Abstract. This research primarily aims to investigate experimentally the mechanical behavior and the crack pattern of the perforated brick block masonry specimens under shear stresses before and after reinforcing procedure with seismic textile. The reinforcing system is a combination between an inorganic matrix made with natural hydraulic lime and white cement and a hybrid multiaxial seismic textile made from Alkali Resistant (AR) glass and polypropylene fibers. The effect of the single-sided application of the sand plaster on the non-strengthened wall specimens is also tested to compare with both reference and retrofitted specimens. The obtained results of the triplet shear tests are given in the manner of force-displacement curves comparatively for all specimens considered without/with textile reinforcement. The shear strength for each triplet specimen investigated is also obtained. The effect of the presence of seismic textile as well as the applied sand plaster on the mechanical behavior of the triplet specimens is emphasized. It is concluded that the shear resistance and the ductility highly influenced by using the seismic textile and relevant plaster.

Keywords: Triplet shear tests · Retrofitting · Seismic textile · Masonry wall
Insulation plaster

1 Introduction

Due to the lack of the ductility, low shear and tensile stress capacity of the masonry components, traditional brick masonry structures have been badly affected in the rural region by the recent strong earthquakes. The cause of most of the deaths and fatal injuries after severe earthquakes is the consequential heavily damage and collapse of the masonry buildings. The earthquake resistant, stability and the ductility capacity of the shear walls of the existing masonry buildings can be increased by employing efficient retrofitting methods. The external application of seismic composite textile to the surface of the walls is one of the most effective methods applied in strengthening masonry buildings. The idea of an earthquake hybrid multiaxial fabric consisting of AR glass and polypropylene fibers have been developed in the Karlsruhe Institute of Technology (KIT) laboratory in the results of a 10-year study [1, 2].

For better understanding the cracking behavior of the masonry units as well as the load-bearing walls of the retrofitted masonry walls under applied loads, many different

© Springer International Publishing AG, part of Springer Nature 2018
S. Fırat et al. (eds.), *Proceedings of 3rd International Sustainable Buildings Symposium (ISBS 2017)*, Lecture Notes in Civil Engineering 7,
https://doi.org/10.1007/978-3-319-64349-6_34

experimental setups and arrangements have been adopted by researchers. The experimental programs involve small test series as well as experiments on small scaled wall specimens and on real sized walls [3, 4].

Within the scope of small scale tests, different shear test methods have been developed to determine the strength parameters and to study of the failure behavior of masonry joints under shear stresses [5–7].

This study is concerned with revealing experimentally the mechanical behavior and the shear performance of the perforated brick block masonry/mortar interface under static vertical loading before and after strengthening with seismic composite fabric. The effect of the single-sided application of the traditional sand plaster on the non-strengthened wall specimens was also investigated by the triplet shear tests to compare with both reference and strengthened specimens.

The obtained test results were given in the manner of force-displacement relationships and shear strength parameters comparatively.

2   Experimental Setup

All experimental tests involved in this research was conducted at the structural mechanics laboratory of Sakarya University. The vertical load-controlled cylinder having loading speed of 1.5 mm/min. is applied perpendicular to the triplet specimens produced to obtain accurate information about the cracking behavior of the two mortar joints under shear stressing.

Triplet shear tests on perforated brick block (Fig. 1) with and without seismic retrofitting were performed. Steel base plate is placed on the middle brick of the triplet specimen for reducing the influence of the bending moment and transmitting the applied vertical load to the mortar joints. The outer bricks are held by two rigid L steel profiles at the edges of the triplet specimens. The results of the experimental test are recorded by an internal inductive displacement transducers integrated in the load cylinder with vertical force capacity of 50 kN.

The technical properties of the perforated brick block are given in Table 1.

Fig. 1. Brick

Table 1. Properties of brick

Dimension (w*l*h)	cm	19*29*13.5
Consumption	pcs/m^2	22–35
Gross dry unit vol. mass	kg/m^3	700
Net dry unit vol. mass	kg/m^3	1800
Pressure strength	N/mm^2	10
Thermal conductivity factor	W/mK	0.32
Fire resistance	–	A1
Tolerance category	–	T1

Each produced test set consists of a triplet of perforated brick blocks shown in Fig. 2 where two specimens were non-strengthening and on the other one was strengthened by using seismic textile and its special mortar. The reinforcing system is a combination between an inorganic matrix (plaster made with natural hydraulic lime and white cement) and a hybrid multiaxial seismic textile (AR glass and polypropylene fibers) shown in Fig. 3. Alkali resistant glass fibers containing a high percentage of zirconia (ZrO$_2$) provide high tensile strength and modulus for the hybrid fabric. The average values of tensile strength in vertical, horizontal and diagonal direction (60° to the vertical) are 2680 N/50 mm, 2100 N/50 mm and 1222 N/85 mm, respectively. The inorganic matrix has the following mechanical properties indicated on the technical datasheet of the product as given in Table 2.

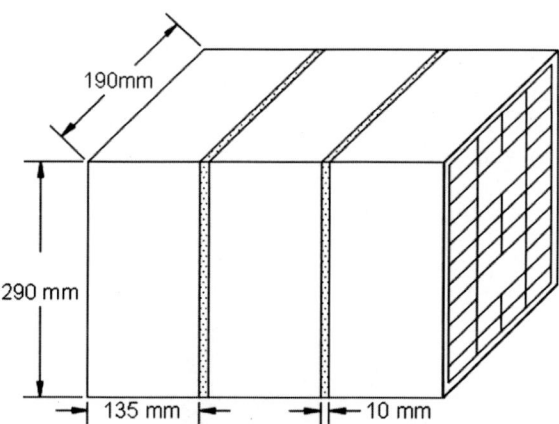

Fig. 2. Triplet specimen

The mortar applied with thickness of 2 cm for the non-retrofitted specimen has the volumetric cement-sand ratio of 1/4. The sand plaster and seismic textile are applied only on single side of the triplet specimens.

The goal of this experimental study is to obtain information on the shear response and the crack pattern of the perforated brick block masonry specimens under shear stresses before and after retrofitting application with seismic textile.

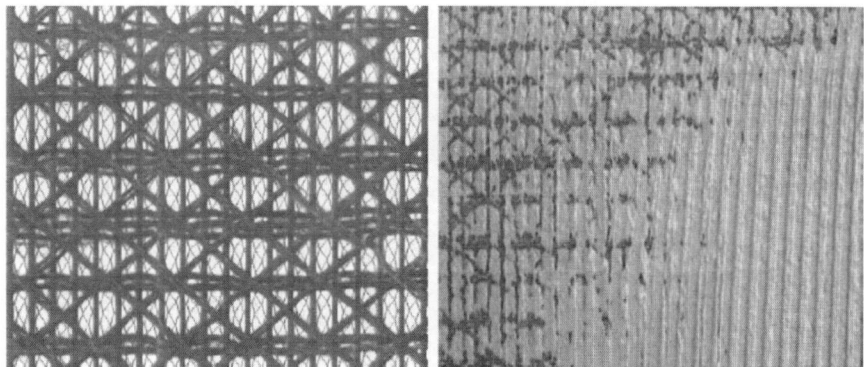

Fig. 3. The hybrid multiaxial seismic textile [1, 2]

Table 2. Properties of the inorganic matrix

Compressive strength (28 days)	14 N/mm^2
Flexural tensile strength (28 days)	4 N/mm^2
Elasticity modules	7500 N/mm^2
Adhesion to the substrate	0.5 N/mm^2

The scheme of the triplet shear test is illustrated in Fig. 4. The relevant parameters used in the triplet shear test are $t_{s1} = 30$ mm, $t_{s2} = 12$ mm and $t_{s3} = 6$ mm, respectively.

All the experimental tests are performed by the test device, which is depicted in Fig. 5.

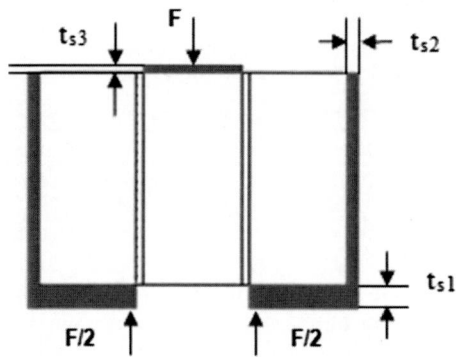

Fig. 4. Scheme of the triplet shear test

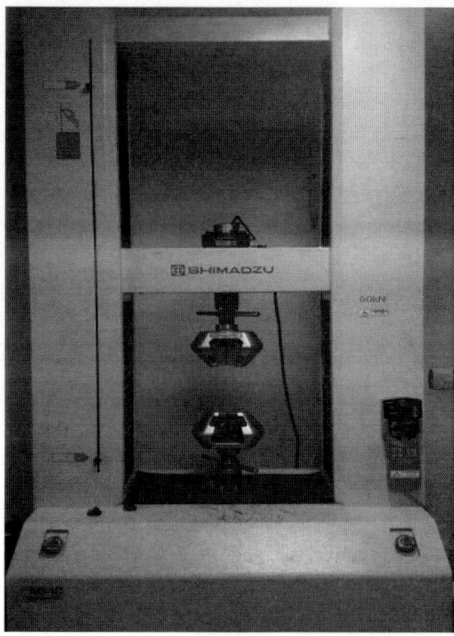

Fig. 5. Test device

3 Experimental Program and Test Results

The shear capacity of the non-retrofitted masonry specimen without applying any sand plaster defined as reference specimen was investigated in the first phase of this research. The Table 3 shows the names of the applied experimental tests.

Table 3. Names of the triplet test specimens

Name	Strengthening situation	Plaster type
D1R00	Reference	Without plaster
D1R01	No/unilateral	Sand plaster
D1G02	Yes/unilateral	Inorganic matrix

Triplet specimen without sand plastered built with masonry perforated brick is coded for test as D1R00. Both Initial and final state of failure the reference specimen under triplet shear test are given in Fig. 6.

The value of the maximum horizontal force of the non-retrofitted specimen without any plaster is achieved as 12,200 N under triplet shear experiment. The obtained force-displacement relationship is illustrated in Fig. 7.

The shear strength f_v for this specimen is calculated with respect to the norm EN 1052-3 [8] as follow:

Fig. 6. Initial state (*left*) and final state (*right*) of D1R00

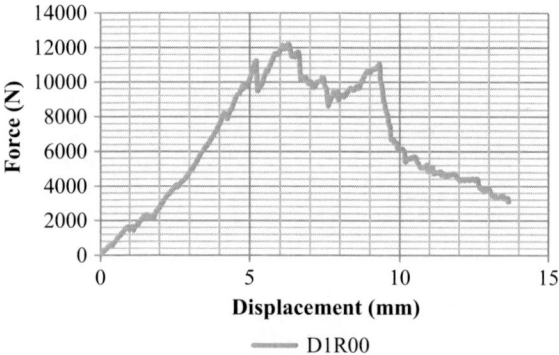

Fig. 7. Force-displacement diagram of the D1R00

$$fv = \frac{F\mathrm{max}}{2A} \quad fv = \frac{12200}{2*(290*190)} = 0.11\,\mathrm{MPa} \qquad (1)$$

where F_{max} is the maximum value of the shear force and A is the cross sectional area of the joint (Eq. 1). In the second stage of this research, the effect of the 2 cm layered sand plaster on the shear capacity of load bearing units of masonry walls are investigated experimentally.

The sand plaster is applied only on single side of the triplet specimen coded as D1R01. The cracking behavior of the central brick under shear stressing is depicted in Fig. 8.

The performance of the unreinforced specimens with sand plaster is given by means of force-displacement relationship (Fig. 9).

In this case, the shear strength f_v is evaluated as 0.260 MPa from the obtained maximum force of 28,700 N.

Fig. 8. Initial state (*left*) and final state (*right*) of D1R01

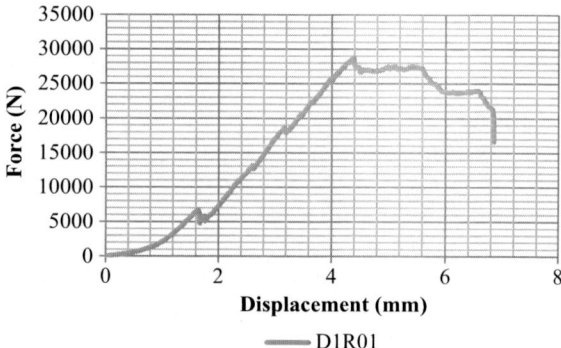

Fig. 9. Force-displacement diagram of the D1R01

In the third stage of this experimental study, the effect of the seismic textile on the shear capacity of the mortar joints is tested. The crack pattern of this third specimen can be seen from the Fig. 10. Hereby, the one-sided retrofitted specimen with 20 mm reinforcing layer is coded as D1G02 for triplet shear test. It has observed from the final state of the experiments shown in Figs. 6 and 8 that all the specimens without retro-fitting had a slide failure along the mortar joints. It is easy to see the strength and the ductility increased by applying of seismic textile.

The value of the maximum force (Fig. 11) for the strengthened specimen is equal to 31,900 N. The strength parameter is estimated as 0.290 MPa for this specimen.

Comparison of the shear strength parameters obtained from the triplet shear test is given in Table 4.

The results of all experimental tests performed were compared in Fig. 12.

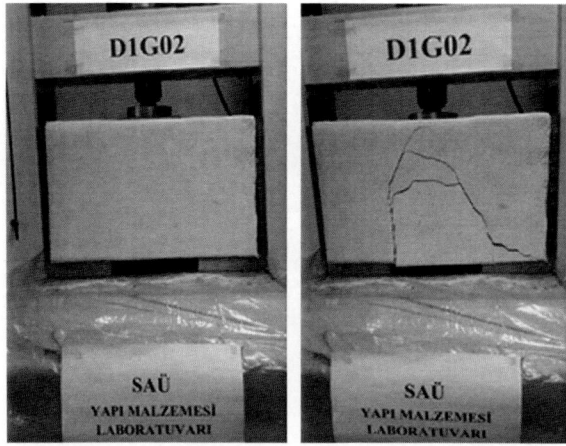

Fig. 10. Initial state (*left*) and final state (*right*) of D1G02

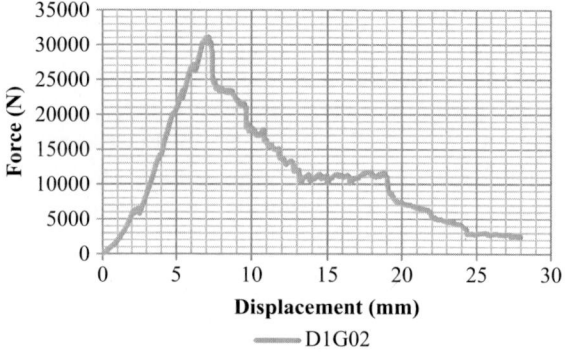

Fig. 11. Force-displacement diagram of the D1G02

Table 4. Shear strength values of three specimens

	D1R00 (MPa)	D1R01 (MPa)	D1G02 (MPa)
f_v	0.111	0.260	0.290
f_{vk}	0.088	0.208	0.232

f_v Shear strength
f_{vk} Characteristic value of the shear strength
$f_{vk} = 0.8f_v$

The damage pattern in the reinforcing system occurred with tilted cracks and with fairly large deep vertical cracks in the outer pumice blocks. When initial cracks occur, the maximum shear force achieved in the strengthened system is about 2.6 times higher than that of the non-strengthened specimen.

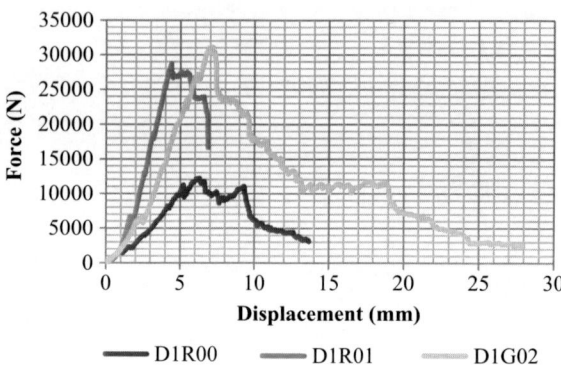

Fig. 12. Comparison of the force-displacement diagrams

4 Conclusions

The triplet shear tests on perforated brick blocks with and without the seismic retro-fitting textile were investigated to evaluate the mechanical behavior and the crack pattern of these specimens. The shear capacity of the non-strengthened specimens with/without applying any sand plaster is also tested in this study.

The results of the experimental study are given with respect to force-displacement curves comparatively for all considered test specimens. The shear strength for each specimen investigated is also obtained. It is concluded that the shear resistance and the energy dissipation highly influenced by using the seismic textile and the relevant plaster.

References

1. Wallner C (2008) Erdbebengerechtes Verstärken von Mauerwerk durch Faserverbundwerk-stoffe Experimentelle und Numerische Untersuchungen. Ph.D. thesis, Karlsruhe Institute of Technology, Karlsruhe
2. Münich JC (2010) Hybride Multidirektionaltextilien zur Erdbebenverstarkerung von Mauer-werk; Experimente und Numerische Untersuchungen mittels eines erweiterten Makromodells. Ph.D. thesis, Karlsruhe Institute of Technology, Karlsruhe
3. Oliveira DVC (2003) Experimental and numerical analysis of blocky masonry structures under cyclic loading. Ph.D. thesis, University of Minho, pp 19–25
4. Rizzo S (2015) Triplet shear tests on pumice blocks BLG19 with and without the seismic retrofitting system Röfix SismaCalce. Final report, Karlsruhe Institute of Technology, Department Reinforced Concrete, Karlsruhe
5. Anglada XR (2014) Shear tests on masonry triplets with different soft layer membranes. B.Sc. thesis, Swiss Federal Institute of Technology, Zürich, pp 9–15
6. Beattie G, Molyneaux TCK, Gilbert M, Burnett S (2001) Masonry shear strength under impact loading. In: 9th Canadian masonry symposium, Fredericton, NB, Canada
7. Lourenço PB, Barros JO, Oliveira JT (2004) Shear testing of stack bonded masonry. Constr Build Mater 18:125–132. doi:10.1016/j-conbuildmat.2003.08.018
8. EN 1052-3 (1996) European norms for methods of test for masonry–Part 3: Determination of initial shear strength

Investigation of Buckling Behavior of FRP-Concrete Hybrid Columns

Ferhat Aydın[(✉)], Tahir Akgül, and Emine Aydın

Sakarya University, Serdivan, Sakarya, Turkey
ferhata@sakarya.edu.tr

Abstract. The use of fiber reinforced plastic (FRP) composite materials in the construction industry increases with each passing day. In recent years, these materials are used as carrier material or reinforcement material in the structure. Glass fiber reinforced plastic (GFRP) profiles are among the most preferred FRP composite materials. GRFP profiles stand out with their high tensile strength, light weight and high corrosion performance. The hybrid use of these materials with classical construction materials offers new advantages. The buckling behavior of the hybrid material formed by placing concrete in plastic state into GFRP box profiles with high tensile strength was investigated in this study. The buckling performance of hybrid columns formed using GFRP profile, normal concrete or reactive powder concrete (RPC) was studied. To this end, normal concrete-GFRP profile and RPC-GFRP profile hybrid columns were produced and buckling testing was performed at different slenderness values. Charts were created after buckling experiments to examine differences in the behavior of the material. As a result of the examinations, the effect of RPC and column slenderness on the buckling performance hybrid columns was identified.

Keywords: Buckling · GFRP · Concrete · RPC · Slenderness

1 Introduction

Fiber reinforced plastic (FRP) composites are preferred due to their high strength and good performance against environmental factors as well as their producibility in different combinations. These new generation of composites are becoming popular due to properties such as high tensile strength, light weight, corrosion performance, resistance against chemicals and electrical insulation [1–3]. However, these materials are not employed in many cases where they can be used instead of other materials because they are not well known by users and designers. FRP composites are estimated to be a good solution in a large part of the available applications [4]. The use of FRPs together with traditional construction materials offer various advantages. These materials are particularly used for strengthening of reinforced concrete construction elements, as profile for carrier building elements and reinforcement in concrete (Figs. 1, 2 and 3).

© Springer International Publishing AG, part of Springer Nature 2018
S. Fırat et al. (eds.), *Proceedings of 3rd International Sustainable Buildings Symposium (ISBS 2017)*, Lecture Notes in Civil Engineering 7,
https://doi.org/10.1007/978-3-319-64349-6_35

Fig. 1. FRP strengthening [5]

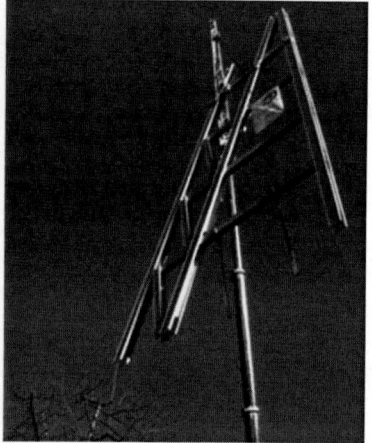

Fig. 2. GFRP eyecatcher building [6]

Fig. 3. FRP bars [7]

There has been many studies on hybrid use of concrete-filled FRP pipe elements or FRP profiles with different dimensions together with concrete [8–16].

Researchers have studied various combinations regarding the hybrid use of concrete and FRP profiles, as can be seen in the literature. Concrete-FRP-steel hybrid construction elements can be seen in Figs. 4 and 5.

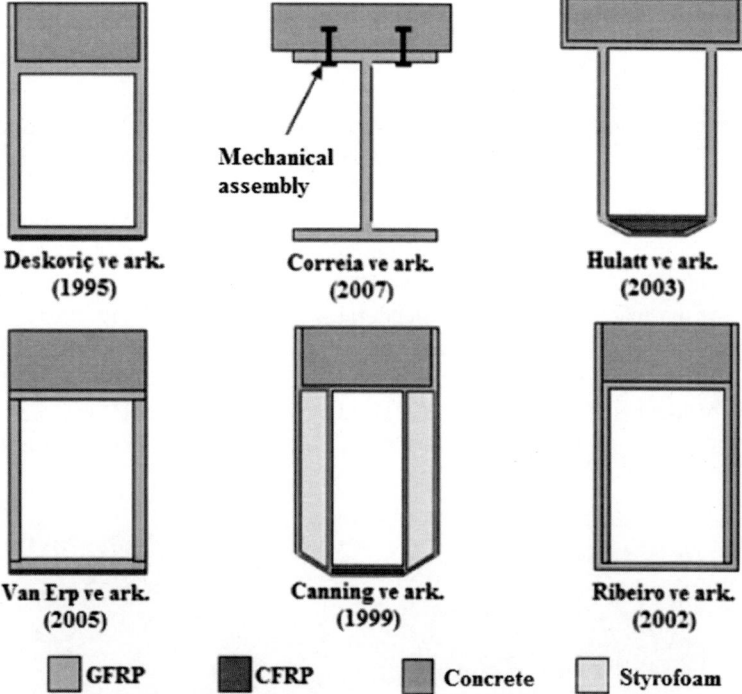

Fig. 4. Hybrid beam sections [17]

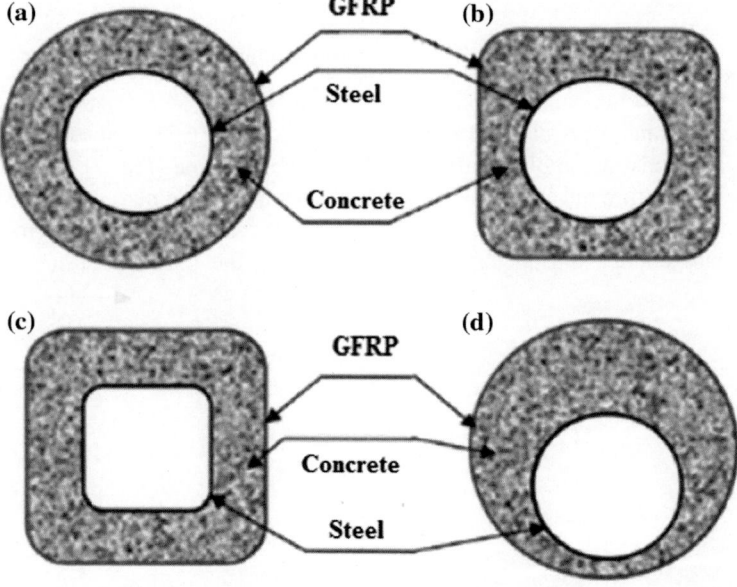

Fig. 5. GFRP-steel-concrete hybrid columns [18]

2   Experimental Studies

2.1   Materials and Method

Various experiments were carried out in order to determine buckling performances of hybrid FRP-concrete columns at different slenderness values. Normal concrete with 25 MPa compressive strength within GFRP profile was used in some of the hybrid columns and RPC concrete with a compressive strength of 200 MPa was used in the remaining samples. The dimensions of the GFRP box profile was 4–75–75 mm. The column lengths of 75–100–150–200–250–320 mm were used to determine the behavior of the material at different slenderness values.

In preparation of hybrid samples, the mixed concrete was filled in GFRP box profile in fresh form and the necessary curing treatment was performed. Once the concrete reached sufficient strength, hybrid columns were cut at aforementioned lengths for testing. Figure 6 shows concrete-filled hybrid columns at different lengths.

Fig. 6. Buckling test specimens

Horizontal and vertical displacements were recorded using data logger after the setup of additional assemblies on the axial pressure machine (Fig. 7).

After completing the experimental setup, the samples were placed and the buckling performance was tested as shown in Fig. 8. Horizontal and vertical displacement values against loads applied during tests were obtained via the Lvdt measuring instrument (Fig. 9).

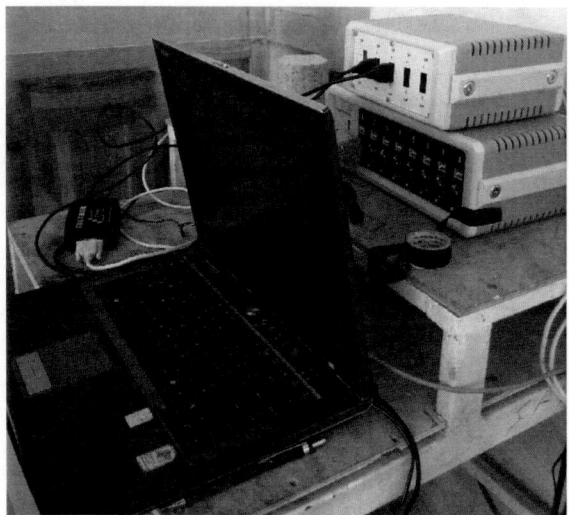

Fig. 7. Test equipment

Fig. 8. Buckling tests

Fig. 9. Horizontal and vertical displacement measurement

2.2 Test Results

As a result of buckling tests, horizontal and vertical displacement values of 75–100–150–200–250–320 mm GFRP-concrete and GFRP-RPC hybrid columns under loading were determined and load-displacement curves were drawn. Horizontal and vertical displacement graphs against loads applied during tests are given in Figs. 10, 11, 12, 13, 14, 15, 16, 17, 18, 19, 20 and 21.

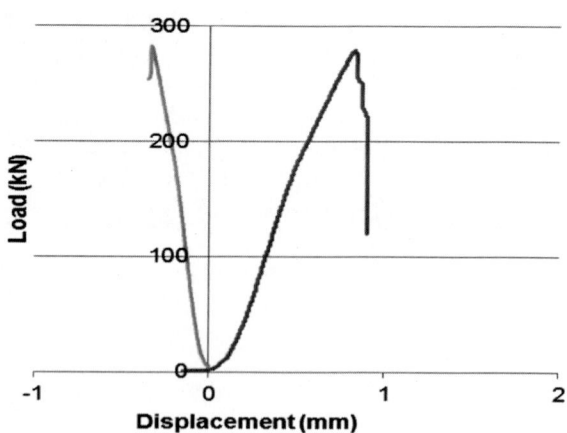

Fig. 10. Hybrid column (normal concrete, L: 75 mm)

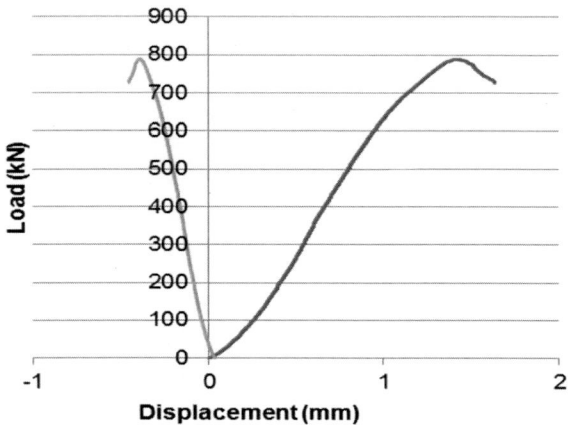

Fig. 11. Hybrid column (RPC, L: 75 mm)

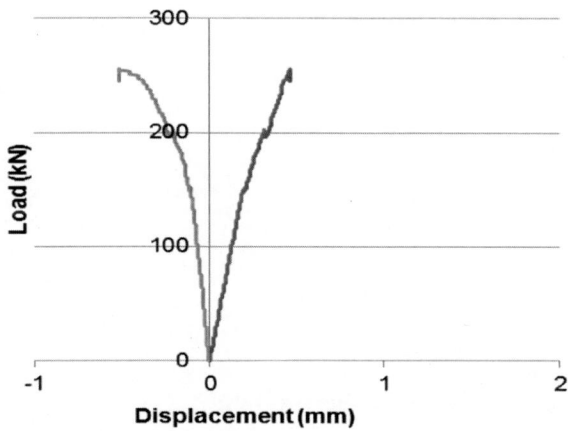

Fig. 12. Hybrid column (normal concrete, L: 100 mm)

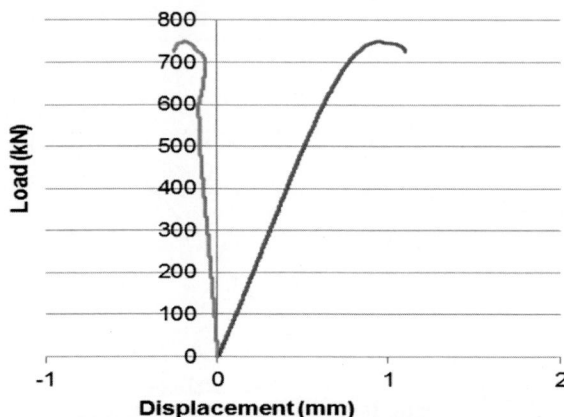

Fig. 13. Hybrid column (RPC, L: 100 mm)

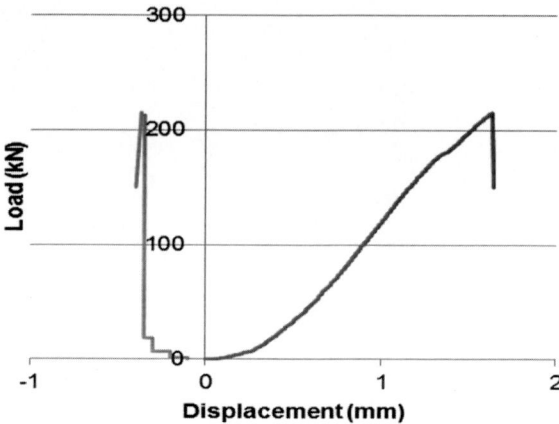

Fig. 14. Hybrid column (normal concrete, L: 150 mm)

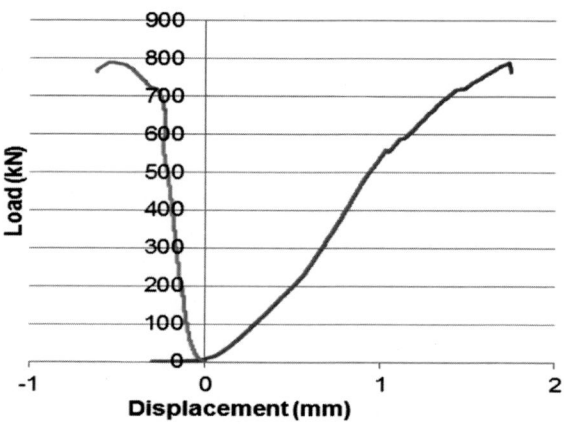

Fig. 15. Hybrid column (RPC, L: 150 mm)

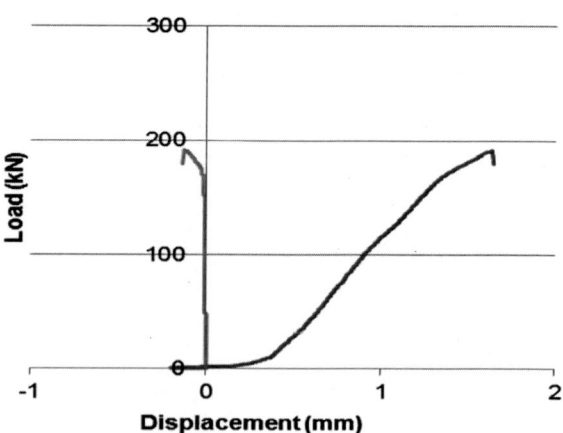

Fig. 16. Hybrid column (normal concrete, L: 200 mm)

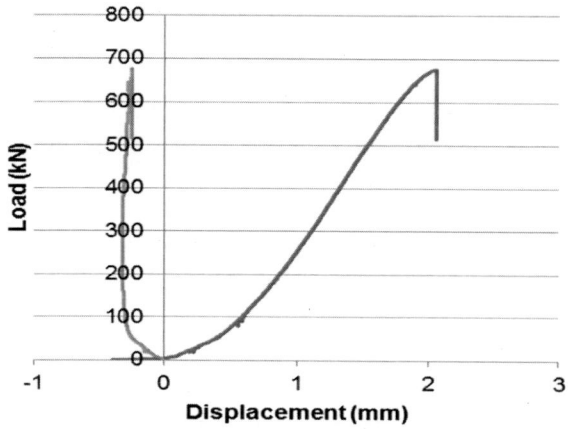

Fig. 17. Hybrid column (RPC, L: 200 mm)

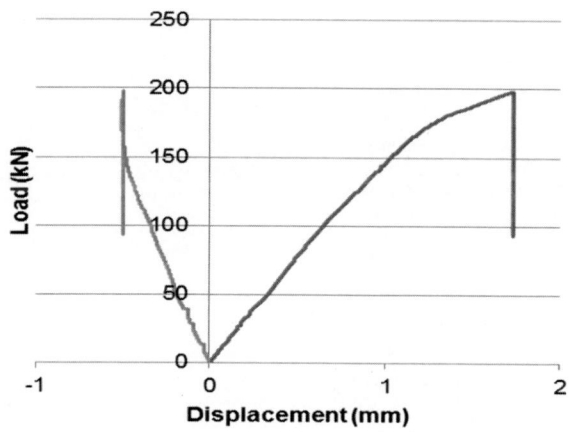

Fig. 18. Hybrid column (normal concrete, L: 250 mm)

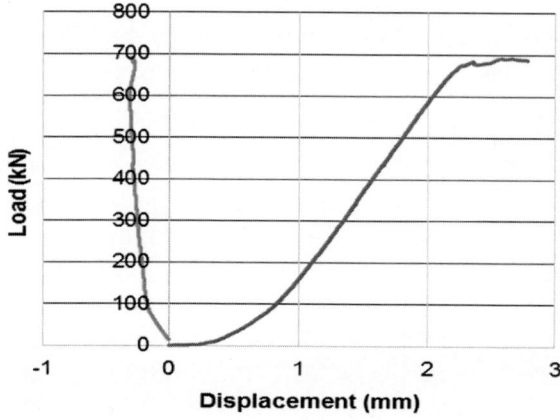

Fig. 19. Hybrid column (RPC, L: 250 mm)

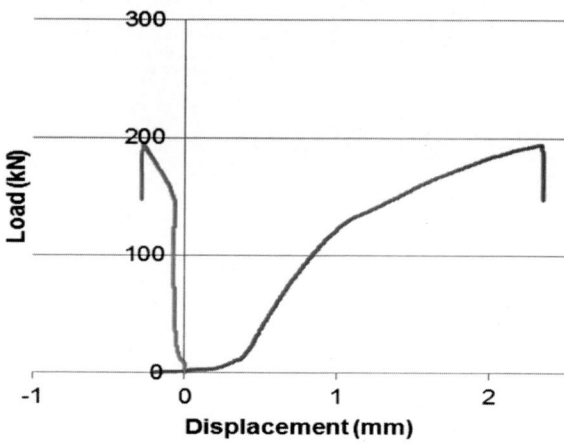

Fig. 20. Hybrid column (normal concrete, L: 320 mm)

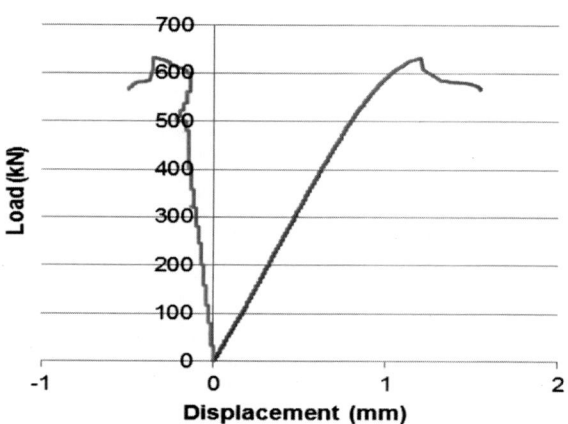

Fig. 21. Hybrid column (RPC, L: 320 mm)

Strength and slenderness values obtained for each hybrid column from graphs are given in Tables 1 and 2.

Table 1. Slenderness and stress values for hybrid columns with normal concrete

Length (mm)	P (kN)	A (mm^2)	σ (N/mm^2)	I (cm^4)	Lk = 1*L	λ
75	283	5625	50.3	263.7	7.5	3.46
100	253	5625	44.9	263.7	10	4.62
150	221	5625	39.3	263.7	15	6.93
200	195	5625	34.7	263.7	20	9.24
250	198	5625	35.2	263.7	25	11.55
320	197	5625	35.0	263.7	32	14.78

Table 2. Slenderness and stress values for hybrid columns with RPC

Length (mm)	P (kN)	A (mm²)	σ (N/mm²)	I (cm⁴)	Lk = 1*L	λ
75	795	5625	141.3	263.67	7.5	3.46
100	740	5625	131.5	263.67	10	4.62
150	793	5625	141.0	263.67	15	6.93
200	675	5625	120.0	263.67	20	9.24
250	695	5625	123.6	263.67	25	11.55
320	632	5625	112.4	263.67	32	14.78

75 × 75 mm hybrid columns filled with normal concrete reached a strength of 50.3 N/mm² at 75 mm, which corresponds to a slenderness of 1. The strength decreased with increasing slenderness and dropped to 35 N/mm² for 32 mm hybrid columns. The strength decreased by approximately 30% compared to the initial value with increasing slenderness.

The strength of 32 mm RPC concrete hybrid columns decreased from 141.3 to 112 N/mm² at a slenderness value of 1. The loss in strength for these columns was approximately 21%.

The stress-slenderness diagram of hybrid columns was drawn and the slenderness equation and the R² value were found from this diagram (Figs. 22 and 23).

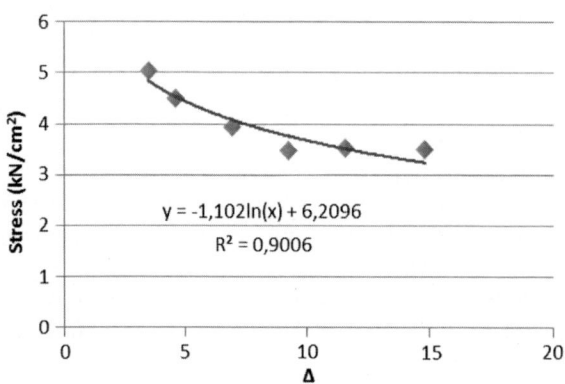

Fig. 22. Stress-slenderness graphic (normal concrete)

An examination of the strength-slenderness graph for the hybrid columns with normal concrete shows that the R² value was 0.90 and the logarithmic curve equation of y = −1.102ln(x) + 6.2096 was obtained with increasing slenderness.

The strength-slenderness graph for the hybrid columns with RPC concrete showed that the R² value was 0.70 and the logarithmic curve equation of y = −1.783ln (x) + 16.416 was obtained. It was found that the hybrid columns with normal concrete had a more compatible behavior at different slenderness values.

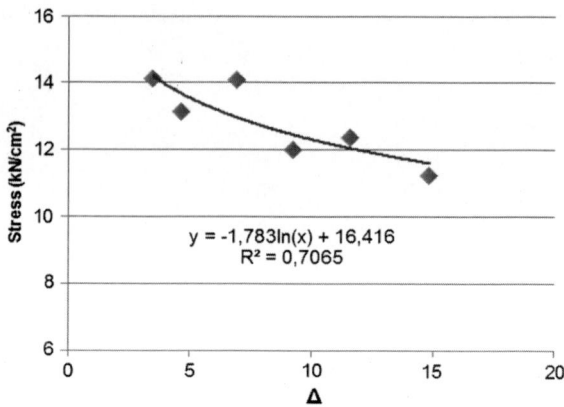

Fig. 23. Stress-slenderness graphic (RPC)

3   Conclusions and Recommendations

The results obtained from buckling testing of hybrid columns with normal concrete and RPC concrete are as follows:

- The hybrid column system offers many advantages such as permanent mold for concrete, hear and water insulation, corrosion strength and improved mechanical performance.
- The strength of 32 mm RPC concrete hybrid columns decreased from 50.3 to 35 N/mm^2 at a slenderness value of 1. The strength of hybrid column at slenderness of 1 taken as reference, the strength decreased with increasing slenderness by 11–22–31–30–30% respectively.
- The strength decreased from 141.3 to 112.4 N/mm^2 for hybrid columns with RPC concrete. The decrease in strength compared to the hybrid column at a slenderness of 1 was 7–0.2–15–13–20.5%.
- A good fit was observed in stress-slenderness graphs of the hybrid columns with normal concrete. The R^2 value was 0.90 and the logarithmic curve equation was $y = -1.102\ln(x) + 6.2096$.
- The R^2 value of the hybrid columns with RPC concrete was 0.70 and the logarithmic curve equation was $y = -1.783\ln(x) + 16.416$.
- The strength of the RPC concrete hybrid columns with high strength increased and the decrease in strength of columns with increasing slenderness occurred only slightly for RPC columns.
- The hybrid construction systems formed by combining traditional construction materials and new generation composites improve inadequate aspects of available construction materials. The use of these new generation composites as alternative materials may provide a solution for many problems related to construction materials.

References

1. Aydın F (2011) Investigation of mechanic performance of hybrid structural element produced using glass fibre reinforced plastic (GFRP) composite and concrete. Ph.D. thesis, Sakarya University, Science Institute, Sakarya
2. Aydın F, Sarıbıyık M (2013) Investigation of flexural behaviors of hybrid beams formed with GFRP box section and concrete. Constr Build Mater 41:563–569
3. Aydın F (2016) Effects of various temperatures on the mechanical strength of GFRP box profiles. Constr Build Mater 127:843–849
4. Cripps A (2002) Fiber reinforced polymer composites in construction, construction. Industry Research & Information Association (CIRIA)
5. https://www.luckett-farley.com/Blog/Article/119/FRP-Concrete-Strengthening-101. 1 Jan 2017
6. http://cclab.epfl.ch/page-13730-en.html. 1 Jan 2017
7. http://www.aslanfrp.com/. 1 Jan 2017
8. Mirmiran A, Shahawy M (1997) Behavior of concrete columns confined by fiber composites. J Struct Eng 123:583–590
9. Fam AZ, Rızkalla SH (2001) Confinement model for axially loaded concrete confined by circular FRP tubes. ACI Struct J 98(4):251–461
10. Becque J, Patnaık AK, Rızkalla SH (2003) Analytical models for concrete confined with FRP tubes. J Compos Constr 7(1):31–38
11. Yu T, Wong YL, Teng JG, Dong SL, Lam ESS (2006) Flexural behavior of hybrid FRP-concrete-steel double-skin tubular members. J Compos Constr ASCE 10(5):443–452
12. Gautam B, Matsumoto T (2009) Shear deformation and interface behaviour of concrete-filled CFRP box beams. Compos Struct 89:20–27
13. Correıa JR, Branco A, Ferreıra JG (2009) Flexural behaviour of multi-span GFRP-concrete hybrid beams. Eng Struct 31:1369–1381
14. Wenlxıao L, Zhıshen W (2004) Flexural performance of newly developed hybrid FRP concrete beams. In: FRP composites in civil engineering, CICE, pp 819–826
15. Nordin H, Taljstena B (2003) Testing of hybrid FRP composite beams in bending. Compos B 35:27–33
16. Hulatt J, Hollaway L, Thorne A (2003) The use of advanced polymer composites to form an economic structural unit. Constr Build Mater 17:55–68
17. Leo B (2009) Design of a fibre-reinforced polymer (FRP) bridge. University of New South Wales at the Australian Defence Force Academy. Second Lieutenant (Singapore Armed Forces), School of Aerospace, Civil & Mechanical Engineering, final thesis report
18. Teng JG, Yu T, Wong YL (2004) Behavior of hybrid FRP-concrete-steel double-skin tubular columns. In: FRP composites in civil engineering, CICE, pp 811–818

An Investigation of Concrete Strength of Hybrid Construction Materials Under the Effect of Heat

Ferhat Aydın[✉], Metin İpek, and Kutalmış Akça

Sakarya University, Serdivan, Sakarya, Turkey
ferhata@sakarya.edu.tr

Abstract. The hybrid use of concrete and Glass Fiber Reinforced Plastic (GFRP) profiles in construction material technologies in recent years offers many new opportunities. Filling fresh concrete into GFRP profile allows for obtaining superior advantages compared to component materials. The purpose of this study is to investigate the change in the strength of concrete in hybrid material exposed to high temperatures. To this end, hybrid compressive samples were prepared by filling GFRP composite box profiles with concrete and plain concrete samples were prepared for comparison. The cubic samples were exposed to high temperatures at 25–200–400–600–800 °C. Strength losses of the plain concrete and the concrete in the hybrid material, which had the same dimensions, under the effect of heat were determined.

Keywords: GFRP · Concrete · Heat · Hybrid materials
Compressive strength

1 Introduction

In recent years, Fiber Reinforced Plastic (FRP) composites, which are increasingly used as building materials, have many positive qualities. These materials are popular especially due to their corrosion performance, lightness and high tensile strength and can be produced in different fiber types. These materials are primarily produced as fabric to strengthen structures, as column or beam to be used as support components and finally as FRP reinforcement material to be used instead of steel reinforcement. Among FRPs, the use of GFRP composites in the form of profile is more common compared to other fiber types due to their inexpensiveness.

The subject of hybrid construction components prepared by filling FRP profiles or pipes with concrete attracts the interest of many researchers and several studies have been conducted on the subject in recent years [1–6]. Scientific studies show that researchers will focus on the hybrid use of traditional construction materials and FRP composites in the upcoming years [7]. Many studies have shown that the hybrid use of FRP composites and traditional materials such as concrete offers solutions to eliminate certain disadvantages of construction materials manufactured solely from FRP [8–11].

The use of concrete within FRP profiles presents many advantages such as area reduction, rigidity, increased strength, prevention of local fractures, improved curing and permeability. Hybrid use may mitigate or completely eliminate the disadvantage arising from the concrete or the GFRP profile. In the traditional concrete production, forming small parts in order to prepare the mold takes a lot of time and causes additional costs. In the hybrid system, on the other hand, a second mold component is not required since the GFRP box profile serves as a mold. Thus, the hybrid system allows for saving time and reducing mold costs largely and this property referred to as permanent form [12–19] in the literature provides great convenience.

Also, regional and local fractures occur in GFRP profiles due to bending [20, 21]. Since GFRP profiles are filled with hardened concrete, these local fractures are reduced. Thus, the hybrid material is expected to show a better performance under bending loads. GFRP box profiles do not let external water and moisture in and also prevent the concrete with plastic consistency in the profile from losing its water and moisture, thus provide great advantages for this operation which is vital for the curing of concrete. In this way, the hydration process of the concrete, which is desired to be cured at 100% relative humidity for 28 days, is completed without any issues [21]. Other advantages of GFRP profiles include good impermeability and water-moisture insulation, allowing for production of construction components with smaller dimensions compared to component materials and increased strength and rigidity values.

Apart from these advantages, the purpose of this study is to determine the effect of GFRP profiles, which protects the concrete in hybrid system against external effects and shows high performance particularly against tensile stresses, on the strength of concrete in case of high temperature or fire.

2   Experimental Studies

Concrete, which is a basic construction material, is only used against compressive stresses in material design since it is quite weak against tensile and bending stresses. GFRP composite material, on the other hand, stands out with its high tensile strength. In the hybrid material design produced by filling GFRP box profile with concrete, concrete is expected to handle compressive stresses and GFRP is expected to handle tensile stresses.

This study examines differences in compressive strength of the concrete in hybrid material caused by high temperature. The room temperature was accepted as the reference point and compressive samples kept at 200–400–600–800 °C were tested to calculate weight loss and compressive strength values under the effect of heat.

2.1   Materials and Method

The effects of the hybrid use of concrete in GFRP box profiles having the same dimensions on heat deformation of concrete were investigated experimentally. To this end, plain concrete samples and hybrid samples were produced by filling GFRP box profiles having the same dimensions with fresh concrete. In axial compression tests, plain concrete samples were tested by exposing the samples to temperatures of 25–200–400–600–800 °C, whereas hybrid samples were tested by exposing the samples to same temperatures, taking the concrete in the GFRP profile out and then applying the compression test. Thus, the changes in compressive strength of profile concrete and concrete taken out from GFRP profile were determined after axial compression tests.

The GFRP box profile had the dimensions of $5 \times 74 \times 74$ mm (Fig. 1). Mechanical and physical properties of the GFRP box profile determined as a result of the tests can be seen in Table 1.

Fig. 1. GFRP box profiles

Table 1. Properties of GFRP box profile

Properties	Values
Specific gravity	1.80
Unit weight	1.78 g/cm^3
Tensile strength	550 MPa
Modulus of elasticity (E)	30,000 MPa

Table 2 shows the components of the concrete, which was produced in a single strength class according to TS 802 [22] using only sand and crushed stone grade 1.

Table 2. Concrete mix design

Material	Volume (dm^3)
Aggregates I (5–12 mm)	379
Sand	336
Cement	105
Water	170
Air	10
Total	1000

Hybrid materials were prepared together with the plain concrete in a single strength class. The concrete with plastic consistency was filled in 74 mm cubic molds and the remaining was filled in GFRP box profile. All samples were kept at 23 °C for 28 days for curing. Then, open parts of the hybrid samples were closed using the same material. Hybrid and plain concrete cubes before and after the effect of heat can be seen in Fig. 2.

Fig. 2. Before and after the heat effect of the samples

Figure 3 shows the exposure of the cubic samples to heat in the furnace and Fig. 4 shows the concrete compressive samples.

Fig. 3. Samples exposed to high temperature in furnace

Fig. 4. Concrete samples

Compression tests were applied to plain concrete and concrete in hybrid samples exposed to different temperatures including the room temperature (Fig. 5).

Fig. 5. Breaking of the samples

2.2 Test Results

Compression tests were performed after heating plain concrete and hybrid samples up to temperatures of 25–200–400–600–800 °C. Weight and strength losses after the performance of the tests were determined. Table 3 shows the results of the compression test performed at room temperature, which was accepted as the reference point.

Table 3. Compressive strength (25 °C)

Sample	Compressive strength (kg/cm^2)
1	278.65
2	220.08
3	268.03
Average	255.58

The average compressive strength was found to be 255.58 kg/cm^2 in tests performed at room temperature, approximately 25 °C.

Table 4 shows weight losses and Table 5 shows compressive strengths of plain concrete and concrete in hybrid samples at 200 °C.

Table 4. Weight Losses (200 °C)

Sample	Hibrit (%)	Plain (%)
1	1.41	1.65
2	0.31	1.55
3	1.11	1.84
Average	0.94	1.68

Table 5. Compressive strength (200 °C)

Sample	Hybrid (kg/cm^2)	Plain concrete (kg/cm^2)
1	248.98	267.46
2	230.76	276.32
3	298.46	269.75
Average	259.40	271.17

The weight loss of concrete in hybrid samples exposed to 200 °C was 0.94% and the compressive strength was 259.40 kg/cm^2. These values were 1.68% and 271.17 kg/cm^2 for plain concrete. Compared to the reference concrete, no strength loss occurred in both plain concrete and hybrid samples. In contrast, slight increases were observed due to hot curing (Fig. 6).

Table 6 shows weight losses and Table 7 shows compressive strengths of samples exposed to 400 °C.

The weight loss was 7.66% for concrete in hybrid samples and 7.13% for plain concrete at 400 °C. Compared to 200 °C, the weight loss increased 8 times for plain concrete and 4 times for hybrid samples. The compressive strength of concrete in hybrid samples was 304.60 kg/cm^2 and the compressive strength of plain concrete was 250.62 kg/cm^2. The comparison with the reference concrete at room temperature can be seen in Fig. 7. Compared to the reference concrete, the strength loss was 2% for plain concrete and the increase in strength was 19% for concrete in hybrid samples at 200 °C.

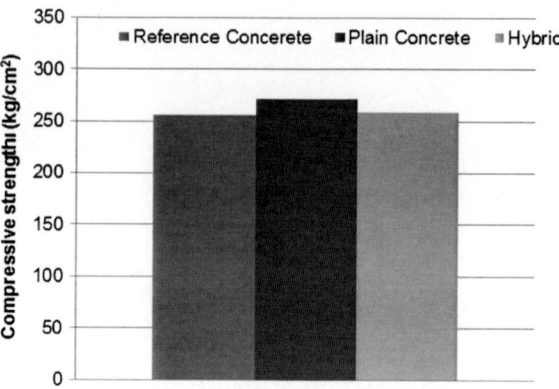

Fig. 6. Comparison of compressive strengths (200 °C)

Table 6. Weight losses (400 °C)

Sample	Hybrid (%)	Plain (%)
1	11.48	6.91
2	4.97	7.33
3	6.53	7.14
Average	7.66	7.13

Table 7. Compressive strength (400 °C)

Sample	Hybrid (kg/cm^2)	Plain concrete (kg/cm^2)
1	279.34	255.78
2	302.63	286.65
3	331.82	209.42
Average	304.60	250.62

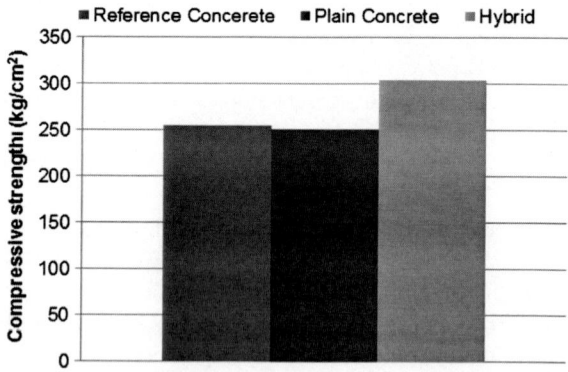

Fig. 7. Comparison of compressive strengths (400 °C)

The GFRP profile on the sample surface was deformed in hybrid material after keeping the sample in the furnace at 400 °C for 1 h, but the profile managed to protect the concrete inside (Fig. 8).

Fig. 8. Samples after 400 °C heat

The temperature was raised to 600 °C during experiments and weight loss values shown in Table 8 and compressive strength values shown in Table 9 were obtained. With the increase in the temperature, the GFRP profile completely was deformed, the matrix of the profile burned and only glass fibers remained. Thus, the weight loss value was calculated only for plain concrete samples due to deformed surface in hybrid samples (Fig. 9).

Table 8. Weight losses (600 °C)

Sample	Plain (%) 600 °C
1	8.08
2	8.74
3	8.12
Average	8.31

Table 9. Compressive strength (600 °C)

Sample	Hybrid (kg/cm^2)	Plain concrete (kg/cm^2)
1	257.78	210.58
2	229.72	222.22
3	227.27	234.81
Average	238.25	222.53

Fig. 9. Samples after 600 °C heat

The weight loss of plain concrete was 8.31% at 600 °C. The weight loss at 600 °C increased approximately 5 times compared to the weight loss at 200 °C.

After exposure to 600 °C, the compressive strength of concrete in hybrid samples was 238.25 kg/cm² and the compressive strength of plain concrete was 222.53 kg/cm². The strength loss rate increased for plain concrete samples with increasing temperature (Fig. 10). The strength loss of concrete in hybrid samples was about 7% at 600 °C, whereas this value was 13% for plain concrete.

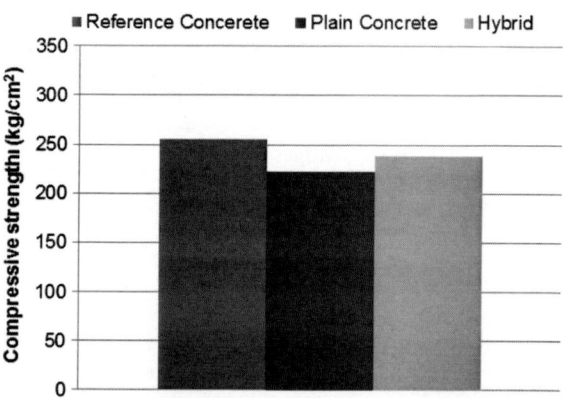

Fig. 10. Comparison of compressive strengths (600 °C)

The temperature was raised to 800 °C and weight loss values shown in Table 10 and compressive strength values shown in Table 11 were obtained. Figure 11 shows the final state of materials once the temperature reached 800 °C. GFRP profile matrix completely burned and only glass fibers remained. Considerable losses were observed in both plain concrete and hybrid samples since the concrete was no longer protected at high temperatures (Fig. 11).

Table 10. Weight losses (800 °C)

Sample	Plain (%) 800 °C
1	7.68
2	10.84
3	10.45
Average	9.66

Table 11. Compressive strength (800 °C)

Sample	Hybrid (kg/cm^2)	Plain concrete (kg/cm^2)
1	142.66	164.03
2	139.47	140.04
3	146.01	149.73
Average	142.71	151.26

Fig. 11. Samples after 800 °C heat

The weight loss of plain concrete was found to be 9.66% at 800 °C. The weight loss at 400 °C increased approximately 6 times compared to the weight loss at 200 °C.

After the effect of heat at 800 °C, the compressive strength of concrete in hybrid samples was 142.71 kg/cm^2 and the compressive strength of plain concrete was 151.26 kg/cm^2. The comparison between strength values of plain concrete and concrete in hybrid samples can be seen in Fig. 12.

Compared to the reference concrete, the strength loss of plain concrete was found to be 44% at 800 °C. This rate was 44% for hybrid samples. Thus, it should be considered that the concrete exposed to 800 °C may lose almost half of its strength.

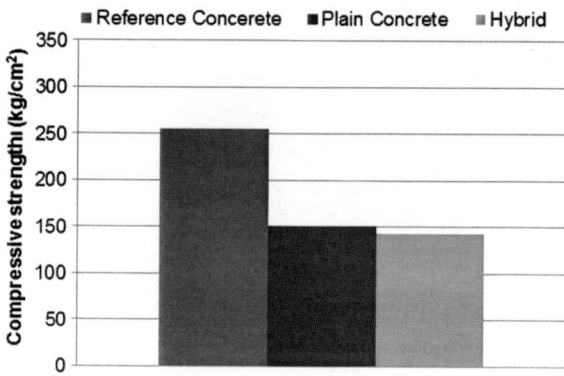

Fig. 12. Comparison of compressive strengths (800 °C)

A slight increase in strength was observed for plain concrete at 200 °C compared to room temperature due to the effect of curing. The compressive strength continued to decrease with increasing temperature. For hybrid samples, the effect of curing was evident at 400 °C and the compressive strength increased. With increasing temperature and burning of the GFRP profile, the concrete lost its protection and considerable strength losses were observed (Fig. 13). The critical temperature value was found to be 200 °C for plain concrete and 400 °C for concrete in hybrid samples. Strength losses increasingly continued after the temperature values mentioned.

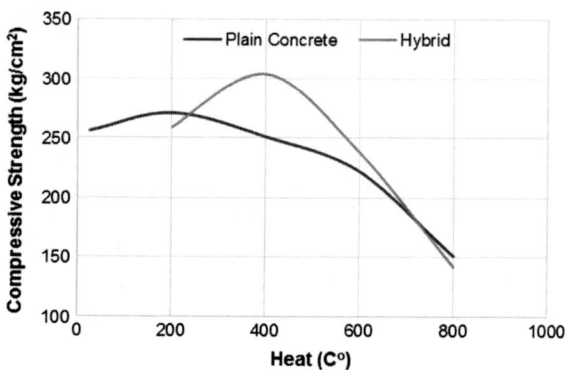

Fig. 13. Decrease in strength under heat effect

3   Conclusions and Recommendations

The results obtained in this study investigating the compressive behavior of hybrid compressive components formed by filling square-shaped GFRP box profiles with concrete at high temperatures are summarized below:

- The weight loss of concrete in hybrid samples was calculated to be 0.94% at 200 °C and 7.66% at 400 °C. The weight loss of concrete in hybrid samples at 400 °C increased approximately 8 times compared to the weight loss at 200 °C.
- The weigh loss of plain concrete was 1.68% at 200 °C, 7.13% at 400 °C, 8.31% at 600 °C and 9.66% at 800 °C. The weight loss increased with increasing temperature as expected. Compared to the weight loss at 200 °C, plain concrete's weight loss increased 4.2 times at 400 °C, 5 times at 600 °C and 5.7 times at 800 °C.
- The average compressive strength of the reference concrete was 255.58 kg/cm^2. Plain concrete's compressive strength was 271.17 kg/cm^2 at 200 °C, 250.38 kg/cm^2 at 400 °C, 222.53 kg/cm^2 at 600 °C and 151.26 kg/cm^2 at 800 °C. The strength loss in plain concrete was 13% at 600 °C and 41% at 800 °C.
- The hybrid material's average compressive strength was 259.40 kg/cm^2 at 200 °C, 304.60 kg/cm^2 at 400 °C, 238.25 kg/cm^2 at 600 °C and 142.71 kg/cm^2 at 800 °C. The strength loss of concrete in hybrid samples was about 7% at 600 °C. Compared to the reference concrete, the strength loss of concrete in hybrid samples was 44% at 800 °C.
- It was determined that GFRP profile matrix completely burned at temperatures over 400 °C and only glass fibers remained. The concrete was no longer protected in hybrid samples once the temperature reached 800 °C. Considerable losses were observed in both plain concrete and hybrid samples since the concrete was no longer protected at high temperatures.
- The critical temperature value at which plain concrete started to lose strength was 200 °C, whereas this value was 400 °C for hybrid samples. GFRP profile delays the strength loss of concrete at low temperatures. Similar strength losses are observed for both concrete types at temperatures over 400 °C.

References

1. Mirmiran A, Shahawy M (1997) Behavior of concrete columns confined by fiber composites. J Struct Eng 123:583–590
2. Fam AZ, Rızkalla SH (2001) Confinement model for axially loaded concrete confined by circular FRP tubes. ACI Struct J 98(4):251–461
3. Becque J, Patnaık AK, Rizkalla SH (2003) Analytical models for concrete confined with FRP tubes. J. Compos Constr 7–1:31–38
4. Yu T, Wong YL, Teng JG, Dong SL, Lam ESS (2006) Flexural behavior of hybrid FRP-concrete-steel double-skin tubular members. J Compo Constr ASCE 10(5):443–452
5. Ozbakkaloglu Togay (2013) Compressive behavior of concrete-filled FRP tube columns: assessment of critical column parameters. Eng Struct 51:188–199
6. Mostafa F, Genda C (2016) Compressive behavior of FRP-confined concrete-filled PVC tubular columns. Compos Struct 141:91–109
7. Hong WK, Kım HC, Yoon SH (2002) Experiment of compressive strength enhancement of circular concrete column confined by carbon tubes. KCI Concr J 14(4):19–144
8. Schaumann E (2008) Hybrid FRP-lightweight concrete sandwich system for engineering structures. Ph.D. thesis

9. Aydın F (2011) Investigation of mechanic performance of hybrid structural element produced using glass fibre reinforced plastic (GFRP) composite and concrete. Ph.D. thesis, Sakarya University, Science Institute, Sakarya, Turkey
10. Aydın F, Sarıbıyık M (2013) Investigation of flexural behaviors of hybrid beams formed with GFRP box section and concrete. Constr Build Mater 41:563–569
11. Aydın F (2016) Effects of various temperatures on the mechanical strength of GFRP box profiles. Constr Build Mater 127:843–849
12. Keller T, Schaumann E, Vallée T (2007) Flexural behavior of a hybrid FRP and lightweight concrete sandwich bridge deck. Compos A 38(3):879–889
13. Hall J, Mottram J (1998) Combined FRP reinforcement and permanent formwork for concrete members. J Compos Constr 2(2):78–86
14. Canning L, Hollaway L, Thorne AM (1999) An investigation of the composite action of an FRP/concrete prismatic beam. Constr Build Mater 13:417–426
15. Rıbeıro MCS, Tavares CML, António JMF, Marques AOT (2002) Static flexural performance of GFRP-polymer concrete hybrid beams. Key Eng Mater 230–232:148–151 (Advanced Materials Forum I)
16. Tianhong L, Peng F, Lieping Y (2006) Experimental study on FRP-concrete hybrid beams. In: Third international conference on FRP composites in civil engineering (CICE 2006), Miami, Florida, USA, December 13–15
17. Fam A, Schnerch D, Rızkalla S (2005) Rectangular filament-wound glass fiber reinforced polymer tubes filled with concrete under flexural axial loading: experimental investigation. J Compos Constr ASCE 9(1):25–33
18. Hamdy MM, Radhouane M (2010) Flexural strength and behavior of steel and FRP-reinforced concrete-filled FRP tube beams. Eng Struct 32:3789–3800
19. Mirmiran A, Shahawy M, Samaan M (1999) Strength and ductility of hybrid FRP-concrete beam-columns. ASCE J Struct Eng 125(10):1085–1093
20. Aydın F, Sarıbıyık M (2010) Compressive and flexural behavior of hybrid use of GFRP profile with concrete. In: International symposium on sustainable development (ISSD 2010), Sarajevo, Bosnia and Herzegovina
21. Aydın F, ve Sarıbıyık M (2011) Investigation of cure effect in hybrid use of GFRP box profiles with concrete. e-J New World Sci Acad 6(4), Article Number: 1A0211:991–1000
22. TSE 802 (2009) Design of Concrete Mix, Turkish Standards Institute, Turkey

Electrical Curing Application on Cement-based Mortar with Different Stress Intensity

Tayfun Uygunoğlu[1]([✉]), İsmail Hocaoğlu[1], and İlker Bekir Topçu[2]

[1] Civil Engineering Department, Engineering Faculty, Afyon Kocatepe
University, 03200 Afyonkarahisar, Turkey
uygunoglu@gmail.com
[2] Civil Engineering Department, Engineering Faculty, Eskişehir Osmangazi
University, 26480 Eskişehir, Turkey
ilkerbt@ogu.edu.tr

Abstract. In this study, the effects of different stress intensity on the electrical resistivity of hardening (setting) cement-based mortar were investigated. In experiments, four different mould (5 × 5 × 10 cm, 5 × 5 × 15 cm, 5 × 5 × 20 cm and 5 × 5 × 25 cm) were used. Different combinations of stress intensity (7.5, 15, 22.5 and 30 V) were used on mixtures. The measurements were done at room temperature. Electrical resistivity and setting time of the specimens with and without electric current application were also investigated. Moreover, specific setting hydration temperature was measured by using thermocouple. As a result, electric current application can be used for obtaining to rapid setting time on the mortar with high volume stress intensity.

Keywords: Electrical resistivity · Mortar · Stress intensity · Hydration time

1 Introduction

Concrete's gaining strength at an early time and it's being used immediately are the results of developing industry [1]. In order to reach these results, there are several cure methods and through these methods, the concrete can get early strength. However, the accelerated cure methods which are applied during pouring the concrete is very limited. Depending upon the hydration time, accelerated cure at the buildings in which ready mixed concrete used is important in terms of the project's being carried out in a predicted time since concrete compressive strength reaches the target level in a short time [2]. Moreover, the prediction of the level of strength in an early time is very crucial in terms of building's both performance and economy. Concrete is not conductive but it can show a conductive feature until its final setting time since it includes water. Furthermore, additive minerals in concrete can change in terms of electrical resistivity of concrete. Different methods have been developed to measure electrical conductivity of mortars and various researches and applications which examine cement's micro structure developments have been carried out [3]. In monitoring of the entire hardening process of a cement paste, electrical resistivity can be used as a parameter. The durability and service life of cement-based materials strongly depends on its transport properties [4]. Electrical conductivity process occurs primarily due to

© Springer International Publishing AG, part of Springer Nature 2018
S. Fırat et al. (eds.), *Proceedings of 3rd International Sustainable
Buildings Symposium (ISBS 2017)*, Lecture Notes in Civil Engineering 7,
https://doi.org/10.1007/978-3-319-64349-6_37

ion transport through the pore solution in a cement-based system and it is an important parameter to study the hydration process of cement pastes at early stages [4, 5]. During the hydration of cement paste after mixing, calcium (Ca) and hydroxyl (OH) ions go into solution within the first 10 min [6]. Then, little happens except for a slow pre-cipitation of semi-crystalline calcium silicate hydrate gel (C–S–H) while the calcium and hydroxyl ion concentrations continue to rise slowly [7, 8]. The degree of hydration describes the process of hydration, and directly relates to the fraction of the hydration products or porous structure in a hydration system in cement-based materials [9]. The variation of resistivity as function of time can indeed reflect internal changes of the pore solution of cement paste with time [10]. As it is well known, the hydration process in cement paste results in the formation of C–S–H, calcium CH, ettringite and other compounds. During hydration, the capillary pores in hardening cement paste are gradually filled up with hydration products and the solid phases form a rigid microstructure with increasing strength. Then, electrical resistivity of cement paste increases with time [11].

Most of the articles in the literature deal with the electrical resistivity of cement paste. Some work has been published on various aspects of electrical measurement of setting and hardened cement paste. Xiao et al. used electrical resistivity measurement and Marsh cone used it for providing a fast way to evaluate and select suitable superplasticizer in concrete [4]. In another study, Xiao et al. [12] measured the electrical resistivity of the concrete mixes and the pore solution within the mixes by a non-contact electrical resistivity apparatus. They reported that electrical resistivity measurement reflects the hydration kinetics of fresh concrete. Abo El-Enein et al. [13] investigated the electrical conductivity of concrete containing silica fume. The results of their study showed that the electrical resistivity can be used as an indication for the setting char-acteristics of pastes made with and without silica fume. Tamaás et al. [14] have shown that the time dependence of the electrical response of cement pastes is complex and influenced by a number of variables such as water content, temperature, clinker fine-ness, gypsum and admixture content. Topçu et al. [4] applied current to cement pastes which have different water/cement ratio (0.40, 0.45, 0.50, 0.55), different mineral admixtures (fly ash, silica fume, and blast furnace slag) and different ratio (0, 10, 20, 30%). As a result, applying electric current can be used for obtaining rapid setting time on the cement paste with high volume mineral admixture [4]. Backe et al. [3] researched the relation between conductivity, porosity, cement chemistry and ion content. According to their theory when porosity increases, conductivity increases too and thus they concluded that it is in relation with hydration degree. As a result, when setting time becomes shorter stripping time becomes shorter too [15]. This research is important for the structures which require rapid repair or fast production such as prefabricate.

2 Experimental Study

In experimental study crushed sand have been used in order to show the impact of electrical resistivity on mortars. Specific gravity of crushed sand is 2.64 and the maximum grain size is 4 mm. In the experiments, CEM 1 42.5 R cement which is suitable to TS EN 197-1 [16] standards have been used.

2.1 Production of Specimen

The mortars whose water/cement ratio is 0.70 are produced with cement dosage of 350. The 5 lt capacity mixer is used to mix admixture. Then the mortars put in electrically isolated wooden moulds whose sizes are 5 × 5 × 10, 5 × 5 × 15, 5 × 5 × 20 and 5 × 5 × 25 (in cm). Unit volume component of 350 dosage mortar has shown in Table 1.

Table 1. Component of mortar per cubic meter

Sample Code	Cement (kg/m^3)	Water (lt/m^3)	Crushed sand (kg/m^3)
M350	350	175	2104

Initially for the mixture of mortars, aggregate and cement were used for 1 min in order to have dry mixture. Then almost 2/3 of mixture water were added to the admixture Finally chemical admixture in the remainder (1/3) of water were put into the admixture and the process of mixing were continued for 3 min. Prepared mortars were put into wooden moulds which has different sizes as 5 × 5 × 10, 5 × 5 × 15, 5 × 5 × 20 and 5 × 5 × 25 (cm × cm × cm). Then some of mortars were applied to 7.5, 15, 22.5 and 30 V stress by direct current (DC) power- supply source during 24 h and some of them were used to control specimen at laboratory conditions. For applying stress intensity to specimens, two copper electrodes were used and put in head end of moulds.

2.2 Preparation of Experimental Setup

For applying stress intensity to the specimens a DC power-supply source Hioki mark which has 30 canal and 60 V capacity has been used. Firstly power supply was connected to a port ("+" pole and "−" pole separately), then data logger was connected to the port. Then direct connection from the port to the specimens was made to apply DC current. In order to measure the temperature of specimens to which stress applied, K type thermocouple were connected. Finally the data from data logger were saved and their graphics were drawn. A view of mortar which electrical current applied to has shown in Fig. 1.

3 Results and Discussions

For measuring temperature of mortars, K type thermocouple was used and in every 2 min the temperature values were recorded by data logger. The comparison of specific hydration temperature of hardening specimens which 7.5, 15, 22.5 and 30 V stress are applied in terms of the setting time is shown in Fig. 1.

From the beginning of hydration time, temperature of cement based mortars have been measured and saved. The mortars whose sizes are 5 × 5 × 10 and 5 × 5 × 15 (in cm) are chosen because the temperature is higher when it is compared with other size of mortars. It is observed that as the stress intensity increases, temperature values get lower, except for 5 × 5 × 10 (Fig. 2).

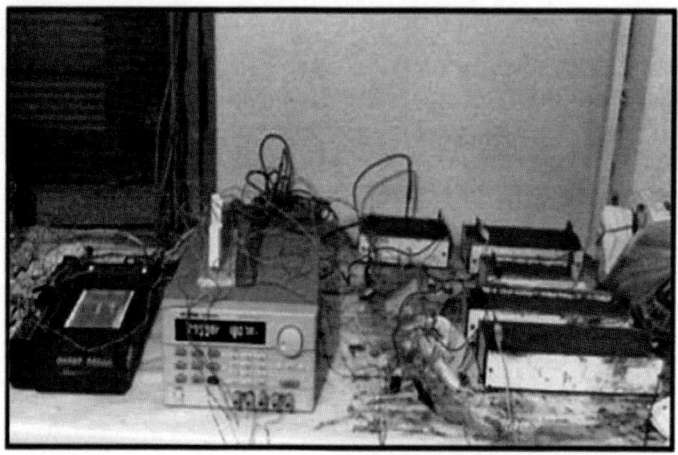

Fig. 1. A view of mortar which electrical current applied to

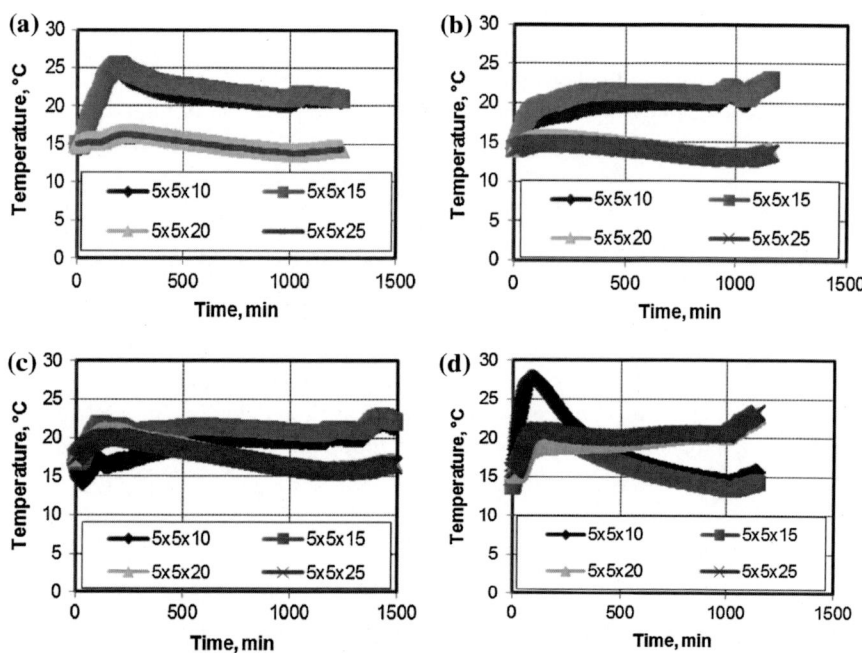

Fig. 2. Specific hydration temperature of mortars (**a** 7.5 V, **b** 15 V, **c** 22.5 V, **d** 30 V)

The conductor between two electrodes in mortar mixture was initially the water in the mixture which is in the liquid phase. At the beginning, all the pores in the mixture have connections among themselves. Therefore, the electrolysis between the electrodes has arisen from the free water of these pores. However, hydrated products occur in the course of time because of the reaction of the main components in the cement paste with water, and the pores which were once connected to each other have been separated. Accordingly, electrical resistivity of cement paste increase in time and when the cement completes its hardening time, its electrical resistivity amount reaches the maximum value. Consequently, the electrical resistivity of cement is directly associated with the hydration of the cement. The mortar having different size have been investigated during the time until the setting time of the mortar has taken (one day at the laboratory having suitable conditions). The resistivity results are presented in Fig. 3 for different voltage.

The conduct of electrical current on the mortar is directly associated with distance of plates in mortar. Due to the electrical conductivity of mortar, it can be seen in Fig. 3 that the mortar which has low distance (5 × 5 × 10 cm) conducts much more electrical current when compared to high distance plates (5 × 5 × 25 cm) in mortar. Therefore, the temperature of low distance mortars was increased and mortar hydrated faster than mortars with high distance. Consequently, resistivity has higher values in small size moulds (5 × 5 × 10 cm) than high distance plates (5 × 5 × 25 cm). When the distance of plates is increased, the resistivity values of mortar were reduced under all the voltage values. On the other hand, the voltage has higher effect on low distance of plate than high distance. The relative compressive strength values were also determined after 28 days of electrical curing process. It can be seen in Fig. 4.

It was interestingly found that when the superposed graphics are examined, it is seen that when the stress intensity is increased (up to 15 V) and the sizes of specimens are made taller (5 × 5 × 20 and 5 × 5 × 25), compressive strength of the specimens are getting higher.

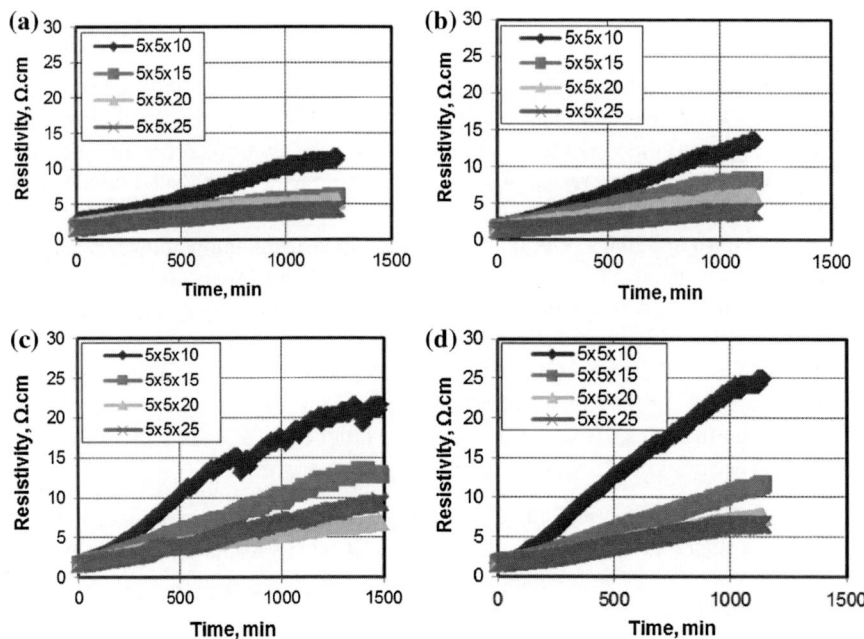

Fig. 3. Electrical resistivity of hardening mortars (**a** 7.5 V, **b** 15 V, **c** 22.5 V, **d** 30 V)

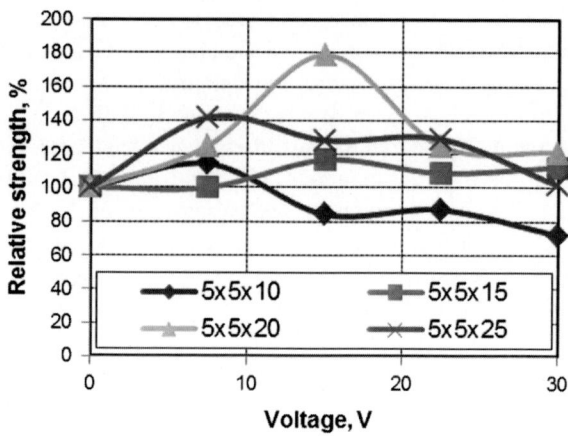

Fig. 4. Relative compressive strength of mortar after DC electrical curing

4 Conclusions

In this study, the effect of stress intensity as a result of applying 7.5, 15, 22.5 and 30 V stress at $5 \times 5 \times 10$, $5 \times 5 \times 15$, $5 \times 5 \times 20$ and $5 \times 5 \times 25$ (in cm) sizes have been investigated. Several conclusions can be drawn from this study:

- When the stress intensity applied to mortars increase (7.5–30 V), specific hydration temperature becomes higher.
- As electrodes approach each other, electrical resistivity becomes higher due to increasing of temperature. The resistivity has higher values in small size moulds ($5 \times 5 \times 10$ cm) than high distance plates ($5 \times 5 \times 25$ cm). When the distance of plates is increased, the resistivity values of mortar were reduced under all the voltage values.
- It can be interpreted that a relationship between measuring temperatures related with hydration time of mortars and resistivity. The electrical resistivity of cement paste increase in time and when the cement completes its hardening time, its electrical resistivity amount reaches the maximum value.
- It is concluded that, 15 V stress application and 20 cm plate distance is the optimum values for the highest strength. When the stress intensity is increased (up to 15 V) and the sizes of specimens are made taller ($5 \times 5 \times 20$ and $5 \times 5 \times 25$), compressive strength of the specimens are getting higher.

Consequently, DC electrical curing of cement based mortar can be modified by the applying of voltage at 15 V with distance of 20 cm. Furthermore, the hydration duration of mortar can be accelerated as a result of applying electric current.

Acknowledgements. Authors thank Afyon Kocatepe University Institute of Natural and Applied Sciences and Coordinatorship of Scientific Research Projects for their support to this study as a doctoral thesis Project with Number of 16.FEN.BİL.43.

References

1. McCarter WJ, Chrisp TM, Starrs G, Blewett J (2003) Characterization and monitoring of cement-based systems using intrinsic electrical property measurements. Cem Concr Res 33:197–206
2. Whittington HW, McCart J, Forde MC (1981) The conduction of electricity through concrete. Mag Concr Res 33:48–60
3. Backe KR, Lile OB, Lyomov SK (2001) Characterizing curing cement slurries by electrical conductivity. SPE Drill Complet 16(4):201–207
4. Topçu İB, Uygunoğlu T, Hocaoğlu İ (2012) Electrical conductivity of setting cement paste with different mineral admixtures. Constr Build Mater 28:414–420
5. Schwarz N, DuBois M, Neithalath N (2007) Electrical conductivity based characterization of plain and coarse glass powder modified cement pastes. Cem Concr Compos 29:656–666
6. Wei X, Li Z (2006) Early hydration process of Portland cement paste by electrical measurement. J Mater Civ Eng 18(1):99–105
7. Buenfeld NR, Newman JB (1987) Examination of three methods for studying ion diffusion in cement pastes, mortars, and concrete. Mater Struct 20:3–10
8. Li Z, Xiao L, Wei X (2007) Determination of concrete setting time using electrical resistivity measurement. J Mater Civ Eng 19(5):423–427
9. Levita G, Marchetti A, Gallone G, Princigallo A, Guerrini GL (2000) Electrical properties of fluidified Portland cement mixes in the early stage of hydration. Cem Concr Res 30:930–932
10. Rajabipour F, Weiss J (2007) Electrical conductivity of drying cement paste. Mater Struct 40:1143–1160
11. Koleva DA, Copuroglu O, Breugel KVG, Ye G, Wit JHW (2008) Electrical resistivity and microstructural properties of concrete materials in conditions of current flow. Cem Concr Compos 30:731–744
12. Xiao L, Li Z (2008) Early-age hydration of fresh concrete monitored by non-contact electrical resistivity measurement. Cem Concr Res 38:312–319
13. Abo El-Enein SA, Kotkata MF, Hanna GB, Saad M, El Razek MMA (1995) Electrical conductivity of concrete containing silica fume. Cem Concr Res 25:1615–1620
14. Tamaás FD, Farkas E, Vörös M, Roy DM (1987) Low-frequency electrical conductivity of cement, clinker and clinker mineral pastes. Cem Concr Res 17:340–348
15. Morsy MS (1999) Effect of temperature on electrical conductivity of blended cement pastes. Cem Concr Res 29:603–606
16. EN 197-1/A3 (2010) Cement—part 1, Compositions and conformity criteria for common cements. *TSE Ankara, Turkey, 2010 (in Turkish)*

Effect of Particle Size Distribution on the Sintering Behaviour of Fly Ash, Red Mud and Fly Ash-Red Mud Mixture

Sedef Dikmen[1(✉)], Zafer Dikmen[1], Gülgün Yılmaz[2], and Seyhan Fırat[3]

[1] Physics Department, Science Faculty, Anadolu University, Eskişehir, Turkey
sdikmen@anadolu.edu.tr
[2] Porsuk Vocational School, Anadolu University, Eskişehir, Turkey
gulgunyilmaz@anadolu.edu.tr
[3] Department of Civil Engineering, Faculty of Technology, Gazi University,
Ankara, Turkey
sfirat@gazi.edu.tr

Abstract. In this study, the characterization and microstructure observation of the sintering behaviour of F (from Seyitömer Thermal Power Plant) and C (from Soma Thermal Power Plant) types fly ashes, red mud and red mud–fly ash mixtures at 4 and 8 h milling times with a mortar type ball are investigated. The particle size distribution (PSD) was carried out by laser technique. The PSD analysis revealed that the particle size decreased with increasing milling time. After milling, the samples were pressed into a cylindrical form without using any additive or binder. The formed samples were sintered at 1050 °C. Phase evolutions and microstructures of the formed samples were carried out by using X-ray diffraction method (XRD) and scanning electron microscope (SEM) method, respectively. The water absorption and compressive strength were tested to determine some physico-mechanical properties. The results of mineralogical analysis as well as the microstructural observations reveal that phase transformations occurred during the sintering treatments. The results showed that at the chosen optimum conditions (F type fly ash-red mud mixture with 1:1 ratio, sintering temperature of 1050 °C and milling time of 4 h), the best characteristics of sintered glass ceramic could be obtained. Values obtained from mechanical testing showed that the fly ash–red mud sintered ceramic materials had increased strength compared to other sintered ceramic materials. The sintered materials from red mud have the highest water absorption value.

Keywords: Fly ash · Red mud · Sintering behaviour · Microstructure
Mechanical properties

© Springer International Publishing AG, part of Springer Nature 2018
S. Fırat et al. (eds.), *Proceedings of 3rd International Sustainable Buildings Symposium (ISBS 2017)*, Lecture Notes in Civil Engineering 7,
https://doi.org/10.1007/978-3-319-64349-6_38

1   Introduction

The disposal of waste material is an issue of ever-increasing concern. Recently, fly ash and red mud are regarded as hazardous industry waste that causes environmental pollution all over the world. The objective of effectively utilize these industrial wastes is to reduce raw material consumption, thus minimizing disposal crisis and treatment costs. These waste materials obtained from the alumina industries and thermal coal-fired power plants.

Red mud, also known as bauxite residue, is a side-product of the Bayer process during the treatment of bauxite with sodium hydroxide for alumina extraction. This slurry contains 15–40% solids and is highly alkaline (pH 11–12.5) [1–2]. Its major constituents are crystalline hematite (Fe_2O_3), which gives it its reddish appearance, boehmite, sodalite, quartz, alumina silicate, with a minor presence of calcite, goethite and gibbsite [3]. Depending upon the quality of the raw material processed and alumina extraction efficiencies, between 1 and 2 tonnes of red mud is generated per ton of alumina produced. Every year more than 120 million tonnes of red mud is discharged to the landfill in the world, which has brought major problem for environmental protection. For instance, only in Eti Aluminium Plant (Seydisehir, Turkey) approximately 260,000 tonnes of red mud generated annually. The red mud landfill which has 10 million m^3 capacity has been currently almost filled up. Although 6 million m^3 capacity is available in another pond, it seems appropriate to consider for a new one in this region [4]. However, only a small amount of this waste about 1% is re-used in the metallurgical as a source for iron and various other purposes [5]. Globally, although there have been multiple attempts to utilize these residues, less than 2% of residues are currently re-used or further processed. Consequently, the vast majority of the residue remains in disposal areas all over the world at present. In recent years, many researchers have been focused on utilization of red mud as the recycling material to produce building materials (such as bricks, glass ceramics, concrete, cements and composite materials) [6–10], coagulants [11], catalysts as agents for neutralizing acidic waste [12] and metal extraction [13] etc.

Similarly, fly ash, a type of solid waste that results from coal fired thermal power plants, which is classified as class F and C ashes according to ASTM C618. The ash containing more than 70 wt% SiO_2 + Al_2O_3 Fe_2O_3 and being low in lime are defined as class F, while those with a SiO_2 + Al_2O_3 + Fe_2O_3 content between 50 and 70 wt% and high in lime are defined as class C. The annual production of fly ash is 900 million tonnes globally that makes its disposal a great challenge [14]. Even though, about 15 million tons of ashes per year are generated in Turkey (production is expected to reach 50 million tons per year by 2020), only small percentage of them are being utilized for application as an ingredient in cement and various other construction products. Many researches have been performed in order to find out application, or process for the disposal of fly ash as an alternative to another industrial resource. The industrial applications of fly ash are grouped into following main categories: construction materials (as a cement additive and as a concrete aggregate) [15], geotechnical application (for asphalt and pavement filler materials) [16–19], agricultural [20], water pollution [21], composite and sintered materials (glass, glass-ceramic and ceramic)

[22], and geopolymer [23]. Among these, especially glass-ceramic materials have shown some promise, which converts industrial wastes product by sintering at high temperatures.

Glass ceramics, which have unique mechanical properties such as high mechanical strength, good dimensional stability and abrasion resistance, are not only suitable for replacing more traditional materials in many applications, but also they can be used in entirely new fields where no alternative material can satisfy the technical demands [24]. Traditionally, the glass-ceramics have been made from pure raw materials, and the products are expensive. Manufacturing glass ceramics with solid waste could recycle the waste and produce commercial products. Red mud contains valuable mineral resources such as CaO, Al_2O_3, SiO_2, Fe_2O_3, and TiO_2. The chemical composition of red mud is quite suitable for producing glass ceramics. Recently, many researches are carried out on producing glass-ceramics with red mud. The mineralogical composition of red mud is particularly close to fly ash. The $CaO–SiO_2–Al_2O_3$ glass ceramic has been obtained from mixing red mud and fly ash successfully [25]. However, there are several researches on red mud and fly ash reuse. As one of the few works, Samal et al. [26] explored the suitability of using red mud and fly ash in order to produce ceramic bodies with satisfactory mechanical properties. In the previous study, the suitability of fly ash with red mud for the development of paving blocks has been evaluated [27]. To utilize both these wastes together by using lead-zinc mine tailings, Liu et al. carried out a research [28] to develop a process for the manufacture of lightweight-foamed ceramics. Those researches, focused in development of fly ash with red mud mixtures in order to obtain commercial building materials, have some advantages, e.g., conservation of resources, which means lower raw materials cost and greater environmental benefits due to use of the waste materials in the process.

The aim of this study is recycling of F (F-FA) and C (C-FA) type fly ashes from two different thermal power plants and red mud (RM) investigating the suitability of glass-ceramics using raw material based on milling, powder compaction and at relatively high temperatures of 1050 °C without any additives or nucleating agents.

2 Experimental Procedures

2.1 Raw Materials

Two different fly ashes were obtained from Seyitömer (Kütahya) and Soma (Manisa) Thermal Power Plants, in Turkey. Additionally, the red mud (slurry) was provided from the Eti Aluminium Inc. Seydişehir, Turkey. Red mud sample was dried in oven at 100 °C for 24 h. A Bruker S8 Tiger XRF spectrometer in standardless mode under helium atmosphere with 18 mm mask was used for determining the chemical composition. Powder samples were loaded into an XRF sample cup to make the measurement by Li2B4O7 fusion bead oxide method. For the measurement a 4 KW, Rh anode X-ray tube was used to generate X-rays. The methods "Best analysis" and "Oxides" were chosen for the analysis. SpectraPlus Eval2 V2.2.454 was used for data interpretation.

The particle size analysis of red mud and fly ashes was carried out using a laser particle size analyzer (Malvern-Mastersizer 2000). The phase composition of the raw materials was performed with a X-ray diffraction (Bruker, D8 Advance), using CuK_α radiation (k = 1.54 Å) at 40 kV and 20 mA, with a 2θ range of $10°–70°$. The samples were scanned with a step of $0.02°$ (2θ). A software package program was utilized to analyze the resulting patterns. Field emission-scanning electron microscopy (FE-SEM) analyses were performed in a ZEISS EVO 50 EP, at 10 kV to observe the changes in microstructure properties of all samples. Table 1 indicates that the main chemical components of the fly ashes are SiO_2 and Al_2O_3, and the other significant amounts of oxides are Fe_2O_3, CaO, MgO and SO_3. Loss on ignition (LOI) of the fly ash mainly results from the unburned carbon content. According to ASTM C618, which considers basically the chemical components, fly ash from Seyitömer ($SiO_2 + Al_2O_3 + Fe_2O_3$ 70 wt%) can be classified as type F (F-FA) and fly ash from Soma ($SiO_2 + Al_2O_3 + Fe_2O_3$ is between 50 and 70 wt%) as type C (C-FA).

Table 1. Chemical composition and some physical properties of fly ashes and red mud (as received)

Component	Fly ash (Seyitömer)	Fly ash (Soma)	Red mud
Chemical composition (wt%)			
SiO_2	53.41	36.37	14.10
Al_2O_3	19.41	19.24	24.55
Fe_2O_3	10.90	4.92	31.91
CaO	4.44	33.52	1.58
MgO	4.64	1.96	0.34
K_2O	2.13	1.44	0.39
Na_2O	0.79	0.30	11.06
TiO_2	0.69	0.50	4.65
MnO	0.12	0.06	0.03
Cr_2O_3	0.11	0.16	0.16
SO_3	0.64	3.34	0.24
Others	0.16	–	0.02
LOI	2.56	1.29	10.94
Physical properties			
Specific surface area (m^2/g)	4.87	3.74	18.32
Density (g/cc)	1.88	2.02	3.05
Particle size (μm)			
d(10)	16.63	3.09	0.35
d(50)	66.72	37.7	1.58
d(90)	96.03	128.6	56.85

As shown in Table 1, Fe_2O_3 and SiO_2 are the major components of the red mud (RM). The both chemical and particle size analysis results are in generally agree with previous studies in literature [3, 4].

2.2   Preparation of Sintered Samples

The raw samples were ground using laboratory ball-mill (FRITCH-Pulverisette 6). First, 40 ml of dry powder was measured with a graduated cylinder. Later, this powder was wet milled in a zirconium mortar containing 40 balls of zirconium for 4 and 8 h, at a water-to-ash ratio of 2. After the grinding process is completed, about 10 ml of sample was taken from each milled suspansion and their average particle size distributions were determined. The milled suspension was filtered and it was dried in an oven at 100 °C for 12 h. The cylindrical shaped samples having a diameter of 13 mm and height of 14–16 mm were produced by uniaxially pressing under the load of 10 MPa using an automated hydraulic press. The pressed samples of F-FA, C-FA, red mud and red mud with fly ashes mixtures (1:1) were placed inside a ashing furnace (Nobertherm) and sintered in air atmosphere at 1050 °C. The heating rate was 6 °C/min and a dwell time at the maximum temperature of 1 h for all samples. After sintering was over, all samples were cooled down in the furnace and then the sintered samples were characterized with techniques like XRD and SEM.

2.3   Characterization of the Sintered Samples

X-ray diffraction (XRD) technique was used to determine the mineralogical changes in crystalline phases of all the sintered samples. SEM analyses were performed to observe the changes in microstructure properties of sintered samples by varying milling time and temperatures. The fractured surfaces of sintered samples were coated with Au prior to examination. Compression tests were performed on cylindrical specimens using an automated Instron 5581 50 kN Tensile Test Machine (Tailored Test).

3   Result and Discussion

3.1   Characterization of Raw Materials

Table 2 shows the particle size distribution of raw materials of two different fly ashes and red mud. As shown Table 2, milling caused a reduction in particle size throughout the range. Even during the 4 h milling process, D(0.9) reduced from 92.03 to 41.7 μm and median particle size changed from 66.72 to 1.96 μm while D(0.1) had a reduction from 16.63 to 1.04 μm for F-FA. These are all large reductions considering the percentage size reductions. In the fine milling process of fly ash and red mud, the network structure would be destroyed by mechanical force, resulting in the breaks of chemical bonds such as Si-O-Si, Al-O-Al and Si-O-Al.

The main components of F-fly ash (Seyitömer) are SiO_2, Al_2O_3 and Fe_2O_3, which constitutes about 84% and C-fly ash (Soma) is SiO_2, Al_2O_3 and CaO, which constitute 94% of total composition. The red mud is Fe_2O_3, SiO_2 and Al_2O_3, which constitute 89% of total composition.

XRD diffraction pattern of as-received, F-fly ash, C-fly ashes and red mud are shown in Fig. 1. In F-fly ash, the main mineralogical phases were identified as quartz (SiO_2) and magnetite (Fe_3O_4). In addition, peaks of anorthite ($CaAl_2Si_2O_8$) and

Table 2. Particle size distribution of milled materials

Milling time	Particle size (μm)	F-FA (Seyitömer)	C-FA (Soma)	Red mud
4 h	d(0.1)	1.04	0.73	0.26
	d(0.5)	2.80	2.46	5.02
	d(0.9)	6.63	4.87	36.85
	d(0.1)	0.70	0.42	0.08
8 h	d(0.5)	1.96	2.65	3.48
	d(0.9)	4.70	3.87	21.82

anhydrite (Calcium sulfate, $CaSO_4$) were found, mostly overlapping with each other (Fig. 1a). In C-fly ash, the main peaks were quartz and free lime (CaO). Few minor peaks of dolomite ($CaMg(CO_3)_2$) and mullite (Aluminum silicate, $Al_{24}Si_{84}O_{204}$) were also reported (Fig. 1b). From the XRD and XRF results, there are some differences between the chemical and mineral compositions of F-FA and C-FA. The XRD data of red mud displays a complex pattern indicating a mixture of metal oxides and complex minerals (Fig. 1c). The dominant phase in red mud is hematite (Fe_2O_3) displaying features at 18.83°, 35.6°, 40.9°, 49.5°, 54.1°, 62.5°, and 64.0°. Besides hematite, quartz, calcite ($CaCO_3$), and ilmenite ($FeTiO_3$) were also present in the as received form of red mud.

Figure 2a–c are SEM micrographs of the fly ashes samples, showing predominantly spherically shaped powder particles, a typical particle morphology for fly ash powders. F-FA comprises individual particles and agglomerates consisting of spherical or irregular shaped particulates, which were also expressed by a former study [29–31]. These spherical particles are glasses produced from melt droplets during incineration of the coal. Some separate hollow spheres were observed in the range between 2 and 30 μm, many of which have diameters less than 15 μm (Fig. 2b). The spherical particles might in turn be hollow sphere filled with smaller spheres, Fig. 2c. In fact, C-FA similar to F-FA based on the morphological structure. However, the shape of C-FA particles are more spherical than that of F-FA and its particle size varies between about 5 and 100 μm. That is, a wide variation in particle size was observed. This result is compatible with the particle size distribution of C-FA given in Table 1. Whereas the particle size varies between 3 and 128 μm, the average particle size is about 38 μm. Such different in the fly ashes of microstructure are ordinary due to variety of fired coal mineralogy and chemistry, coal particle size of and coal combustion conditions of the power plants.

The SEM morphology of the red mud at 30 K magnification is given in Fig. 2d. The red mud composed of very fine particle size where the particle size ranges from 1 μm to over a few microns. Red mud has a variety of different morphology, some individual particles look like flake and prism.

XRD patterns of glass ceramic samples produced from F-FA by grinding treatment are given Fig. 3. At 1050 °C, the anorthite component of the F-FA disappeared. With sintering temperature and increasing milling time up to 8 h, the quartz peak intensities decreased due to signify damages in crystal structure and increasing the amount of amorphous phase.

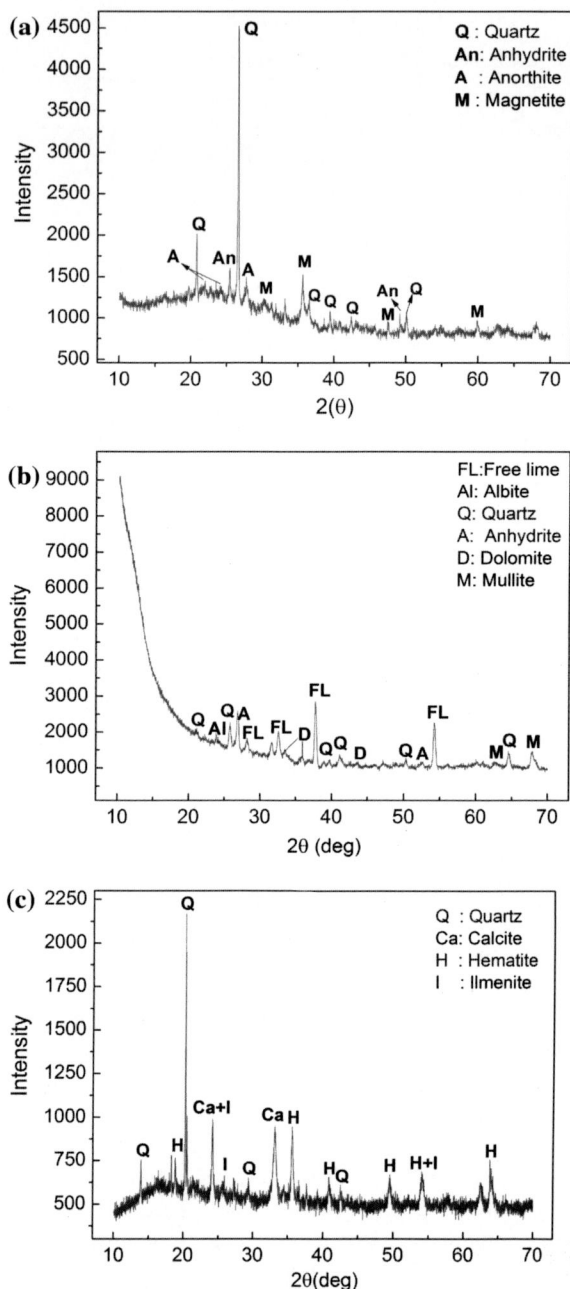

Fig. 1. XRD patterns of raw materials **a** F-FA, **b** C-FA, **c** RM

Fig. 2. Scanning electron micrograph of raw materials **a** F-FA-Mag:300x, **b** F-FA-Mag:6Kx, **c** C-FA-Mag:3Kx, and **d** RM-Mag:20Kx

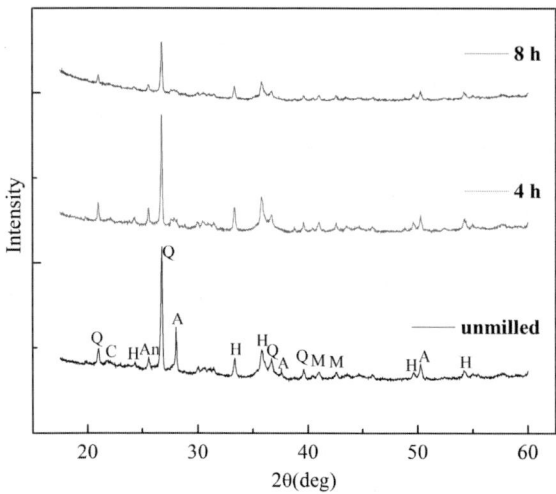

Fig. 3. X-ray diffraction pattern of unmilled, milled for 4 and 8 h of F-FA at 1050 °C

XRD analysis reveal that the samples C-FA, RM, F-FA/RM and C-FA/RM sintered at 1050 °C, Fig. 4. RM and F-FA/RM samples are hematite structure. More quartz peaks in structure are seen in XRD pattern of C-FA/RM sample. RM sample shows the best crystalline. This result is compatible with SEM image of C-FA/RM given in Fig. 6d.

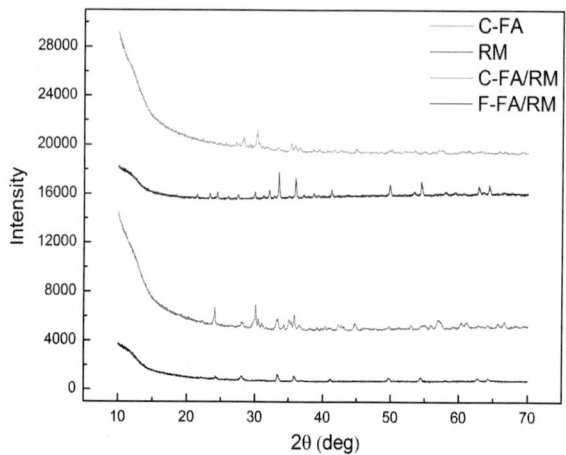

Fig. 4. X-ray diffraction pattern of milled C-FA, RM, F-FA/RM and C-FA/RM for 8 h, at 1050 °C

The microstructures of unmilled, 4 and 8 h milled and sintered F-FA samples at 1050 °C temperatures were examined by SEM (Fig. 5a–c). The surface of the sample milled for 8 h is rough and homogeneus microstructure. Increasing the milling time, a much density sample that has a much smoother fracture surface. The unmilled F-FA sample was clearly showed that spherical pores were found in the formation.

SEM micrograph of the sintered C-FA the presents of a different morphology than original form (Fig. 6a). There are tiny crystals of about 1 μm size in a glassy structure and the spherical particles have been disappeared. This is due to the fact that when calcite and dolomite break down into CaO and MgO, the Ca and Mg are incorporated into the new mineral phases, principally in high temperature silicates such as gehlenite, wollastonite and diopside, which tend to obstruct the melting process [26].

Figure 6b shows the surface morphology of the sintered red mud. The darker phases represent the quartz, whereas the lighter region represents the magnetite, very small size of iron particles throughout the microstructure. The fired specimens show a homogenous microstructure composed by isolated pores, irregular shapes, and size crystals and ceramic matrix surrounding the former phases. It was observed that red mud contains very irregular spherical particles whereas treated fly ash (F-FA/RM and C-FA/RM mixture) have agglomerated structure. These agglomerated structures may be due to binding of particles with the organic compounds (Fig. 6c, d). The main difference is the amount and size of pores (Fig. 5d) which are more in red mud with 50 wt% of fly ash mixtures. The irregular crystals were composed mainly of quartz,

(a)   **(b)**

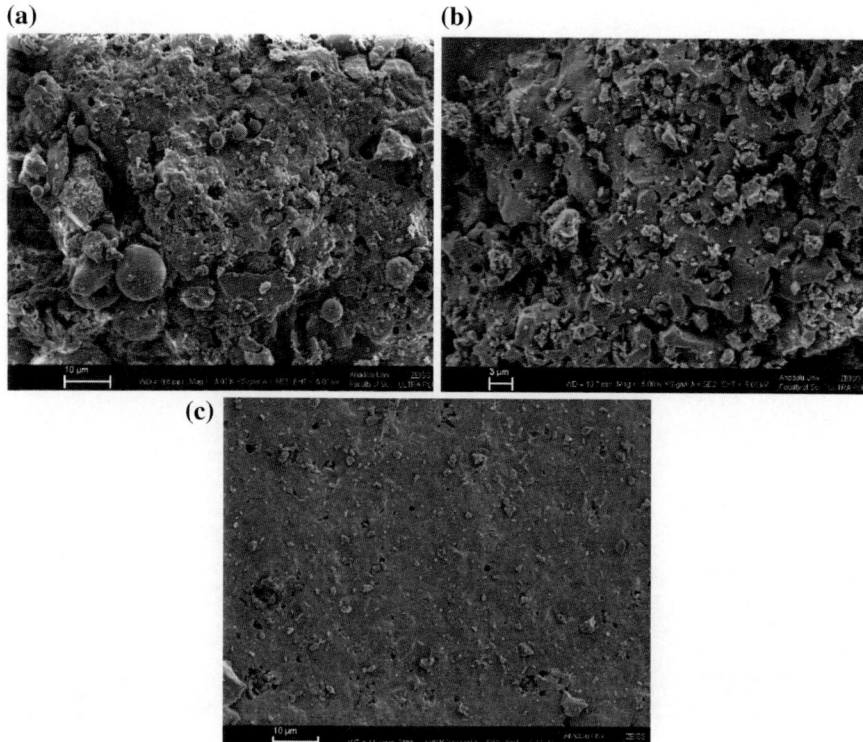

Fig. 5. Scanning electron micrograph of sintered materials at 1050 °C **a** unmilled F-FA, **b** 4 h milled F-FA, **c** 8 h milled F-FA

which has been identified by the main crystalline phase. Darker regions are quartz crystals. Small lighter region particles are related to glassy phase distributed throughout the ceramics [26].

Compressive stress tests showed that increasing the milling time higher strength the samples. The best compressive stress values were obtained for F-FA/RM sample specimen. The results in shown Table 3.

Glass ceramics prepared from the F-FA/RM mixture without adding additive and binder with a pressure strength of 155.30 MPa have been observed to have good specifications in comparison with the literature [23] found that the highest pressure strength of fly ash-based geopolymer bricks containing sodium hydroxide/sodium silicate additives at various ratios was 47.6 MPa which was considerably higher than the limit values given in national and international standards. The reason why C-FA has lower compressive strength than F-FA is that C type fly ash reacts with alkaline chemicals very quickly due to the high content of calcium oxide components. For this reason, experiments with C type of fly ash appears to be stabilized when glass ceramics are produced, but within a few hours, capillary cracks are formed. They can break down before and after the heat treatment. Another possible reason might be the presence of high levels of free calcium oxide in the C type fly ash.

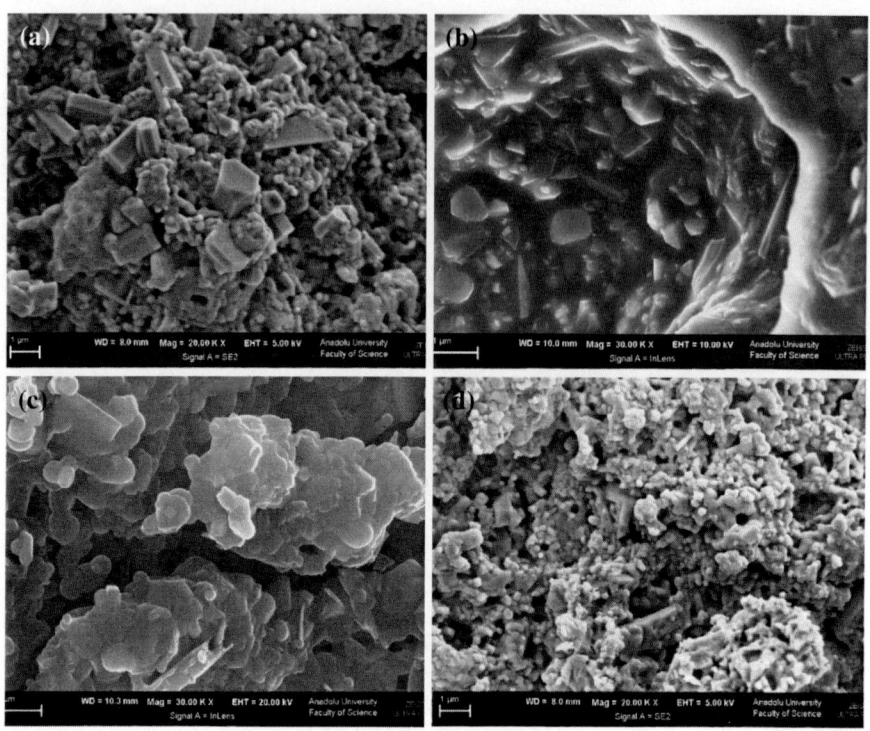

Fig. 6. Scanning electron micrograph of sintered materials at 1050 °C and milled during 8 h **a** C-FA, **b** RM, **c** F-FA/RM mixture, **d** C-FA/RM mixture

Table 3. Compressive strength of sintered materials at maximum compressive load

Sintered samples	Milling time (h)	Compressive stress (MPa)
F-FA	4	0.441
	8	0.688
C-FA	4	0.211
	8	0.256
RM	4	108.61
	8	120.20
F-FA/RM mixture	4	147.87
	8	155.30
C-FA/RM mixture	4	63.47
	8	86.08

4   Conclusions

The characterization in terms of both phase evolution during sintering and microstructure for milling time have been investigated on the ceramic bodies obtained from F type fly ash, C type fly ash, red mud and red mud-fly ashes mixtures.

The mineralogical evolution of F-FA examined by XRD revealed that quartz and magnetite are the main crystalline phase in the original material. In C-fly ash, the main peaks were quartz and free lime. Few minor peaks of dolomite and mullite were also reported. The pattern of red mud displays a complex pattern indicating a mixture of metal oxides and complex minerals. The dominant phase in red mud is hematite. Besides hematite, quartz, calcite and sodalite were also present in the as received form of red mud. In addition, new phases such as sodium aluminium silicate, calcium aluminum silicate, hematite, hercynite appear as a result of different chemical reaction between red mud and fly ash during sintering. The glassy phase is formed due to the partial dissolution of quartz and combination of other minor phases during sintering process. Increasing milling time gives to reduce a partial dissolution of quartz and therefore, the amount of amorphous phase increases. The silicate bonded phases developed at sintering temperature 1050 °C in the ceramic bodies red mud and fly ash mixtures. Scanning electron microscopy showed a homogenous microstructure composed of isolated pores, irregular shape, size crystal and ceramic matrix surrounding the formed phase. Detailed characterization and microstructure investigation opens up the possibilities of processing the materials in sintering stage for future materials production.

References

1. Borges AJP, Hauser-Davis RA, Oliveira TF (2011) Cleaner red mud residue production at an alumina plant by applying experimental design techniques in the filtration stage. J Clean Prod 19(15):1763–1769
2. Li LY (1998) Properties of red mud of tailings produced under varying process conditions. J Environ Eng 124(3):254–264
3. Liu Y, Naidu R (2014) Hidden values in bauxite residue (red mud): recovery of metals. Waste Manag 34:2662–2673
4. Arslan S, Ucbeyiay H, Çelikel B, Baygül M, Avcu S Demir GK (2015) Eti aluminum red mud characteristics and evaluation of dewatering performance. In: Bauxite residue valorization and best practices conference-2015. KU Leuven, Belgium (5–7/10/2015)
5. Kurtoglu SF, Soyer-Uzun S, Uzun A (2016) Tuning structural characteristics of red mud by simple treatments. Ceram Int 42(15):17581–17593
6. Liu S, Guan X, Zhang S, Xu C, Li H, Zhang J (2006) Sintering red mud based imitative ceramic bricks with CO$_2$ emissions below zero. Mater Lett 191:1–224
7. Yang J, Zhang D, Hou J, He B, Xiao B (2008) Preparation of glass-ceramics from red mud in the aluminum industries. Ceram Int 34(1):125–130
8. Liu RX, Poon CS (2016) Utilization of red mud derived from bauxite in self-compacting concrete. J Clean Prod 11(2):384–391
9. Kang SP, Kwon SJ (2017) Effects of red mud and alkali-activated slag cement on efflorescence in cement mortar. Constr Build Mater 133:459–467
10. Abbasi SM, Rashidi A, Ghorbani A, Khalaj G (2016) Synthesis, processing, characterization, and applications of red mud/carbon nanotube composites. Ceram Int 42 (15):16738–16743
11. Zhao Y, Zhang L, Ni F, Xi B, Xia X, Peng X, Luan Z (2011) Evaluation of a novel composite inorganic coagulant prepared by red mud for phosphate removal. Desalination 273(2–3):414–420

12. Feng Y, Wu D, Liao C, Deng Y, Zhang T, Shih K (2016) Red mud powders as low-cost and efficient catalysts for persulfate activation: pathways and reusability of mineralizing Sulfadiazine. Sep Purif Technol 167:136–145

13. Kumar S, Kumar R, Bandopadhyay A (2006) Innovative methodologies for the utilization of wastes from metallurgical and allied industries. Resour Conserv Recycl 48(4):301–314

14. Zhang M, Zhao M, Zhang G, Mann D, Lumsden K, Tao M (2016) Durability of red mud-fly ash based geopolymer and leacing behavior of heavy metals in sulphiric acid solutions and deionized water. Constr Build Mater 124:373–382

15. Ersoy B, Kavas T, Evcin A, Başpınar S, Sarıışık A, Önce G (2008) The Effect of BaCO₃ addition on the sintering behaviour of lignite coal fly ash. Fuel 87:2563–2571

16. Blissett RS, Rowson NA (2012) A review of the multi-component utilisation of coal fly ash. Fuel 97:1–23

17. Fırat S, Yilmaz G, Cömert AT, Sümer M (2012) Utilization of marble dust, fly ash and waste sand (Silt-Quartz) in road subbase filling materials. KSCE J Civ Eng 16(7):1143–1151

18. Fırat S, Comert AT (2011) Curing time effects on CBR of stabilized kaoline with fly ash, lime and cement. J Fac Eng Archit Gaz 26(4):719–730

19. Öntürk K, Fırat S, Vural İ, Khatib JM (2014) Uçucu kül ve mermer tozu kullanarak yol altyapısının iyileştirilmesi", J Polytech 17(1) (Special Issue):35–42

20. Pandey VC, Singh N (2010) Impact of fly ash incorporation in soil systems. Agric Ecosyst Environ 136:16–27

21. Li F, Wu W, Li R, Fu X (2016) Adsorption of phosphate by acid-modified fly ash and palygorskite in aqueous solution: experimental and modeling. Appl Clay Sci 132:343–352

22. Barbieri L, Lancellotti I, Manfredini T, Queralt I, Rincon J (1999) Design, obtainment and properties of glasses and glass-ceramics from coal fly ash. Fuel 78:271–276

23. Zeybek O (2009) Fly ash based geopolymer brick production. Master of Science Thesis, Anadolu University, Graduate School of Sciences, Civil Engineering Program, Eskisehir, Turkey (Advisor: O. Arioz)

24. McMillan PW (1979) Glass-ceramics, 2nd edn. Academic Press, London

25. Liu W, Yang J, Xiao B (2009) Review on treatment and utilization of bauxite residues in China. Int J Miner Process 93:220–231

26. Samal S, Ray AK, Bandopadhyay A (2015) Characterization and microstructure observation of sintered red mud-fly ash mixture at various elevated temperature. J Clean Prod 101:368–376

27. Kumar A, Kumar S (2013) Developtment of paving blocks from synergistic use of red mud and fly ash using geopolymerization. Constr Build Mater 38:865–871

28. Liu T, Tang Y, Li Z, Wu T, Lu A (2016) Red Mud and fly ash in corporation for lightweight foamed ceramics using lead-zinc mine tailings as foaming agent. Mater Lett 183:362–364

29. Yılmaz H (2015) Characterization and comparison of leaching behaviors of fly ash samples from three different power plants in Turkey. Fuel Process Technol 137:240–249

30. Erol M, Genc A, Ovecoglu ML, Kucukbayrak S, Tapyik Y, Yucelen E (2000) Characterization of a glass-ceramic produced from thermal power plant fly ashes. J Eur Ceram Soc 20:2209–2214

31. Yılmaz G (2012) Structural characterization of glass–ceramics made from fly ash containing SiO_2–Al_2O_3–Fe_2O_3–CaO and analysis by FTIR–XRD–SEM methods. J Mol Struct 1019:37–42

Key Success Factors Impacting the Success of Innovation in UAE Construction Projects

Mansour Faried[(⊠)], Malak Saad, and Khalid Almarri

BUiD, Dubai, United Arab Emirates
{2014133025, 2014133069}@student.buid.ac.ae,
khalid.almarri@buid.ac.ae

Abstract. *Purpose* Several Former researches has been concerned with the Innovation concept, however, to the extent of authors knowledge, there are less literature concerned the application of Innovation concept in construction projects in United Arab of Emirates, with a clear identification of the exact role of each member of the participants in UAE innovation process, The purpose of this research is to investigate and identify the critical factors which influence and enable innovation in UAE construction projects. Also, identify responsible party for each factor application. *Design/methodology/approach* The research took form of qualitative approach first, which rely on critical analysis of relevant literatures focused on innovation in construction projects and similar industries, either in UAE or similar Contexts, which have been found to be applicable in UAE construction projects, aiming to identify the critical factors influencing the innovation in construction industry in UAE, and extend the research to classify the ownership of each factor among all parties involved in the process. Then, the findings have been examined using the quantitative approach though a designed survey to validate the finding and measure the Impact of each factor on the success of UAE innovative construction projects, as well as the correlation and regression between the different identified factors with each other. *Findings* According to the study results, three major players has been identified as a main influencer to the innovation process in UAE construction project which are: Government, R&D institutions & organizations. Also a set of ten factors categories has been identified as a critical for innovation success which are: Support of R&D investment, Commitment, Vision & Policies, advanced procurement systems, Leadership style, market oriented researches, Flexibility of experts' movement, knowledge exchange, In-house testing facilities and Decentralized university system. Application of those ten factors categories found to be different base on the scope of each player of the three key players which resulted into thirty key success factors for innovation in UAE construction projects. Furthermore, the study identified the impact strength of each factors in order to identify the scope of each key player to focus and involve as a form of scope matrix. Also, types of innovation has been identified as the "technical innovation" and "organizational innovation", the study result suggest that both of them required to be developed simultaneously in order to deliver the targeted "successful innovation in UAE construction project" *Research limitations/implications* The identified factors are considered as a main factors but not an all-inclusive list, further research may add more factors which might be directly or indirectly impact the innovation in UAE construction projects. *Practical implications* The concluded list of critical factors influencing the innovation in UAE construction projects with its scope demarcation matrix will help the major

© Springer International Publishing AG, part of Springer Nature 2018
S. Fırat et al. (eds.), *Proceedings of 3rd International Sustainable Buildings Symposium (ISBS 2017)*, Lecture Notes in Civil Engineering 7,
https://doi.org/10.1007/978-3-319-64349-6_39

players involved in innovation process to recognize the contribution expected from each, and then can be considered as a base for any socio-economic initiatives can be introduced in present of near future to support the process of innovation in UAE construction projects. *Originality/value* The study generate its value through integration of partial knowledge gained by previous various study into a comprehensive focused analysis on the innovation in construction projects in the country of UAE, which to extent of authors knowledge, never been discussed that precisely before.

Keywords: Innovation · Construction · UAE · Vision & policies R&D investment

1   Introduction

Background

The construction sector is the main economy sector in gulf region, where more than $570 billion is the budget of the construction projects awarded since 2005 only in gulf region. Meanwhile UAE market is the major player in the gulf region and that was really obvious when the gulf region market impacted after Dubai real state crisis in 2009 [1]. The call for innovation was essential for the UAE construction market in order to survive and sustain with stable improvement and growth, which will have positive influence to attract the foreign investment to the country [2].

In fact the call for innovation mentioned above had full support from the country government because it was in the same direction with the strategic plan and leadership vision of H.H Sheikh Mohammed Bin Rashid Al Maktoum. Moreover, it was declared that the year of 2015 will be "Innovation Year". Also, the government declared its willingness to achieve its goal for innovation future of UAE as the most innovative nation in the world after only seven years [3].

Therefore, organizations will have to improve the ability and competence to introduce projects that are efficient of producing innovative products and services, which is robust to the above mentioned risks. Such firms need to review how to consider, introduce, design and apply its projects to provide successful innovation.

The Study Question

What are the factors could enhance the innovation of construction industry in UAE?

The Aim of the Study

The study is aim to spot on the critical factors which impact the innovation in UAE construction projects, which could help the construction firms and specialists to enhance the construction innovation within the country.

The Objectives of the Study

1. To conduct a literature review to introduce the concept of innovation and its impact on general practice.
2. To determine the factors which impact the innovation in construction.

3. To determine the critical factors which impact the innovation in construction business in the UAE.

To propose a framework for UAE construction practitioners for improving the innovation process within the project life cycle.

2   Literature Review

2.1   Definition of Construction Innovation

As per Ozorhon [4] Innovation is complex and dynamic even if it is not a technical innovation. Innovation is the application of the new brilliant ideas at any aspect or field [5]. In regards of construction industry context, Construction industry is totally different than the manufacturing and service industries because of the project base type of construction business, which makes each project unique. That also, reflects on the innovation behaviour [6]. Construction innovation focus on the project and considering the achievement of its goals, the process of construction innovation covers many of project elements such as, human resources, cost, management, problem solution and implementation etc. [7].

2.2   The Characteristics of Construction Innovation

(1) Project-based cooperative innovation; there are two levels of cooperation; project and strategic. In other words, there is cooperation between parties involved in the project and there is long-term cooperation between different construction organizations in the industry [4].
(2) Integrative innovation; the integration in Construction is essential not only for technological aspects but also needed from society, management, etc. [8].
(3) Open innovation; due to the diversity of construction project activities and phases there are many parties involved such as; contractors, suppliers, consultants, etc. every party of these parties consider as a source for innovation that makes the construction innovation a kind of open innovation [9].
(4) Dynamic innovation; the construction projects are complex, dynamic, and nonlinear, so that the construction innovation has the same characteristics, because each project is unique than the others [8].

2.3   Historical Development of Innovation Concept in UAE

Last decade has witnessed a huge impact on innovation due to technology rapid development. Moreover, the worldwide market became more competitive and fast growing, the innovation power has been focus on demand rather than the production especially after the widely use of the internet [10].

Among the Gulf region countries, The United Arab Emirates (UAE) is the country featured with fast growing and innovative economy market. Also, UAE's innovation is share 30% of business activities [11]. Although the UAE is classified as hydrocarbon-rich country, the government realized the importance of economy transformation from relying on hydrocarbon exports to diversity in income sources and sustainable economy in order to have long-term economic development, so the government set up its vision 2021 with ambition goals [12]. In October 2014, His Highness Sheikh Mohammed bin Rashid Al Maktoum, Vice President and Prime Minister of the UAE, has announce that there are seven development sectors named; renewable energy, transport, education, health, technology, water and space, will be taken the focus under the government innovation strategy of the same year [3]. In November 2014, the UAE government has approved the title of 2015 to be the year of innovation, also the government gave the direction to establish the National Innovation Committee (NIC), it will gather the UAE's efforts and supervise the process to lead the country toward the top of the list of innovative countries over the world with challenged duration of seven years only to achieve its goal [3]. In Feburary 2015 a seven-dimension program announced by the government, South Korea was the example of innovative country that UAE aim to implement especially in education sector, schools and universities [12].

Major funds in UAE has directed to serve key industries, for example hydrocarbon services, semiconductor, space, clean energy and construction, in the same line of UAE vision to take the lead of knowledge-based economy [13]. According to Al-Jundi [14], UAE enterprises, especially construction business, required innovation. Notwithstanding, some examples of construction firms have effectively innovating and challenging in the worldwide markets, most of small and medium construction organizations does not have a high level of innovation while the international firms have more innovation as the small and medium firms usually operate in traditional manner [3]. One of the ways to push for innovation is the international competition. Although, to push the innovation in construction business it is a must to reflect the firm's business model. Plus, the general climate to make the innovation possible [15]. As per Blayse et al. [16] the construction innovation should be seen as 'product system' innovation, this perspective includes all the parties usually involves in construction industry such as; contractors, consultants, suppliers, manufactures, clients and technical support providers.

2.4   Widely Used Forms of Innovation

Abbot et al. [17] has defined three types of innovation; firstly, product innovation or technical innovation, which related to specific construction products and could be drawn-out to include the whole construction system. Moreover, design of the product, product material and manufacturing process of the product also considered in the product innovation. Secondly, procurement innovation or organizational innovation, which related to the way that the project was procured, new forms of contract and supply chain management are considered as samples of procurement innovation. Thirdly, process innovation, which related to the process itself and how the work is going to be achieved. Although, the type of information for steel frame building and cast in situ concrete building can be similar. But the participating firms and supply chain could be totally different [18]. The study will focus only on the first and second types of innovation as the process innovation was found more related to the manufacturing rather than construction. Abbot and Behrens [19] has proposed five levels of innovation related to technical and construction organizations as below:

- Generic innovations, which is aiming to have a new paradigms like the improvement of the concrete characteristics.
- Epochal innovations, it is a sub-degree of generic innovations. It is about improvement of certain sector of the activity.
- Altering innovations, it is promote to alter at the high level of the organization.
- Entrenching innovations, it is only a modification to the existing methodology in align with the strategic direction.
- Incremental innovation, it is how to use the same collected input to improve and enhance the output.

2.5   Key Success Factors of Innovation in UAE Construction Projects

The following section will brief the identified factors which counted responsible of the success of innovation in UAE construction project through deep review of relevant literatures

- Support of R&D investment

A positive and strong relationship has been found between the development of national innovation and level of local R&D in that particular country [20]. Government reports Conducted in recent years indicated that low interest in R&D Investment in UAE is one of the key factors which negatively impact the innovation in Local construction market [21]. Accordingly, reducing its contribution share in economic growth in the country [22]. Schilirò [3] has promoted that Link between the innovation in UAE and investment on R&D and highlighted that it is not only a matter of the investment amount on R&D, but also on the correct deployment of that investment to be aligned with the innovation strategy. Supporting of R&D investment needs to be done by government's through different mechanisms like rewards system, tax relief … etc. [23]. Also, through organization by allocation of certain amount of Business development for in-house R&D [24]. As well as, R&D institutes independent institutes and universities which needs to find marketing mechanisms to raise funds for supporting their research's [25].

- Commitment

Commitment has been identified in various literatures as very crucial factor for the success of construction industry innovation in UAE. Since the Innovation process will require to go through various administrative and technical challenges, the commitment of organization top management on innovation will enhance the integration between team members and support them to come over those challenges [26]

That commitment will require extending from the government and authority leaders through the main organization leaders up to supply chain leaders and managers [27].

Rahman et al. [28] highlighted the importance of leader commitment on promoting the team work concept, trust, conflict resolution and respect, which all considered as enabler for innovation.

- Vision & policies

Innovation in UAE has been initiated and strongly backed by the vision of the Country leadership, as the government has realized the role of innovation to reform the economy into a sustainable economy, which relies on wide range of knowledge rather than consumable resources [12, 14]. Recently on late of 2014, the Government of UAE under leader ship of H.H Sheikh Mohammed Bin Rashid al Maktoum has lunched the "National Innovation strategy "which determined the sectors to be under focus of Innovation and development, such sectors includes the transport, renewable energy, space, health, water & education, eventually most of those sectors require the support of construction industry which consequently will call for innovation in construction industry itself [3], Such vision is translated in reality into a set of polices which include new regulations, direction of investment, engorgement of private sector and modification of education curriculums to include motivation concept in its core and methods [29].

- Advanced procurement systems

Procurement systems has been identified as a major factor affecting the innovation in construction activities. Blayse and Manley [22] highlighted that the adoption of non-traditional procurement and contracting method such as partnership, PPP, joint ventures ... etc. Abuelmaatti and Ahmed [30] facilitate the cooperation and problem solving through un-traditional solutions. Akintoye et al. [31] have seen the innovation process in construction more successful when the integration between the various stakeholders and supply chain became more effective, and the advanced procurement system will be a vital enabler for such effective integration.

Albaloushi and Skitmore [32] focused suppliers and Supply chain management (SCM) in the context of construction projects in UAE. They have concluded that the SCM will positively facilitate the collaboration among project parties when strategic and innovative procurement concepts have been adopted, which will unify the goals and will keep the team focus on innovation targets.

- Leadership style

Leadership style found to be linked to organization innovation ability according to various literatures, teams leaders play an essential role in fixing and implementing the require policies and procedures to achieve innovation ([33]; Fig. 1).

There are two main styles of leadership identified in literatures: "transactional style" which developing the relationship between the team members and their leader through clear communication on what exactly is required from them, with cutting edge objectives and goals [34], such style of leadership found with positive impact on team innovation when the emotion factor among team member is low, and it has a negative impact when the emotion factor goes up.

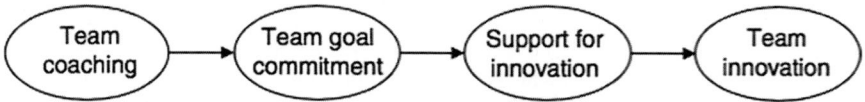

Fig. 1. Team leader role in achieving team innovation [33]

The other style of leadership is the "transformational style" which can be defined as motivating team members to go beyond their nominated capacity through changing their thoughts and perceptions [35]. Such style found to be opposite to previous style, as it will support and enable team innovation abilities whenever the emotional factors are high, while if the emotion factor goes down the innovation opportunity will be shut down. [36].

- Market oriented researches

A detailed study conducted by Alam [37], to explore the relationship between the firm innovation and firm market orientation, the study was a quantitative research based on a questioner survey and the study results suggested a strong positive relationship between firm innovation capacity and firm market orientation, furthermore, the study promoted the customer orientation as a competition strategy to survive in high competitive environment, and the innovative solution which will be generated accordingly will be highly expected by market as it represents its real needs.

That study is supported by another study conducted by Al-Abd et al. [38] but with more focus on UAE context, and they have suggested that organization and research centres need to be more focus in developing and responding to market demands in order to enhance innovation system in UAE.

- Flexibility of experts movement

Movement of skilled labours, specialist and experts between organizations will allow new solution to generate through mixture of different working cultures and knowledge [39]. Alinaitwe et al. [20] stated that policies and regulation which hindering the flexibility of specialist movements between companies are considered as a barrier for national innovation in UAE.

The same concept has come across the study conducted by Randeree [39] which aims to measure the perception of expats workers towards their organizations in UAE, the result finding mention that although the skilled labour movement restrictions would be positively support organization stability, however it negatively affect skilled labours motivation and innovation ability.

- Knowledge exchange

Shortage of professionals, experts and skilled labours are always counted as main challenge for construction industry in UAE, and has been marked as one of major risks responsible for projects delay [40]. UAE government has realized that fact, and they have launched a strategy in February 2015 composed of seven initiatives to support the national innovation in the country, among those initiatives are to held annual exhibitions for innovation by the participation of schools, universities, government entities and private sectors. Those exhibition will be a potential attraction of global experts to follow and participate. Another, initiative was about holding training programmes for potential innovators by world leading experts under the fund of UAE government [3].

Same concept will be achieved through encouragement of foreign investment in innovative construction projects in UAE. Especially, for world industry leaders, who will be having their own R&D centres, and they will be coming to country along with their knowledge and experts, a live example to that is Masdar City in Abu Dhabi [41]

- In-house testing facilities

An important study carried out by Castellani et al. [42] attempting to measure the impact of the distance between R&D institution and their testing laboratories on the value of the R&D process output, and the results indicated that although common culture, language, and even religious back ground are strongly impacting the output, and according to data collected from more than 6000 R&D firm located in 59 country; close geographic proximity of R&D institute and the testing laboratory will positively enhance the R&D process output [42].

Al Hallami et al. [43] have come across the same topic while analyzing the challenges and obstructions that R&D centers in UAE are currently facing, and according to their analysis, the have recorded the shortage of local and in-house testing facilities as main barriers affecting the construction innovation development in UAE [43].

- Decentralized university system

Creating a decentralized high education system in UAE has been identified by Al-Abd et al. [38] as part of comprehensive strategy proposed to be adopted by UAE government to support local innovation development.

The relationship between the decentralization concept and innovation has been explained by Pamuk et al. [44] as the decentralization concept will reflect diversity in cultures, priorities, and solution methodology, which consequently will enable innovation development. And since Universities are considered as main R&D institution where Innovation will be generated and Founded [25], decentralizing the university systems will ensure broader and faster innovation development [38].

3   Discussion

By analysing the data collected above through the Literature review, we can identify that there are three main key players held responsible for innovation in UAE construction projects, those key players are:

(1) Government,
(2) R&D institutions (i.e. universities, independent research centres … etc.)
(3) Construction organizations (i.e.: clients, developers, consultants, contractors and suppliers).

Those three key players are in charge of ten factor categories identified as enabler for innovation in UAE construction project.

Each of the ten numbers of factors categories found to be with different application base on the scope of each player of the three major players and its relation to the factor categories. Accordingly the 10 factors categories could be translated into thirty factors impacting the Innovation in UAE construction project (ten factors categories, each contain three factors as the application of factor category from the perspective/involvement of each player of the three key players identified above). As per the following table:

According to the identified factors and their grouping via Table 1 above, the following hypotheses can be concluded:

Table 1. Key success factors of innovation in UAE construction projects

Factor category	Government factors	Organization factors	R&D institute factors
Support R&D investment	Support investment on R&D through Government initiatives	Budget allocation for R&D within organization financial plans	R&D institutes self-funding strategies
Reference	[3, 20–25]		
Commitment on innovation	Government commitment for innovation	Organizations top management commitment to innovation adoption	R&D institutes development of new procurement systems
Reference	[26–28]		
Visions & policies	Support innovation through Government vision and policies	Alignment of organization strategy with government innovation policy	R&D institutes leadership style
Reference	[3, 12, 14, 29]		
Advanced procurement system	Government policy acceptance of new procurement system	Organizations generation of new procurement systems	R&D institutes; Flexibility of experts movement
Reference	[22, 30–32]		
Leadership style	Government leadership style	Construction organizations leadership style	R&D institutes establishment of in-house testing facilities
Reference	[34–36]		
Market oriented research	Government response to market needs	Customer oriented organization strategy	Universities diversity in curricular and education methodologies
Reference	[37, 38]		
Flexibility of experts movements	Flexibility experts movement within labour law	Organization human resource policies	Allowance for experts movement between R&D institutes
Reference	[20, 39]		
International knowledge exchange	Cross governments knowledge exchange	Cross organization knowledge exchange	Cross universities knowledge exchange

(*continued*)

Table 1. (*continued*)

Factor category	Government factors	Organization factors	R&D institute factors
Reference	[3, 40, 41]		
In-house testing facilities	Government support for local testing facilities	Organization dependency on local testing facilities	R&D institutes development of new procurement systems
Reference	[42, 43]		
Decentralized university system	Government Education system support for Decentralization	Organization acceptance of different education curricular graduates	R&D institutes leadership style
Reference	[25, 38, 44]		

H1: government key success factors group are significantly support the success of innovation in UAE construction projects

H2: construction organization key success factors group are significantly support the success of innovation in UAE construction projects

H3: R&D Institutes key success factors group are significantly support the success of innovation in UAE construction projects

H4: Integration of Key success factors groups of Government, organization and R&D institute will result in higher success rate of innovation in UAE construction projects.

The above Hypotheses will be tested for validity and significance in the next section using quantitative research approach.

4 Methodology

This section of the paper represents the verification of the Hypotheses through analysing the data collected via designed Questionnaire (Appendix 1).

This methodology have been selected and design to measure practical applicability of theoretical research finding within two dimension in actual projects known as innovative projects, first dimension is measure the occurrence of each key success factor in the project, and second dimension is to measure the success rate of innovation in the project. And through analysing the relationship between both dimensions, then the relationship between variables can be concluded. And since actual data collection is base of analysis, then the survey questionnaire method with projects experts found to be the most appropriate method for this research.

The survey has two sections; the first section is designed to collect personal information to represent the candidate suitability for the survey in terms of Age, Level of education, years of experience ... etc. And the second section is designed to collect data related to the research parameters. Second section is composed of 4 scales, first 3 scales are collecting data about the Independent variables (critical success factors of innovation in UAE construction projects) and the last scale is measuring the dependant variable (success of innovation in UAE construction project).

The relative occurrence of the identified thirty critical success factors has been assessed using Likert rating system on five levels starting from "strongly agree" which represent the strong occurrence up to strongly disagree which represent the complete absence of the factor. Candidate has been selected from participants in various Innovative projects, their organization also has a verity between contractors, clients and consultant.

A pilot test has been carried oud on a sample of 5 persons, and according to the pilot result a minor adjustment has been made to the questionnaire and then floated to participants for survey.

Questionnaire has been circulated by means of hard copy handed over by hand, or electronically by means of email for some participants who is currently located out of country.

A total of forty (40) invitations to survey have been sent out, and total of twenty eight (28) filled questionnaire returned with a response rate of 70% which is considered acceptable [45].

The results has loaded on SPSS statistics software, and the items has been grouped in four factors, first factors is combining the items which is related to government (independent variable), second factor is combining the items related to Organization (independent variable), third factor is combining items related to R&D institutes (independent variable), and last factors is combining the items that measure the success of innovation in UAE Construction project (dependent variable). The following tests has been carried out with its correspondent values:

(a) Reliability test

There were thirty factors identified in the literature review as key success factors for innovation of construction projects in UAE. However, the reliability test was done also for the same factors. Cronbach's alpha technique was applied to measure the internal consistency and was found with value of 0.891 (Table 2) which is above 0.7 and accordingly the collected data for the identified factors could be considered as highly reliable [46].

Table 2. Reliability statistics using alpha method

Reliability statistics	
Cronbach's alpha	N of items
0.891	34

The test has been repeated for all items not including factors nor global using the half split Model, the value of the half split resulted was 0.823 (Table 3) which is very close to the alpha value, and accordingly confirm the high reliability of the collected data [46].

Table 3. Reliability statistics using half split method

Reliability statistics			
Cronbach's alpha	Part 1	Value	0.809
		N of items	17
	Part 2	Value	0.828
		N of items	17
	Total N of items		34
Correlation between forms			0.702
Spearman-Brown coefficient	Equal length		0.825
	Unequal length		0.825
Guttman Split-Half coefficient			0.823

(b) Factor analyses

Principal components analysis (PCA) has done for 30-question survey related to independent variables to figure out the KSFs of innovation in construction projects on 28 individuals. The suitability of PCA has evaluated before start the analysis. Correlation analyses was done and found that all factors have one or more correlation coefficient more than 0.3. PCA found out five components that had correlation bigger than one and which translated 26.9, 13.4, 11.6, 8.1 and 4.2% of the total variance, respectively. Visual check of the screen plot presented that five components could be taken [47]. Moreover, the five-component taken encountered the interpretability criterion. However, there are six factors has been eliminated out of total thirty factor concluded from the literature review.

The five-component taken explained 59.9% of total variance. Rotated technique was run to explain the survey statistics, the rotation technique represent the 'simple structure' [48]. The explanation of the information was according to the factors pointed out from the questionnaire which was expected to have strong loadings of government and R&D institutes items on Component 1, government, organization and R&D institute's items on Component 2 and 5, organization items only on Component 3 and government and organization items on Component 4. Component loadings of the rotated technique can be shown on Table 4.

• Factor Group 1

That Principal component represent 26.9% of the total variance of key success factor. There are seven factors in that group as below:

– Government policy acceptation of new procurement system; it is important for government to adopt non-traditional method of procurement system such as Joint Venture, Partnership ... etc. in order to enhance the innovation in construction industry [22].
– Government leadership style and R&D institutes leadership style; Leadership style is essential for any organization aims to improve its innovation ability through the well understanding of teams leaders to the implementation of polices that make the team achieve the required innovation [33].

Table 4. Rotated component matrix

Rotated component matrix

	Component				
	1	2	3	4	5
Non-traditional procurement system used in the project were accepted by law/codes and authorities	0.636	0.206	0.003	0.192	0.255
Leadership style of governmental leaders positively impacted innovation in my project	0.662	0.075	0.173	0.162	0.002
Project representatives participated in international events sponsored by government (exhibition/seminars/conferences)	0.622	0.102	0.378	0.072	0.449
Education system in UAE accept diversity and allow for decentralization	0.708	0.213	0.214	−0.055	−0.112
R&D institutes in UAE are able to raise fund for their research through marketing strategies	0.756	0.022	−0.056	0.025	−0.172
Leadership style of R&D institutes involved in my project supported the teamwork and facilitated the innovation environment	0.570	0.461	−0.241	−0.407	0.008
R&D institutes policies allows for experts transfer and movement within local market	0.658	0.223	−0.051	0.037	−0.043
Government initiative supported the Investment in R&D related to innovation in my project	0.345	0.616	0.227	−0.144	0.150
My organization strategy is inline with government innovation policy	0.295	0.849	−0.053	0.150	−0.047
R&D institutes in UAE are working on evaluation and analyzing of the new procurement systems	0.295	0.849	−0.053	0.150	−0.047
R&D institutes in UAE their own in-house testing facilities required to carry on their researches	−0.030	0.630	0.401	0.293	0.070
There are diversity & variety Universities' curricular and education methodologies in UAE	−0.032	0.673	0.143	−0.127	−0.241
There is an allocated budget within my organization for R&D and new technologies researches.	0.184	−0.108	0.575	0.056	−0.086
Leadership style for my project leaders facilitate the teamwork and enhanced personal output of team members	0.214	0.206	0.556	−0.146	0.049
My organization adopt and implement a customer oriented strategy	0.227	0.167	0.662	−0.072	−0.008
Human resource policies within the project organization strictly follow labor laws.	−0.235	0.040	0.717	−0.108	0.147
Participating organizations were more preferring to carry out the required tests in local facilities	−0.200	0.140	0.596	−0.197	−0.346
Authorities and government monitor and response to market demands and challenges	0.033	−0.043	0.164	−0.521	0.351

(*continued*)

Table 4. (*continued*)

Rotated component matrix

	Component				
	1	2	3	4	5
Locally generated test results were accepted and accredited by local authorities	0.344	0.156	0.028	*0.815*	0.121
My organizations top management commit to innovation adoption & support innovation trials	0.285	0.052	0.368	−0.609	0.199
There was an innovative procurement system applied and used in my project (fully or partially for some packages)	0.344	0.156	0.028	*0.815*	0.121
The Innovation in my project is in line with government general vision and policy	0.008	−0.065	0.410	−0.145	*0.649*
My Organization is welcoming and accepting the Hiring of different education curricular graduates	0.005	0.280	0.285	−0.128	*−0.767*
Local & international events allows knowledge exchange between universities & R&D institutes	0.339	−0.130	0.128	0.200	*−0.622*
Governmental authorities committed to success of innovation in my project	**0.404**	**0.378**	**0.240**	**−0.005**	**−0.010**
Current labours law allowed the project to acquire the required experts and specialist from local market	**0.385**	**0.105**	**0.288**	**−0.491**	**0.058**
There is a possibility of knowledge exchange between organizations in local or international market	**0.137**	**0.470**	**0.144**	**0.421**	**0.438**
R&D institutes in UAE committed to promote innovation in Construction industry	**0.438**	**0.378**	**0.185**	**−0.320**	**−0.276**
Strategy of R&D institutes in UAE is in line with government innovation policy	**0.198**	**0.343**	**0.483**	**0.258**	**0.375**
Researches made by R&D institutes in UAE are tailored to fit construction market needs	**0.430**	**0.468**	**0.032**	**0.349**	**0.165**

- Cross governments knowledge exchange; UAE government realized the importance of knowledge exchange. Therefore, it has encourage the international cooperation and holding of international exhibtation related to construction innovation [3].
- Government Education system support for Decentralization; the UAE government understand the importance of decentralization of education. In order to enhance the construction innovation in UAE, so it was a part of the country strategy to support the innovation [38].
- R&D institutes self-funding strategies; R&D institutes and universities are required to find the means to fund its researches for innovation. It is important to have its financial strategy to encourage the researchers to work on innovation topics [25].
- R&D institutes; Flexibility of experts movement; the movement of professionals and experts between R&D institutes will let new ideas to come through mixture of diverse working cultures and knowledge [39].

- Factor Group 2

That Principal component represent 13.4% of the toatl variance of key success factor. There are five factors in that group as below:

– Support investment on R&D through Government initiatives; there is a positive connection between the investment on the R&D and the level of the country innovation [20].
– Alignment of organization strategy with government innovation policy; UAE government has an innovation policy and strategy which needs to be implemented by all public and private construction firms to achieve the required innovation in construction industry [3].
– R&D institutes development of new procurement systems; Procurement systems recognized as a main factor impacting the innovation in construction industry. Therefore, the development of new procurement system will be highly required [30].
– R&D institutes establishment of in-house testing facilities; the availability of in-house testing facilities will help the R&D institutes to accelerate the process and reduce the time required to conduct such experiments [42].
– Universities diversity in curricular and education methodologies; the diversity in curricular will allow many education establishments to participate and contribute in the education system in the country which will enable and enhance the innovation development [44].

- Factor Group 3

That Principal component represent 11.16% of the total variance of key success factor. There are five factors in that group as below:

– Budget allocation for R&D within organization financial plans; each organization is required to allocate certain amount of its budget for innovation development, In order to be able to compete with the other firms in construction industry [24].
– Construction organizations leadership style; the teams leaders have a role to motivate the team members to go afar their nominated capacity through changing their views and opinions [35].
– Motivating team members to Customer oriented organization strategy; there is a positive relationship between customer oriented concept and the firm's innovation, the customer oriented concept will allow the construction firms to survive in UAE competitive market [37].
– Organization human resource policies; the restrictions on movement of skilled workforce is impacting negatively the individual's motivation for innovation [39].
– Organization dependency on local testing facilities; the construction organization would save a lot of project time by rely on the local testing facilities, which will enable the innovation [42].

- Factor grouping 4

That Principal component represents 8.1 of the total variance of key success factors. There are four factors in that group as below:

– Government response to market needs; the government needs to pay more attention in responding to market requirements to enable and enhance the construction innovation in UAE [38].

– Government support for local testing facilities; the shortage of local testing facilities is UAE is defined as a barrier for innovation. Therefore, the government is required to support such facilities in order to enhance the innovation in the country [43].

– Organizations top management commitment to innovation adoption; the obligation of firm top management on innovation will increase the incorporation between team individuals and support them to come over implementation challenges [26]

– Organizations generation of new procurement systems; the Supply chain management of construction projects in UAE will certainly enable the association amongst project parties once innovative procurement system take place, which will combine the efforts to achieve the innovation targets [32].

- Factor grouping 5

That Principal component represents 4.2 of the total variance of key success factors. There are four factors in that group as below:

– Support innovation through Government vision and policies; the government has recognized the importance of innovation to transform the economy from normal to a sustainable economy, that depends on diversity of knowledge instead of consumable resources [14],

– Organization acceptance of different education curricular graduates; acceptance of the diversity in curricular will allow many individuals to join the organizations which will enable and enhance the innovation development [44]

– Cross universities knowledge exchange; Universities are to held international seminars and exhibitions in terms of innovation, which could be attraction point for innovation leaders all over the world to share their knowledge [3].

(c) Correlation:

A Pearson's product-moment correlation has been done to measure the relationship between the key success factor for innovation in construction in projects and the successful innovation. The test has run in two stages;

- **First stage:** before factor reduction procedure in order to examine the literature review hypotheses, the correlation test has been carried out between the dependent variable group (Group DV) and the three main groups of Independent variables: governmental factors group, organizational factors group and R&D institutes factors group. Each group represents once of the identified Hypotheses H1, H2 & H3.

- **Second stage:** after factor reduction procedure in order to measure the five group of factors with the successful innovation in construction projects. Initial analyses displayed the association to be linear with all variables normally dispersed, as evaluated by Shapiro-Wilk's test ($p > 0.05$), and there were no outliers. There was a significant positive correlation between group factor number one 0.010, group factor number two 0.001, group factor number three 0.026 and group factor number four 0.010, While there was a weak positive correlation with group factor number five 0.230 (refer to Table 5).

Table 5. Correlation test

Correlations

		Gov	Org	RnD	G1	Gr2	G3	G4	G5	G-DV
Gov	Pearson correlation	1	0.558**	0.681**	0.874**	0.516**	0.387*	0.698**	0.256	0.563**
	Sig. (2-tailed)		0.002	0.000	0.000	0.005	0.042	0.000	0.188	0.002
	N	28	28	28	28	28	28	28	28	28
Org	Pearson correlation	0.558**	1	0.575**	0.407*	0.582**	0.903**	0.454*	0.539**	0.585**
	Sig. (2-tailed)	0.002		0.001	0.032	0.001	0.000	0.015	0.003	0.001
	N	28	28	28	28	28	28	28	28	28
RnD	Pearson correlation	0.681**	0.575**	1	0.758**	0.833**	0.262	0.427*	0.390*	0.650**
	Sig. (2-tailed)	0.000	0.001		0.000	0.000	0.177	0.023	0.040	0.000
	N	28	28	28	28	28	28	28	28	28
G1	Pearson correlation	0.874**	0.407*	0.758**	1	0.487**	0.186	0.619**	0.188	0.478*
	Sig. (2-tailed)	0.000	0.032	0.000		0.009	0.344	0.000	0.339	0.010
	N	28	28	28	28	28	28	28	28	28
G2	Pearson correlation	0.516**	0.582**	0.833**	0.487**	1	0.309	0.276	0.160	0.611**
	Sig. (2-tailed)	0.005	0.001	0.000	0.009		0.110	0.156	0.415	0.001
	N	28	28	28	28	28	28	28	28	28
G3	Pearson correlation	0.387*	0.903**	0.262	0.186	0.309	1	0.259	0.434*	0.419*
	Sig. (2-tailed)	0.042	0.000	0.177	0.344	0.110		0.184	0.021	0.026
	N	28	28	28	28	28	28	28	28	28
G4	Pearson correlation	0.698**	0.454*	0.427*	0.619**	0.276	0.259	1	0.120	0.478*
	Sig. (2-tailed)	0.000	0.015	0.023	0.000	0.156	0.184		0.544	0.010
	N	28	28	28	28	28	28	28	28	28
G5	Pearson correlation	0.256	0.539**	0.390*	0.188	0.160	0.434*	0.120	1	0.230
	Sig. (2-tailed)	0.188	0.003	0.040	0.339	0.415	0.021	0.544		0.238
	N	28	28	28	28	28	28	28	28	28
G-DV	**Pearson correlation**	**0.563****	**0.585****	**0.650****	**0.478***	**0.611****	**0.419***	**0.478***	**0.230**	**1**
	Sig. (2-tailed)	**0.002**	**0.001**	**0.000**	**0.010**	**0.001**	**0.026**	**0.010**	**0.238**	
	N	**28**	**28**	**28**	**28**	**28**	**28**	**28**	**28**	**28**

**Correlation is significant at the 0.01 level (2-tailed)
*Correlation is significant at the 0.05 level (2-tailed)

- **The result of stage 1:** found that there was a statistically significant relationship between government, organization and R&D factors as independent variables and successful project innovation as dependant variable. Correlation was significant at the 0.01 level for all of three groups of factors. Therefore, we can accept the literature review hypotheses [49].
- **The result of stage 2:** the total factors included in Group1,2,3&4 could be considered as critical factors impacting innovation, which the this test result couldn't proof the correlation between the factors in Group 5 and the dependant variable. Accordingly the factors in group 5 could be excluded from this test result.
- The results of both stage 1 and stage 2 correlation test can be seen combined in Table 5 above for reference.

(d) Regression:

Regression test has been conducted on the survey parameters to confirm the correlation and predict the relation development between the dependant variable (success of Innovation) and the various independent variables (KSF for innovation success) as per the following steps:

Firstly, A linear regression between the global independent variable (all thirty factors included) and the dependent variable (all items related to dependant factors) was run and found that the global independent factors could statistically significantly predict successful innovation in construction projects in UAE, $F (22,184) = 14.40$, $p < 0.0005$ the global independent factors accounted for 12.9% of the explained

Table 6. Regression test—global IV with DV

ANOVA[a]

Model	Sum of squares	df	Mean square	F	Sig.
1					
Regression	25.306	1	25.306	22.184	0.000[b]
Residual	29.659	26	1.141		
Total	54.964	27			

[a]Dependent variable: groupDV
[b]Predictors: (constant), globalIV

variability in successful innovation in UAE (As per Table 6).

Secondly, A linear regression was run consequently between the dependent variable and each group of the factor groups number 1, 2, 3, and 4 and found that the ANOVA F value for each consequently as the Following: 7.682 (Group1), 15.492 (Group2), 5.547 (Group3) & 7.709 (Group4)—as per Tables 7, 8, 9 and 10.

From the above analysis, since the F value of the regression between the dependant variable and the Global in table number 6 above found higher than the value of regression between the dependant variable and each of the Independent variable groups. It could be understood that the ultimate rate of Innovation success will be

Table 7. Regression test—group1 with DV

ANOVA[a]

Model	Sum of squares	df	Mean square	F	Sig.
1					
Regression	12.536	1	12.536	7.682	0.010[b]
Residual	42.429	26	1.632		
Total	54.964	27			

[a]Dependent Variable: group DV

Table 8. Regression test—Group2 with DV

ANOVA[a]

Model	Sum of squares	df	Mean square	F	Sig.
1					
Regression	20.522	1	20.522	15.492	0.001[b]
Residual	34.442	26	1.325		
Total	54.964	27			

[a]Dependent Variable: group DV
[b]Predictors: (Constant), Group2

Table 9. Regression test—Group3 with DV

ANOVA[a]

Model	Sum of squares	df	Mean square	F	Sig.
1					
Regression	9.664	1	9.664	5.547	0.026[b]
Residual	45.300	26	1.742		
Total	54.964	27			

[a]Dependent Variable: groupDV
[b]Predictors: (Constant), Group3

Table 10. Regression test—Group4 with DV

ANOVA[a]

Model	Sum of squares	df	Mean square	F	Sig.
1					
Regression	12.570	1	12.570	7.709	0.010[b]
Residual	42.394	26	1.631		
Total	54.964	27			

[a]Dependent Variable: group DV
[b]Predictors: (Constant), Group4

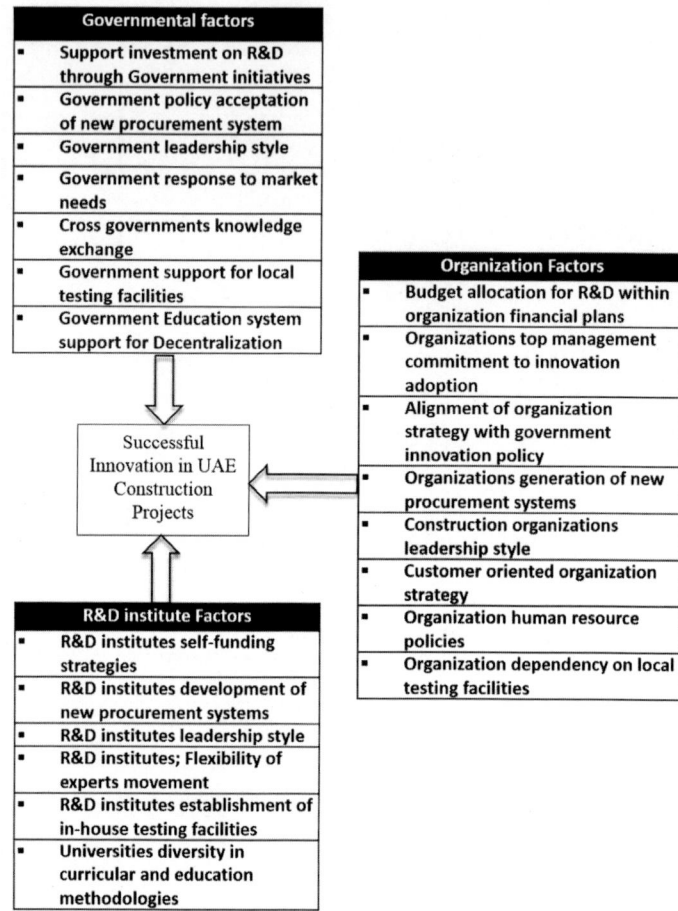

Fig. 2. Conceptual model

achieved when all the identified factors are implemented and integrated together. Which by logic prove and verify the validity of the hypotheses Number *H4* (Fig. 2).

5 Conclusions

The above analysis has concluded the following results:

- Data collected are highly reliable and could be trusted for study
- The tested hypotheses *H1, H2* and *H3* found valid due to high correlation between each group of factors and the success of innovation (Table 5)
- The tested Hypothesis *H4* found valid due to the achievement of Higher Anova F value when testing the regression with the global Independent factor (which represent the whole factors groups) comparing to individual Anova F value which

achieved when testing the regression of each group of factors with the success of innovation. (Tables 6, 7, 8, 9 and 10)

- The factors rotated transformation using the CP dimensions reduction method and further correlation test has highlighted the significant impact of total number of twenty one factors out of thirty factors tested. As total of nine factors can be ignored in this test (Six factors could be ignored due to low loading value and three factors can be ignored due to non-correlation with the dependant variable.
- The final segregated twenty one factors can be again grouped in three groups as per the following conceptual model:

6   Recommendations

According to previous finding and Conclusion, this section will provide a set of recommendation which can be taken as guide line for any adopted practice concerned with enabling the Innovation in UAE construction projects:

- Government should support the Investment in R&D via various Initiatives, the formal system should be flexible enough to accept new forms of procurements and contracts, government also need to be very well responsive to market needs.
- Government should organize the events which allow for knowledge exchange like competitions, and conferences on the governmental level with other countries.
- Government should support the establishment of laboratories and testing facilities locally, and concerned authorities should be willing to accept and accredit the locally generated test reports.
- The education system within the country should allow for diversity and decentralization.
- The Construction Organization in UAE should allocate the sufficient fund to support R&D investment within organization and project budget, and top management of the organization should commit on the Innovation success.
- Construction organizations need to align their strategy with governmental innovation vision, they should generate and accept new forms of procurement and their human resource policies should support the experts' rotation within the local market.
- Organization need to be customer oriented and operating base on market demand
- R&D Institutes and Universities in UAE should develop their marketing strategies to raise funds to support their researches and their system should allow for expert rotation within the local market
- Universities in UAE should develop research to evaluate and assess the new procurement methods, and they should have their own testing facilities or at least support and rely on the local testing facilities.

References

1. Neaime S (2012) The global financial crisis, financial linkages and correlations in returns and volatilities in emerging MENA stock markets. Emerg Mark Rev 13(3):268–282
2. Erogul M (2013) Entrepreneurial activity and attitude in the United Arab Emirates. Innov Manag Policy Pract 2159–2186
3. Schilirò D (2015) Innovation in small and medium enterprises in the United Arab Emirates. Int J Soc Sci Stud 3(5)
4. Ozorhon B (2013) Analysis of construction innovation process at project level. J Manag Eng 29(4):455–463
5. Ozorhon B, Abbott C, Aouad G (2014) Integration and leadership as enablers of innovation in construction: a case study. J Manag Eng 30(2):256–263
6. Widén K, Olander S, Atkin B (2013). Links between successful innovation diffusion and stakeholder engagement. J Manag Eng
7. Liu H et al (2014) Identification of critical success factors for construction innovation: from the perspective of strategic cooperation. Front Eng 1(2):202–237
8. Li BC (2010) Engineering innovation: break through the barriers and avoid the traps. Zhejiang University Press, Hangzhou
9. Wang MJ, Zhang ZS (2011). Analyse of the formation and operation mechanism for technological innovation network of major construction projects. Chin Eng Sci
10. Hacklin F, Battistini B, Von-Krogh G (2013) Strategic choices in converging industries. MITSloan Manag Rev 55(1):64–73
11. Schwab K (ed) (2013) The global competitiveness report 2013–2014: Full data edition, Geneva, World Economic Forum
12. Schiliro D (2013) Diversification and development of the United Arab Emirates'economy. J Appl Econ Sci (JAES) 2(24):228–239
13. El-Sokari H, Van Horne C, Huang Z, Al-Awad M (2013) Entrepreneurship. An Emirati Perspective, Abu Dhabi, Khalifa Fund for Enterprise Development & Zayed University
14. Al-Jundi S (2012) Economic diversification in the United Arab Emirates = التنويع الاقتصادي في الإمارات العربية المتحدة
15. Suliman A (2013) Organizational justice and innovation in the workplace: the case of the UAE. J Manage Dev 32(9):945–959
16. Blayse AM, Manley K (2004) Key influences on construction innovation. Constr Innov Inf Process Manag 4(3):143–154
17. Abbot C, Jeong K, Allen S (2006) The economic motivation for innovation in small construction companies. Constr Innov Inf Process Manag 6(3):187–196
18. Kagioglou M, Cooper R, Aouad G, Sexton M (2000). Rethinking construction: the generic design and construction process protocol. Eng Constr Archit Manag 7(2):141–153
19. Abbot-Smith K, Behrens H (2006) How known constructions influence the acquisition of other constructions: the german passive and future constructions. Cogn Sci 30(6):995–1026
20. Alinaitwe HM, Widén K, Mwakali J, Hansson B (2007) Innovation barriers and enablers that affect productivity in Uganda building industry. J Constr Dev Ctries 12(1):59–75
21. Bygballe L, Ingemansson M (2014) The logic of innovation in construction. Ind Mark Manag 43(3):512–524
22. Blayse A, Manley K (2004) Key influences on construction innovation. Constr Innov Inf Process Manag 4(3):143–154
23. Aouad G, Ozorhon B, Abbott C (2010) Facilitating innovation in construction: Directions and implications for research and policy. Constr Innov Inf Process Manag 10(4):374–394

24. Manley K (2008) Implementation of innovation by manufacturers subcontracting to construction projects. Eng Constr Archit Manag 15(3):230–245

25. Muscio A, Quaglione D, Vallanti G (2013) Does government funding complement or substitute private research funding to universities? Res Policy 42(1):63–75

26. Mollaoglu-Korkmaz S, Swarup L, Riley D (2013) Delivering sustainable, high-performance buildings: influence of project delivery methods on integration and project outcomes. J Manage Eng 29(1):71–78

27. Dainty A, Millett S, Briscoe G (2001) New perspectives on construction supply chain integration. Supply Chain Manag Int J 6(4):163–173

28. Rahman M, Kumaraswamy M, Ling F (2007) Building a relational contracting culture and integrated teams. Can J Civ Eng 34(1):75–88

29. Miniaoui H, Schilirò D (2017) Innovation and entrepreneurship for the diversification and growth of the gulf cooperation council economies. Bus Manage Stud 3(3):69

30. Abuelmaatti A, Ahmed V (2014) Collaborative technologies for small and medium-sized architecture, engineering and construction enterprises: implementation survey. J Inf Technol Constr 19:210–224

31. Akintoye A, Goulding J, Zawdie G (2012) Construction innovation and process improvement. Wiley-Blackwell, Chichester

32. Albaloushi H, Skitmore M (2008) Supply chain management in the UAE construction industry. Int J Constr Manag 8(1):53–71

33. Rousseau V, Aubé C, Tremblay S (2013) Team coaching and innovation in work teams. Leadersh Organ Dev J 34(4):344–364

34. Bass B, Avolio B, Jung D, Berson Y (2003) Predicting unit performance by assessing transformational and transactional leadership. J Appl Psychol 88(2):207–218

35. Yukl G (1999) An evaluation of conceptual weaknesses in transformational and charismatic leadership theories. Leadership Q 10(2):285–305

36. Pieterse A, van Knippenberg D, Schippers M, Stam D (2009) Transformational and transactional leadership and innovative behavior: The moderating role of psychological empowerment. Journal of Organizational Behavior 31(4):609–623

37. Alam MM (2014) Market orientation and innovation: are they related concepts? Int J Trends Econ Manag Technol (IJTEMT), 3(6)

38. Al-Abd Y, Mezher T, Al-Saleh Y (2012). Toward building a national innovation system in UAE. In: 2012 Proceedings of PICMET'12: Technology Management for Emerging Technologies. IEEE, pp. 2086–2099

39. Randeree K (2014) Organisational justice: migrant worker perceptions in organisations in the United Arab Emirates. J Bus Syst Gov Ethics 3(4):59–69

40. Motaleb O, Kishk M (2013) An investigation into the risk of construction projects delays in the UAE. Int J Inf Technol Proj Manag 4(3):50–65

41. Madichie N (2011) IRENA—Masdar City (UAE)—exemplars of innovation into emerging markets. Foresight 13(6):34–47

42. Castellani D, Jimenez A, Zanfei A (2013) How remote are R&D labs? Distance factors and international innovative activities. J Int Bus Stud 44(7):649–675

43. Al Hallami MO, Van Horne C, Huang VZ (2013) Technological innovation in the United Arab Emirates: process and challenges. Transnatl Corp Rev 5(2):46–59

44. Pamuk H, Bulte E, Adekunle AA (2014) Do decentralized innovation systems promote agricultural technology adoption? Experimental evidence from Africa. Food Policy 44:227–236

45. Baruch Y (1999) Response rate in academic studies-a comparative analysis. Hum Relat 52 (4):421–438

46. Norusis MJ (1992) SPSS for windows: advanced statistics release 5. SPSS Incorporated, Upper Saddle River
47. Cattell R (1966) The scree test for the number of factors. Multivar Behav Res 1(2):245–276
48. Thurstone L (1947) The calibration of test items. Am Psychol 2(3):103–104
49. Norušis M (2006) SPSS 15.0 guide to data analysis. Prentice Hall, Upper Saddle River

Analysis of Strengthened Composite Beams Under Flexural Stress

Emre Ercan[1(✉)], Bengi Arısoy[1], Ali Demir[2], and Anıl Özdemir[3]

[1] Department of Civil Engineering, Faculty of Engineering, Ege University,
İzmir, Turkey
{emre.ercan,bengi.arisoy}@ege.edu.tr
[2] Department of Civil Engineering, Faculty of Engineering, Celal Bayar
University, Manisa, Turkey
ali.demir@cbu.edu.tr
[3] Department of Civil Engineering, Faculty of Technology, Gazi University,
Ankara, Turkey
anilozdemir@gazi.edu.tr

Abstract. A steel-concrete composite beam is composed of a steel beam and concrete slap connected with shear connectors. Composite beams are highly efficient structural members in order to load carrying capacity because the tension component of the force pair originating from bending is carried by the steel profile and the compressive component is carried by the concrete slab in composite beams. In this study, numerical and experimental analysis of steel-concrete composite and strengthened steel-concrete composite beams is presented. In experimental study, one strengthened and one un-strengthened specimens are produced and tested in bending. Strengthening is produced applying carbon fiber reinforced polymers sheet to the lower flange of the steel beam and using steel fiber reinforced concrete in the concrete slab. Specimens are tested under four-point loading test. During the tests, load, deflection and strain values are collected by data acquisition system. In numerical study, the finite element models of the steel-concrete composite beams are generated and analyzed by Atena-GiD program. Comparison and evaluations are made in terms of strength, applicability, stiffness and energy consumption about the steel-concrete composite beams with carbon fiber reinforced polymers and steel fibered concrete for both numerical and experimental results.

Keywords: Composite beam · CFRP · Steel fibered concrete
Strengthened composite beam · Four-point loading test

1 Introduction

The bearing system that produced with reinforced concrete slabs and the steel beam connected with shear elements is called steel-concrete composite beam (Fig. 1). These systems are more economical than steel beams that bear load alone by the concrete slab that sits freely on them. This is because, in composite beams, the tension component of the force pair originating from bending is carried by the steel profile and the compressive component is carried only by the concrete slab or jointly with a portion of the

© Springer International Publishing AG, part of Springer Nature 2018
S. Fırat et al. (eds.), *Proceedings of 3rd International Sustainable Buildings Symposium (ISBS 2017)*, Lecture Notes in Civil Engineering 7,
https://doi.org/10.1007/978-3-319-64349-6_40

steel component. Therefore, a steel profile that is weak to buckling is relieved entirely or to a large extent from carrying the compressive component of bending.

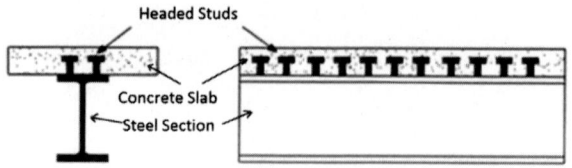

Fig. 1. Typical steel-concrete composite beam [2]

Two types of composite beams can be fabricated, fully and partially; a fully-composite beam has a sufficient number of shear connectors (headed studs) that prevent the slip between the concrete and steel beam, after concrete crushing and steel beam yielding. In a partially-composite beam, shear connectors fail before the concrete crushing under compression and, the slip between the concrete and steel beam would occur. Existing rules for the designing of composite structural members are given in detail in the literature [1, 2].

In the literature, there have been many studies associated with composite beams. Studies are typically based on strengthening concrete and steel with a variety of materials. Some of the studies are summarized below:

In the study conducted by Çetin and Yelgin [3], composite beams were reinforced with CarboDur plate elements. The behavior of reinforced composite beams was investigated experimentally and numerically and the results were compared. A simple beam experimental setup was established with the concrete slab at the bottom and the steel profile at the top of the beam, and the load was applied from the middle point of the beam because the behavior of the composite system in the negative moment region was investigated in the study.

Değerli [4] formed a composite beam with a reinforced concrete slab and steel profiles in his research and used bolts for shear connection. In the study, composite behaviour, the separation of the slab from steel profile and the effect of loaction of bolts at different distances and crack formation were investigated. The beams were loaded with a single load from the middle point. It was confirmed that the bolts were very good shear elements.

Gedik [5] stated in his study that enabling steel-concrete composite beam behaviour provided significant reductions in construction costs of multi-story steel buildings. In his study, the study was conducted analytically. The obtained results are compared with available experimental results.

Teng et al. [6] investigated strengthened steel-concrete composite beams with carbon fiber reinforced polymers CFRP under bending experimentally and numerically. In the study they explained the moment capacity increase and the location of de-bonding between CFRP sheets and steel under heavy loads.

In the study conducted by Ağcakoca and Aktaş [7], it was aimed to determine the necessary CFRP amount to ensure designed behavior in steel-concrete composite beams. In the study, they obtained reinforced beams by applying an HM-CFRP strip to the bottom flange of the steel beam that is forming a composite profile and the beam was subjected to a four-point loading test. The collapse occurred with the rupture of HM-CFRP strips.

As mentioned above CFRP materials are usually bonded to steel beams to increase the elastic stiffness and ultimate capacity of steel-composite beams but there is limited research into comparisons between experimental and numerical analysis of steel-concrete composite beams in which both using strengthened concrete and steel beam. In this study, experimental and non-linear Finite Element Method (FEM) analysis results of un-strengthened and strengthened steel-concrete composite beams were compared.

2 Material Method

The 3000 mm steel-concrete composite beam is tested in a four-point loading test to investigate the effect of CFRP and steel fibered concrete on bending capacity of beam and compare it with numerical solutions. Headed studs are used to provide bonding between concrete slab and steel beam. Steel beam is IPE 200 (S275). Effective reinforced concrete slab width is calculated as beff = 80 cm; reinforced concrete slab thickness is calculated as d = 10 cm [1, 2]. Technical drawing of the composite beam is given in Fig. 2. Strengthening of the composite beam is ensured by CFRP bonding to the bottom of the lower flange of the steel.

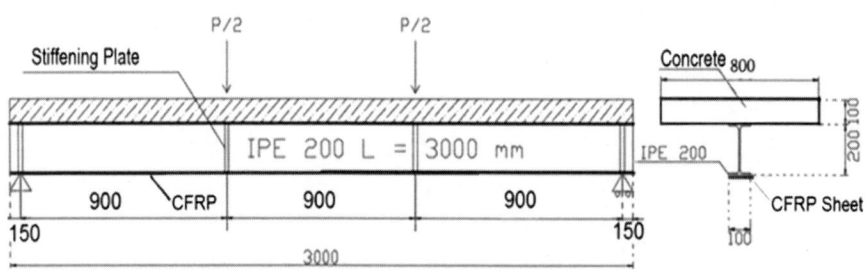

Fig. 2. Drawing of test specimen steel-concrete composite beam

2.1 Mechanical Properties of Concrete and Steel Fibered Concrete

The C25 concrete mixtures were prepared in the factory and brought to the laboratory by a trans-mixer. Strengthening of concrete established by adding 3 cm hooked steel fibres. The proportion of steel fibres was taken 2% by volume. Cube and cylindrical samples were taken while concrete casting and tested on the same day as the experiments done [8]. Test results are presented in Table 1.

Table 1. Mechanical properties of concrete

Property	Un-strengthened	Strengthened
Uniaxial strength (MPa)	28.26	29.3
Splitting tensile strength (MPa)	2.8	3.4
Modulus of elasticity (MPa)	30,026	29,896

2.2   Mechanical Properties of CFRP

Strengthening materials is CFRP sheets, brand named as SikaWrap 300C, and epoxy Sikadur 330. Two layers of CFRP are applied to bottom surface of the lower flange of the steel beam. Mechanical properties of CFRP and the epoxy resin are presented in Table 2.

Table 2. Mechanical properties of CFRP and epoxy

Properties of unidirectional CFRP	Remarks of SikaWrap 300C
Fiber orientation	0°
Areal weight (g/m2)	300 ± 10
Density (g/m3)	1.78×10^{-6}
Thickness (mm)	0.166
Tensile strength (MPa)	3900
Elastic modulus (MPa)	230,000
Ultimate tensile strain (%)	1.5%
Properties of resin	Remarks of Sikadur 330
Tensile strength (MPa)	30
Elastic modulus (MPa)	3800
Properties of Lamina (1 mm for each layer)	Remarks of Sika
Elastic modulus (GPa)	33

2.3   Preparation of the Composite Beam

Shear between upper flange of steel beam and concrete slab was calculated to determine number of headed studs (shear connectors) in order to have full bonding. The biggest shear force that may occur in the beam before it reaches yielding was calculated as 190 kN in computer analysis. Headed shear studs were calculated under a shear force of 200 kN to ensure not to have slip between concrete and steel beam during the experiment [1, 2, 9, 10]. According to this, 75 mm in length and 19 mm in diameter headed studs were chosen and welded on IPE200 steel profile with 280 mm distance. Stiffening plates were welded to the web of IPE 200 profile from both sides on the supports and at loading point (Fig. 3). Mounting stiffening plates, headed studs and the details of support system drawings are presented in Fig. 4. Then, a $100 \times 800 \times 3000$ mm concrete mold was prepared, a 2.5 cm concrete cover was created for welded wire fabric Q131/131 and concrete casting was done by trans-mixer. Finally, CFRP sheets were bonded to the lower flange of the IPE 200 (Fig. 5). The CFRP strengthening procedure includes removing rust from steel with sand blasting, application of priming adhesive layer Sikadur 330 and applying two layers of the SikaWrap 300C sheets. All applications were performed at room temperature and specimens were cured for at least 28 days under laboratory conditions before testing, [10].

Fig. 3. Headed studs

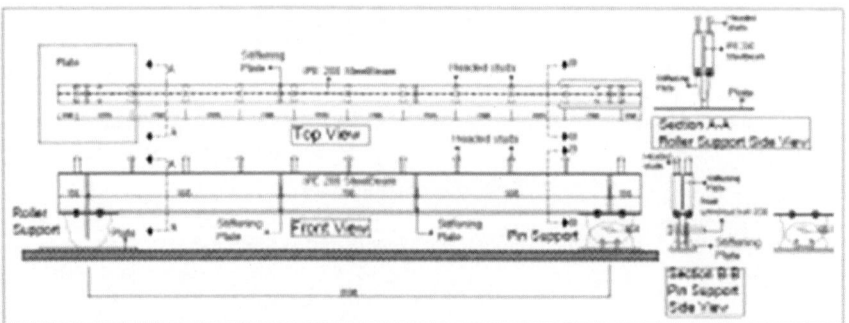

Fig. 4. Drawing of supports, headed studs and stiffening plates on IPE 200 profile [10]

2.4   Experimental Setup

The 3000 mm beam was loaded as in Fig. 6, at the four-point loading test. The loading was cyclic and, load was increased 25 kN in each cycle. Incremental cyclic loading was continued until failure of the beam.

In the experimental setup, displacement controlled 500 kN loading capacity hydraulic loading system and 8-channel Testbox 1001 data acquisition system was used. A total of three potentiometers were placed to the 1/4th and half of the beam. Strain-gauges were placed on the upper and lower flanges of the middle of the IPE 200 profile to measure the strain values (Figs. 6 and 7). Thus, it could be identified when

Fig. 5. Mold, concrete casting and CFRP strengthened steel concrete composite beam

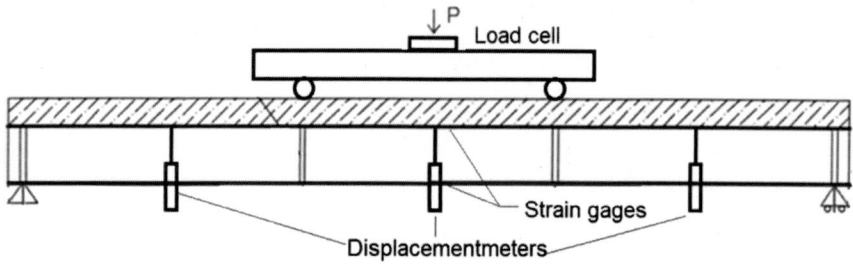

Fig. 6. Schematic view of test set-up

Fig. 7. Steel-concrete composite beam test set-up

the yielding could have occurred in the steel beam. Load value was obtained with the help of the 500 kN load cell. The data that received from all sensors were transferred to the computer with the Testbox1001 data acquisition system which has a sampling rate 8 S/s.

2.5   Finite Element Model of the Composite Beams

Finite element model of the composite beam was prepared in the Atena-GiD program [11]. The bonding between concrete and steel beam is constituted via "fixed contact" as explained in Atena program documentation Par4–6 [11, 12] to ensure fully-composite beam model. The finite element model of the un-strengthened beam consists of 27,818 tetrahedral elements and 7597 nodes (Fig. 8). CFRP are assigned as linear elements to the flange of the steel beam as explained in the Atena program documentation part 11 for CFRP strengthened model [12]. The strengthened model has 300 additional line elements and 315 nodes that represent CFRP (Fig. 9). The material properties of concrete obtained as a result of the tests were attained to the program. Material properties given by the manufacturers were used for steel and CFRP material. Displacement values obtained in the experimental results were given as input data for the numerical model. The Newton–Raphson method for non-linear analysis and LU algorithm for the solution of the matrix team were chosen, and the models were analysed.

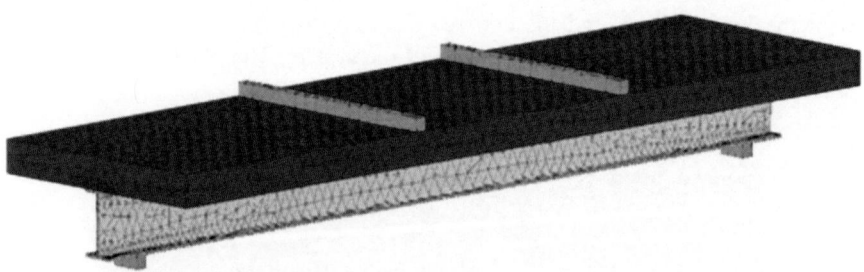

Fig. 8. FE model of the un-strengthened composite beam

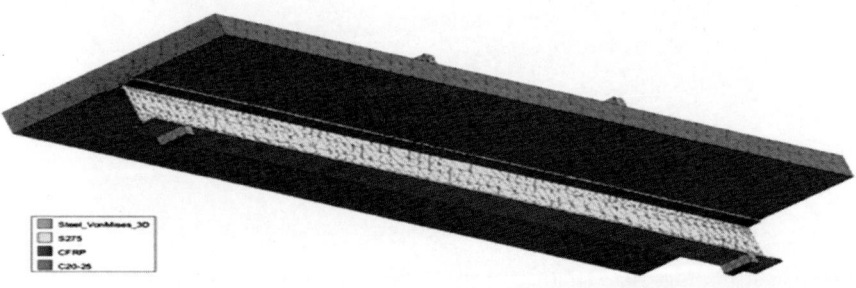

Fig. 9. FE model of the strengthened composite beam

3   Results

3.1   Four-Point Loading Test Results of Un-strengthened Sample

Under increased cyclic loading, the first shear cracks occurred, in the concrete slab, in the 9th load cycle at 225 kN. In the points where the load was applied the deflection value was observed as 6.4 mm. Permanent deformations occurred after the 10th load cycle at 250 kN and the rigidity of the system and slope of the load-displacement curve began to decrease. A significant yielding in the steel beam was observed at the 320 kN load and after this, the amount of deflection increased (Fig. 10). Shear cracks expanded slowly in the region near the area of the applied load (Fig. 11). Brittle fracture in concrete was not observed until the failure of the system. The obtained load deflection graph is presented in Fig. 14.

3.2   Four-Point Loading Test Results of Strengthened Sample

Permanent deformations occurred after the 10th load cycle at 250 kN and the rigidity of the system and slope of the load-displacement curve began to decrease. The first micro shear cracks occurred in the 11th load cycle at 275 kN. In the points where the load was applied, the deflection was observed as 7.7 mm. A significant yielding in the steel beam was observed at the 378 kN load, because at this load, the CFRP began to separate from the steel and the CFRP ruptured with a noise (Fig. 12). After this, the amount of deflection increased. Shear cracks expanded suddenly in the region near the area of the load applied, and the concrete slab crushed (Fig. 13). Brittle fracture in concrete was not observed until the failure of the system. The obtained load deflection is presented in Fig. 14.

Fig. 10. Deformed shape of the un-strengthened sample

Fig. 11. Shear cracks near the loading point of un-strengthened sample

Fig. 12. Rupture of CFRP in strengthened sample

3.3   Finite Element Analysis of the Composite Beam

The maximum deflection-total load value graphics was obtained from the FEM analyses and is presented in Fig. 15 for both models. The crack map and principle stress contour graphs obtained for un-strengthened and strengthened composite beams as a result of the analysis and the stress distribution are given in Figs. 16 and 17. As seen in

Fig. 13. Shear cracks near the loading point

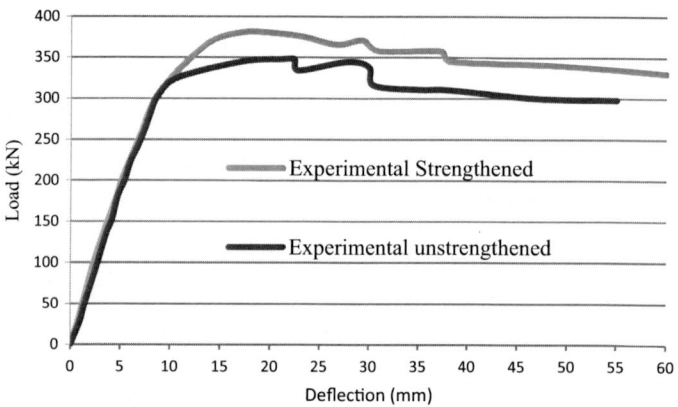

Fig. 14. Incremental circular load-deflection graph of strengthened composite beam

the figures, the yielding value 275 MPa for S275 steel was exceeded in the lower flange of the steel for both types of beams. Concrete was crushed by exceeding average compressive value of 29.8 MPa. The maximum failure load of the strengthened model was 380 kN that is higher than failure load of the un-strengthened model that is 350 kN. However, the rigidity of the strengthened beam was a little higher.

4 Discussion

The linear regions of the load-deflection graphs that were obtained as a result of Nonlinear FE analysis and experiment were overlapped on linear region and very close in non-linear region. The experimental rigidity of strengthened and un-strengthened beams showed more rigid behavior when it was outside the linear region than FE

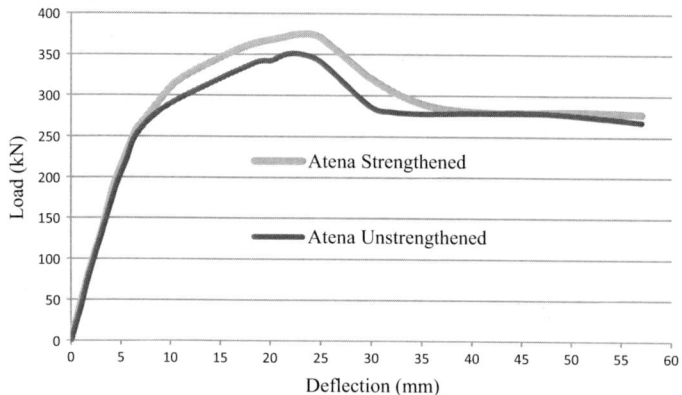

Fig. 15. Load-deflection curves obtained from Atena-GiD FE program

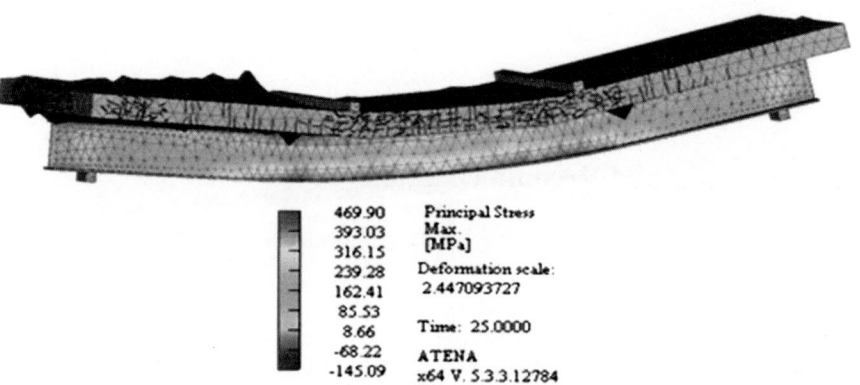

Fig. 16. Principle stress contour and cracks in un-strengthened composite beam

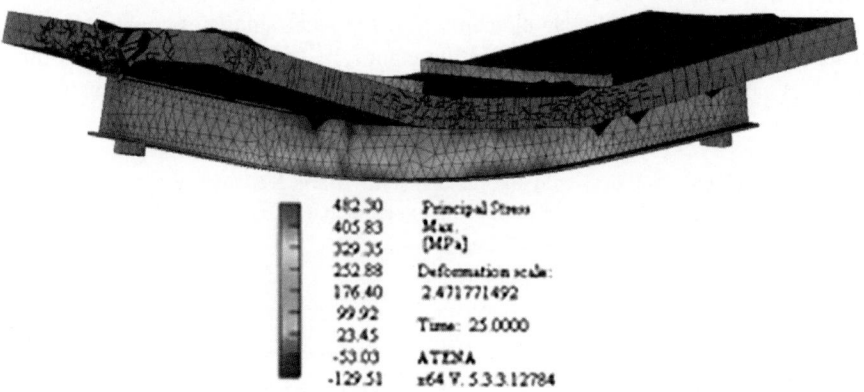

Fig. 17. Principle stress contour and cracks in strengthened composite beam

analysis results. The maximum load of the experimental and FE analysis results are similar. The reason for the rigidity increase is that the modulus of elasticity of CFRP is 230 GPa and that of steel is 200 GPa, that they are very close each other. A detailed comparison of the results is presented in Fig. 18.

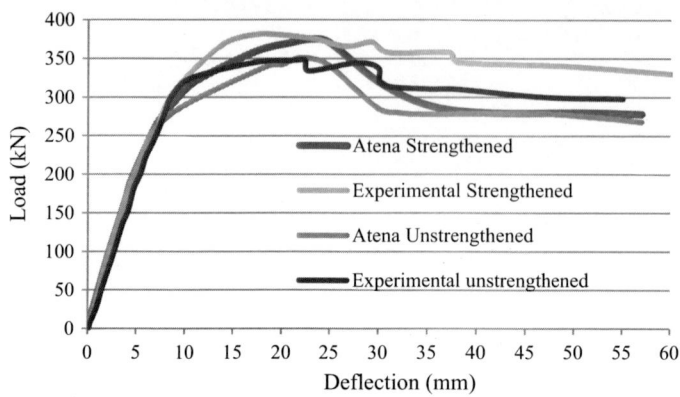

Fig. 18. Load-deflection curves obtained from numerical model and experimental study

5   Conclusions and Recommendations

As a result of the experiments performed in this study the following conclusions and recommendations were obtained.

Experimental and numerical results of the composite beams have given the same flexural rigidities in the linear region. However, experimental results indicate that samples exhibit larger rigidity in non-linear region. Maximum load-carrying capacities are similar for FE and experimental results.

CFRP have a small effect on rigidity but an 8% increase is observed in the maximum load capacity. Increasing the proportion of CFRP will increase the maximum load carried. Steel fibered concrete decrease number of cracks in the concrete slab.

Embedding CFRP as linear elements into the steel simulates maximum loading capacity very well. However, delamination cannot be seen. For this reason, if there is no delamination, finite element analysis gives considerably good solution for CFRP strengthened composite beams.

Acknowledgements. Author would like to thank Mr. Barış KURAL, Mr. Sahir CİLLO and Mr. Selçuk ORAL for their support during the experimental study.

References

1. Arda TS, Yardımcı N (2000) Çelik Yapıda Karma Elemanların Plastik Hesabı. Birsen Yayınevi, İstanbul
2. Çelik Yapıların Tasarım ve Yapım Kuralları (2016) Ministry of Environment and Urbanisation. Ankara, Turkey
3. Çetin Ö, Yelgin N (2002) Karbon Lifleri ile Takviye Edilmiş Kompozit Kirişlerin Negatif Moment Bölgesindeki Davranışlarını Deneysel ve Teorik Olarak İncelenmesi. Sakarya Univ J Sci 6(1):129–134
4. Değerli A (2010) Structural behavior in negative moment of composite beams which are combined with bolts. Sakarya University, Institute of Natural Sciences, The Degree of Master Thesis, Sakarya, Turkey
5. Gedik E (2008) Modelling of full and partially shear connected composite beam with finite element method. Gazi University, Graduate Scholl of Natural and Applied Sciences, The Degree of Master Thesis, Ankara, Turkey
6. Teng JG, Fernando D, Yu T (2015) Finite element modelling of bonding failures in steel beams flexurally strengthened with CFRP laminates. Eng Struct 86:213–224
7. Ağcakoca E, Aktaş M (2012) Defining development length of HM-CFRP in composite I-section strengthened with HM-CFRP. e-J New World Sci Acad 7(2):47–59
8. Erdoğan TY (2013) Concrete. METU Press, Ankara
9. Requirements for Design and Construction of Reinforced Concrete Structures (2000) TS 500, Turkish Standards Institution, Ankara, Turkey
10. Özyılmaz E (2016) Experimental study of composite beams which is strengthened with CFRP under bending and shear forces. Ege University, Graduate Scholl of Natural and Applied Science, The Degree of Master Thesis, İzmir, Turkey
11. Atena Version 5.3 (2016) Červenka Consulting S.R.O., Prague, Czechia
12. Pryl D, Červenka J (2016) ATENA program documentation. Červenka Consulting S.R.O, Prague

Wave Propagation and Vibration Isolation in Soils

Erkan Çelebi[(⊠)]

Department of Civil Engineering, Faculty of Engineering, Sakarya University,
Sakarya, Turkey
ecelebi@sakarya.edu.tr

Abstract. The goal of this study is to deal with the modelling and calculation of the ground-borne vibrations and the effect of the wave propagation on foundation response with emphasis on vibration screening systems. Herein, ANSYS FEM software program was used to simulate vibratory source induced wave propagation and dynamic foundation-soil interaction problem. Extensive parametric investigations on the screening performance of open and in-filled trench barriers have been done for both active and passive isolation cases. The obtained results are important for structural engineers since provide practical information and methods about the dissipation of the strong vibration energy.

Keywords: Wave propagation · Vibration isolation · Dynamic load
Finite element analysis · Absorbing boundary

1 Introduction

In problems associated with the soil-structure interaction and wave propagation it is significant to know well the dynamic parameters and geotechnical engineering characterization related to the underlying soil conditions of the engineering constructions to be investigated of structural vibrations and also to have efficient software available for the computational modelling of soil-structure coupled system, in which the site properties are required as input.

The Rayleigh wave plays an important role on the general behavior of the surface wave propagation and on transmitting vibration energy because of its simplicity and the close relationship of its velocity to the shear-wave velocity for soil materials. To minimize the effects of strong ground vibrations induced by human made activities on the nearby structures, vibration isolation measures are placed especially into poor soils having bearing capacity problem. Because of low cost and without great difficulty to construct, both open trench and concrete barriers can be useful methods in practical civil engineering applications for reductions of ground-borne vibrations.

Depending on the rapid development of the numerical analysis techniques with the significant progress of computer technologies, there has recently been a remarkable scientific and engineering interest in evaluating the adverse effects of strong ground-borne vibrations on environmental structures. In the past three decades, numerous notable research studies based on numerical, analytical and experimental methods focused on the shielding performance of both open trench and solid wave

© Springer International Publishing AG, part of Springer Nature 2018
S. Fırat et al. (eds.), *Proceedings of 3rd International Sustainable
Buildings Symposium (ISBS 2017)*, Lecture Notes in Civil Engineering 7,
https://doi.org/10.1007/978-3-319-64349-6_41

barriers considered under the dynamic behavior of soil-foundation-structure coupled system [1–4]. In order to realistically idealize the unbounded region, various modeling methods such as the finite difference method, finite element method (FEM), thin layer-flexible volume method, boundary element method (BEM) and their coupling procedures have been conducted for the analysis of interaction problems. Many researchers have primarily dealt with the development of several modeling techniques to efficiently simplify the analysis of the wave propagation problems in unbounded soil media and to understand the screening mechanism of wave isolation barriers [5–9].

In this study, the computational simulation of ground vibrations generated by dynamic load source and the effect of the wave propagation on foundation response including the screening performance of open and in-filled trench barriers was directly performed by employing 2D finite element model by assuming plane-strain condition with Drucker–Prager failure criterion for plastic deformations and soil yielding. Herein, the infinite soil medium was considered as a finite region by using absorbing boundaries.

2   Computational Model of Soil-Foundation Interaction Problem

To assess the response of the vibrating soil–foundation system under periodic stationary loads and to reveal the effect of the installed trench and concrete wall barriers on wave propagation, parametric investigations for both passive and active screen cases have been performed in the frequency domain by using two dimensional (2D) finite element (FE) model under plane-strain conditions within the frame of ANSYS software package [10] (Fig. 1).

To simulate adequately the response of the local soil medium to cycling loading conditions and to evaluate the contribution of non-linear behavior of the soil on the foundation vibration in the finite element procedure, the mechanical properties of the underlying soil media is considered by an undrained elastic-plastic Drucker–Prager model (Table 1).

In the considered problem the underground is regarded as a homogeneous layer on the top of rigid bedrock. Viscous-spring dynamic artificial boundaries (combin 14) are used at both right and left-hand side of the 2D infinite soil region to realistically dissipate energy due to incoming waves. The bottom of the soil layer is idealized by rigid boundary conditions.

In this numerical example, despite of the geometrical damping having main effect, 2.5% of the critical damping is chosen as the material damping value. The main material parameters for underlying soil are given as shear wave velocity of $c_s = 220$ m/s, total unit weight of $\rho_s = 15$ kN/m^3 and Poisson's ratio of $v = 0.42$. The cohesion and friction angle is to be assumed as $c = 2$ kN/m^2 and $\theta = 24°$, respectively. The properties of the concrete wall barrier and considered foundations are taken as $c_s = 2400$ m/s, $\rho_s = 24$ kN/m^3 and $v = 0.2$. The FE mesh size is defined as the lateral extent of $L_f = 45$ m and the total depth of $H_f = 10$ m. In the case of vibration isolation, the depth H_t and its width B_t of the wave barrier are taken as 8 m and 1 m, respectively for achieving an ideal vibration mitigation under financially and structural feasibility.

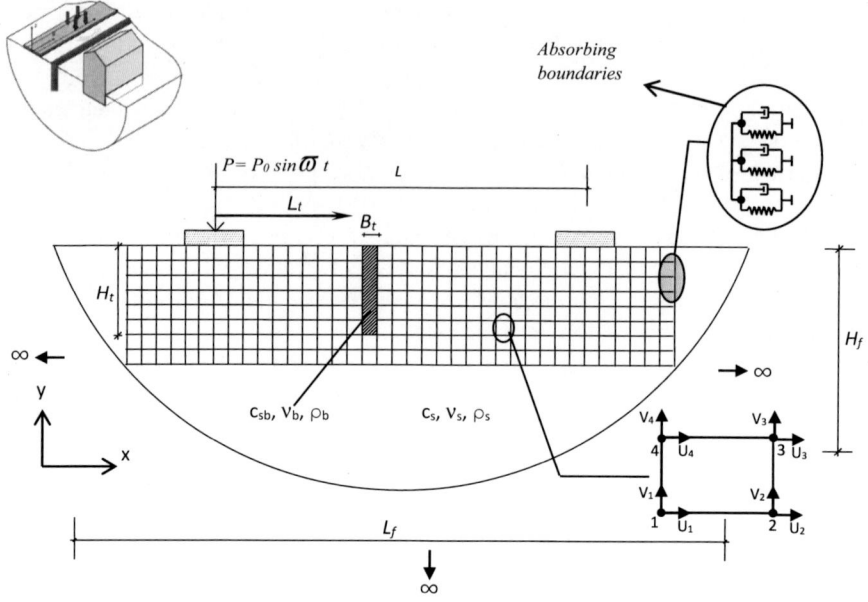

Fig. 1. Problem statement and the mathematical model

Table 1. Constitutive material models in ANSYS [10]

Name	Yield criterion	Flow rule	Hardening rule	Material response
Bilinear isotropic hardening	von Mises/Hill	Associative	Work hardening	Bilinear
Bilinear kinematic hardening	von Mises/Hill	Associative	Kinematic hardening	Bilinear
Multilinear isotropic hardening Nonlinear isotropic hardening	von Mises/Hill von Mises/Hill	Associative Associative (Prandtl-Reuss equations)	Work hardening Kinematic hardening	Multilinear Bilinear
Multilinear kinematic hardening	von Mises/Hill	Associative	Kinematic hardening	Multilinear
Nonlinear kinematic Hardening	von Mises/Hill	Associative	Kinematic hardening	Nonlinear
Anisotropic	Modified von Mises	Associative	Work hardening	Bilinear, each direction (tens, and comp.) different
Drucker-Prager	Mises-dependence on hydrostat.stress	Associative or non-associative	Non	Elastic-perfectly plastic

The distance L_t between the trench and the foundation to be subjected to dynamic load is 5 m in the case of active isolation. The clear distance between two concrete surface foundations is 24 m. The dynamic source on the foundation is simulated by a fixed sinusoidal point load for forcing amplitude of 2500 kN in a frequency range of practical importance of 0–100 Hz. Two dimensional four-node rectangular elements (plane 42) are used in the FE mesh generation of the soil region. This type of element consists of two translational degrees of freedom at each node under plain strain condition.

3   Numerical Results

The effect of depth, material stiffness and location of the wave barrier on the vibration screening efficiency under plane strain conditions has been investigated by employing FE analysis in the frequency domain. For numerically investigations, the observation points have been chosen centrically above the rigid foundations.

When the wave barrier is constructed nearby the dynamic load, such application is named as active isolation. If the wave barrier is placed away from the applied load but nearby the foundation of the building to be protected from incoming waves, such far field isolation is named as passive isolation. At first, screening effects of installing rectangular open trench and concrete trench barriers have been compared with the original site for the active isolation case. The wave propagation patterns of the transmitted waveform and the influence on the vibration amplitudes in the case of open (Fig. 2c) and concrete in-filled trenches (Fig. 2d) are obviously different to the case of no trench as shown in Fig. 2a.

Comparing the vibration shielding performance of open trench with concrete wall, the open trench type wave barrier having with ideal construction depth (here $H_t = 8$ m) gives better results with regard to obstruct the incoming waves. Furthermore, as shown in Fig. 2 when comparing (a) and (b), the depth of the trench have significant effect on the transmitted waveform and vibration reduction.

The discrepancy of the screening performance and transmitted waveforms depends on dynamic properties of propagating wave which occur after hitting an obstacle such as reflection, refraction and diffraction varied with the in-filled material stiffness of the trench type wave barriers.

The comparison of the wave propagation patterns of the transmitted waveform presented in Fig. 3 for both wave barriers demonstrates again that an open trench is more effective on blocking the energy waves. In the same manner, open trench barrier gives the best isolation effect in the passive isolation case. Because of traveling a longer propagation path surrounding the trench barrier, there is a certain amplitude reduction in the incoming waves from the vibratory source.

The observation point on the center of the foundation under loading is defined as A_1, the other foundation protected from the strong vibrations is specified as A_2 for active isolation. Vertical vibration amplitudes of the observation points on the considered foundations in the case of active isolation applied by both open and concrete in-filled trench barrier are comparatively summarized in Table 2.

Table 3 presents the amplitudes of vertical foundation vibrations for passive isolation case.

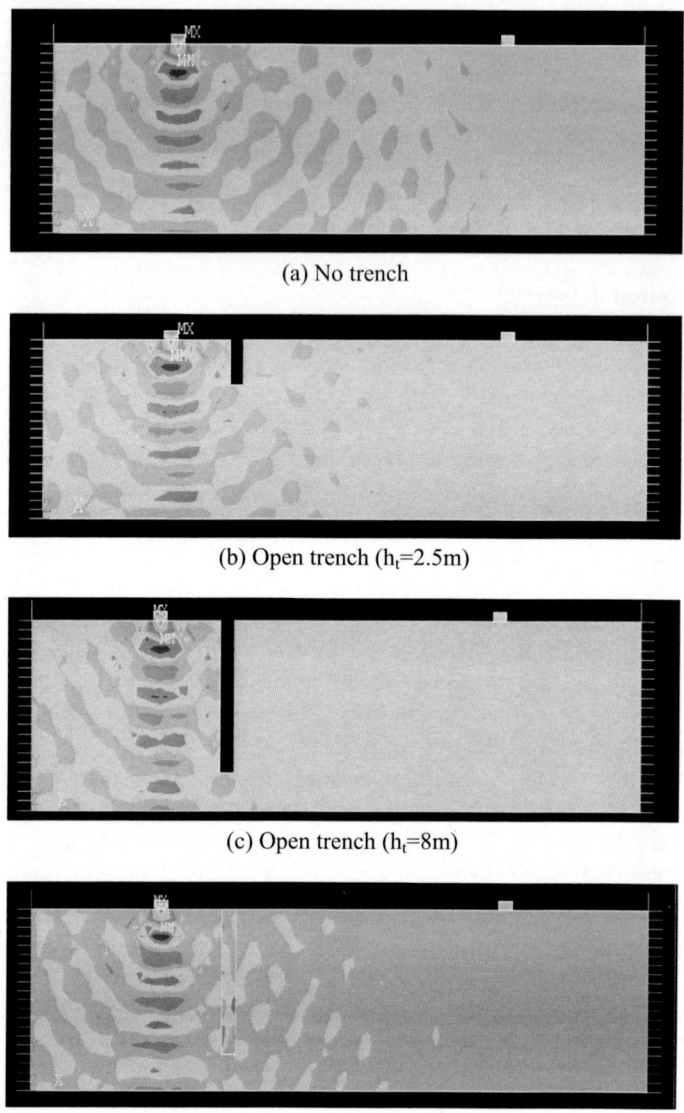

(a) No trench

(b) Open trench (h_t=2.5m)

(c) Open trench (h_t=8m)

(d) Concrete- infilled trench (h_t=8m)

Fig. 2. Wave propagation pattern for active isolation by installing open trench and concrete wall barrier

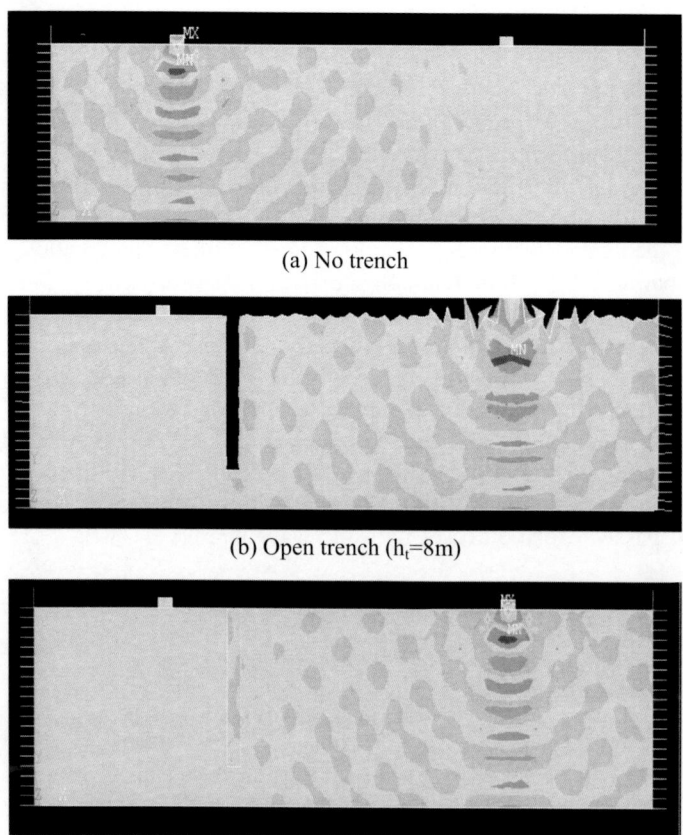

(a) No trench

(b) Open trench (h_t=8m)

(c) Concrete-infilled trench (h_t=8m)

Fig. 3. Wave propagation pattern for passive isolation by installing open trench and concrete wall barrier

Table 2. Response of the foundations for active isolation

Vertical displacement (mm)	Without trench	Open trench (H_t = 8 m)	Concrete wall (H_t = 8 m)
A_1 point	23.3	31.3	26.2
A_2 point	10.7	2.81	6.83

Table 3. Response of the foundations for passive isolation

Vertical displacement (mm)	Without trench	Open trench (H_t = 8 m)	Concrete wall (H_t = 8 m)
A_1 point	23.3	23.6	23.2
A_2 point	10.7	2.88	6.87

4   Concluding Remarks

In this study, computational simulation of the soil-foundation interaction problem and the effects of wave propagation on the response of wave barriers were directly accomplished by employing 2D finite element model under plane-strain condition with Drucker–Prager failure criterion for plastic deformations and soil yielding. Herein, viscous-spring dynamic artificial boundaries were considered for modeling of the infinite soil medium. The effects of depth, material stiffness and location of the trench type wave barrier on the vibration isolation were numerically investigated.

Applying open or in-filled trench type wave barriers such as concrete wall can mitigate the foundation vibrations significantly. The use of an open trench is more effective than using an in-filled trench. The installing depth and location of the trench type barriers plays an important role on the screening performance.

The most efficient vibration method is to place an open trench in the close vicinity to the foundation to be protected (passive vibration isolation). On the other hand, there is a no considerable amplification in the response of the load applied foundation for the investigated passive isolation cases by using wave barriers.

References

1. Beskos DE, Dasgupta G, Vardoulakis IG (1986) Vibration isolation using open or filled trenches, part 1: 2-D homogeneous soil. Comput Mech 1(1):43–63
2. Klein R, Antes H, Le Houedec D (1997) Efficient 3D modelling of vibration isolation by open trenches. Comput Struct 64(1–4):809–817
3. Ulgen D, Toygar O (2015) Screening effectiveness of open and in-filled wave barriers: a full-scale experimental study. Constr Build Mater 86:12–20
4. Çelebi E, Fırat S, Beyhan G, Çankaya I, Vural I, Kırtel O (2009) Field experiments on wave propagations and vibration isolation by using wave barriers. Soil Dyn Earthq Eng 29:824–833
5. Leung KL, Vardoulakis IG, Beskos DE, Tassoulas JL (1991) Vibration isolation by trenches in continuously non-homogenous soil by the BEM. Soil Dyn Earthq Eng 10(3):172–179
6. Ahmad S, Al-Hussaini TM (1991) Simplified design for vibration screening by open and infilled trenches. J Geotech Eng 117(1):67–88
7. Celebi E, Fırat S, Cankaya I (2006) The effectiveness of wave barriers on the dynamic stiffness coefficients of foundations using boundary element method. Appl Math Comput 180:683–699
8. Andersen L, Nielsen SRK (2005) Reduction of ground vibration by means of barriers or soil improvement along a railway track. Soil Dyn Earthq Eng 25(7–10):701–716
9. Zoccali P, Cantisani G, Loprencipe G (2015) Ground-vibrations induced by trains: Filled trenches mitigation capacity and length influence. Constr Build Mater 74:1–8
10. Ansys Inc. Theory reference—Ansys manual, http://www.ansys.com

Concrete Strength Variation Effect on Numerical Thermal Deformations of FRP Bars-Reinforced Concrete Beams in Hot Regions

Ali Zaidi[1(✉)], Aissa Boussouar[1], Kaddour Mouattah[1],
and Radhouane Masmoudi[2]

[1] Structures Rehabilitation and Materials Laboratory (SREML), Civil
Engineering Department, University of Laghouat, Laghouat, Algeria
a.zaidi@mail.lagh-univ.dz
[2] Civil Engineering Department, University of Sherbrooke, Sherbrooke, QC,
Canada
Radhouane.Masmoudi@USherbrooke.ca

Abstract. The steel corrosion phenomenon could reduce the durability and the serviceability of concrete structures reinforced with steel bars. Moreover, the repair cost of these structures is very expensive. Consequently, it seems necessary to substitute steel bars by fiber reinforced polymer (FRP) bars, in concrete structures, because of their high properties, particularly, their excellent corrosion resistance and high tensile strength-to-weight ratio. Nevertheless, the use of FRP bars in concrete structures, built in hot regions, may cause splitting cracks within concrete at the interface of FRP bars-concrete, and eventually the failure of the concrete cover. This paper presents a nonlinear finite element investigation using ADINA software to analyze the effect of concrete strength variations on thermal deformation distributions in the concrete cover surrounding glass FRP (GFRP) bars for reinforced concrete beams under high temperatures up to 70 °C. The main results show that the concrete strength variation has no big influence on the transverse thermal deformation of FRP bars-reinforced concrete beams for thermal loads less than the cracking thermal load ΔT_{cr}, producing the first radial cracks in concrete at the FRP bar-concrete interface, varied from 20 to 35 °C depending on the ratio of concrete cover thickness to FRP bar diameter (c/d_b) and the compressive concrete strength f_c' varied from 1 to 3.2 and 25 to 90 MPa, respectively. However, for thermal loads greater than ΔT_{cr}, the transverse thermal deformations decrease with the increase in the concrete strength. Comparisons between analytical and numerical results in terms of thermal deformations are presented.

Keywords: Concrete cover · GFRP bar · Concrete strength variation
Thermal deformation · Numerical simulation

© Springer International Publishing AG, part of Springer Nature 2018
S. Fırat et al. (eds.), *Proceedings of 3rd International Sustainable
Buildings Symposium (ISBS 2017)*, Lecture Notes in Civil Engineering 7,
https://doi.org/10.1007/978-3-319-64349-6_42

1 Introduction

The use of fiber reinforced polymer (FRP) in civil engineering constructions is becoming as an effective solution to eradicate the steel corrosion problem. Furthermore, FRP bars have a high tensile strength, low density, and high stiffness. However, the transverse thermal behavior of FRP bars embedded in concrete should be well understood. The thermal incompatibility between FRP bars and concrete in the transverse direction may cause radial cracks within concrete at the interface of FRP bars-concrete under high temperatures, and consequently, the reduction of the durability and the serviceability of concrete structures. Although, many researches were carried out on thermal behavior of concrete structures reinforced with FRP bars taking into account different parameters such as the concrete cover thickness, FRP bar diameter, temperature variation, humidity, shape of concrete elements, spacing between FRP bars, and combined temperature and mechanical loadings [1–7]. The nonlinear numerical analysis of concrete strength variation effects on the distribution of thermal deformations in the concrete cover surrounding FRP bars, when the confining action of concrete is asymmetric, need more investigation. This paper presents a nonlinear finite element study, using ADINA software, to analyze transverse thermal deformations in concrete beams reinforced with glass FRP (GFRP) bars submitted to high temperatures up to 70 °C varying the compressive concrete strength f'_c from 25 to 90 MPa and the ratio of concrete cover thickness to FRP bar diameter (c/d_b) from 1 to 3.2.

2 Analytical Models Background

The transverse thermal deformations are predicted using analytical models established by Zaidi and Masmoudi [4], Masmoudi et al. [3], Aiello et al. [2], and Rahman et al. [1] for a concrete element reinforced with FRP bars submitted to a temperature variation ΔT. The radial pressure P exerted by FRP bar on concrete cover at the interface of bar-concrete when the temperature increases is due to the difference between the transverse coefficient of thermal expansion of FRP bars and concrete. This pressure is given by the following equation:

$$P = \frac{(\alpha_t - \alpha_c)\Delta T}{\frac{1}{E_c}\left(\frac{r^2+1}{r^2-1} + v_c\right) + \frac{1}{E_t}(1 - v_{tt})} \tag{1}$$

where $r = b/a$ is the ratio of concrete cylinder radius ($b = c + d_b/2$) to FRP bar radius ($a = d_b/2$); E_c is the modulus of elasticity of concrete; v_c is Poisson's ratio of concrete; α_c is the coefficient of thermal expansion of concrete; E_t is the modulus of elasticity of FRP bar in the transverse direction; v_{tt} is Poisson's ratio of FRP bar in the transverse direction and α_t is the transverse coefficient of thermal expansion of FRP bar.

The transverse thermal deformation in concrete (ε_{ct}) and in FRP bar (ε_{ft}), at the interface of FRP bar-concrete, due to the radial pressure P and the temperature variation ΔT, are given by the following equations:

$$\varepsilon_{ct}(a) = \frac{P}{E_c}\left(\frac{r^2+1}{r^2-1} + v_c\right) + \alpha_c \, \Delta T \tag{2}$$

$$\varepsilon_{ft}(a) = \alpha_t \, \Delta T - \frac{(1-v_{tt})}{E_t}P \tag{3}$$

The transverse thermal deformation of concrete $\varepsilon_{ct}(b)$, at the external surface of concrete cover of prismatic concrete beams reinforced with FRP bars, due to the radial pressure P and the temperature variation ΔT, is given by:

$$\varepsilon_{ct}(b) = \frac{2P}{E_c(r^2-1)} + \alpha_c \Delta T \tag{4}$$

3 Nonlinear Numerical Investigation

3.1 Reinforced Concrete Beams Description

The detail of concrete beams reinforced with glass FRP (GFRP) bars used in this study is presented in Fig. 1. The cross-sections of concrete beams ($b_1 \times h$) were 76×100, 100×125, and 100×150 mm^2. The ratio of concrete cover thickness to FRP diameters c/d_b was varied from 1 to 3.2. The concrete used in this study has been considered to have a non-linear behavior. The concrete properties such as the compressive concrete strength (f'_c), the tensile concrete strength (f_{ct}), the modulus of elasticity (E_c), are reported in Table 1. The coefficient of thermal expansion α_c, Poisson's ratio v_c, concrete density γ_c, were equal to $(11.6 \pm 2.1) \times 10^{-6}/\,°C$; 0.17; 2.4 g/cm^3, respectively. The elastic modulus and the tensile strength of concrete have been evaluated using equations recommended by ISIS Canada (2007) [8].

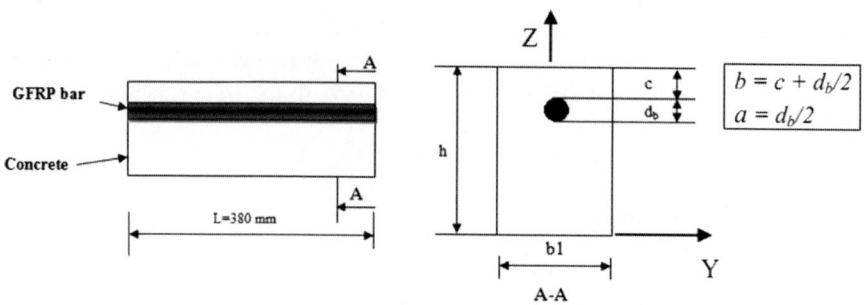

Fig. 1. Modeled concrete beam reinforced with GFRP bar

The mechanical properties of the GFRP bars determined experimentally by Zaidi and Masmoudi [4] are reported in Table 2. GFRP bars had a linear elastic behavior. The average values of the transverse and the longitudinal coefficients of thermal

Table 1. Mechanical properties of concrete

Compressive strength, f'_c (MPa)	Tensile strength, f_{ct} (MPa)	Modulus of elasticity, E_c (MPa)
25	3.0	24,944
30	3.3	26,621
35	3.5	28,164
40	3.8	29,601
45	4.1	30,951
50	4.2	32,227
55	4.4	33,441
60	4.6	34,601
65	4.8	35,713
70	5.0	36,784
75	5.2	37,817
80	5.4	38,816
85	5.5	39,784
90	5.7	40,724

Table 2. Mechanical properties of GFRP bars used

Bar diameter, d_b (mm)	Ultimate tensile strength, f_{fu} (MPa)	Longitudinal modulus of elasticity, E_l (GPa)	Longitudinal poisson's ratio, v_{lt}
9.5	627 ± 22	42 ± 1	0.28 ± 0.02
12.7	617 ± 16	42 ± 1	
15.9	535 ± 9	42 ± 1	
19.1	600 ± 15	40 ± 1	
25.4	N/a	N/a	

expansion (CTE) of GFRP bars for five FRP bar diameters tested were found to be equal to $\alpha_t = 33 \times 10^{-6}/°C$ and $\alpha_l = 9 \times 10^{-6}/°C$, respectively. While the modulus of elasticity (E_t) and Poisson's ratio (v_{tt}) of GFRP bars in the transverse direction were found to be equal to 7.1 GPa and 0.38, respectively.

3.2 Finite Element Model

A nonlinear numerical model was established using ADINA software, for a concrete rectangular cross-section reinforced with GFRP submitted to a temperature variation, to investigate the concrete strength variation effect on the distribution of transverse thermal deformations in concrete cover and GFRP bars (Fig. 1). The cross-section of concrete has been modeled by means of two dimensional plane stress elements since the axial deformations are constant. The finite element analysis has been performed only for the half of the concrete rectangular cross-section because of the symmetric of the cross section with respect to z-z axis. The temperature variation (ΔT) increased with

an increment of +5 °C up to 70 °C was applied statically over all the cross-section of
the GFRP bar-reinforced concrete beam. The meshing of both concrete and GFRP bar
was carried out using triangular elements with 6 nodes, as shown in Fig. 2. GFRP bar
material was modeled to have a linear elastic behavior, however the concrete was
assumed to have a nonlinear behavior (the concrete and elastic models defined by
ADINA were used for concrete and GFRP, respectively). A perfect bond was considered at the interface between GFRP bar and concrete. These both materials share the
same nodes at the interface. In this study, splitting cracks appear when the maximum
tensile stress in the circumferential direction reaches the tensile strength of concrete.

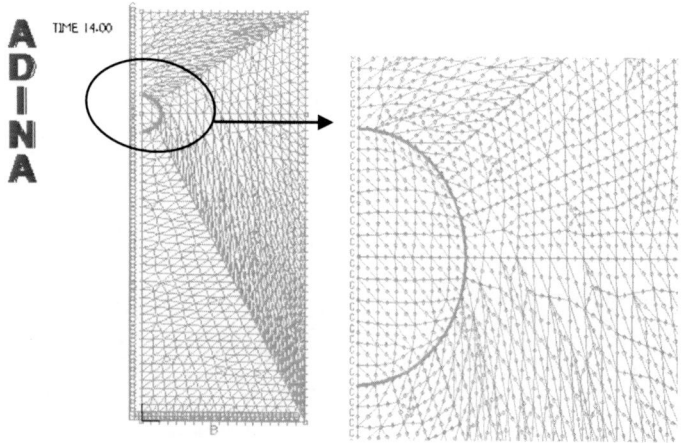

Fig. 2. Meshing of half of cross-section of concrete beam reinforced with GFRP bar

3.3   Numerical Results and Discussions

Figures 3 and 4 present transverse thermal strain curves as a function of the temperature variation and the compressive concrete strength, at the interface of FRP
bar-concrete, for concrete beams reinforced with GFRP bars having a ratio of concrete
cover thickness to FRP bar diameter (c/d_b) equal to 1.0 and 3.2, respectively. These
figures show that the thermal strain curves are linear until the cracking thermal load
ΔT_{cr}, producing the first radial cracks in concrete at the FRP bar-concrete interface
(Fig. 5), varied from 20 to 35 °C depending on the ratio c/d_b and the compressive
concrete strength f'_c varied from 1 to 3.2 and 25 to 90 MPa, respectively. From ΔT_{cr}
thermal loads, the strain curves become nonlinear and exhibit important values because
of the development of a circular crown of radial cracks in the concrete surrounding
GFRP bar, as shown in Fig. 5. Also, it can be noted that the concrete strength variation
has no big influence on the transverse thermal strains for thermal loads less than ΔT_{cr}.
However, for thermal loads greater than ΔT_{cr}, the transverse thermal strains decrease
with the increase in the concrete strength.

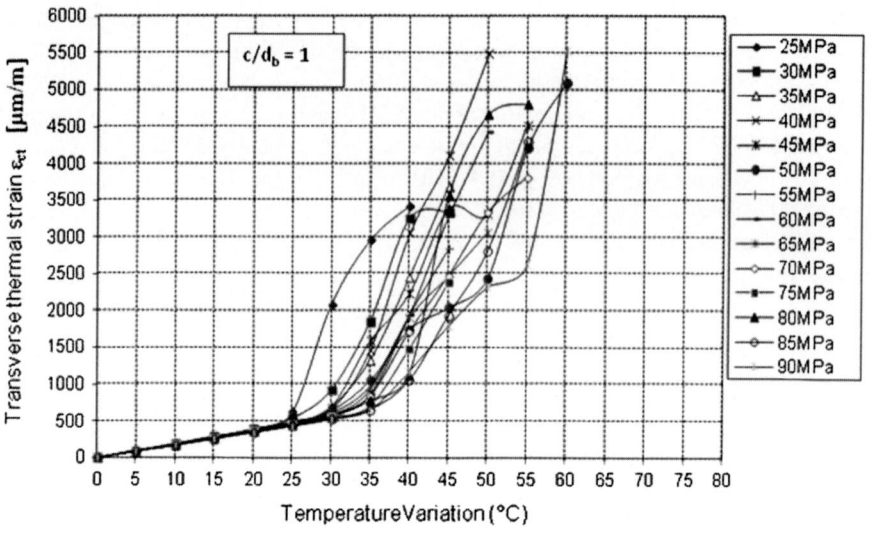

Fig. 3. Transverse thermal strains at the interface for concrete beams reinforced with GFRP having $c/d_b = 1$ and different concrete strength f_c' varied from 25 to 90 MPa

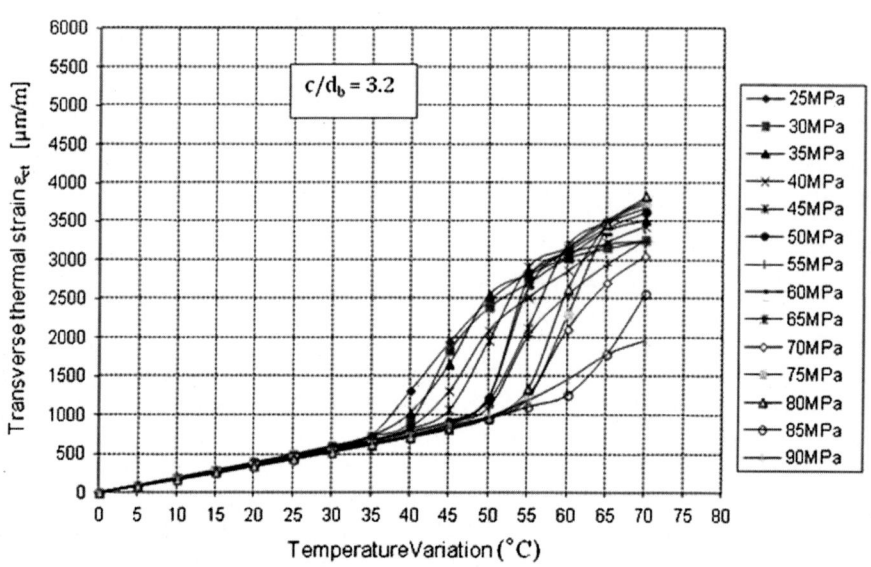

Fig. 4. Transverse thermal strains at the interface for concrete beams reinforced with GFRP having $c/d_b = 3.2$ and different concrete strength f_c' varied from 25 to 90 MPa

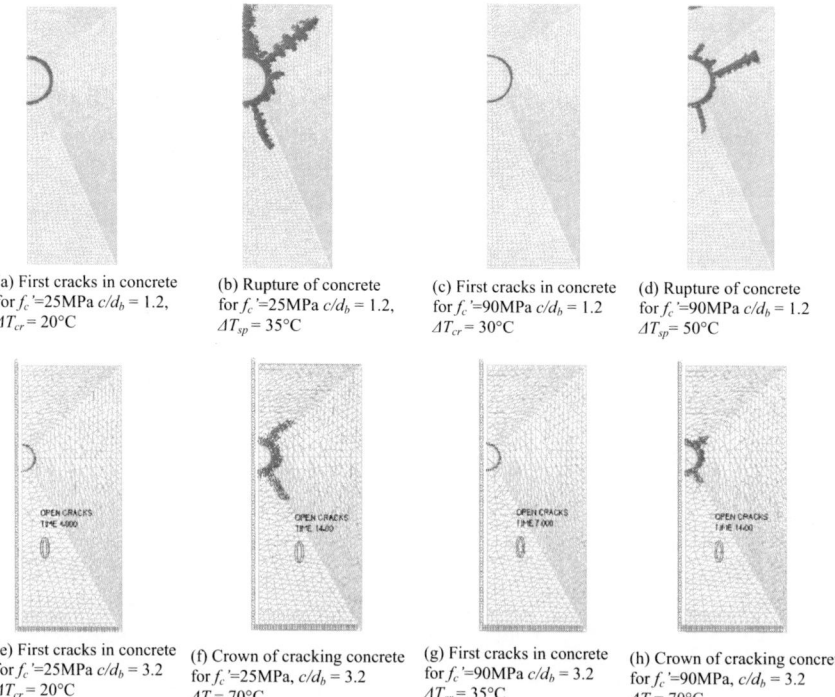

(a) First cracks in concrete for f_c'=25MPa c/d_b = 1.2, ΔT_{cr} = 20°C

(b) Rupture of concrete for f_c'=25MPa c/d_b = 1.2, ΔT_{sp} = 35°C

(c) First cracks in concrete for f_c'=90MPa c/d_b = 1.2 ΔT_{cr} = 30°C

(d) Rupture of concrete for f_c'=90MPa c/d_b = 1.2 ΔT_{sp}= 50°C

(e) First cracks in concrete for f_c'=25MPa c/d_b = 3.2 ΔT_{cr} = 20°C

(f) Crown of cracking concrete for f_c'=25MPa, c/d_b = 3.2 ΔT = 70°C

(g) First cracks in concrete for f_c'=90MPa c/d_b = 3.2 ΔT_{cr} = 35°C

(h) Crown of cracking concrete for f_c'=90MPa, c/d_b = 3.2 ΔT = 70°C

Fig. 5. Typical cracking concrete patterns for GFRP bars-reinforced concrete beams having c/d_b = 1.2 and 3.2; f_c' = 25 and 90 MPa

4   Comparisons Between Numerical and Analytical Results

Figures 6 and 7 show the comparison between analytical and numerical curves in terms of transverse thermal strains at the interface of FRP bar-concrete as a function of the temperature variation and the typical compressive concrete strength (25 and 90 MPa) for a ratio of c/d_b = 1 and 3.2, respectively. From these figures, it can be seen that the transverse thermal strains values predicted from the numerical model, at FRP bar-concrete interface, are in good agreement with the analytical results until the thermal loads values (ΔT_{cr}), producing the first cracks in concrete at the interface, ranging from 20 to 35 °C (Fig. 5), depending on the compressive concrete strength (f_c') and the ratio c/d_b. From these cracking thermal loads ΔT_{cr}, the non-linear numerical model exhibits high values because of the presence of cracks in the concrete which has not been considered in the analytical model based on the theory of elasticity. The same observation can be noted for transverse thermal strains at external surface of concrete cover for concrete beams having a ratio c/d_b = 1 (Fig. 8) where the thermal strain curves predicted from numerical and analytical models are almost linear and similar until the thermal loads, producing the failure of concrete cover or profound radial

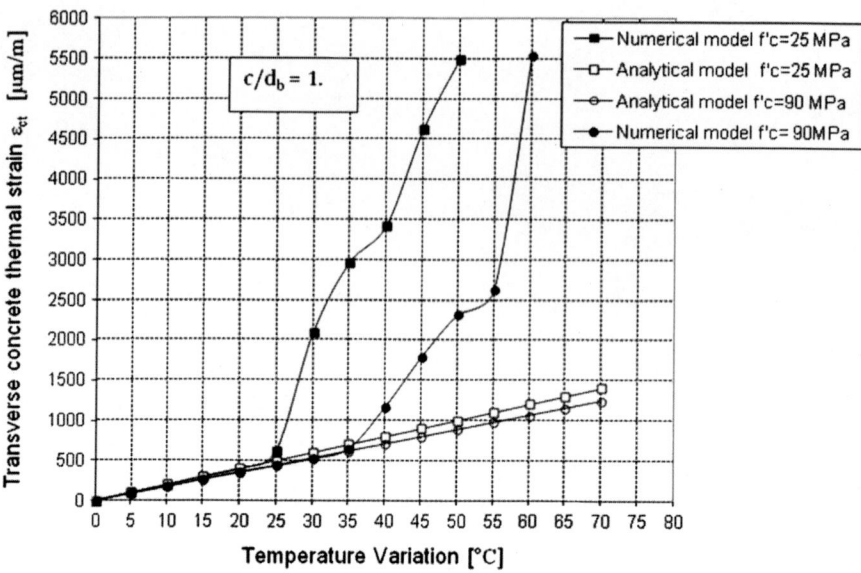

Fig. 6. Transverse thermal strains at FRP bar/concrete interface for prismatic concrete beams reinforced with GFRP bar having $c/d_b = 1$ varying concrete strength f_c'—comparison between numerical and analytical results

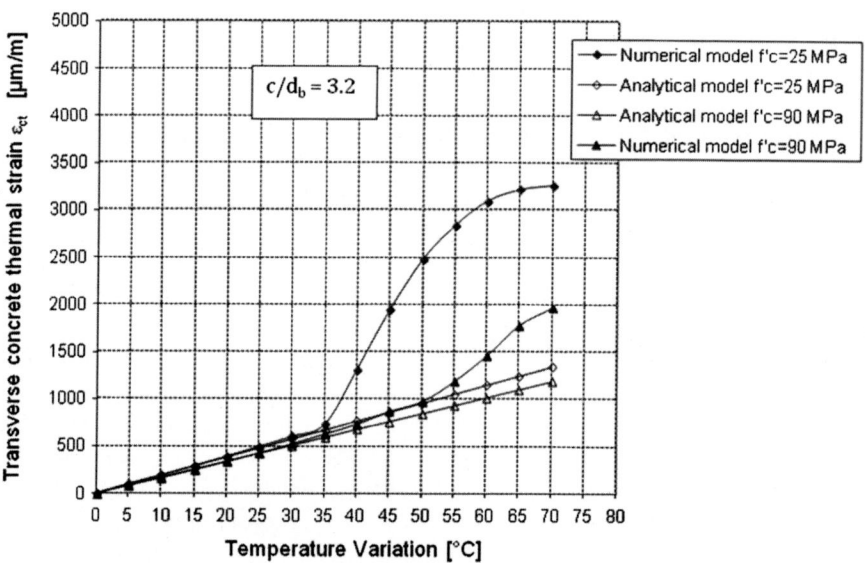

Fig. 7. Transverse thermal strains at FRP bar/concrete interface for prismatic concrete beams reinforced with GFRP bar having $c/d_b = 3.2$ varying concrete strength f_c'—comparison between numerical and analytical results

cracks, ranged from 30 to 50 °C corresponding to the compressive concrete strength of 25 and 90 MPa, respectively, as shown in Fig. 8.

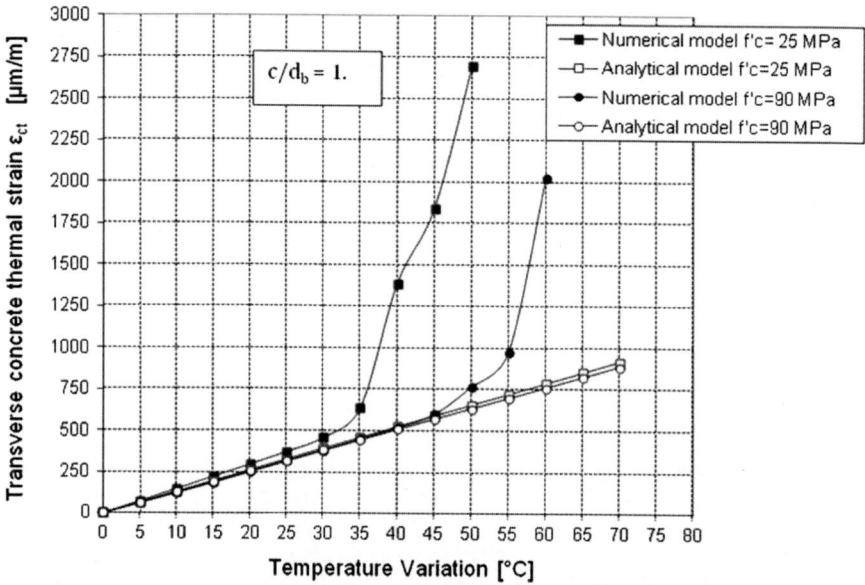

Fig. 8. Transverse thermal strains at external surface of concrete cover for prismatic concrete beams reinforced with GFRP bar having $c/d_b = 1$ varying concrete strength f'_c—comparison between numerical and analytical results

5   Conclusions

- The transverse thermal strain curves are linear until the cracking thermal load ΔT_{cr}, producing the first radial cracks in concrete at the FRP bar-concrete interface, varied from 20 to 35 °C depending on the ratio of concrete cover thickness to FRP bar diameter c/d_b and the compressive concrete strength f'_c varied from 1 to 3.2 and 25 to 90 MPa, respectively. From ΔT_{cr} values, the thermal strain curves become nonlinear and present high values because of the development of the circular crown of radial cracks in the concrete surrounding GFRP bar.
- The concrete strength variation has no big effect on the transverse thermal strains for thermal loads less than ΔT_{cr}. However, for thermal loads greater than ΔT_{cr}, the transverse thermal strains decrease with the increase in the concrete strength.
- The transverse thermal strains values predicted by the numerical model, at FRP bar-concrete interface, are in good agreement with the analytical results until the thermal loads values (ΔT_{cr}), producing the first cracks in concrete at the interface, ranging from 20 to 35 °C, depending on the compressive concrete strength (f'_c) and the ratio c/d_b. From these cracking thermal loads ΔT_{cr}, the non-linear numerical

model exhibits important values because of the presence of cracks within concrete which has not been considered in the analytical model based on the theory of elasticity.

- The transverse thermal strains, at external surface of concrete cover of GFRP bar-reinforced concrete beams, predicted from numerical and analytical models are almost similar until the thermal loads, producing the failure of concrete cover or profound radial cracks, ranged from 30 to 50 °C corresponding to the compressive concrete strength of 25 and 90 MPa, respectively.

Acknowledgements. The writers would like to acknowledge the support of the Civil Engineering Department of Sherbrooke University (Canada) and also the Structures Rehabilitation and Materials Laboratory (SREML) of Laghouat University (Algeria). The opinion and analysis presented in this paper are those of the authors.

References

1. Rahman HA, Kingsley CY, Taylor DA (1995) Thermal stress in FRP reinforced concrete. In: Proceedings of annual conference of Canadian society for civil engineering, Ottawa, pp 605–614
2. Aiello MA, Focacci F, Nanni A (2001) Effects of thermal loads on concrete cover of fiber reinforced polymer reinforced elements: theoretical and experimental analysis. ACI Mater J 35:332–339. doi:10.14359/10402
3. Masmoudi R, Zaidi A, Gerard P (2005) Transverse thermal expansion of FRP bars embedded in concrete. J Compos Constr 9:377–387. doi:10.1061/(ASCE)1090-0268(2005)9:5(377)
4. Zaidi A, Masmoudi R (2007) Effect of concrete cover thickness and FRP-bars spacing on the transverse thermal expansion of FRP bars. In: 8th international symposium on fiber reinforced polymer reinforcement for concrete structures. University of Patras, Department of Civil Engineering, Patras, Greece
5. Zaidi A, Masmoudi R, Bouhicha M (2013) Numerical analysis of thermal stress-deformation in concrete surrounding FRP bars in hot region. Constr Build Mater 38:204–213. doi:10.1016/j.conbuildmat.2012.08.047
6. Zaidi A, Mouattah K, Masmoudi R, Hamdi B (2015) Finite element modeling of fiber reinforced polymer bars embedded in prismatic concrete beams under high temperatures. J Reinf Plast Compos 34:315–328. doi:10.1177/0731684415571191
7. Bellakehal H, Zaidi A, Masmoudi R, Bouhicha M (2014) Behavior of FRP bars-reinforced concrete slabs under temperature and sustained load effects. Polymers 6:873–889. doi:10.3390/polym6030873
8. Intelligent Sensing for Innovation Structures (2007) Reinforcing concrete structures with FRP. Design manuals no 3, ISIS-Canada, Winnipeg, Manitoba, Canada, pp 41–43

A Study on the Use of Advanced Nondestructive Testing Methods on Histroric Structures

Rukiye Tuğla[1(✉)], Rüya Kılıç Demircan[2], and Gökhan Kaplan[3]

[1] Abana Sabahat-Mesut YILMAZ Vocational Schools Kastamonu, Kastamonu University, Kastamonu, Turkey
rkykockar@gmail.com
[2] Civil Engineering Department, Gazi University, Ankara, Turkey
[3] Kastamonu University, Kastamonu Vocational Schools, Kastamonu, Turkey

Abstract. Historic structures are social and cultural heritage for the land they are built on. To protect this heritage and to preserve it for the generations to come are among the most important responsibilities of nations. However numerous historical structures are found in Turkey, the awareness around nondestructive maintenance processes which will keep the identity of the structure intact is not developed as required. Historical structures require preservation, and necessary maintenance and reinforcement measures to be taken in order to survive damages arising from several reasons. Such work must be designed by multidisciplinary professionals and must be applied adhering to the original form of the structure. Among the limiting factors involved in an intervention to a structure is the knowledge on the strength and material properties of the structural elements used. Structural elements of a historical building must be subjected to measurements based on a number of assumptions made for parameters such as compressive strength, shear strength and elasticity modules. As a result of practices based on such assumptions, historical structures may face unnecessary repair and reinforcement processes. With the advancement made in technology now it is possible to obtain the necessary data from historical buildings without causing damage or causing minimal damage using nondestructive or semi-destructive test methods. Therefore, it is important and necessary to use and popularize the use of nondestructive testing methods for historical structures. This study investigates the following nondestructive testing methods and reports on the results obtained in a comparative manner: concrete rebound hammer method; flat-jack method; penetration resistance method; ultrasonic method; impact-echo method; and magnetic and electrical methods.

Keywords: Nondestructive testing methods · Historic structures · Restoration

1 Introduction

It is of utmost importance to protect and ensure they are safely offered to the generations to come in order to maintain the data for the development of the humanity. These structures offer information on the building techniques used, and the culture,

© Springer International Publishing AG, part of Springer Nature 2018
S. Fırat et al. (eds.), *Proceedings of 3rd International Sustainable Buildings Symposium (ISBS 2017)*, Lecture Notes in Civil Engineering 7,
https://doi.org/10.1007/978-3-319-64349-6_43

beliefs and sociologic structure of that period. It is important for the humanity to further such an information flow [10].

The main purpose of maintenance and restoration of historic cities must involve historic and cultural sustainability, protection of the identity of such historic structures for the generations to come, assessment of historic structures in the structure inventory, and maintenance of the traditional settlement model. Thus, extensive protective efforts are of importance for the sustainability of physical and social structure in a historic environment where change and transformation are underway [17].

In addition to the protective efforts, repairment and strengthening are also important aspects for the survival of historic structures. Repairment and strengthening is an area of expertise and must be conducted by qualified people using right materials, equipment and techniques. It is a must to gather information about the condition of the structure before initializing repairment and strengthening process. If damaged, the reasons behind the damage must be investigated and these reasons must be eliminated. The strength of structural material, material properties and physical conditions must be known. And such data must be obtained without damaging the structure. Historic structural elements must be subjected to measurements for parameters such as compressive strength, shear strength and elasticity module. As a result of practices based on such assumptions, historical buildings may face unnecessary repair and reinforcement processes. In order to eliminate such damage, it is possible to obtain the necessary data from historic structures without causing any damage or causing minimal damage with the use of nondestructive or semi-destructive test methods. Among these tests are infrared thermal imaging, superficial hardness test, in situ pressure test, ultrasound test, radioactive method, penetration resistance, and endoscopic method.

This study aims to communicate information about the damages in historic structures and their reasons and the techniques used in order to obtain information about structures before the repairment-strengthening work. A number of examples are provided from previous use of test techniques in historic structures.

2 Damages in Historic Structures and Their Reasons

Historic structures are the most important sources of information about the era they were built. Built with the technology and materials available in a period of the past and defying the damage of the years, historic structures are exposed to damages today due to a number of reasons which may lead to their destruction. It is of utmost importance to know the reasons behind such damages in terms of the repairment and strengthening efforts. Among these damages are;

Damages Due to the Ground: Impaired or heterogeneous soil strength leads to movement of the structural elements in time which result in visible failures in the structure [1].

Ground Movements Such as Earthquakes: often able to merely bear the load of pressure strains, historic structures are damaged extensively due to earthquakes which create a great tensile stress on the structure. Such damages may cause even demolition when necessary measures are not taken.

Fires, Wars, and Vandalism: However fire's impact on stone or brick buildings is limited when compared to wooden buildings, fire impairs the architectural properties of the structure while high temperatures during a fire hazard lead to cracks and collapses affecting the material properties. Historic structures are receiving extensive damage during wars as the destructive power of modern warfare increases. This kind of a damage may even lead to complete destruction of a historic structure. Vandalism is an intentional act of destruction and is a commonly used term to define destruction of monuments by random people [1]. Drawing graffitis on historic monuments, littering, and causing destruction, which are commonly seen in historic structures today are examples of vandalism.

Air Pollution and Traffic: A number of polluting gasses and solutions damage the facades of buildings resulting in wear and loss of material properties in time. Any vibrations due to vehicle traffic and their impact on the structure's foundation lead to damages in historic structures.

Long-Term Natural Causes: Historic structures may be damaged under different natural conditions, if necessary measures are not taken. Among these natural causes are temperature differences, freeze-thaw cycles, exposure to continuous humidity.

Structural Design Faults: If there are any dimensional faults in the carrier systems of the buildings due to their original design, in other words, if carriers such as walls, pillars, supports, etc. are not able to carry vertical and horizontal loads, then serious damages may occur. A wall yields when it is unable to support cross-sectional loads, and when this is the case for the pillars, it may lead to dislocation of arches, vaults, and domes it supports, which in return may result in collapse [1].

Structural Material's Strength Loss: Deterioration of the structures accelerate when the materials used (stones, bricks, adobe bricks, wood, etc.) are not of good quality. For example, wear of the material is accelerated when the clay content of the stones used in construction increases. In the case of brick structures, structural strength increases when the bricks used are cured properly. Damages such as rapid wear, breaking off, indent formation, surface losses, disintegration, etc. occur when poor quality bricks are used.

Poor Workmanship: The combination of proper construction technique and good workmanship along with good quality materials is ideal. Any faults such as using insufficient amounts of binding material and use of faulty techniques used in bonding may lead to damages in masonry buildings.

Sinking, settling, cracks, dislocations observed in historic structures due to afore-mentioned reasons are especially common in masonry structures. Masonry structures are heavy and rigid structures due to the material properties. However their compressive strength is relatively higher, they offer low strength in terms of tensile strength and shear stress. Cracks, dislocations and disintegration may be observed in the structure due to above-mentioned effects. Cracks in the structure and their location give us the information on the stress distribution needed to deduct the reasons behind such damages. Knowing the reason behind the formation of cracks is of utmost importance in order to take appropriate repairment and strengthening decisions.

Sinking and/or settling of bearing elements is common in historic structures due to the differences in the ground. Different kinds of sinking and settling in bearing walls are manifested in inclined cracks which generally occur in the wall plane. Which side the sinking and/or settling is inclined to can easily be defined if the direction of the crack is detected [26].

Any covering elements such as dome, vaults and aches are exposed to joint dislocations in time due to ground movements and earthquakes when they are not able to withstand the tensile stress created by the inclination moments. These structures may be demolished if necessary measures are not taken.

Repairment and strengthening of historic structures is a field of expertise. Information such as material properties, mechanical properties and physical conditions must be obtained before any restoration effort is initiated for a structure damaged due to the reasons listed above. When obtaining such data, the condition of the structure must be considered. Today's technology made it easier to obtain such information. It is now possible to obtain the desired information with no or minimal damage to the structure using nondestructive or semi-destructive test methods.

3 Nondestructive Testing Methods

3.1 Infrared Thermal Imaging Method

Infrared thermal imaging is commonly used for efforts related to fuel and energy preservation. Today, this method is used extensively in fields such as military, industry, medicine, meteorology, architecture and engineering [30]. Moreover, many scientists have preferred this method for their studies especially focusing on the detection of materials and structural issues in historic buildings. This method is used to detect especially heat loss, humid regions, gas escape, thermal bridge and the status of thermal insulation [14, 18, 28, 30, 31]. In addition, infrared thermal imaging is also used for failures in historic structures, rainwater and waste water drainage system issues, and structural cracks [29]. Thermal cameras are used for infrared thermal imaging (Fig. 1).

Fig. 1. Examples of thermal cameras

This nondestructive test method which uses thermal cameras provides thermal map of the material used in accordance with the color scales at threshold value defined with the detection of the radiation emitted by the material. Thus, the area to be examined is defined. This method allows for the detection of thermal issues, humid areas, gas escape, heat bridges, etc. in the areas examined [18, 31]. In other words, problematic/unproblematic areas of the structure being investigated are revealed in a nondestructive manner [15, 16, 29, 30].

The emissivity coefficient must be known (generally between 0.90 and 0.95 for structural material) when examining a structure using infrared thermal imaging method. Lower or higher emissivity coefficients is a factor influencing the accurate temperature measurement with thermal cameras [21, 23]. Thermal cameras are also affected by other parameters along with emissivity. Among these parameters are the environmental temperature of the place thermal camera is used, wind velocity of the environment, relative humidity, the distance between the camera and the object to be tested, the angle of the camera, the time and season the test will be conducted [6, 30]. Infrared thermal imaging is used in three different methods. The first one is used to detect the heat transfer from a medium where a heater is used to the outer medium; while the second is used where there is a heat flow between two surfaces with different ambient temperatures. Finally, it is also used to inspect the heat differences with the application of heat radiation to both sides of the object where there is no temperature differences or heaters in place [29, 30]. It is possible to obtain both qualitative and quantitative data using thermal cameras. Analysis can be conducted using relevant software [30].

3.2 Superficial Hardness/Concrete Test Hammer Method

One of the oldest nondestructive test methods, concrete test hammer method is used to measure the hardness of the material. Hardness is one of the most important mechanical properties when it comes to distinguish materials. Hardness is the resistance of the material against an object which tries to penetrate the material surface. Superficial hardness was developed by Ernst Schmidt in 1948 and is recognized as Schmidt test hammer after the scientists name [32]. The working principle of this method involves the measurement of the rebound of a spring-loaded mass impacting on a surface. The hardness of the material increases as the rebound increases [12].

3.3 Flat-Jack Method

Also known as in-situ stress test, the Flat-jack method allows for identification of a number of mechanical properties of a structural wall. Compressive strength, elasticity modulus and poisson ratio are among the mechanical properties measured with the flat-jack method [19]. Flat-jack testing tool allows for the measurement of the change of length (Δl, mm) in response to the power (P, kN) applied to an element as per ASTM C1196–92 [5]. Flat-jack test mechanism involves two measuring methods. The first method uses a double flat-jack and is able to define compressive strength, elasticity modulus and poisson ratio; while the second method uses a single flat-jack and is able to define the current compressive stress of the wall [19]. The mechanism involves a compressor, manometers, flat-jacks, comparators and pins. Flat-jacks are used in order to apply compression to the surface, while comparator measures the replacement and pins locate the comparator [5, 12].

3.4   Ultrasound Test Method

Ultrasound velocity test method is a nondestructive method which gives information on the quality of the concrete, its internal structure, its porosity, and its compressive strength, cracks and the depth and orientation of these cracks when necessary correlations are applied. It operates with the principle of the measurement of ultrasonic pulse velocity through a material. It is calculated using the equation below (Eq. 1) [8, 32].

$$V = L/\Delta T \tag{1}$$

Here V is ultrasonic pulse velocity (km/s), while L is the length of the material and ΔT is the difference in time for ultrasonic pulse travel.

In the ultrasound test, material surface is contacted with two piezoelectric transducers without any void between them and the surface. The first transducer sends the ultrasound waves and the second received these ultrasound waves. The time of transmission and velocity of the ultrasound waves are then measures [24].

If the density of the material is poor and/or the material has cracks, then the diffusion of the soundwaves, therefore, the ultrasound pulse velocity is low [12]. When the material is robust, in other words, when it has reduced amount of pores and it offers a better strength, then the ultrasound pulse velocity is higher. However, this test is not sufficient by itself to define the strength and must be combined with other techniques.

3.5   Radioactive Method

Radioactive test method was developed in 50s in order to examine materials. This test mechanism consists of an electromagnetic radiation source and a sensor. The sensor measures the time required for the radiation to reach the other side of the material. When the sensor is in the form of a special photographic film then it is called radiography, and when the sensor converts the radiation into electrical waves then it is called radiometry [9, 12].

3.6   Penetration Resistance Method

This is a test method operates with the principle of the measurement of penetration depth of a probe or a nail shot at the concrete using a gun developed in the USA in 1964. Windsor probe is similar to the concrete test hammer. It is used in order to have an idea about the strength of the concrete and identifies the concretes resistance against penetration. This test method is affected by the aggregates used in the concrete due to its operating principle [32].

3.7   Endoscopic Method

The elements of the bearing system used in structures are rather large and hard to investigate. This is a method preferred when the material used in a bearing system is to be defined where such material cannot be visually inspected or identified using a borehole sample. This method involves drilling a hole of 1 cm in diameter in the structure and obtaining images of the structure with the use of a cable and a camera attached to it, thus, making it possible to identify the material used. This method is especially important for the diagnosis of the bearing system of historic structures [3].

3.8   Ground Penetrating Radar Method

In ground penetrating radar method, electromagnetic waves are sent through a medium, the times between receiver and transmitter are recorded and the target area is scanned. Thus, it is possible to reveal unknown characteristics due to the physical discontinuity of the medium. This method does not allow for the exploration of mechanical properties of the material, however, it is able to define the physical characteristic which cannot be identified with visual inspection or with the use of borehole sample [13].

4   Examples of Application

1. Tavukçuoğlu et al. [28] investigated the Şengül Bath, a fifteenth century Ottoman structure, using infrared thermal imaging in combination with microclimatic measurements in order to explore the thermal problems in the material due to wrong repairments along with the material properties and thermal properties of Historic Turkish Baths.
2. Çiçek [11] published a postgraduate thesis, titled "A Study on the Thermal Performances of Historic Turkish Baths", which focused on the thermal and physical properties of Historic Turkish Baths, the original microclimate properties of the structure, thermal insulation properties and detection of thermal issues using infrared thermal imaging. The study used nondestructive methods such as microclimatic analyses, infrared thermal imaging, thermal transfer analysis methods and vapor transfer analysis methods and it was supported with lab analyses in order to assess the results obtained.
3. Akevren [2] conducted a postgraduate study to make it possible for investigating historic stone walls without destructing the structure in order to improve the use of infrared thermal imaging under field conditions. The study focused on the Cenabi Ahmet Pasha Mosque located in Ankara and built in sixteenth century. Material decay of the current structure, humidity problem and cracks formed in the structure were investigated using infrared thermal imaging and ultrasound test methods. It was found that infrared thermal imaging allows the researcher to distinguish the depths of the cracks.
4. Noland et al. [22] made a study on the masonry structures available in the USA and used flat-jack test method in order to obtain information on such structures and assessed the results.

5. Qinglin and Xiuyi [25] improved the thick flat-jack equipment with the high capacity of dislocation and used it on the soft wall material found in China.
6. Kuran and Dabanlı [19] employed the flat-jack method for the Sultan Alaaddin Mosque located in the Inonu district of Eskişehir and defined the compressive strength and elasticity modulus of this structure. Figure 2 shows the application of the flat-jack test for Alaaddin Mosque.

Fig. 2. Flat-jact test conducted by Kuran and Dabanlı

Test results showed that the compressive strength of the masonry wall is 1.5 MPa. In the interior of the mosque, compressive strength of the brick arch located in the north of the sanctuary was also 1.5 MPa [19].

7. Simões et al. [27] employed flat-jack test for an old masonry structure (circa eighteenth century) located in Lisboa. The test mechanism of the flat-jack test is given in Fig. 3.

Flat-jack test results showed that the compressive strength of the wall is 0.91 MPa. According to the Secant method, the elasticity modulus is found to be 1817.4 MPa and the poisson ratio is measured as 0.31.

8. Bianco [7] has conducted investigations of the Saint Domenico Monastery located in Naples using the endoscopy method. Figure 4 shows the images from the endoscopy conducted in the monastery. Existence of macro cracks was identified as a result of the endoscopy conducted in the internal parts of the structural element.
9. Assunçao et al. [4] employed Ground Penetrating Radar and tomography methods on the columns of Hospital de Sant Pau i la Santa Creu located in Barcelona and built in 1905. The ground penetrating radar image of the structure is given in Fig. 5.

As seen in Fig. 5, there are vertical radar lines on the column. There were strong and irregular reflections between the radar lines. Figure 5b shows the change in the stress distribution on the exterior part due to the brick thickness. On the other hand, due to the asymmetrical surface distribution of the internal surface parts of the bricks, the reflection curves here are even more irregular. However the exterior surface of the

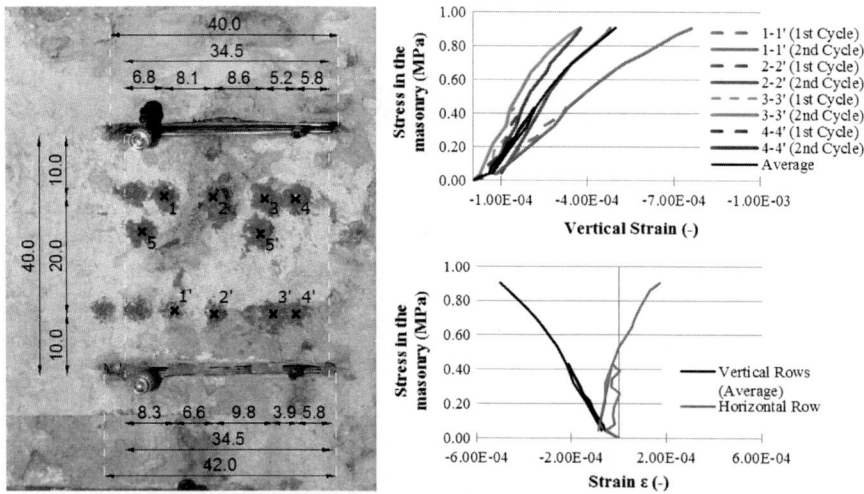

Fig. 3. The flat-jack test mechanism used by Simões et al. and the relevant stress–strain curves

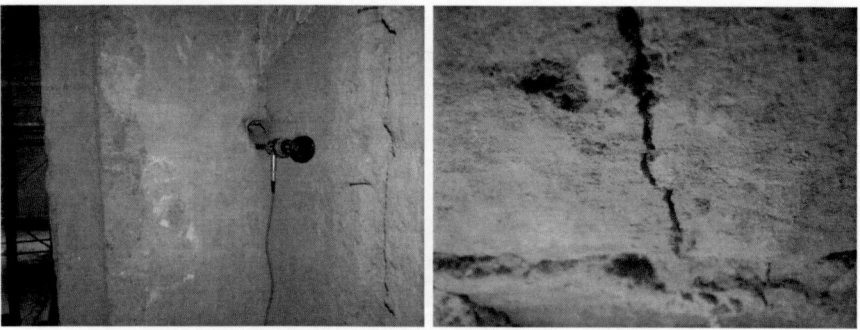

Fig. 4. Damage assessment using the endoscopy method

column seems to be regular, it was found that the bricks available in the exterior surface are not that uniform. Figure 5c shows that the middle parts of the column have a more regular structure. It is believed that the reason behind such a regularity is the existence of metallic material [4].

Figure 3 shows the radar images of the column in horizontal plats of 3 cm thickness. As seen in Fig. 6, the middle parts of the column have a more homogenous structure. The results of the radar and tomography tests showed the existence of metallic material in the middle parts of the column. The velocity was higher in these parts. Existence of metallic material in the middle part of the column was visually inspected with radar lines and was proved as seen in Fig. 7 [4].

10. Maierhofer and Roellig obtained tomographic images from masonry walls using IR cameras. Figure 8 shows the tomographic images of the cracks formed on the wall surface in a stepwise manner. Light and dark areas are detected on the image. Light

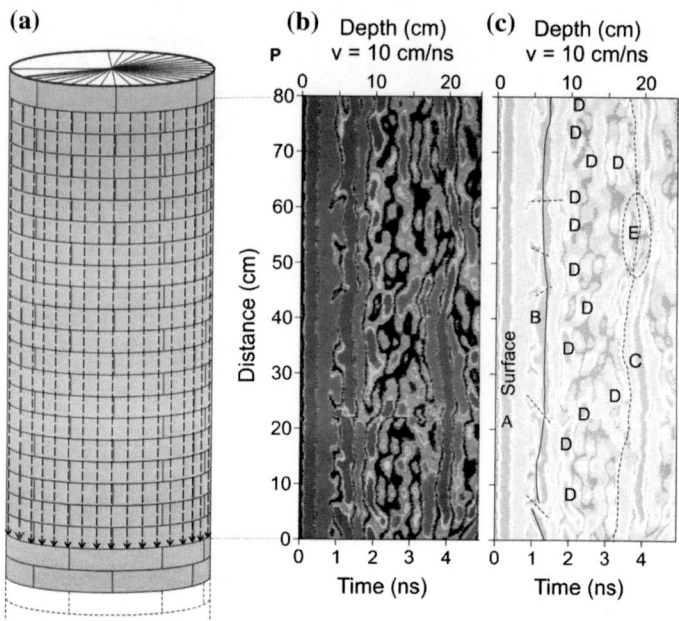

Fig. 5. Ground penetrating radar test conducted on Hospital de Sant Pau i la Santa Creu [4]

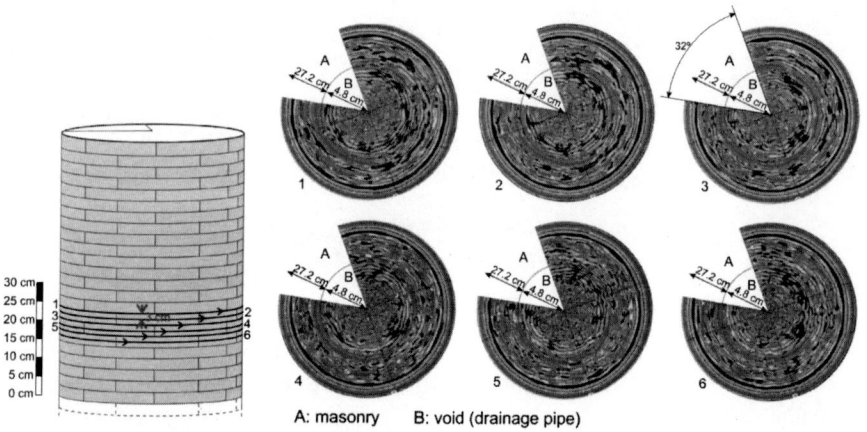

A: masonry   B: void (drainage pipe)

Fig. 6. Ground penetrating radar test conducted on the column and relevant radar diagrams [4]

areas indicate the rocks used in the wall construction while the dark areas are the joints. It was found that cracks are formed in the joints between bricks/rocks [20].

Wall joints of the northern part of the church were investigated using tomographic images (Fig. 9). The image in the middle explores the joint properties with respect to the changes in the temperature. It was found that joints have very high and low

(a) (b) (c)

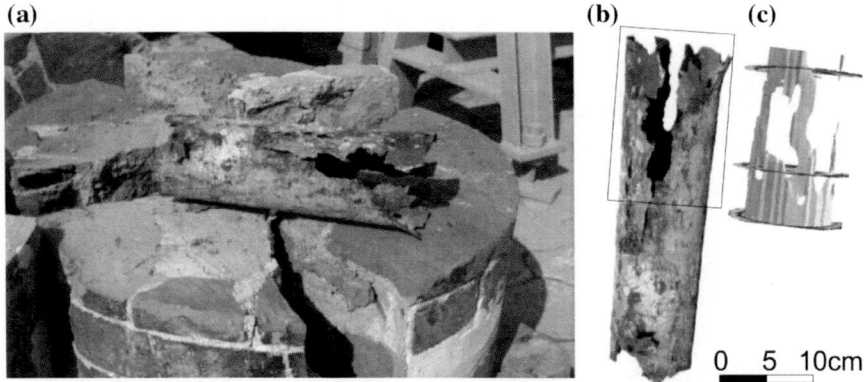

Fig. 7. Metallic material available in the middle parts of the column [4]

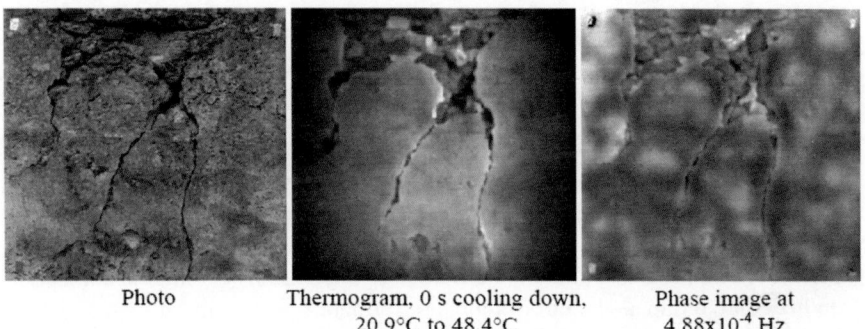

Photo   Thermogram, 0 s cooling down,   Phase image at
                              20.9°C to 48.4°C 4.88×10^{-4} Hz

Fig. 8. Damage assessment using IR camera [20]

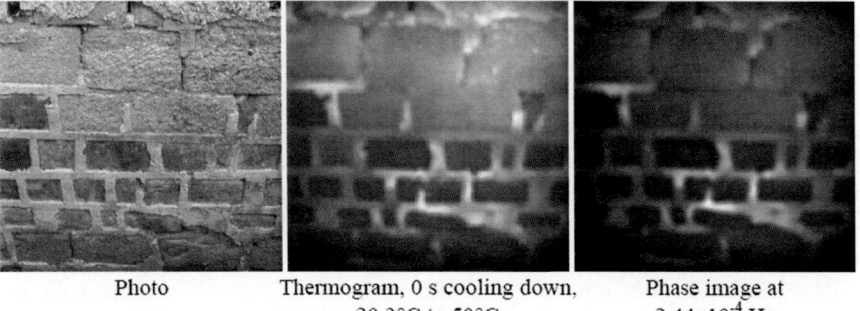

Photo   Thermogram, 0 s cooling down,   Phase image at
                              20.3°C to 50°C 2.44×10^{-4} Hz

Fig. 9. Damage assessment in the brick wall using IR camera [20]

temperature values. The white joints indicate that the material do not fill the gap between brick fully and that it is where the material loss will start [20].

As shown in Fig. 10, the humidity content of the wall surface can be defined using tomographic methods. The dark areas available in the tomographic images show that the humidity content is higher [20]. A water movement was found to affect the structure from the ground as seen in Fig. 10.

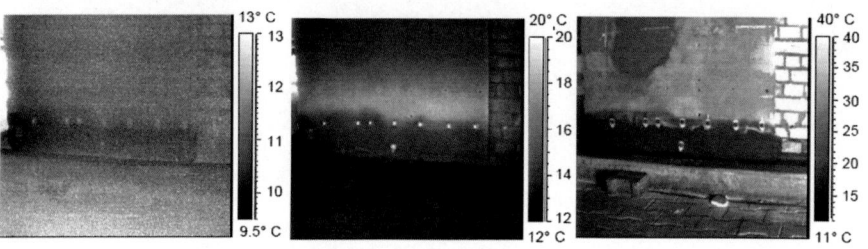

Fig. 10. Damage assessment using tomographic methods [20]

5 Results

It is an important concept for nations to preserve historic monuments and to conduct necessary repairment and strengthening practices in order for their survival for the generations to come. Damaged structures must be analyzed before any repairment-strengthening application takes place. The reasons behind the damage must be studied and eliminated and technical information about the structure must be available. Such information can be obtained without damaging the structure or altering its original conditions.

Concrete hammer test and ultrasound test methods are now insufficient nondestructive test methods given the technological advancements. Therefore, the use of advanced nondestructive test methods are gaining widespread use in the structural investigations. Among the advanced nondestructive test methods are flat-jack, ground penetrating radar, endoscopy, infrared thermal imaging and tomography methods.

Flat-jack test is a very strong nondestructive test method used for identifying the mechanical properties of historic structures. Mechanical properties such as compressive strength, elasticity modulus and poisson ratio can be measured with the flat-jack method. It is possible to detect the existence of metallic material in the bearing elements using radar images. The use of IR cameras for tomographic images allows for data on the cracks and humidity content of the structural elements. Endoscopy method, on the other hand, does not offer detailed information when compared to the other methods.

This study investigated the nondestructive test methods used in the assessment of historic structures and examined the previous studies. The results of this study showed that it is possible to obtain data on the material decay, invisible cracks formed in the structure, humidity problems and thermal properties along with values such as elasticity modulus, poisson ratio and compressive strength using nondestructive test methods.

Thus, it shows that it is possible to conduct repairment/maintenance practices on the historic structures, which are the representatives of the historic identity, preserving their natural conditions and without causing any further damage.

References

1. Ahunbay Z (1996) Tarihi Çevre Koruma ve Restorasyon. YEM yayınları. İstanbul, pp 23–45 (Turkish)
2. Akevren S (2010) Non-destructive examination of stone masonry historic structures–quantitative Ir thermography andultrasonic testing. Yüksek Lisans Tezi, Orta Doğu Teknk Üniversitesi Fen Bilimleri Enstitüsü, Ankara, pp 1–148
3. Aköz F, ve Yüzer N (2001) Tarihi Yapılarda Malzeme Özelliklerinin Belirlenmesinde Uygulanan Yöntemler. Yıldız Teknik Üniversitesi, İstanbul (Turkish)
4. Assunçao SS, Gracia VP, Caselles O, Clapes J, Salinas V (2014) Assessment of complex masonry structures with GPR compared to other non-destructive testing studies. Remote Sens 2014(6):8220–8237. https://doi.org/10.3390/rs6098220
5. ASTM C 1196–92 (1997) Standard test method for in situ compressive stress within solid unit masonry estimated using Flatjack measurements
6. Balaras CA, Argiriou AA (2002) Infrared thermography for building diagnostics. Energy Build 34:171–183
7. Bıanco A (2012) Endoscopic analysis supporting issues of historic stratigraphic investigations: the case history of Saint Domenico Monastery in Naples-Italy. In: 18th World conference on nondestructive testing Durban South Africa
8. BS 1881: Part 203 (1986) Recommendations on the non-destructive testing of concrete in the form of plain, reinforced and prestressed test specimens present components and structures by the measurement of ultrasonic pulse velocity
9. Carino NJ (1991) Nondestructive testing of concrete: history and challenges, concrete technology past, present, and future. In: Proceeding of V. Mohan Malhotra symposium
10. Çavuş M (2011) Tarihi Yapılarda Üst Örtülerin Çelik Malzeme İle Sağlamlaştırılmasının Sonlu Elemanlar Yöntemiyle Modellenmesi. Doktora Tezi Gazi Üniversitesi-Fen Bilimleri Enstitüsü Ankara (Turkish)
11. Çiçek P (2009) Thermal performance assessment of historical Turkish Baths. Yüksek Lisans Tezi, Orta Doğu Teknk Üniversitesi Fen Bilimleri Enstitüsü, Ankara, pp 1–160
12. Erköseoğlu G (2012) Kültürel Mirasın Depremden Korunması Ve Turizme Kazandırılması Amacı İle Uygulanabilecek Güçlendirme Teknikleri. Uzmanlık Tezi, T.C. Kültür ve Turizm Bakanlığı Yatırım ve İşletmeler Genel Müdürlüğü, Ankara (Turkish)
13. Gioloti A, Proteco S (2001) Experiences on georadar applicability on masonry structures. RILEM TC 177-MDT workshop on on-site control and non-destructive evaluation of masonry structures, Mantova, Italy
14. Grinzato E, Vavilov V, Kauppinen T (1998) Quantitative infrared thermography in buildings. Energy Build 29:1–9
15. Grinzato E, Bison PG, Marinetti S (2002) Monitoring of the ancient buildings by the thermal method. J Cult Herit 3:21–29
16. Kandemir Yucel A, Tavukcuoglu A, Saltik EN (2007) In situ assessment of structural timber elements of a historic building by infrared thermography and ultrasonic velocity. Infrared Phys Technol 49:243–248

17. Kılıç Demircan R, Kaplan G, Gültekin AB (2016) Analysis of restoration mortars used for strengthening of historical buildings in the context of sustainability criteria. SBE İstanbul
18. Koçkar R (2012) Tuğla duvarlardaki ısıl özelliklerin ve ısıl sorunların kızılötesi ısıl görüntüleme ve sıcak kutu yöntemleriyle incelenmesi. Yüksek Lisans tezi, Gazi Üniversitesi Fen Bilimleri Enstitüsü, Ankara (Turkish)
19. Kuran F, Dabanli Ö (2012) Tarihi Yığma Yapıların Mekanik Özelliklerinin Yerinde Yapılan Flat-Jack (Yassı Kriko) Deneyi İle Belirlenmesi. Restorasyon, pp 180–187
20. Maierhofer C, Roellıg M (2009) Active thermography for the characterization of surfaces and interfaces of historic masonry structures. In: NDTCE'09, non-destructive testing in civil engineering, Nantes, France
21. Maldague XPV (2001) Theory and practice of infrared technology for nondestructive testing. Wiley, New York
22. Noland JL, Atkinson RH, Schuller MP (1990) A review of the flat-jack method for nondestructive evaluation. In: Proceedings of the nondestructive evaluation of civil structures and materials, Boulder, USA
23. Ocana SM, Canas GI, Requena IG (2004) Thermographic survey of two rural buildings in Spain. Energy Build 36:515–523
24. Postacıoğlu B (1981) Cisimlerin Yapısı ve Özellikleri-İç Yapı ve Mekanik Özellikler. Cilt 1, İTÜ Matbaası, İstanbul (Turkish)
25. Qinglin W, Xiuyi W (1988) The evaluation of compressive strength of brick masonry in-situ. In: 8th International brick/block Mas. conference, Dublin, Ireland
26. Sesigür H, Çelik OC, Çılı F (2007) Tarihi yapılarda taşıyıcı bileşenler, hasar biçimleri, onarım ve güçlendirme. İstanbul Bülten 89:10–21
27. Simões A, Gago A, Lopes M, Bento R (2012) Characterization of old masonry walls: flat-jack method. In: 15 WCEE Lisboa
28. Tavukcuoglu A, Cicek P, Grinzato E (2008) Thermal analysis of an historical Turkish bath by quantitaive IR thermography. Quant Infrared Thermogr J 5(2):151–173
29. Tavukçuoğlu A, Akevren S, Grinzato E (2010) In-situ examination of structural cracks at historic masonry structures by quantitative infrared thermography and ultrasonic testing. J Mod Opt 57(18):1779–1789
30. Titman DJ (2001) Applications of thermography in non-destructive testing of structures. NDT&E Int 34:149–154
31. Tuğla R, Tavukçuoğlu A, Arslan M (2013) Examination of thermal properties and failures of brick walls by the use of infrared thermography and hot box method. International conference & exhibition on Application of efficient & renewable energy technologies in low cost buildings and construction, Ankara, Turkey, pp 180–199
32. Yaman İÖ (2010) Betonda Tahribatlı/Tahribatsız Muayene Metodları. Ders Notu Orta Doğu Teknik Universitesi Ankara (Turkish)

Studying the Historical Structure Damage Due to Soil Hazards and Examination of Applied Repairment-Strengthening Techniques

Rüya Kılıç Demircan[✉] and Pınar Sezin Öztürk Kardoğan

Civil Engineering Department, Technology Faculty, Gazi University, Ankara,
Turkey
ruyakilic86@gmail.com, sezinozturk@gazi.edu.tr

Abstract. Historical structures which are like a bridge between past and future are the social and cultural heritage. One of the most important responsibilities of society is that this heritage must be preserved and transferred to future generations safely. Historic structures were damaged by natural hazards, originating from soil, earthquakes and human impacts. These structures of lives can be extended to make repairment and strentghening. Prestudy must be done to decide type of the repairment and strentghening. Firstly, historical structure situation must be specified and determined hazards assessment after that the type of the hazards must be stated so repairing and strengthening must be done. These processes should be done to historical structures by experts in a way to least interference. Soil damage can affect negatively to historical structure. In this study, soil damage for historical structures was approached. Settlement, swelling, segregation, rupture, crack and sliding are hazards due to soil effects. These damages which include in the specific historical structure are viewed and then, recommended retrofit and reparing works are researched.

Keywords: Historical structures · Damage due to soil · Repairment-strengthening

1 Introduction

Any formation built in the past by the people to meet their needs and sustain their lives is now carrying a historic value for us. One of the most important is the historic structures. Historic structures are significant cultural assets which reflect the archeological, aesthetic, economic, politic, architectural and technical properties of the land they are built on. To protect and maintain such a cultural heritage is the responsibility of the nation.

With the destructive effects of the time, historic structures are being destroyed and most commonly demolished due to several factors such as natural causes, human impact, etc. It is possible to leave our cultural heritage to the generations to come with the restoration efforts of specialists such as maintenance-reinforcement, provided that necessary measures are taken.

One of the most important issues concerning historic structures is the damage to the foundation soil of the structure. Issues such as sinking/settling, changing groundwater levels, decaying foundation piling, adverse effects of the neighboring groundwork, insufficient drainage, and earthquakes lead to different types of damages in structures.

© Springer International Publishing AG, part of Springer Nature 2018
S. Fırat et al. (eds.), *Proceedings of 3rd International Sustainable
Buildings Symposium (ISBS 2017)*, Lecture Notes in Civil Engineering 7,
https://doi.org/10.1007/978-3-319-64349-6_44

Cracks in the covering elements such as dome, arch, and vault and columns and load-bearing walls, joint dislocations, uneven sinking/settling are the reflections of the aforementioned foundation damage on the structure.

Such cases require the investigation of a specialist in order to define the reasons behind the damage. Suitable repairment-strengthening technique must be used in order to eliminate the causes of the damage.

This study addresses the ground-related damages on the elements of the historic structures and gives examples of these damages using well-known historic structures. It then discusses the suitable repairment-strengthening efforts for such buildings.

2   Damages in Historic Masonry Structures

Historic structures survive until today being damaged by several reasons such as earthquakes, ground-related damages, structural material losing its strength, long-term natural causes, abandonment of the structure, improper use and wrong restoration practices, fire, war, vandalism, air pollution and traffic. Some types of the damages due to aforementioned factors can be listed, as follows.

2.1   Damages in Load-Bearing Walls

Masonry structures are heavy and rigid structures. Thus, masonry structures receive the power of an earthquake to the fullest. Tensile and non-ductile behavior under pressure leads to sudden sinking/settling of the structure without having a significant plastic strain. Weak connections between the walls and the covering system may cause the damage occurring at a weak point of the structure to easily spread therefore causing destruction. Large window and door openings and wall arrangement being unsymmetrical lead to additional stress concentration which increases the damage [1].

Vertical loads and earthquake loads are compensated by load-bearing walls in masonry structures. In general, tensile strength of the wall material and shear strength of the mortar used are lower in masonry structures. The most important causes of damage are cracks, dislocation and disintegration due to the tensile strength caused by the impact of the earthquake building shear stress on the walls.

Crack formation is one of the most common types of damages forming on load-bearing walls. Knowing the reason behind the formation of cracks is of utmost importance in order to take appropriate repairment-strengthening decisions. Cracked spots and their distribution on the structure will give an idea about the reason behind the formation of cracks and stress distribution of the structure [2].

Another important type of damage is the one caused by the wall's movement shifting its axis. Cracks as well as shifting axis give indications in terms of the movement of the load-bearing system. Another type of wall damage is the sinking/settling of the wall due to the differences in the soil or large differences in the vertical loads among the wall. Different kinds of sinking/settling in walls are manifested in inclined cracks which generally occur in the wall plane in Fig. 1. Which side the sinking/settling is inclined to can easily be defined if the direction of the crack is detected [3].

Fig. 1. Cracks visible on the main outer wall of a historic mosque [4]

2.2   Damage in Arches and Vaults

Vaults are made of consecutive arches and any damage in the arches also applies for the vaults. As the masonry material are not strong enough to absorb the tensile strengths caused by the effect of bending moment, the arches need to be built in a way to form only compression stress. A bending moment of zero at all profiles defines the arch shape. This shape is the inverted version of the catenary which shows the natural load transfer. Catenary is the shape of a chain or flexible cable hung from both ends freely under the gravitational effect. Shaped like the inverted catenary under its own weight, the bending moment of the arch at all of its profiles is zero and there is no tensile force forming at none of these profiles. İnverted catenary is the ideal shape for an arch under its own weight. Figure 2 shows two different arches, one low and one high, represented using catenaries. Built in such manner, these arches are rather strong and there are no disadvantages if they have thin profiles. As the shape of the arch draws away from the inverted catenary tensile strength starts to build on the arch, which is translated into cracks and deformations.

Joint dislocation can be observed to a degree in the arches and vaults of historic structures on the arch/vault axis or perpendicular to this axis. The reason behind this is the reduced thrust of the arch/vault spring line in time. Figure 3 shows the joint dislocations in a damaged vault.

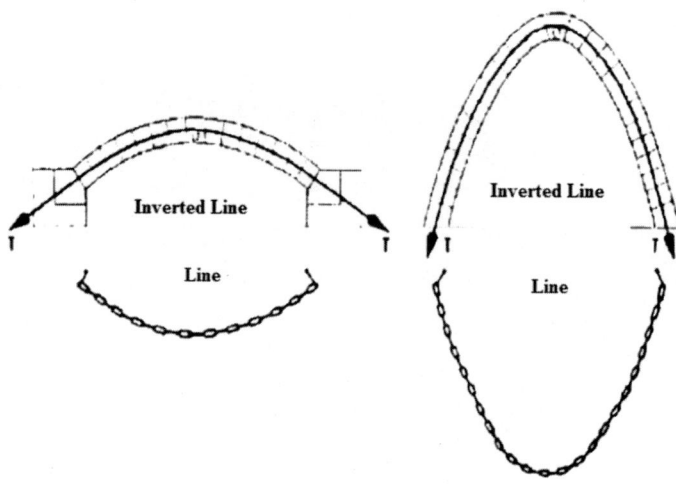

Fig. 2. Catenaries and their inverted arches, low and high [2]

Fig. 3. Joint dislocations in a vault [3]

The location of the spring line in an arch, the weight of the arch, the shape of the arch curve, width of the arch, width of the wall or pedestal the arch stands on, impact of the load on the arch and the condition of the strainer which prevents the arch to spread, all have an impact on the stability of the arch, therefore, on the arch deformations [2].

Damages in the arches and vaults are observed when the vertical and horizontal loads on the arch profile reach the load carrying capacity, when tensile strength builds on the arch profiles, when strainer bars are corroded losing their functionality, and when the structural bearing of the strainers are moved or dislocated [4], in Fig. 4.

(**a**) Buckling (Kitchens of Edirne Palace)   (**b**) Rupture (Kitchens of Edirne Palace)

(**c**) Buckling (Vasat Atik Ali Pasha Mosque)

Fig. 4. Damage types in iron strainers [3].

2.3   Dome Damages

Formed in a way to rotate an arch around its vertical axis, vertical loads of a dome are transferred to the bottom of the dome starting from the keystone and gradually transferring to the neighboring stones. The perpendicular force of gravity on individual stones is transferred to the neighboring stones in a crosswise manner. Thus, the load at the bottom of the dome has two components, horizontal and vertical. Damage in a dome is most commonly a result of this horizontal force. Vertical component of the load is transferred to the load-bearing elements such arches, walls, etc., while horizontal force is offset by supports and strainers in order to prevent the dome from spreading. It is possible to counter the horizontal force with think main walls while it is also possible to use thinner walls when counterweight towers are used. This horizontal force creates tensile strength on the horizontal plane and shear stress on the vertical plane of the frame which prevents the dome to spread acting as a support for the dome.

Tensile stress will build on the dome when the frame cannot carry the load of the dome. Domes are units under pressure and any tensile stress which may build on the dome may lead to crack formation on the vertical plane, therefore, damaging of the dome. Figure 5 shows the crack formation from the top to bottom of the dome due to the tensile stress built on the dome [2].

Fig. 5. Dome crack [2]

3   Ground Survey and Identification of Soil Properties

It is very important to look into the soil properties along with the structure itself when examining the historic structures. In the assessment of soil parameters, it is necessary to first conduct a number of surveys and observations on the land and then identify soil properties with lab tests. It will be the proper approach to interpret the structure and damages after identifying the soil properties of historic structures.

Samples must be taken from observation holes and drills, the groundwater level must be defined and a number of tests must be conducted on the soil in order for a classification of the soil the structure is built on and identification of soil properties [5].

4 Ground Damages in Historic Structures

Possible damages related to the ground in historic masonry structures can be listed, as follows:

- Sinking/settling
- Damages due to changing groundwater levels
- Damages due to decaying of foundation piles
- Damages due to excavations in the vicinity of the structure
- Damages due to insufficient drainage
- Damages caused by earthquakes [5].

Settlement
Settlement in a structure can occur due to various factors such as underground erosion, structural collapse of soil, thermal changes, frost heave, vibration and shocks, landslide, creep and mining subsidence, besides the inherent variable soil conditions [6]. In the Fig. 6, settlements are showed schematically.

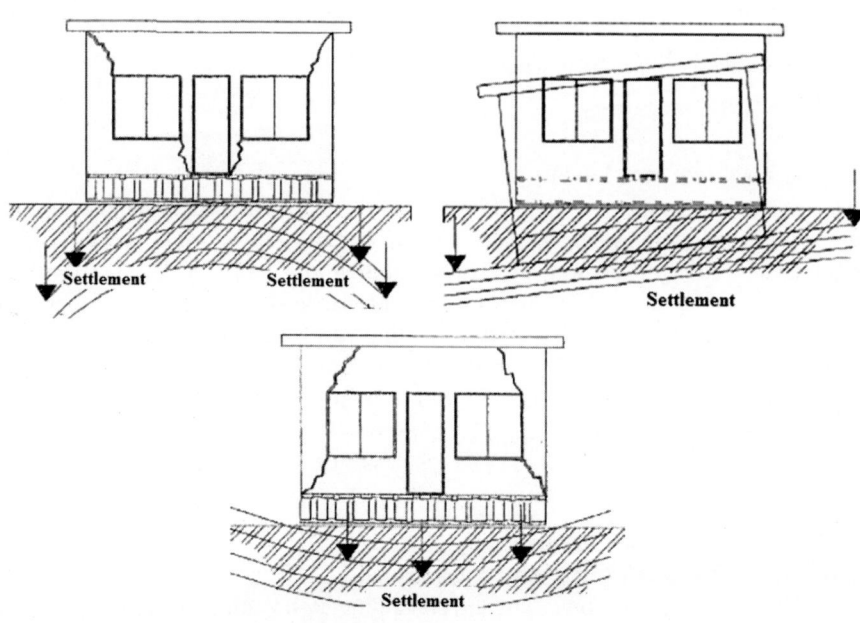

Fig. 6. Schematic of settlement examples [7]

Even sinking/settling usually does not compromise the stability and lifecycle of a structure and causes no damage due to sinking [8].

In the past, there were some examples of settlement of historical building such as Naziresha's Mosque. Naziresha's Mosque is the only well-preserved historical structure in Elbasan, Albania in Fig. 7 [9].

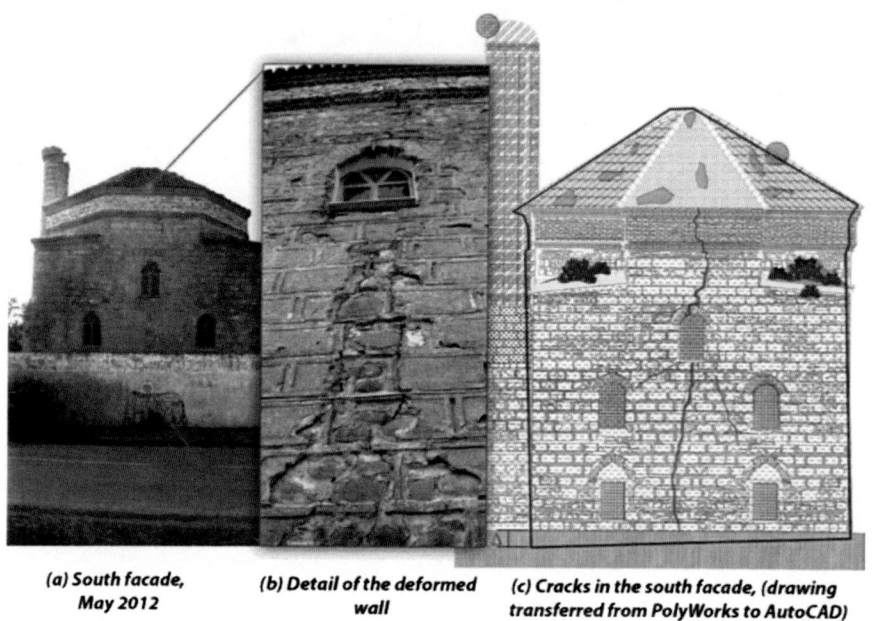

(a) South facade,
May 2012

(b) Detail of the deformed
wall

(c) Cracks in the south facade, (drawing
transferred from PolyWorks to AutoCAD)

Fig. 7. Cracks due to differential settlement (1 = 18 cm) in the south façade of Naziresha's Mosque [9]

Changing Groundwater Levels

Due to excavations in the vicinity of historic structures, decreasing groundwater levels lead to decreased pore water pressure and as a result soil particle weight will be increased, thus, soil in the lower levels of ground will be subjected to increased loads [5]. Sinking/settling may occur due to the additional load. Damages and cracks will occur in the structure when the sinking/settling is above the threshold, in Fig. 8. Nevertheless, decreasing groundwater levels also cause retreat cracks in the ground.

Resulting decrease in the volume will lead to significant sinking/settling. Pore water pressure increases with the increasing groundwater level, soil gets soft and the shear strength of the soil is compromised. The result is insufficient bearing capacity of the soil which may lead to sinking/settling.

Decaying of Foundation Piles

Historic structures used to be built on a foundation system involving wooden piles driven in the soil where the soil is soft and containing high volumes of water [5]. As the wooden piles do not contact with air and water, they are not subject to decay. In case of

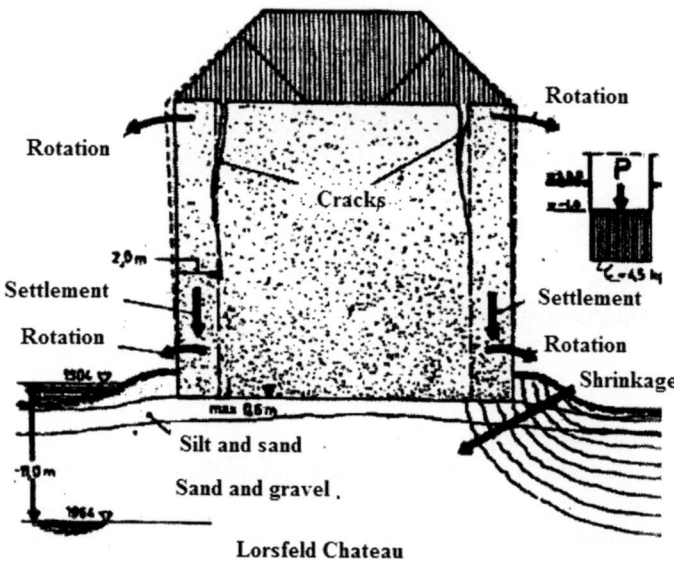

Fig. 8. Retreat occurring due to decreased groundwater levels [10]

decreasing groundwater levels due to external factors piles are exposed to air which may lead to decay and sinking/settling may occur due to decaying piles, in Fig. 9.

Excavations in the Vicinity of the Structure

Excavations around the historic structures may cause dislocation and damages. Any void around the foundation due to excavation disrupts the balance of the foundation causing structural damages in the structure and the foundation system.

Insufficient Drainage

In cases where the carrier soil layer is made of clay and silt, the water captured in the soil will lead to transfer of silt and clay particles in the soil due to decreased permeability if there is no sufficient drainage when the soil is wet due to rainfall or other reasons and all of these lead to local collapses in the ground [10].

Earthquake Effect

Historic structures are generally heavy and rigid structures. As they are heavy and rigid structures, they receive more earthquake force. Any change in the groundwater levels and/or increased pore ratio due to the impact of the earthquake will lead to cracks in the structure and structural damage.

4.1   Foundation Reinforcement in Historic Structures and Ground Improvement Methods

Improving the foundation and soil behavior of historic structures is one of the methods used to keep the sinking/settling at an acceptable level. These are:

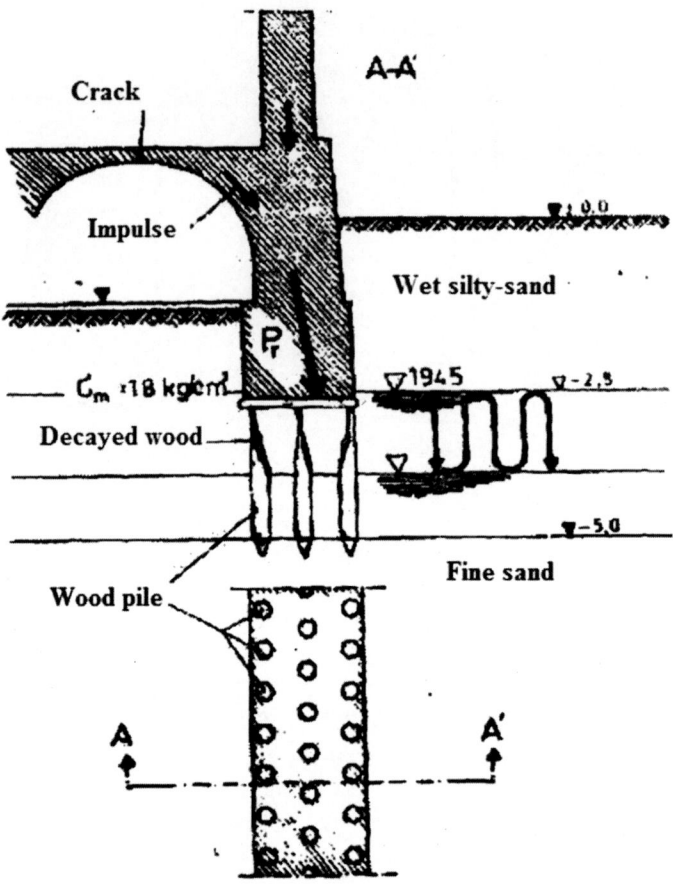

Fig. 9. Partial foundation detail of the Diersfordt Church—Pile Damage [10]

1. Expanding the foundation using traditional foundation strengthening methods and/or deepening foundation in order to improve foundation behavior.
2. Methods based on improving sub-foundation soil; foundation soil is improved using cement injection, chemical injection or jet-grout columns.
3. Methods to transfer the foundation loads to deeper soil layers. Driven piles, poke piles, mini piles, jet-grout columns are used for this purpose [15].

5 Some Historic Structures Which Were Subjected to Ground-Related Damage

Konya Alaadin Mosque

Konya Alaadin Mosque is built on the Alaadin hill. However the mosque was built in the twelfth century, it was enlarged in time. As the mosque towers on a filled soil and due to the increased pore ratio in the soil due to factors such as rainfall, a number of sinking/settling processes have led to cracks in the structure [10]. Vertical deformation due to a number of sinking/settling processes caused cracks in the structure. Figure 10 shows the cracks occurred after these sinking/settling processes. Vertical deformation due to a number of sinking/settling processes confirms the vertical flexural cracks.

However the reinforcement work conducted in 1967 limited some of the settling/sinking processes, ground settling/sinking continues. Concrete strainer located for the purpose of reinforcement limits the vertical deformation, however, shear cracks still form in the structure due to settling/sinking. Considering all the aspects noted above, it was seen that the structure requires a foundation improvement along with foundation reinforcement. As a solution, inside and outside of the mosque were improved with injection and the carrier system was supported using mini piles [10].

Cenab-ı Ahmet Pasha Mosque

The only work of Sinan, the Architect, in Ankara, this mosque and sepulcher reflecting the classic Ottoman architectural style is a masterpiece. It was built in 1565. Designed in a rectangular geometry of 17.2 × 17.8 m, the mosque's dome frame stands directly on a main wall of 1.8 m thickness. Standing on a rectangular plan with its cubic frame and single dome, the mosque's southern wall height is higher than its northern wall height due to the inclined land it was built on [12].

The mosque was built on clay and silt soil. The drought in recent years has changed the groundwater profile at a great extent. Due to this change, a number of sinking/settling processes were observed in the clay and silt soil. The mosque is now closed due to the cracks occurred in the dome and walls as a result of the sinking/settling, Fig. 11. It was recommended too use mini fore piles which will be driven to the rock layer in order to prevent displacement due to settling, and use of a steel sheet frame around the dome pedestal was recommended in order to prevent further formation of cracks in the dome [13].

Konya Mevlana Museum

Passed away in 1273, the sepulcher of Rumi was built in 1274 by Bedrettin Tebrizi, the architect, under the government of Alaaddin Kayserin of Seljuks. The original structure of Rumi's sepulcher included a dome standing on four pedestals while the structure of the current sepulcher was built in the reign of Alaaddin Ali Bey of Karamanids. Historical records provide documents on the restorations implemented by Alaaddin Ali Bey of Karamanids [14].

Analyses showed that some of the pedestals and walls of the museum were settled/sunk in different times which led to crakes in the dome joints, walls, and arches, Fig. 12. The soil on which the museum was built on consists of a mix of clay, silt and sand. An examination of the SPT values of the foundation level, it can be seen that the foundation of the museum is driven into filling soil with varying concentrations. Offering a loess structure, these filling soils have a higher load-bearing capacity when they are dry, while they may lead to several settling/sinking processes when they are saturated with water. The cracks occurred in the museum are a result of such sinking/settling.

Fig. 10. Crack formation due to a number of sinking/settling processes in Konya Alaaddin Mosque [11]

Fig. 11. Wall and dom cracks observed in the Cenab-ı Ahmet Pasha Mosque due to settling/sinking at different times [13]

Consisting of construction techniques and sections belonging to different periods, Rumi Museum has been subjected to ongoing restorations and attempts to improve the structure without foundation improvements proved to be insufficient. A durable improvement and restoration also requires foundation improvement. When foundation and the structure are assessed in combination, it can be seen that methods such as jet-grout will be the ideal choice for improvement of such foundations. However, due to the quality of the museum (including sepulchers, receiving visitors continuously, etc.) such applications are not possible. As an alternative to the foundation improvement, Uretek method can be applied as a new technique. This technique allows for foundation improvement while the museum is still accepting visitors [14].

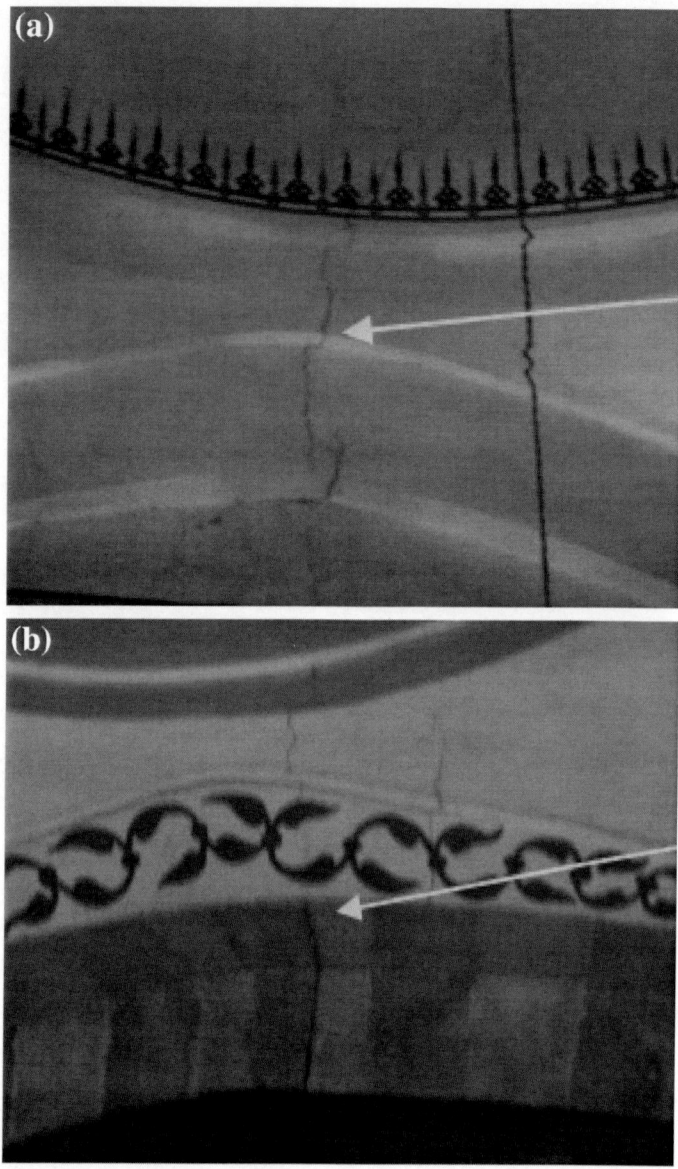

Fig. 12. **a** rear front wall-dome joint crack and **b** front arch-dome joint crack

6   Conclusions

The purpose of maintenance-restoration work on historic structures is to make it possible for countries and nations to leave these historic treasures to the generations to come in the proper way. The actual purpose of these practices conducted on historic structures is to utilize the structure in its natural form and to prolong its lifecycle.

Historic structures are being damaged due to several reasons and many of them are demolished due to neglect. Ground-related damages are one of the most common damages. They lead to cracks, dislocations, and sinking/settling of the structural elements. These damages must be rehabilitated by specialists on time using the right method. Maintenance-reinforcement work must be commenced after protective measures are taken without compromising the current condition of the structure. Sufficient repairment-strengthening must be implemented if there are any damages in the flooring, inner/outer walls and the roof sections of the structure. The soil on which the structure was built needs to be improved using the proper improvement method and the foundation must be reinforced, if necessary. Methods, which will allow for the preservation of the historic structures without compromising their texture, must be selected. The materials and methods selected for the repairment-strengthening processes of the historic structures must be suitable to the historic texture of the structure.

References

1. Celep Z, Kumbasar N (2004) Deprem mühendisliğine giriş ve depreme dayanıklı yapı tasarımı, İstanbul
2. Aköz H (2008) Deprem Etkisi Altındaki Tarihi Yığma Yapıların Onarım Ve Güçlendirilmesi, İTÜ Fen Bilimleri Enstitüsü, Yüksek Lisans Tezi, İstanbul
3. Sesigür H, Çelik OC, Çılı F (2007) Tarihi yapılarda taşıyıcı bileşenler, hasar biçimleri, onarım ve güçlendirme. İstanbul Bülten 89:10–21
4. Bayraktar A, Kocasinan Y, Büyükgökmen G, Kökçü E (2005) Yığma Yapıların Deprem Güvenliğinin Araştırılması Çalıştayı, Sunum Notları, ODTÜ
5. Erköseoğlu G (2012) Kültürel Mirasın Depremden Korunması Ve Turizme Kazandırılması Amacı İle Uygulanabilecek Güçlendirme Teknikleri, Uzmanlık Tezi, T.C. Kültür Ve Turizm Bakanlığı Yatırım Ve İşletmeler Genel Müdürlüğü
6. Naik S, Naik NP, Kandolkar SS, Mandrekar RL (2011) Settlement of structure of a structure. In: Proceedings of Indian geotechnical conference, Paper No: Q-285
7. Arun G (2016) Tarihi Yığma Yapılarda Hasar Teşhisi, TMMOB İnşaat Mühendisleri Odası, Gaziantep Şubesi Eğitim Semineri
8. Mahrebel HA (2006) Tarihi Yapılarda Taşıyıcı Sistem Özellikleri, Hasarlar, Onarım Ve Güçlendirme Teknikleri, İstanbul Teknik Üniversitesi, Yüksek Lisans Tezi, Fen Bilimleri Enstitüsü
9. Yardım Y, Mustafaraj E (2015) Effects of soil settlement and deformed geometry on a historical structure. Nat Hazards Earth Syst Sci 15:1051–1059
10. Namlı M (2001) Tarihi Yapıların Temel Sistemleri ve Temel Takviyesi Yöntemleri, Yüksek Lisans Tezi, İstanbul Teknik Üniversitesi, Fen Bilimleri Enstitüsü, İstanbul
11. https://www.google.com.tr/search?q=konya+alaeddin+camii+%C3%A7atlaklar&sa=X&espv=2&biw=1920&bih=901&tbm=isch&tbo=u&source=univ&ved=0ahUKEwic9vqugN_QAhVDHxoKHUWmBH0QsAQIGA#imgrc=1Y6vk1mBGIA6HM%3A
12. Başkan S (1993) Cenabı Ahmet Paşa Camii, Kültür Bakanlığı/1407 Tanıtma Eserleri/49, 1. Baskı
13. Köseoğlu Ç (2011) Investigation of a damaged historical mosque with finite element analysis, METU, The Degree of Master of Science

14. Yıldız M, Sezer R, Ekinci O, Yıldız E (2011) Mevlana Müzesinde Oluşan Hasar Nedenlerinin Araştırılması ve Uygun Temel Güçlendirme Sisteminin Seçilmesi. ISSN:1306-3111. e-J N World Sci Acad 6(4), Article Number: 1A0249

15. Toğrol E (1994) Temel Takviyesi Yöntemlerine Yeni Bir Bakış, Zemin Mekaniği ve Temel Mühendisliği, V. Ulusal Kongresi, ODTÜ, Ankara, Cilt III. S, pp 887–917

Self-Sensing Behavior Under Monotonic and Cyclic Loadings of ECC Containing Electrically Conductive Carbon-Based Materials

Mustafa Şahmaran[1](✉), Ali Al-Dahawi[2], Vadood Farzaneh[1],
Oğuzhan Öcal[1], and Gürkan Yıldırım[3]

[1] Department of Civil Engineering, Engineering Faculty, Hacettepe University,
Ankara, Turkey
{msahmaran, vadoodfarzaneh, oguzhanocal24}@gmail.com
[2] Department of Building and Construction Engineering, University of
Technology, Baghdad, Iraq
alimk_ceg@yahoo.com
[3] Department of Civil Engineering, Engineering Faculty, Adana Science and
Technology University, Adana, Turkey
gurkanyildirimgy@gmail.com

Abstract. The development of self-sensing (piezoresistivity) feature which is one of the non-structural properties of cementitious composites to be multifunctional is under focus in the present study. This capability is considered as one of the best alternatives to continuously monitor the damage and deformations of infrastructures. The self-sensing behavior of cubic and prismatic specimens under monotonic uniaxial compression and cyclic flexural loadings respectively was investigated and compared with dielectric ECC materials. The results showed that the incorporation of carbon-based electrically conductive materials within the cementitious composites have a significant effect on monitoring the damage and deformation of cement-based materials effectively.

Keywords: Self-sensing · Carbon-based materials · Multifunctional cementitious composites (MCC) · Engineered cementitious composites (ECC)

1 Introduction

Concrete is one of the most commonly used construction materials in the world to build the infrastructure facilities. It can be seen in the residential, industrial, commercial, sports and agricultural facilities etc. It can be regarded that damage resulting from the production of cement as a construction material, which is the main binding material in concrete, of less damage compared to other construction materials such as reinforcing steel. So, concrete is a tool of development and progress in various activities around the world.

After the construction of concrete, it is rare that maintenance is performed on it, and the performance of maintenance generally begins at the later ages when the serviceability index of concrete has reached low levels, which requires large amounts of money in order to raise it by a few if it is possible. The early time sensing for the need of concrete maintenance increases the safety situations and service life of the concrete structures, reduces the amount of funds spent compared to the previous situation, as well as, provides multi-alternative techniques to carry out the maintenance.

As the necessity is the mother of invention, as has been said, a new technique emerged in recent years made it possible to constantly monitor the concrete structures. This is called structural health monitoring (SHM).

Thus, the SHM technology allows the evaluation of the current state and health of public structures, can reveal progressive damage, and permit appropriate processes of repair to be implemented before infrastructures become unsafe and unserviceable. In addition to that, SHM technique has the ability to diagnose the distress position and its intensity within the infected infrastructure. Moreover, by SHM technique, it is possible to count the traffic flow and weight of the vehicles that use either bridges or highways.

The development of new property in concrete for SHM, which enables a continuous monitoring of different distresses inside concrete structures, has begun in concrete-related research in recent years [1]. Consequently, some additional research works have been started to be performed recently to equip cementitious composites such as concrete with non-structural properties in addition to its structural and performance properties. Apart from structural and performance properties of cementitious composites, the most important non-structural property of this kind of materials can be counted as the ability to self-sense damage and strain formation. Cementitious composites can be regarded as "multifunctional" in the cases where the materials exhibit functional characteristics as described above along with the structural and performance properties. In general, structural materials that do not contain any sensor but used for sensor purpose is called self-sensing materials.

In other words, self-sensing (piezoresistivity) is the capability of cement-based materials reinforced by electrically conductive fillers to sense the stress and strain changes (i.e., by monitoring the effects of stress and strain changes on the electrical resistivity) [2]. When the deformation or stress of the piezoresistive cement-based materials is occurred, the contact state between the fillers and the matrix is changed, which affects the electrical resistance of the cement-based materials. Strain, stress, crack and damage can therefore be detected through measuring of the electrical resistance [2–5].

2 Experimental Program

2.1 Materials

Cement is one of the common ingredients that was used in all mixtures. It was an ordinary Portland cement (PC) CEM I 42.5R. Its properties are compatible with the requirements of EN-197-1 [6] and ASTM-C150 [7] Type I cement. The chemical composition and physical properties of Portland cement are provided in Table 1, while the particle size distribution of it can be seen in Fig. 1. Fly ash (FA) of Class-F was used as another ingredient. Chemical and physical properties of FA are presented in

Table 1 in addition to the particle size distribution which is demonstrated in Fig. 1. This type of FA is compatible with the requirements of ASTM-C618 [8]. For all mixtures, normal tap water was used. To reduce the negative impact of air bubbles within the mixtures, a defoamer was used in the production of multifunctional cementitious composites. High range water reducing admixture (HRWRA) was used in all mixtures. It is a polycarboxylic ether based superplasticizer concrete admixture. The product is compatible with ASTM-C494/C494M [9], type F. It has an amber color. Its density is 1.082–1.142 kg/l. In order to enhance the mechanical properties of the fabricated cementitious composites [10] against the different loading scenarios, and then getting the ability to monitor the changes in electrical resistivity more broadly at different load levels, Polyvinyl Alcohol (PVA) fibers of 8 mm length, 39 μm diameter and 1.3 specific gravity were used. The nominal tensile strength and elastic modulus of these fibers were 1620 MPa and 42.8 GPa respectively. Fine silica sand with a maximum grain size of 400 μm, a specific gravity of 2.60 and water absorption capacity of 0.30% was incorporated with a saturated surface dry (SSD) condition. Chemical composition and physical properties of silica sand are provided in Table 1. Moreover,

Table 1. Chemical composition and physical properties of Portland cement, class-F fly ash and silica sand

Chemical composition, %	PC	FA	Silica sand
SiO$_2$	20.77	52.22	38.40
Al$_2$O$_3$	5.55	16.58	10.96
Fe$_2$O$_3$	3.35	6.60	0.81
MgO	2.49	2.10	7.14
SO$_3$	2.49	0.02	1.48
CaO	61.43	7.98	34.48
Na$_2$O	0.19	0.86	0.18
K$_2$O	0.77	1.53	0.86
Loss on ignition	2.20	10.36	3.00
Physical properties			
Specific gravity	3.06	2.10	2.60
Blaine fineness (m^2/kg) [11]	325	269	–
BET (m^2/kg) [12]	–	–	–

particle size distribution is shown in Fig. 1. Chopped pan-based electrically conductive carbon fibers with the length of 12 mm and aspect ratio of 1600 were employed in this study during the fabrication of CF-reinforced cementitious composites. The properties of the fibers can be seen in Table 2. Multi-walled carbon nanotubes were utilized as other conductive materials in the present study. Physical and chemical properties of this product are given in Table 3.

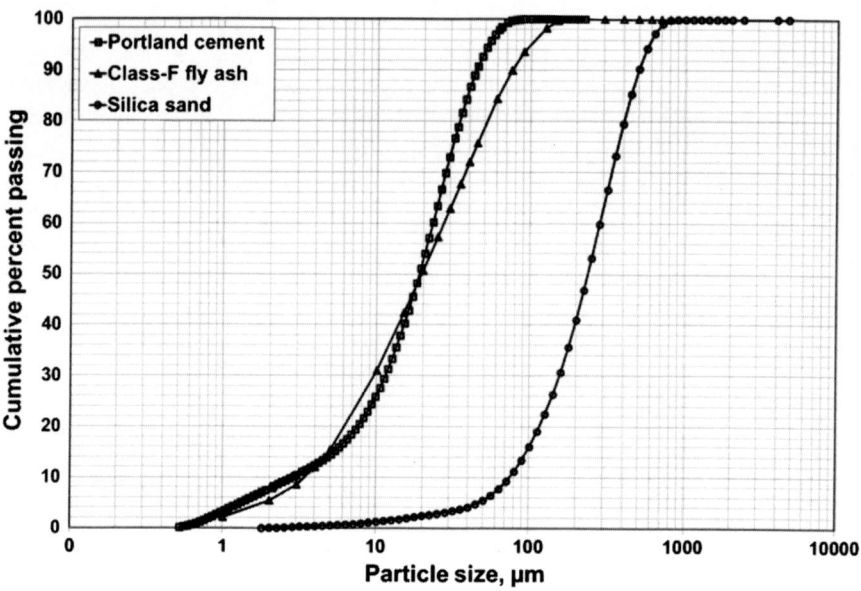

Fig. 1. Particle size distributions of Portland cement, class-F fly ash and silica sand

Table 2. Typical properties of chopped carbon fiber

Typical properties	Unit	Value
Carbon fiber content	%	100
Fiber diameter	μm	7
Fiber unit weight	kg/m³	1700
Fiber length	mm	12
Metal contamination	–	<0.1 g/1000 g
Tensile strength	MPa	4150
Tensile modulus	GPa	252

Table 3. Physical and chemical properties of MWCNT

Physical properties	CNT
Diameter (nm)	~ 10–30
Length (μm)	~ 10–30
Surface area (m²/g)	>200
Chemical properties	
Purity	>90%

2.2 Mixing Methods and Mixture Proportions

The produced mixtures were mixed according to the suggested best mixing procedure of the micro- and nano-scale electrically conductive materials that have been recommended by Al-Dahawi, Öztürk [13]. These mixtures were poured in the oiled molds, placed in their molds at $50 \pm 5\%$ RH, 23 ± 2 °C and covered with plastic sheets for 24 h.

After demolding process, the specimens were, then, moved into isolated plastic bags to be cured at $95 \pm 5\%$ RH, 23 ± 2 °C. To remove the potential effects of the internal moisture on the electrical measurements due to polarization, the specimens were dried in an oven at 60 °C for 24 h before testing [14, 15]. Afterwards, all specimens were tested at room temperature.

During the preparation of mixtures, water to cementitious materials (PC + FA) ratio (W/CM), aggregate to cementitious materials ratio (aggregate/CM) and fly ash to Portland cement ratio (FA/PC) were kept constant at 0.27, 0.36 and 1.2, respectively. The mixtures were reinforced by CF or CNT with the dosages which were recommended by Al-Dahawi, Sarwary [16] to create the electrically conductive network which is responsible for the self-sensing issue. These dosages are 1% by vol. of mixture and 0.55% by wt. of CM for CF and CNT respectively. In addition to the electrically conductive material, the mixtures were, also, reinforced by 2.0% vol. of PVA as recommended by many researchers [3, 10, 17–24]. A defoamer dosage of 0.3% by weight of cementitious materials (PC + FA) was utilized to remove the undesirable air bubbles as recommended by the manufacturer company and proposed by [25]. To get the same workability and know the optimum dosage of HRWRA that should be used, mini slump flow spread tests were carried out on these mixtures.

2.3 Testing Setup

50 mm-cubic specimens were prepared to measure the self-sensing ability of the developed multifunctional cementitious composites under monotonic uniaxial compression. A 2000 kN capacity compression machine was used to carry out the compression test. Two brass plate electrodes of 70 mm in length and 10 mm in width were embedded in the fresh mixtures very close to the top surface of the cubic specimen as can be seen in Fig. 2. To interrupt the possible electrical contact during the tests and minimize the mistakes of the recorded deflection by the loading machine to the maximum extent, insulative plastic covers were inserted between the specimen and compressive loading plates (Fig. 3). The loading rate was 0.36 MPa/s as recommended by ASTM-C109/C109M [26].

A concrete resistivity meter with 2-probe was utilized to measure the electrical impedance changes of the cubic specimens under the effects of the uniaxial compression loads (Fig. 4).

The self-sensing capability of ECC beam specimens with different carbon-based materials was evaluated under four-point bending test with repeated loading applications. To perform the cyclic flexural test, a 100 kN capacity universal testing machine was used. The supports and contact points of the universal testing device were covered with PVC cling film to cut any possible electrical connection with the testing device,

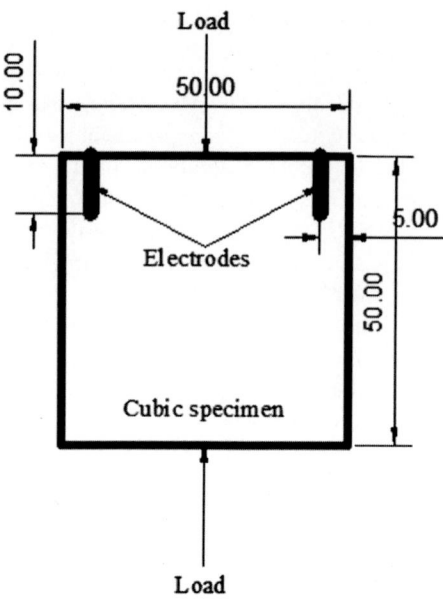

Fig. 2. Front view of the configuration setup for compression test (all dimensions are in mm)

Fig. 3. Closer view of specimen taken during testing under uniaxial compression test

which could affect overall results (Fig. 5). DC source meter with four-probe method was utilized to measure the voltage (by the inner electrodes) due to the applied current (by the outer electrodes) (Fig. 5). Flexural load was applied and released three times until reaching 70% of the modulus of rupture level each time, at a loading rate 0.30 mm/min.

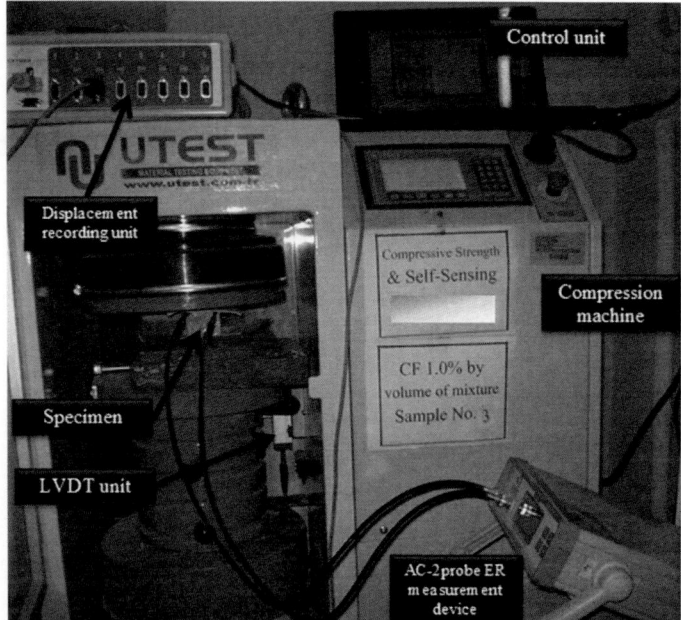

Fig. 4. Uniaxial compression and AC-2probe electrical resistance tests setup

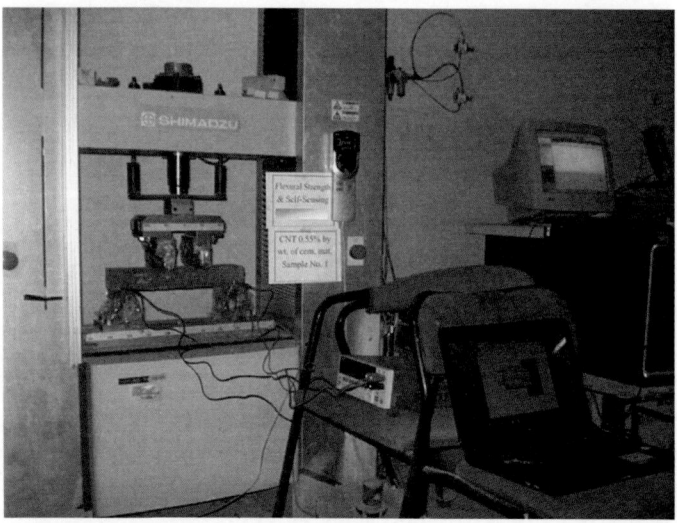

Fig. 5. Four-point flexural test together with DC-4probe electrical test

3   Results and Discussions

Three 28-day-old cubic specimens from each mixture in addition to the reference mixture (without conductive materials) were tested and average results were taken into consideration. Fractional changes in ER measurements (FCER) of different ECC specimens relative to bulk ER (without any loading) ($\Delta\rho/\rho o$) and deformation in compression (mm) were plotted against testing time. Figure 6 shows the self-sensing behavior.

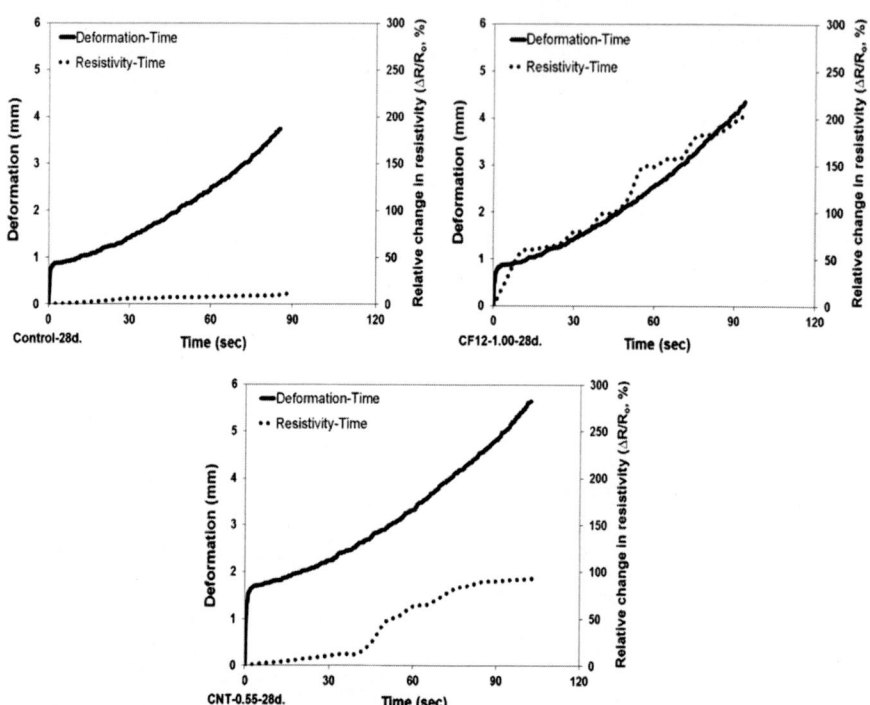

Fig. 6. Mechanical and electrical behavior of ECC specimens with and without carbon-based conductive materials under compressive loading

The figure clearly demonstrates that application of compressive loading caused ER results of ECC specimens to increase, as expected, due to brass electrodes being separated from each other. Control ECC specimens without additional carbon-based material were not responsive to deformation as in the case of CF- and CNT-bearing specimens. The figure shows that ECC specimens with CF were able to sense even relatively small deformation levels, starting at the beginning of loading. The changes in relative ER results were slightly smoother during the initial stages of compressive loading and became more pronounced after a certain level due to the changes in crack characteristics of ECC specimens upon compressive loading. By comparing the FCER at the end of the loading process for the different ECC carbon-based mixtures in addition to the control dielectric mixture, it is easy to see the superiority of CF-bearing specimens with around 200% FCER followed by CNT-bearing specimens with around 100% FCER. The incorporation of electrically conductive materials (CF and CNT) to the ECC mixtures was not only created the self-sensing property to these mixtures but also improved their mechanical properties [13].

Plot related to the self-sensing behavior of ECC specimens with and without (plain) carbon-based materials upon three applications of loading and unloading is shown in Fig. 7. All ECC mixtures produced in this study were able to sense damage in the

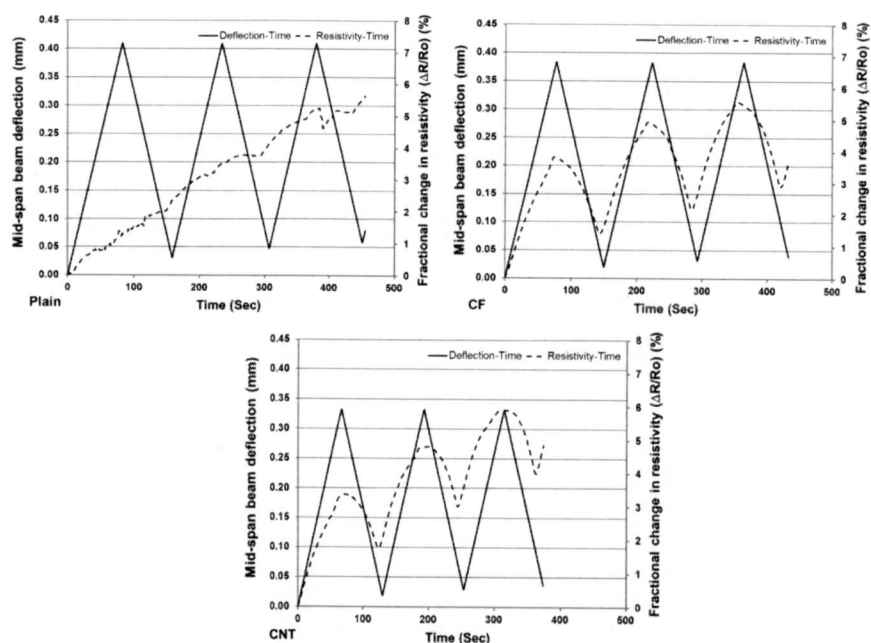

Fig. 7. Self-sensing behavior of ECC specimens subjected to cyclic flexural loading

plastic range. Plain specimens with no carbon-based material additions were not very successful in sensing repetitive unloading, due most probably to the inadequate discharge of specimens upon unloading and polarization effects with the application of DC current [27]. However, the plain specimens were responsive to increased levels of deformation under bending loading; after three cycles of loading and unloading, FCER results reached level about 5.8%. The beneficial effects of carbon-based material use can be clearly visualized by looking at the graphs in Fig. 7. When specimens were strained to certain deformation levels, applications of both loading and unloading were successfully sensed for each repetition. On the other hand, this is not the case with respect to the control specimens without carbon-based conductive materials. Concordantly, upon gradual increases in loading, there were increments in electrical resistivity due to the divergence of conductive filler, while decrements followed upon unloading due to the convergence of conductive fillers. In other words, there was reversibility in the FCER results with loading and unloading. This suggests that to be able to more clearly sense damage in terms of changes in electrical properties, cementitious systems should be highly electrically conductive.

4 Conclusions

The following conclusions have been obtained from the current study:

- The carbon-based materials, CF and CNT, proved themselves worthy in capturing changes in compressive damage.
- All cyclic loading and unloading applications were successfully sensed by ECC specimens with different types of carbon-based materials with the superiority of CF-based ECC specimens; this was not the case for plain specimens, which did not sense repetitive unloading. In other words, there was clear reversibility in the electrical measurements with unloading. This therefore showed that in order to properly detect damage, cementitious systems must be highly electrically conductive.

Acknowledgements. The authors gratefully acknowledge the financial assistance of the Scientific and Technical Research Council (TUBITAK) of Turkey provided under Project: 114R043.

References

1. Chen P-W, Chung D (1993) Carbon fiber reinforced concrete as an electrical contact material for smart structures. Smart Mater Struct 2(3):181
2. Chung DDL (2002) Piezoresistive cement-based materials for strain sensing. J Intell Mater Syst Struct 13(9):599–609
3. Hou T-C, Lynch JP (2005) Conductivity-based strain monitoring and damage characterization of fiber reinforced cementitious structural components. In: Proceedings of SPIE 12th annual international symposium on smart structures and materials, March 6–10. 2005. International Society for Optics and Photonics, San Diego, California, USA
4. Azhari F (2009) Cement-based sensors for structural health monitoring, in civil engineering. The University of British Columbia, Vancouver, p 184
5. Ou J, Han B (2009) Piezoresistive cement-based strain sensors and self-sensing concrete components. J Intell Mater Syst Struct 20:329–336
6. EN-197-1 (2000) Cement-part I: COMPOSITION, specifications and conformity criteria for common cements, in European Standard, p 29
7. ASTM-C150 (2007) Standard specification for Portland cement, in Annual book of ASTM standards, p 8
8. ASTM-C618 (2005) Standard specification for coal fly ash and raw or calcined natural pozzolan for use in concrete, in Annual book of ASTM standards, p 3

9. ASTM-C494/C494M (1999) Standard specification for chemical admixtures for concrete, in Annual book of ASTM standards, p 9
10. Li VC, Wang S, Wu C (2001) Tensile strain-hardening behavior of polyvinyl alcohol engineered cementitious composite (PVA-ECC). Mater J 98(6):483–492
11. ASTM-C204 (2007) Standard test methods for fineness of hydraulic cement by air-permeability apparatus, in Annual book of ASTM standards, p 9
12. Brunauer S, Emmett PH, Teller E (1938) Adsorption of gases in multimolecular layers. J Am Chem Soc 60(2):309–319
13. Al-Dahawi A, Öztürk O, Emami F, Yıldırım G, Şahmaran M (2016) Effect of mixing methods on the electrical properties of cementitious composites incorporating different carbon-based materials. Constr Build Mater 104:160–168
14. Sun M-Q, Liew RJ, Zhang M-H, Li W (2014) Development of cement-based strain sensor for health monitoring of ultra high strength concrete. Constr Build Mater 65:630–637
15. Li H, Ou J, Xiao H, Guan X, Han B (2011) Nanomaterials-enabled multifunctional concrete and structures, in Nanotechnology in civil infrastructure: a paradigm shift. In: Gopalakrishnan K, et al (eds) Springer, Berlin, pp 131–173
16. Al-Dahawi A, Sarwary MH, Öztürk O, Yıldırım G, Akın A, Şahmaran M, Lachemi M (2016) Electrical percolation threshold of cementitious composites possessing self-sensing functionality incorporating different carbon-based materials. Smart Mater Struct 25(10):1–15
17. Kendall A, Keoleian GA, Lepech MD (2007) Materials design for sustainability through life cycle modeling of engineered cementitious composites. Mater Struct 41(6):1117–1131
18. Şahmaran M, Li VC (2007) De-icing salt scaling resistance of mechanically loaded engineered cementitious composites. Cem Concr Res 37(7):1035–1046
19. Şahmaran M, Li VC (2008) Durability of mechanically loaded engineered cementitious composites under highly alkaline environments. Cement Concr Compos 30(2):72–81
20. Li VC, Şahmaran M (2010) Engineered cementitious composites. Transportation research record. J Transp Res Board 2164(1):1–8
21. Şahmaran M, Lachemi M, Hossain KMA, Li VC (2009) Internal curing of engineered cementitious composites for prevention of early age autogenous shrinkage cracking. Cem Concr Res 39(10):893–901
22. Şahmaran M, Özbay E, Yücel HE, Lachemi M, Li VC (2011) Effect of fly ash and PVA fiber on microstructural damage and residual properties of engineered cementitious composites exposed to high temperatures. J Mater Civ Eng 23(12):1735–1745
23. Sahmaran M, Yücel HE, Yildirim G, Al-Emam M, Lachemi M (2014) Investigation of the bond between concrete substrate and ECC overlays. J Mater Civ Eng 26(1):167–174
24. Yucel HE, Guler M, Jashami H, Yaman IO, Sahmaran M (2013) Thin ECC overlay systems for rehabilitation of rigid concrete pavements. Mag Concr Res 65(2):108–120
25. Li-Ping G, Wei S, Qing-Yu C (2010) Multiscale material design and crack path prediction of a polymer-modified cementitious cover layer. Comput Mater Sci 49(1):184–189
26. ASTM-C109/C109M (2005) Standard test method for compressive strength of hydraulic cement mortars (using 2-in. or [50-mm] cube specimens), in Annual book of ASTM standards, p 9
27. Reza F, Batson GB, Yamamuro JA, Lee JS (2001) Volume electrical resistivity of carbon fiber cement composites. ACI Mater J 98(1):25–35

Determination of Favorable Time Window for Infrared Inspection by Numerical Simulation of Heat Propagation in Concrete

Ersan Güray[(⊠)] and Recep Birgül

Department of Civil Engineering, Faculty of Engineering, Muğla Sıtkı Koçman University, 48170 Muğla, Turkey
ersan.guray@mu.edu.tr

Abstract. Every bridge is subjected to a thorough inspection process every other year at most. Nondestructive evaluation techniques, especially noncontact methods, are gaining popularity to take part in structural health monitoring of existing bridges for expediting the inspection process. Infrared thermography is one of the noncontact testing methods; it is based on capturing and processing the thermal gradient on a radiant surface which is highly affected by the ambient environmental conditions. The objective of this study is to numerically search for an appropriate time window to carry out infrared inspections. To this end, a numerical model of a bridge deck with certain initial and boundary conditions was used to numerically obtain temperature differentials at any nodes across the model for a period of 24 h. A delamination with a constant thickness was positioned in the concrete deck. The transient solutions of the nonlinear partial differential equation were obtained by utilizing the finite element method. The numerical results point to afternoon as the most favorable time window to conduct infrared inspections; this result coincides with some of the experimental research found in literature. Additionally, it was shown that the existence of water in the defect greatly affected the heat conduction process.

Keywords: Infrared inspection · Bridge decks · Subsurface defects

1 Introduction

Efficiency of transportation networks is adversely affected by bridges showing signs of deterioration. Various inspection techniques such as hammer sounding and chain drag, core drilling if deemed necessary, are utilized to monitor the soundness of concrete bridges. Since these techniques call for lane closures and experienced personnel, fast and especially noncontact nondestructive evaluation techniques are desired by highway agencies. A recent research report [1] devoted to nondestructive testing techniques to identify concrete bridge deck deterioration lists 14 candidate methods. One of these 14 methods is infrared thermography. Researchers compared the efficiencies of different inspection techniques; one such study by Vaghefi et al. [2] compared the detection accuracies of infrared thermography and chain drag at 10 core sampling locations. Chain drag yielded 80% of damage detection accuracy while infrared thermography located the defects with 40% accuracy. However, another comparative study [3]

© Springer International Publishing AG, part of Springer Nature 2018
S. Fırat et al. (eds.), *Proceedings of 3rd International Sustainable Buildings Symposium (ISBS 2017)*, Lecture Notes in Civil Engineering 7,
https://doi.org/10.1007/978-3-319-64349-6_46

conducted at a different bridge reports that both impact echo and infrared thermography located the subsurface delaminations with 100% accuracy at 8 core sampling locations while chain drag showed 75% accuracy. SHRP2 report [1] states that the accuracy of infrared thermography might differ depending on the time window for conducting the inspection work, ambient environmental conditions and surface conditions of a target object.

Qualified engineers and inspectors have been using infrared thermography since 1980s to locate subsurface flaws and disintegration of concrete in roadways and bridges. Infrared thermography devices capture electromagnetic waves in the infrared wavelength [4]. These waves are radiations due to temperature variations on surfaces of a target object. Basics of infrared thermography are provided as a separate section given below in this study. Expected outcome of an infrared thermography inspection work is to get locations of subsurface voids and delaminations in concrete. Even though this inspection method promises good potential for delamination detection and characterization, it has several downsides as well. It may be able to locate a subsurface defect; however, it is not able to provide the depth information for the existing flaw [5]. In addition, deep delaminations that are beyond the thermal sensitivity of an infrared camera available to an inspector will be missed and impair the quality of the inspection work. Boundary conditions, surface anomalies, patches and stains on the surface of a target object are also known to affect infrared measurements [6]. There is another aspect to consider when conducting infrared inspections; Washer et al. [7] reported that the accuracy of damage detection by infrared thermography is greatly affected due to daily temperature variations, thus it is vitally important to choose a favorable time window to carry out infrared inspections.

Since solar loading is an important parameter to include in the inspection process, ASTM D4788-03 [8] stipulates a minimum of 3-h direct sunshine to create a temperature difference of 0.5 °C and states that weather conditions must include sunshine. Although the ASTM standard does not specify a time window, these requirements lead to a day time inspection process; it is left to the inspection personnel to decide what time window would be most favorable to conduct the inspection. The ASTM standard also prohibits testing if the wind velocity exceeds 50 km/h (30 mph).

In some commercial applications [6], determining an appropriate time window is accomplished by continuously monitoring the target structure and the ambient environmental conditions for a predefined time period. This approach requires extra visits to a target structure; in addition, access to the structure must be provided to perform data collection, this may cause further difficulty to inspection personnel. Washer et al. [7] devised an experiment to investigate the effects of solar loading on the ability of locating subsurface flaws. They reported that there could be a substantial delay between solar loading on a concrete structure and observation of subsurface defects and that intermittent cloud cover could lessen the thermal contrast in the infrared images. The authors concluded that a late afternoon inspection would be essential to offer optimized conditions for deeper defects in a concrete structure. Other researchers [9, 10] reported similar results stating that deeper flaw detection depends very much on ambient environmental conditions as well as size of defects.

2 Scope of the Study

Desired accuracy in detecting subsurface flaws using infrared imaging relies on seizing the best time window to capture the highest temperature gradients on the surface of a target object. Reports of experimental research mostly point to a late afternoon being the most favorable time period for inspection purposes; however, there are also publications with contradictory outcomes. Since these experiments are designed for different purposes, it would not be fair to compare and judge their results since there are too many variables to consider in each of these experiments such as size of defects, locations of flaws in experimental specimens, ambient environmental conditions and the thermal sensitivity of the infrared equipment used in the experiments. Therefore, this study focuses on a numerical analysis rather than an experimental work to numerically search for an appropriate time window to perform infrared inspection for a concrete deck. It is deliberately kept relatively simple not to obscure the main objective of the study. In addition, interested reader can repeat the analysis for his/her own objectives with different weather conditions and material properties.

3 Basics of Infrared Thermography

There are three modes of heat transfer: conduction, convection and radiation. Heat transfer takes place by one or a combination of these three modes. Thermal properties of an object govern how heat transfers from and into that specific object. These thermal properties are known as specific heat, mass density, heat capacity, thermal conductivity and thermal diffusivity [4]. Infrared thermography is related to the radiation mode of heat transfer since it builds on measuring the radiative heat flow. It is well known that any object above absolute zero temperature radiates energy in the infrared zone of the electromagnetic spectrum. These radiations from a target surface can be captured by infrared scanning devices and then converted to temperature values according to the Stephan-Boltzmann law [11].

$$W = \varepsilon\sigma T^4 \tag{1}$$

where W = radiant flux emitted per unit area (W/m^2), σ = Stephan-Boltzmann constant (5.67 $\times$ 10^{-8} W/m^2/K^4), ε = emissivity value of the target (from available tables and charts) and T = absolute temperature of the target (K).

This law states that W, radiant energy, of a target surface is equal to two factors multiplied by the fourth power of the absolute temperature, T, of the target. An infrared device measures W and then calculates T. The emissivity value (ε) of a target is a surface characteristic; it is a constant for a material in question over a given temperature range and under specific measurement conditions. Emissivity value (ε) is the ratio of actual emission from an object to that from a hypothetical source called a "black body" at the same temperature [12]; it is an influential parameter since, as the Stephan–Boltzmann law indicates, the greater the value of emissivity, the higher the radiant flux emitted.

Radiant energy is energy of electromagnetic waves; each electromagnetic wave is defined by its wavelength. An appropriate infrared capturing device must be utilized for a specific measurement task so that the maximum wavelength of the emitted energy is in the measurement range of the device. The Physics law that helps to determine the wavelength of maximum radiation is known as Wien's displacement law.

$$\lambda_m = b/T \tag{2}$$

where λ_m is the wavelength of maximum radiation (μm), b is the Wien's displacement constant (2897 μm/K) and T is the absolute temperature of the target (K).

This equation contains a constant, Wien's displacement constant (2897 μm/K), leaving the other two terms being inversely related to each other. When one of them is specified, the other is easily calculated. This equation is used to relate the surface temperature of a target to the maximum wavelength of the emitted energy from the same surface. As an example, for the 300 K ambient room temperature value, the peak wavelength of infrared radiation would be 2897/300 $\approx$ 10 μm. If the temperature range of a target surface can be reasonably estimated, then Wien's displacement law is utilized to select the right infrared equipment for proper measurements. Appropriate infrared sensors are able to detect those invisible wavelengths and allow us to determine surface temperature of an object [4].

Infrared thermography technique relies on the fact that subsurface flaws affect the rate of heat flow. These subsurface flaws cause temperature differentials on the surface. Measurements of these surface temperature differentials, along with proper analysis and interpretations, help to locate subsurface anomalies. However, there are important environmental factors to consider when the objective is to detect subsurface defects in concrete; these factors are solar loading, wind speed, ambient temperature changes [12] and relative humidity [13]. If the objective is to register an accurate temperature reading of a specific point on a concrete object, then the temperature calculation process of an infrared device must take all these factors into account precisely. On the other hand, if the goal is to depict the temperature differentials on the surface of a concrete object, then it would be sufficient to pick an appropriate time window such that environmental factors are at their optimal levels for the object in question to emit infrared radiation. There are studies focusing on determining favorable time windows experimentally for concrete bridge decks [14, 6, 12]. In these studies, several variables such as ambient temperature, pressure, wind speed and wind directions are measured and analyzed to investigate their effects. They reported that the time of the day is an influential parameter to include in the analyses since solar loading varies throughout the day. The present study explores the versatility of numerical analysis to determine the best time window so that measurable temperature differentials occur on the surface of a concrete bridge deck. These temperature differentials are essential for infrared inspections to reveal satisfactory results leading to capturing subsurface flaws in the concrete deck. Ambient environmental temperature is represented by a sine function to characterize the solar loading; other factors such as humidity and wind speed are assumed to be included in the sine function representation. Any estimated or actual weather data with preferred frequency might easily replace this sine function.

4   Numerical Model of a Concrete Deck with Delamination

The numerical model is constructed for a two dimensional cross-sectional area of a concrete deck. Since a typical deck thickness for a concrete bridge is 200 mm, the height of the model is set to 200 mm. The width of the model is set to be twice as big 400 mm, to keep the computational cost at a minimum as shown in Fig. 1. Side edges

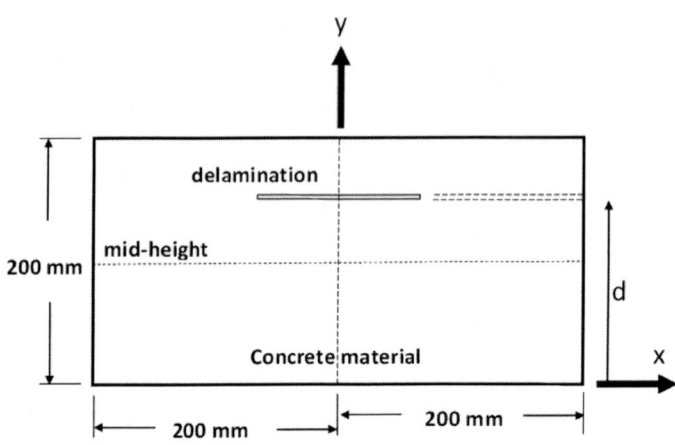

Fig. 1. Model geometry

are also assumed to be heat insulated to provide boundary conditions for the numerical model. The delamination has a thickness (t) of 2 mm, 1% of the deck height; the width is set to be 100 mm so that a big enough delamination can be positioned in the deck and still have a symmetry along the vertical axis. The delamination is positioned with respect to the deck soffit; these positions are denoted by "d" and take arbitrary values of $d = 187$ mm, $d = 177$ mm, $d = 167$ mm and $d = 157$ mm. The numerical results are obtained at the nodes located on the deck surface while the delamination is positioned at these four predetermined locations.

The temperature varies with respect to the time of the day; the ambient temperature (T_a) variation for a period of a day is represented by a sine function as given in Fig. 2. The top surface of the deck is assumed to be warmer than the deck soffit since the top surface is exposed to direct sun light during the day time. The temperature values at different times of the day can be represented by Eq. 1 for the top surface and by Eq. 2 for the deck soffit as given below.

@the top surface : $T_a = 25 - 10\mathrm{Cos}(\pi(t - 240)/720)$ for $0 \leq t \leq 1440$ min   (3)

@the bottom surface : $T_a = 25 - 10\mathrm{Cos}(\pi(t - 480)/720)$ for $0 \leq t \leq 1440$ min   (4)

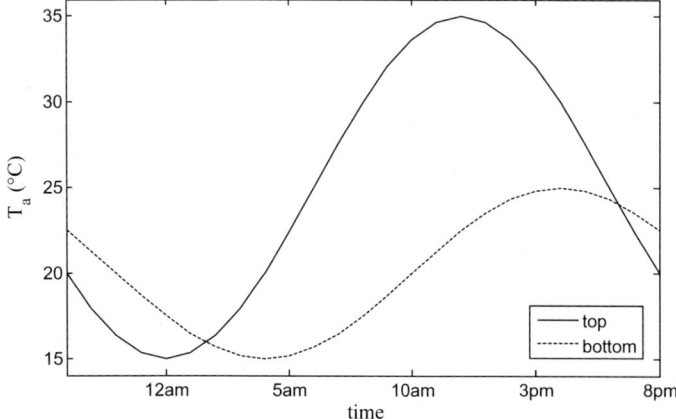

Fig. 2. Ambient temperature values throughout the day at the top and bottom surfaces

Heat energy between air and the deck surfaces is transferred in the form of convection. The convective heat transfer coefficient of the air is taken as (h_c)air = 20 W/m²K which refers to a mild wind with a constant velocity of approximately 1.1 m/s blowing close to the top and bottom surfaces of the deck.

5   Material Properties and Method

This study focuses on a concrete bridge deck that has delaminations. The test cases considered are: (a) concrete deck without any delamination to represent sound concrete, (b) the deck with air-filled delamination, and (c) the deck with water-filled delamination. Water-filled delamination case is included to investigate the effects of the content of a delamination.

Material properties of concrete, water and the air is given in Table 1. Initial temperature distribution is set to 18 °C uniformly throughout the model.

Table 1. Properties of the materials in the deck model

Materials	Heat conductivity (k) [kcal/hr.m° C]	Specific heat (c_p) [J/kg° C]	Density (ρ) [kg/m³]
Concrete	2	880	2500
Water	0.58	4200	1000
Air	0.024	1000	1.2

Although heat energy is transferred in the form of convection between the air and the deck surfaces, heat transfer mode changes to conduction through the concrete deck and the delamination. In the cases of water- and air-filled delaminations, the heat flows only by conduction since the delamination thickness (t) is sufficiently small and the temperature difference is small enough between the top and bottom surfaces of delamination such that convection does not initiate. The governing equation of the problem, to solve the transient temperature variation inside the deck with the delamination, is defined with the well-known heat equation but ignoring the radiation and convection terms such that:

$$\frac{\partial T}{\partial t} = \propto \nabla^2 T \tag{5}$$

where $\nabla^2 = \left(\frac{\partial^2}{\partial x^2} + \frac{\partial^2}{\partial y^2}\right)$ is the Laplacian operator and $\propto = \frac{k}{\rho C_p}$ is the thermal diffu-sivity. The transient solution of the problem is performed on ANSYS Mechanical Workbench. Geometry of the model is defined first in an APDL (ANSYS Parametric Design Language) input file; then, a finite element mesh is generated with rectangular elements subjected to a mesh size from 0.5 to 20 mm as shown in Fig. 3. The model contains approximately 7500 quadrilateral finite elements. As seen in Fig. 3, the finite

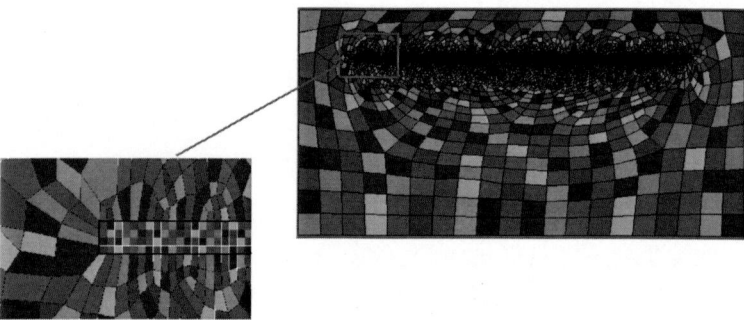

Fig. 3. Finite element mesh for the case when delamination is at $d = 157$ mm

element mesh becomes much denser around the delamination region. Convection boundary condition is imposed to the top and bottom surfaces of the deck; side boundaries are assumed to be heat insulated. Temperature data is collected from the nodes located at the top surface.

6   Numerical Results and Discussion

The results of numerical analyses are presented in the following figures. The first thing to notice in these figures is that the initial temperature which is set to 18 °C at the beginning of each analysis is the same for all cases. Temperature values on the top surface change gradually with respect to the time of the day; delamination locations significantly affect these changes. The analyses conducted cover a period of 24 h since the temperature variation trend becomes obvious in this time period for each case separately.

Variations of the temperature values on the surface of the concrete deck are more pronounced as expected for the case when the delamination is positioned close to the top surface, $d = 187$ mm from the deck soffit. Existence of a delamination close to the top surface causes an extensive change of the temperature variation monitored on the top surface especially for the case of air-filled delamination as shown in Fig. 4. There are clearly observable time periods throughout the 24-h period where the temperature

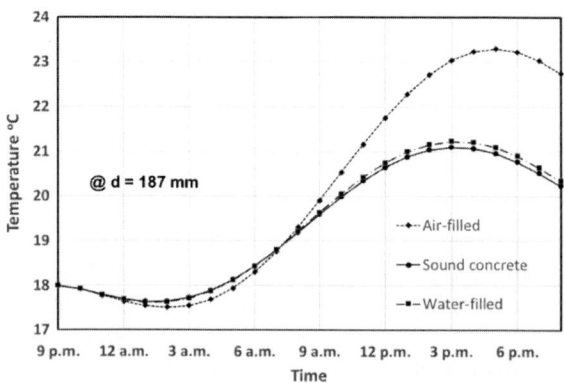

Fig. 4. Temperature variation on the surface when the delamination is at 187 mm

differentials are obvious. Around 7 p.m. the temperature gradient is about 2.5 °C. Commercially available infrared devices are able to detect very low temperature gradients since they have a thermal sensitivity value of 0.04 °C [15]; however, researchers and inspectors prefer somewhere around ten times the thermal sensitivity of the available infrared camera as the threshold level for their measurements [16]. As seen in Fig. 4, almost any day time after 10 a.m. is good to conduct infrared inspections to detect air-filled delaminations because temperature gradients appear to be more than 0.4 °C in this time window. However, when the delamination is filled with water, temperature gradient values are lower than 0.4 °C at any time of the day, causing the inspections not to reveal the subsurface flaw. The reason is that the conductivity of water is higher than that of air; in turn, heat energy is transferred a lot easier. In other words, as a material; the water cannot block the heat transition from passing through the delamination. However, air-filled delamination provides some insulation which results in higher temperature gradients on the surface. Although the values are lower than 0.4 °C for the case of water-filled delamination, the temperature gradients might still be detectable depending on the capability of the available infrared camera; but this would not be acceptable to warrant the adequacy of the inspection.

When the delamination is positioned away from the top surface, $d = 177$ mm, the available time window to conduct infrared inspections reduces to a portion of afternoon, starting from 3 p.m. to capture air-filled delaminations as seen in Fig. 5. As for water-filled delaminations, there is no time window to capture this type of

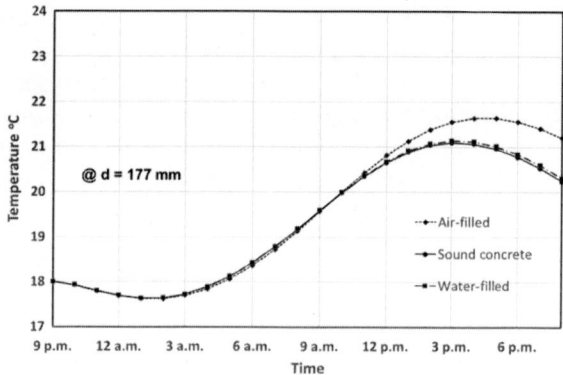

Fig. 5. Temperature variation on the surface when the delamination is at 177 mm

delaminations; in other words, when the delamination is positioned away from the top
surface and filled with water, infrared inspections would not locate the flaw at all.

The obvious temperature variations on the top surface of the concrete deck disappear with deeper delamination locations at $d = 167$ mm and $d = 157$ mm for both
air-filled and water-filled delaminations as shown in Figs. 6 and 7. Commercially
available infrared cameras will not be able to detect these existing flaws whether it is
filled with water or not.

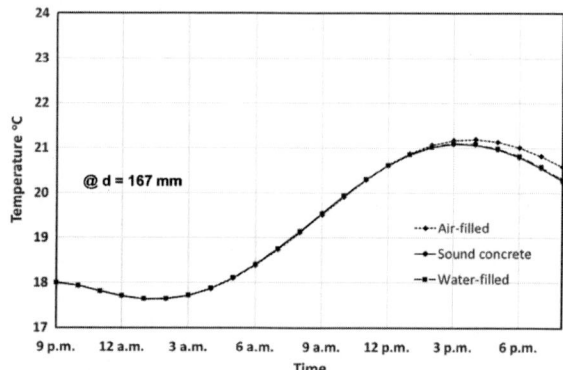

Fig. 6. Temperature variation on the surface when the delamination is at 167 mm

Since the objective is to determine a favorable time window to conduct infrared
inspections on concrete decks, the results of the numerical analyses will be more
informative if presented in a way that focuses on depths of delaminations with respect
to a specific hour of a day. In other words, the information presented in Figs. 4, 5, 6
and 7 should be reexamined on the basis of hours of a day. Consider an instance where
the temperature gradients are more pronounced on the surface of concrete deck around
7 p.m. as shown in Fig. 8. It is clear from the figure that water-filled delamination

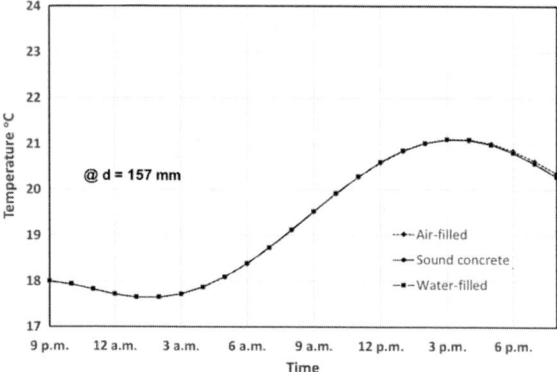

Fig. 7. Temperature variation on the surface when the delamination is at 157 mm

cannot be detected at any depths at this hour of the day since the temperature difference is less than 0.4 °C whereas air-filled delaminations can be detected if the delamination locations are closer to the sop surface; i.e., $d = 177$ mm and $d = 187$ mm. Delaminations located deeper than 170 mm will have a temperature difference of 0.4 °C and be missed by commercially available infrared cameras.

Data presentation similar to Fig. 8 is given below for each hour separately in Fig. 9. The first graph in the graph array shows that the delamination closest to the top surface,

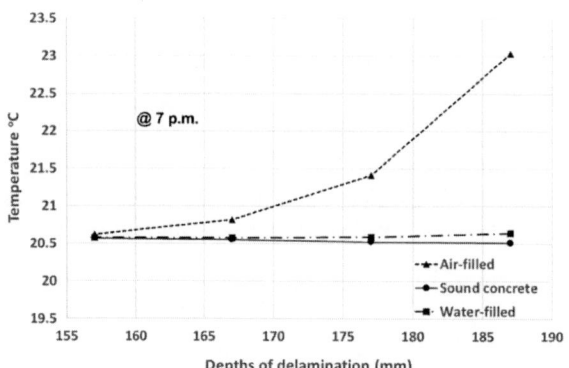

Fig. 8. Temperature gradients on the top surface at 7 p.m.

$d = 187$ mm, becomes detectable at 10 a.m. if it is an air-filled flaw. If the delamination is located at a deeper position, $d = 177$ mm, then it can be detected at 2 p.m. A careful scrutiny of this figure would reveal that the best time window to conduct infrared inspection for the given weather data is between 3 and 7 p.m. This result well coincides with the experimental results found in literature [14, 6, 12, 16].

It should be obvious that graphical results of a preliminary numerical analysis for different weather conditions will guide infrared inspectors to determine when to carry out their inspection works so that they would get the most out of their infrared equipment (Figs. 10, 11, 12, 13, 14, 15, 16, 17 and 18).

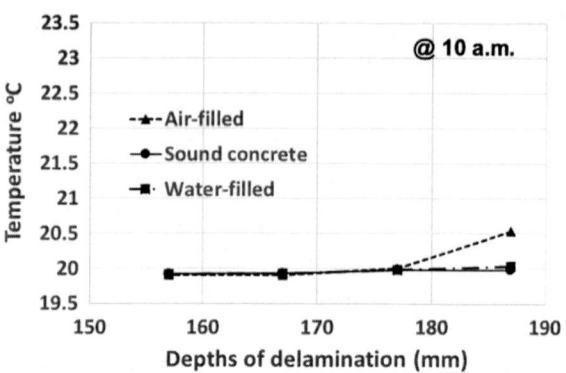

Fig. 9. Temperature gradients @10 a.m.

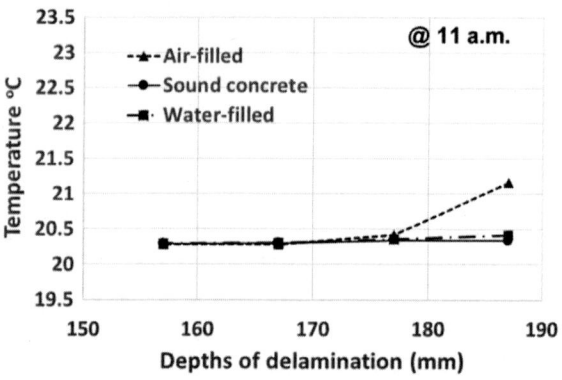

Fig. 10. Temperature gradients @11 a.m.

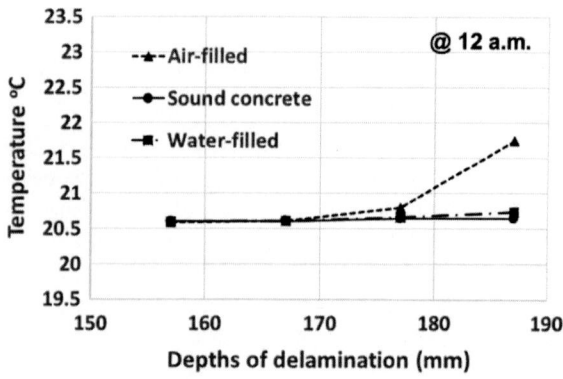

Fig. 11. Temperature gradients @12 a.m.

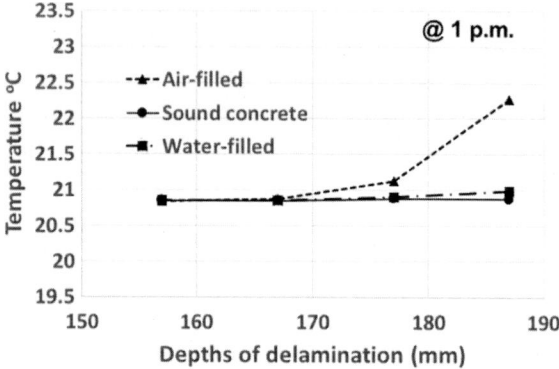

Fig. 12. Temperature gradients @1 p.m.

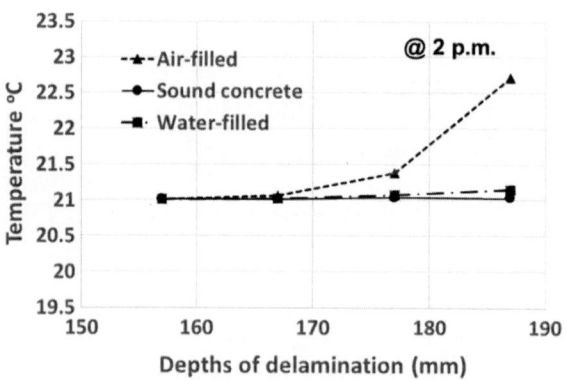

Fig. 13. Temperature gradients @2 p.m.

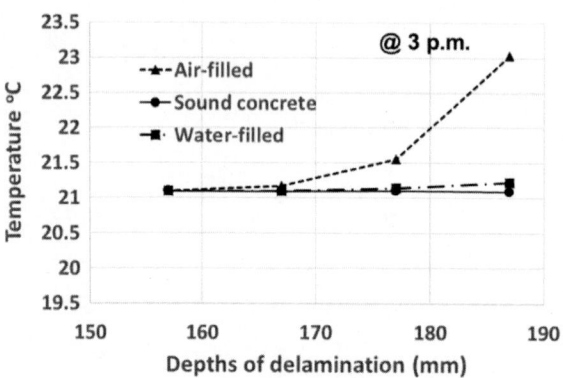

Fig. 14. Temperature gradients @3 p.m.

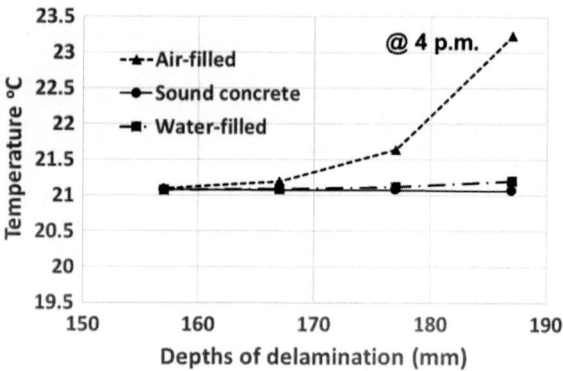

Fig. 15. Temperature gradients @4 p.m.

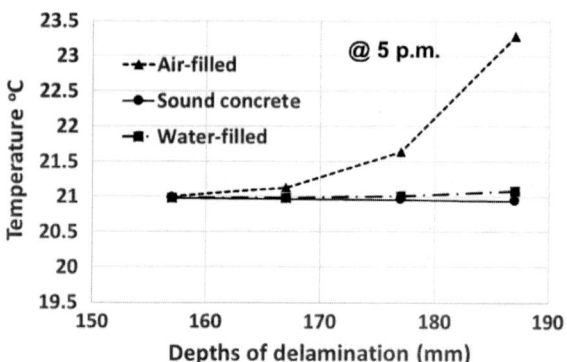

Fig. 16. Temperature gradients @5 p.m.

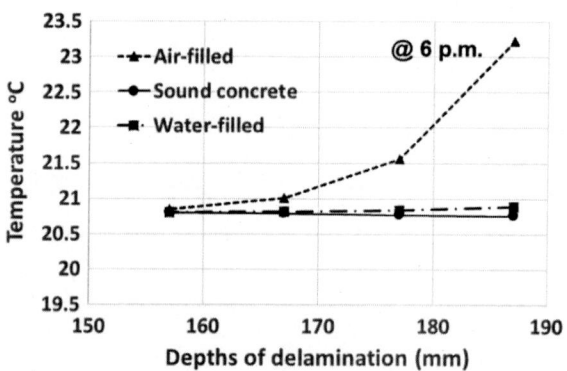

Fig. 17. Temperature gradients @6 p.m.

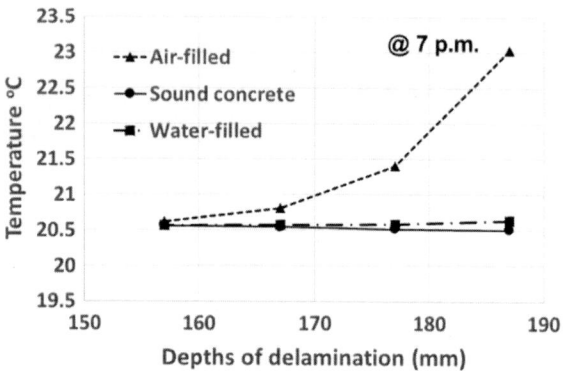

Fig. 18. Temperature gradients @7 p.m.

7   Conclusions

Infrared inspection of concrete bridge decks is a simple and practical nondestructive evaluation technique since small temperature differences on the surface due to sub-surface defects in the deck can be captured by commercially available infrared cameras. The objective of this study is to numerically determine a favorable time window to conduct infrared inspection to get the most out of available infrared equipment. To this end, heat transfer through the deck for a duration of 24 h is investigated when a delamination with a constant thickness is positioned at various depths in the deck. Two cases regarding the delamination are considered: (a) the delamination is air-filled, and (b) the delamination is water-filled. Air-filled delamination clearly presents its existence by creating a temperature gradient on the surface because it creates a discontinuity for the conduction process. On the other hand, water-filled delamination induces temperature differences lower than 0.4 °C; in turn, it becomes difficult for the infrared devices to detect the delamination. The reason for the low values of temperature gradient stems from the fact that the conductivity of water is higher than that of air; thus, heat energy is easily transferred making it difficult to detect the existence of the delamination. It is also observed that both the delamination depths and the time of a day are influential parameters to consider before making a decision regarding when to conduct an infrared inspection work. The results of this numerical analysis showed that the time window between 3 and 7 p.m. is the most favorable duration to perform infrared inspection under the weather conditions specified at the beginning of the study.

References

1. Gucunski N, Imani A, Romero F, Nazarian S, Yuan D, Wiggenhauser H, Shokouhi P, Taffe A, Kutrubes D (2013) Nondestructive testing to identify concrete bridge deck deterioration. SHRP 2 Report, S2-R06A-RR-1, Transportation Research Board, Washington, DC

2. Vaghefi K, Ahlborn TM, Harris DK, Brooks CN (2015) Combined imaging technologies fo concrete bridge deck condition assessment. J Perform Constr Facil 29(4):04014102
3. Oh T, Kee S, Arndt RW, Popovics JS, Asce M, Zhu J (2013) Comparison of NDT method for assessment of a concrete bridge deck. J Eng Mech 139:305–314
4. Kaplan H (2007) Practical applications of infrared thermal sensing and imaging equipment 3rd edn. Society of Photo-Optical Instrumentation Engineers, Bellingham, Washington
5. Meola C, Carlomagno GM, Giorleo L (2004) Geometrical limitations to detection of defect: in composites by means of infrared thermography. J Nondestruct Eval 23(4):125–132
6. Washer G, Fenwick R, Bolleni N (2010) Effects of solar loading on infrared imaging o subsurface. J Bridge Eng 15(4):384–390
7. ASTM D4788-03 (2007) Standard test method for detecting delaminations in bridge deck. using infrared thermography. American Society of Testing Materials
8. Yehia S, Abudayyeh O, Nabulsi S, Abdelqader I (2007) Detection of common defects i concrete bridge decks using nondestructive evaluation techniques. J Bridge Eng 12:215–22.
9. Ahlborn TM, Shuchman RA, Sutter LL, Harris DK, Brooks CN, Burns JW (2013) Report bridge condition assessment using remote sensors. United States Department of Trans portation Research and Innovative Technology Administration. Project No DT0S59-10-H-00001
10. Starnes MA, Carino NJ, Kausel EA (2003) Preliminary thermography studies for quality control of concrete structures strengthened with fiber-reinforced polymer composites J Mater Civ Eng 15(3):266–273
11. Minkina W, Dudzik S (2009) Infrared thermography: errors and uncertainties. Wiley Chichester
12. Washer G, Fenwick R, Bolleni N, Harper J (2009) Effects of environmental variables or infrared imaging of subsurface features of concrete bridges. Transp Res Rec J Transp Re: Board 2108:107–114
13. Zhang J, Gupta A, Baker J (2007) Effect of relative humidity on the prediction of natura convection heat transfer coefficients. Heat Transf Eng 28(4):335–342
14. Matsumoto M, Mitani K, Çatbaş FN (2013) Bridge assessment methods using image processing and infrared thermography technology: on-site pilot application in Florida. In Proceedings of transportation research board; 92nd annual meeting, Washington, DC
15. FLIR website. http://www.flir.com. Accessed 1 Dec 2015
16. Washer G, Fenwick R, Bolleni N (2009) Development of hand-held thermographic inspection technologies. Organizational Results Research Report, September 2009 OR10-007, Submitted to MoDOT

Determination of Self-Healing Performance of Cementitious Composites Under Elevated CO$_2$ Concentration by Resonant Frequency and Crack Opening Measurements

Süleyman Bahadır Keskin[1(✉)], Kasap Keskin Özlem[1],
Gürkan Yıldırım[2], Mustafa Şahmaran[3], and Özgür Anıl[4]

[1] Department of Civil Engineering, Engineering Faculty, Muğla Sıtkı Koçman
University, Muğla, Turkey
{sbkeskin, okasapkeskin}@mu.edu.tr

[2] Department of Civil Engineering, Engineering Faculty, Adana Science and
Technology University, Adana, Turkey
gurkanyildirimgy@gmail.com

[3] Department of Civil Engineering, Engineering Faculty, Hacettepe University,
Ankara, Turkey
sahmaran@hacettepe.edu.tr

[4] Department of Civil Engineering, Engineering Faculty, Gazi University,
Ankara, Turkey
oanil@gazi.edu.tr

Abstract. Global warming is a phenomenon that incontrovertibly affects daily lives of human beings in almost all aspects. Definitely, construction industry, especially concrete as most commonly used construction material, is not exempt from the effects of global warming. Nevertheless, there is a lack of information on how the change in atmospheric conditions influences self-healing behavior of cementitious materials. This research examines the impact of increased CO$_2$ concentrations in the atmosphere on the self-healing capability of cementitious materials in terms of resonant frequency and crack opening measurements. For this purpose, to clearly disclose the effect of tremendous increase in the environmental CO$_2$ concentration as a result of global warming, Engineered Cementitious Composites (ECC) which possess advanced intrinsic self-healing capability were employed. For this purpose, sound and pre-cracked ECC specimens containing fly ash and ground granulated blast furnace slag were tested by resonant frequency after 28 days of initial curing up to 28 + 90 days with 15 days intervals and crack openings were observed for each testing age. Moreover, in order to accelerate the capture of CO$_2$ from the environment, a third ECC mixture was prepared by adding Ca(OH)$_2$ to the ECC mixture incorporating fly ash. The results showed that CO$_2$ present in the environment can improve the self-healing behavior of ECC mixtures, which is a promising finding in terms of environmental concerns. Possibility of capturing and decreasing the CO$_2$ from the atmosphere by self-healing mechanism will make the ECC a more environmentally friendly construction material additional to its superior technical properties.

Keywords: High CO$_2$ environment · Cementitious composites · Self-healing

© Springer International Publishing AG, part of Springer Nature 2018
S. Fırat et al. (eds.), *Proceedings of 3rd International Sustainable
Buildings Symposium (ISBS 2017)*, Lecture Notes in Civil Engineering 7,
https://doi.org/10.1007/978-3-319-64349-6_47

1 Introduction

Cracking in structural members is an important phenomenon on account of service life given the fact that durability of cementitious materials is directly or indirectly related with the issues regarding crack formation. Recently, many researchers have been focused on developing cementitious materials for structural use that possess high durability and extended service life. Although it is a known fact that achieving crack free cementitious material is almost impossible, it has been well documented by the researchers that materials such as Engineered Cementitious Composites (ECC) can be considered as crack free due to their tight crack widths even under extreme loading conditions. In many occasions it has been shown that ECC with self-controlled multiple and micron sized tight crack formations in cracked state have better or at least comparable transport properties with respect to conventional sound concrete [1] This property of ECC not only increases the service life but also can decrease the repair and maintenance costs of the structures for a sustainable infrastructure.

Along with the extended service life and lower repair and maintenance costs multiple micron sized crack formation in ECC brings about superior mechanical performance especially under tensile and flexural loading. ECC possesses compressive strength values ranging between 30 and 100 MPa and under direct tensile load, it can reach up to 8% strain with a strain hardening behavior [2]; yet, when tensile properties are evaluated under flexural loads, it shows flexural hardening behavior [3]. Nevertheless, in many studies an intrinsic property of ECC which further decrease the need for repair and maintenance has been emphasized namely self-healing. Self-healing is a phenomenon that has attracted many researchers especially in the near past those pioneered the development of different special techniques such as using hallow fibers, microencapsulation, expansive agent and bacteria [4]. However, in order to artificially initiate the self-healing process these methods necessitates the use of an external material that is not typically used in concrete production. On the other hand besides those aforementioned, an important self-healing mechanism which does not require any additional material to process namely intrinsic or autogenous self-healing is potentially inherent in all cementitious materials as the source of the self-healing is the ongoing reaction of the previously unhydrated cementitious materials and the reactions of some hydration products with water and atmospheric CO_2. It has also been shown by researchers that the crack widths significantly affect intrinsic self-healing as small cracks have higher tendency to close by self-healing [5]. Due to the high amounts of cementitious materials incorporated in ECC and the multiple and tight crack widths that remain generally under 100 μm [6]. ECC is a perfect example of intrinsically self-healing cementitious materials. Self-healing properties of ECC have been revealed many times by numerous researchers and it is still a hot topic among materials scientist.

For the last several decades, there is a continuously growing attention placed on reducing the emission of greenhouse gases as their detrimental effect on climate change is obvious. In this context, CO_2 emission which constitutes three quarters of the total greenhouse gas emission seems to be primarily responsible for the climate change [7]. Concentration of CO_2 in the atmosphere increased to 390 ppm from 280 parts per million (ppm) between preindustrial period and 2010 which corresponds to an increase of 39% [8]. Furthermore it is predicted that the concentration of CO_2 may rise up to 1000 ppm by the year 2100 [9].

For any type of ECC two main mechanisms of intrinsic self-healing are encountered in the literature. One of these mechanism is the healing of cracks by ongoing hydration in which the preformed cracks are filled with the hydration products of the previously unhydrated material. This type of self-healing is the dominant mechanism when pozzolanic materials such as fly ash are used in ECC production with Portland cement. The other mechanism is the carbonation in which the preformed cracks are filled with calcium carbonates as calcium hydroxide formed during the hydration reactions of Portland cement react with the CO_2 in the atmosphere or as the Ca^{2+} ions leached from hydration products react with carbonates or bicarbonates those formed a result of reaction with CO_2 and water. Carbonation mechanism can be observed in ECC mixtures produced with either fly ash or ground granulated blast furnace slag (S), however it is the dominant mechanism for the ECC produced with S [10–14].

In this respect, although carbonation is a potential cause of deterioration in reinforced structures, it has been shown that the same carbonation mechanism is beneficial in terms of self-healing of cracks in ECC. In this paper, the effect of CO_2 rich environment on the self-healing behavior and capability of different ECC mixtures containing fly ash and S is investigated in terms of changes in resonant frequencies and crack widths. In addition to the conventional ECC mixtures produced with FA and S, since it was previously shown by Yildirim et al. [15] that additional hydrated lime can potentially enhance the self-healing behavior of fly ash incorporated ECC through carbonation in order to increase the environmental CO_2 uptake and enhance self-healing simultaneously, as a third ECC mixture, additional hydrated lime was also incorporated in ECC mixture containing fly ash.

2   Experimental Program

In order to investigate the effect of CO_2 rich environment on the self-healing capability, three different ECC mixtures were prepared with different pozzolanic materials (PM). ECC 1 and ECC 2 were produced with Class-F fly ash (FA) and ground granulated blast furnace slag (S), respectively with a PM to Portland cement (PC) ratio of 1.2 by weight. Additional to these two mixtures, ECC 3 was produced by adding hydrated lime (CH) to ECC 1 mixture by 5% of total weight of cementitious materials (CM). For all ECC mixtures, besides the PMs, CEM I 42.5 R type ordinary PC, silica sand with a maximum aggregate size of 0.4 mm, water, polycarboxylic-ether based superplasticizier (SP) and polyvinyl alcohol (PVA) type fibers with 39 μm diameter, 8 mm length and 1610 MPa nominal tensile strength were used. Amount of SP was adjusted to achieve adequate fiber distribution in each ECC mixture type. The mixture proportions and the chemical and physical properties of PC, FA and S are given in Tables 1 and 2, respectively.

Table 1. Mixture proportions for different ECCs

Mix ID	PC	W/CM	Sand/CM	FA/PC	S/PC	CH (kg/m^3)	PVA (kg/m^3)
ECC 1	1	0.27	0.36	1.2	–		26
ECC 2	1	0.27	0.36	–	1.2		26
ECC 3	1	0.27	0.36	1.2	–	62	26

Table 2. Chemical and physical properties of PC, FA and S

Chemical composition	PC	FA	S
CaO	61.43	3.48	35.09
SiO_2	20.77	60.78	37.55
Al_2O_3	5.55	21.68	10.55
Fe_2O_3	3.35	5.48	0.28
MgO	2.49	1.71	7.92
SO_3	2.49	0.34	2.95
K_2O	0.77	1.95	1.07
Na_2O	0.19	0.74	0.24
Loss on ignition	2.20	1.57	2.79
Physical properties			
Specific gravity	3.06	2.10	2.79
Blaine fineness (m^2/kg)	325	269	425

Cylindrical specimens with dimension Ø100 × 200 mm were prepared from each ECC mixture to be tested for self-healing evaluation with and without CO_2 environment. After keeping 24 h at 50 ± 5% RH and 23 ± 2 °C, all the specimens were demolded and placed in isolated plastic bags at 95 ± 5% RH and 23 ± 2 °C and cured till the end of 28 days. At the end of 28 days, Ø100 × 200 mm cylindrical specimens were cut into Ø100 × 50 mm pucks by a diamond blade saw. Half of the specimens were kept sound while the other half were pre-loaded up to 70% of the ultimate deformation under splitting tensile test to introduce the microcracks. Ultimate deformation capacities of mixtures were determined as the average of results obtained by testing four Ø100 × 50 mm puck specimens from each ECC mixture under splitting tensile test at 28 days. Ultimate deformation capacities were 1.7, 1.6 and 1.5 mm for ECC 1, ECC 2 and ECC 3, respectively.

After the application of pre-loading and initial reference measurements at 28 days, half of the total pre-cracked and sound specimens from each ECC mixture type were placed in a CO_2 incubator that can keep the environmental conditions constant at 90 ± 5% RH, 50 ± 5 °C and 3% CO_2 level. The other half of pre-cracked and sound specimens from each mixture were also placed in another environmental chamber at 90 ± 5% RH and 50 ± 5 °C; the only difference is the lack of high CO_2 concentration. Tests for self-healing evaluation were applied on the specimens at 28 + 15, 28 + 30, 28 + 45, 28 + 60, 28 + 75 and 28 + 90 days. As a result, the changes in the properties of different ECC mixtures with time were obtained for both sound and pre-cracked specimens with and without CO_2 application.

Self-healing performance of ECC samples were evaluated by Resonant Frequency (RF) and crack width measurements. Accordingly, for RF test 4 sound and pre-cracked set of specimens were used for each mixture and average of measurements obtained was recorded. For crack characterization at least 8 specimens were used and the crack widths were measured by using a crack microscope.

RF test method was applied in accordance with ASTM C 215 [16]. Impact was generated with steel balls having three different diameters (8, 12, 14 mm), and RF value (in terms of Hertz) was given by the device after analyzing the frequency spectrums of impacts generated. The measurements were done perpendicular to cracks in pre-cracked specimens and along a premarked direction in sound specimens.

3 Results and Discussion

Resonant frequency in terms of self-healing evaluation, was only used in limited number of studies. Each ECC mixture used in this study is separately dealt in terms of self-healing performance. Before the discussion of the test results it should be noted that this study focuses mainly on the effect of incidental carbonation caused by the increase in environmental CO_2 concentration. For this reason, relative humidity, temperature and CO_2 concentration to which the specimens were exposed during the experimental study were set forth in order not to effect the carbonation in an unnatural manner. Thus, no effort was given to accelerate the formation or increase the amount of carbonation products by trading off the nature of the reactions. Under normal environmental conditions it is known that during carbonation reactions, CO_2 reacts with both $Ca(OH)_2$ and CSH gel, however it is showed that CO_2 reacts with CSH only after all $Ca(OH)_2$ present in the pore solution is used in carbonation reaction [17] and the rate of carbonation is mainly dependent on two main factors namely humidity and CO_2 concentration. Inside the cementitious composite, CO_2 should be able to reach the reactants of carbonation reaction which is more attainable in a dry environment since the diffusion of CO_2 is faster in air than in water, [18] however as carbonation reactions need water, humidity is also required [19]. In this respect, it is shown that the optimum relative humidity for carbonation is between 50 and 70% which is the main reason in most of the publications relative humidity in between these values is generally used [20]. However in this research, as the main concern is self-healing and since intrinsic self-healing requires water, a relative humidity of 90% is implemented to evaluate self-healing and carbonation simultaneously. As for CO_2 concentration, when the amount of CO_2 is higher in the environment, the rate of carbonation is expected to be higher. Although, increasing CO_2 concentration seems to help augment carbonation effect on self-healing, in the literature it is stated that the nature of carbonation reaction differentiates from the normal CO_2 environment of 0.03% when a threshold CO_2 concentration is exceeded. It is demonstrated that when the CO_2 concentration increases to 3% (which is the value used in this study) the characteristics of CSH gel does not differ much when compared to 0.03% natural CO_2 concentration. Furthermore, as the concentration increases to 10 and 100%, CSH gel vanishes after carbonation which totally alters the nature of cementitious composite [21]. In addition to this it was also shown that there is consensus between results of carbonation tests conducted under natural and 4% CO_2 concentration asserting that this type of accelerated test can be used to evaluate carbonation effect at long term [22]. For this reason, in order not to compromise natural self-healing process and to stay inside the logical limits, CO_2 concentration is fixed to 3% in this study.

Results obtained from RF test for each ECC mixture at all testing days (28, 28 + 15, 28 + 30, 28 + 45, 28 + 60, 28 + 75 and 28 + 90) are given in Figs. 1, 3 and 5 in normal and CO_2 rich environment conditions.

3.1 ECC Mixtures Containing Fly Ash

As shown in Fig. 1, resonant frequency results are affected majorly by crack formation rather than CO_2 amount in the atmosphere. This behavior is normal as resonant frequency is mainly affected by dynamic modulus and the mechanical performance of the

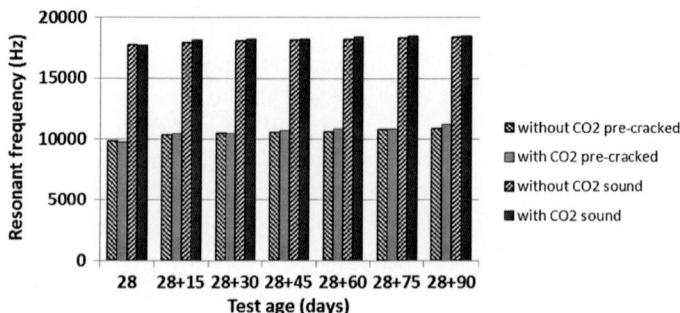

Fig. 1. Resonant frequency results of ECC 1 specimens

material rather than materials porosity and pore chemistry. However when the resonant frequency values for each testing age is considered to compare the effect of CO_2 rich environment, it is observed that after initial curing of 28 days, at almost all ages of testing, for both sound and pre-cracked specimens, specimens cured under rich CO_2 environment reached higher resonant frequency values. Higher resonant frequency values may be indicators of advanced hydration as hydration is the main parameter that affects the mechanical performance. Additional surface carbonation may also be effective on the resonant frequency in the case of CO_2 rich environment.

In order to better understand the effect of the CO_2 concentration, a diagram was plotted in Fig. 2. In this diagram vertical axis represents the initial crack opening formed on the specimens at the age of 28 days while the horizontal axis represents the

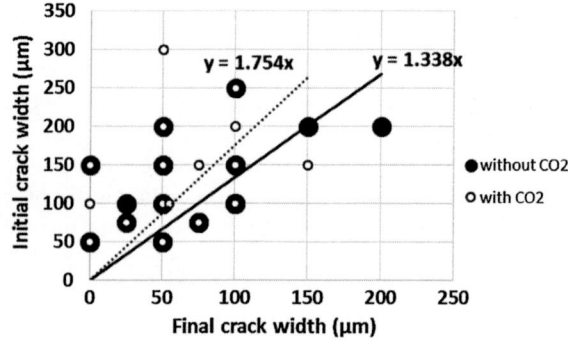

Fig. 2. Crack closure diagram for ECC 1 specimen

final crack width 90 days after the initial 28 days period. However in this diagram some data points overlap which may vitiate the impact of the overlapping points. To overcome this, data points were marked by circles with different diameters and also for each set of data a trendline was also plotted so that both the number of cracks and the amount of crack closure could be taken into account together. In this diagram data points flocculating closer to the vertical axis or a trendline with a higher slope indicates that the cracks are closed or healed better. As seen from Fig. 2, for pre-cracked specimens, crack closure may also be attributed to the increase in resonant frequency values which indicates a better self-healing performance. When percent change in resonant frequency with respect to 28 days resonant frequency values of each specimen is considered, it is clearly seen that for sound specimens cured under atmospheric CO_2 concentration, the increase is measured 3.3% while for those cured under 3% CO_2 concentration it is 4.5%. On the other hand, for pre-cracked specimens these ratios are 10.4 and 15.4% for atmospheric and rich CO_2 environment, respectively. This gap between the results is clear evidences of the beneficial effect of CO_2 on self-healing.

3.2   ECC Mixtures Containing Slag

Resonant frequency test results revealed that there is no significant difference among the sound specimens cured under normal and high CO_2 concentrations. In the case of sound specimens, surface carbonation probably may not contribute to mechanical performance which is different than in the case of ECC 1 specimens, owing to the fact that ECC 2 specimens already possess developed mechanical performance before surface carbonation. As mentioned previously, resonant frequency is influenced by crack formation extensively rather than curing duration and conditions. On the other hand, the extent of the variation in the test results upon pre cracking varies largely depending on the mineral admixture used. For instance, pre cracking yielded a drop in the resonant frequency of about 11,600 and 7950 Hz for ECC 2 and ECC 1 specimens, respectively. Although all specimens were pre-loaded to 70% of their ultimate deformation, different mixtures may have different crack characterizations. This issue is discussed earlier by Sahmaran et al. [23] and it was concluded that slag bearing ECC specimens possess larger crack widths with smaller number of cracks while the fly ash bearing ones possess smaller crack widths with large number of cracks which is also consistent with this study as ECC 1 specimens have an average crack width of 112 μm, while ECC 2 specimens possess an average crack width of 133 μm. The substantial decline in resonant frequency upon pre loading in ECC 2 specimens can be attributed to large crack widths resulted upon loading.

As seen from Fig. 3, although curing under CO_2 rich environment has almost no influence on sound specimens, this is not the case for pre cracked specimens. It is clear that pre cracked ECC 2 specimens achieved higher resonant frequency values upon curing under high CO_2 concentration. This improvement in self-healing capability is also evident in crack closure diagrams provided in Fig. 4 as crack closure performance under CO_2 rich environment is slightly better. This enhancement can be attributed to faster and further carbonation reactions attained by abundance of CO_2 that results in

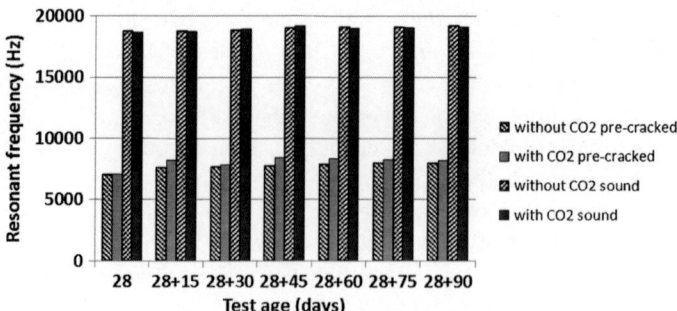

Fig. 3. Resonant frequency results of ECC 2 specimens

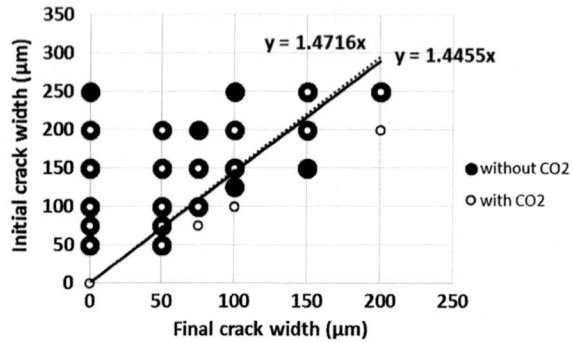

Fig. 4. Crack closure diagram for ECC 2 specimens

plugging of the cracks with $CaCO_3$. When compared to ECC 1 mixtures, crack closure performance upon CO_2 rich curing is seen to be less pronounced in ECC 2 mixtures.

3.3   ECC Mixtures Containing Fly Ash and Calcium Hydroxide

Specimens containing fly ash with additional $Ca(OH)_2$ are expected to attain higher maturity and enhanced mechanical performance through additional CSH gel formation as the pozzolanic capacity of the system is increased, whereas the naturally formed $Ca(OH)_2$ through hydration reactions may not be enough to react with all fly ash available in the system especially for system that contains more than 30–40% fly ash [24]. In addition, high amounts of $Ca(OH)_2$ in the cementitious system may result in breakdown of fly ash particles yielding them to be more reactive through exposing the inner silicates for reacting to form CSH [25]. Moreover inside the cementitious matrix, $Ca(OH)_2$ addition provides an improved particle size distribution and structure through microfiller effect as it fills the gaps between the cementitious materials.

According to the resonant frequency test results provided in Fig. 5, upon pre-cracking average resonant frequency of sound ECC 3 specimens decreased 9285 Hz as a result of crack formation while the amount of drop is calculated 7960 Hz

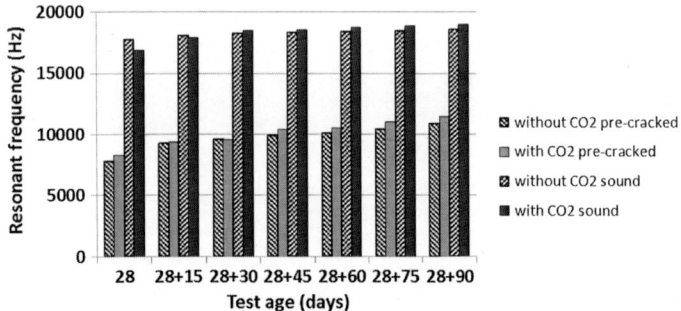

Fig. 5. Resonant frequency results of ECC 3 specimens

for ECC 1 specimens. This discrepancy due to additional $Ca(OH)_2$ is an outcome of its positive effect on hydration which yields higher matrix fracture toughness and thus larger crack widths. For sound specimens, after 90 days of curing in moist cabinet, average increases of 4.9 and 12.3% in the resonant frequency results are measured for specimens cured under atmospheric and 3% CO_2 concentrations, respectively. In order to evaluate the effect of additional $Ca(OH)_2$, if this values are compared with the ECC 1, changes in the resonant frequency values are found 3.3 and 4.5% for atmospheric and high CO_2 concentrations, respectively implying that regardless of curing condition additional $Ca(OH)_2$ yields better mechanical performance in which high CO_2 concentration is also beneficial. This kind of behavior is also observed for pre-cracked specimens, $Ca(OH)_2$ addition resulted in 39.4% escalation in the results for atmospheric and 38.7% for high CO_2 concentration. These close results suggest that there is no effect of CO_2 concentration on the results, although significant improvements are observed for pre-cracked ECC 1 specimens. However for ECC 1 specimens, average percent increases are measured around one third of those measured for ECC 3 specimens which reveals that $Ca(OH)_2$ addition has a tremendous influence on self-healing performance in terms of resonant frequency test. This improvement is attributed to the fact that additional $Ca(OH)_2$ fortifies self-healing through additional CSH formation rather than carbonation, as CO_2 concentration which may only benefit self-healing through $CaCO_3$ precipitation shows no or insignificant effect. In addition to this, crack

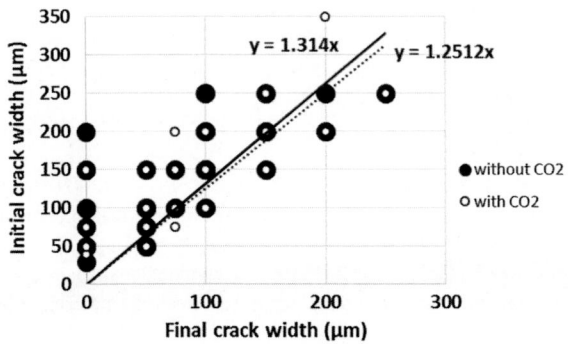

Fig. 6. Crack closure diagram for ECC 3 specimens

closures in Fig. 6 shows that crack widths closures are worse than ECC 1, in ECC 3 possibly because of the larger crack width occurred during pre-loading and self-healing through CSH gel formation rather than carbonation effect.

4 Conclusions

This study represents the results of an experimental investigation that aim to reveal the possible effect of day by day accelerating global warming and greenhouse gas emission in the broad framework by focusing specifically on the effect of high concentrations of CO_2, the major greenhouse gas, in order to project the long term impact on the self-healing performance of Engineered Cementitious Composites. The following conclusions are drawn from the study.

Sound specimens of ECC 1 mixture exhibit higher values of resonant frequency upon CO_2 rich curing. Self-healing performance also improved through CO_2 rich environment as a result of carbonation which in turn also supported by the crack closure rates.

Crack closure rates of pre-loaded ECC 1 specimens showed that cracks visually healed better under high CO_2 concentration.

Resonant frequency result of ECC 2 specimens were lower compared to ECC 1 specimens mainly due to larger crack widths encountered upon pre-loading. Clear benefit of CO_2 rich curing is also revealed by resonant frequency results and also supported with the crack closure rates.

Sound ECC 3 specimens yielded higher resonant frequency results while this results became more pronounced with high CO_2 concentration. For pre-cracked specimens although curing condition did not influence the results much, $Ca(OH)_2$ addition yielded extreme self-healing performance most probably due to additional CSH formation.

Acknowledgments. The authors gratefully acknowledge the financial assistance of the Scientific and Technical Research Council (TUBITAK) of Turkey provided under Project: MAG-112M876.

References

1. Şahmaran M, Li VC (2010) Engineered cementitious composites: can composites be accepted as crack-free concrete? Transp Res Rec 164:1–8
2. Li VC (2002) Advances in ECC research. SP 206-23 ACI Special Publication on Concr. Mater Sci Appl, pp 373–400
3. Sahmaran M, Lachemi M, Hossain KMA, Ranade R, Li VC (2009) Influence of aggregate type and size on ductility and mechanical properties of engineered cementitious composites. ACI Mater J 106:308–316
4. Wua M, Johannesson B, Geiker M (2012) A review: self-healing in cementitious materials and engineered cementitious composite as a self-healing material. Constr Build Mater 28:571–583

5. Van Tittelboom K, Gruyaert E, Rahier H, De Belie N (2012) Influence of mix composition on the extent of autogenous crack healing by continued hydration or calcium carbonate formation. Constr Build Mater 37:349–359
6. Sahmaran M, Li VC (2009) Durability properties of micro-cracked ECC containing high volumes fly ash. Cem Concr Res 39:1033–1043
7. Huaman RNE, Jun TX (2014) Energy related CO_2 emissions and the progress on CCS projects: a review. Renew Sustain Eng Rev 31:368–385
8. Lal R (2013) Soil carbon sequestration SOLAW. Backgr Thematic Rep. http://www.fao.org/fileadmin/templates/solaw/files/thematic_reports/TR_04b_web.pdf
9. Stewart MG, Wangb X, Nguyen MN (2011) Climate change impact and risks of concrete infrastructure deterioration. Eng Struct 33:1326–1337
10. Neville AM (2002) Autogenous healing: a concrete miracle. Concr Int 24:76–82
11. Lepech MD, Li VC (2009) Water permeability of engineered cementitious composites. Cem Concr Compos 31:744–753
12. Ozbay E, Şahmaran M, Lachemi M, Yücel HE (2013) Self-healing of microcracks in high volume fly ash incorporated engineered cementitious composites. ACI Mater J 110:33–44
13. Sahmaran M, Yildirim G, Erdem TK (2013) Self-healing capability of cementitious composites incorporating different supplementary cementitious material. Cem Concr Compos 35:89–101
14. Zhang Z, Qian S, Ma H (2014) Investigating mechanical properties and self-healing behavior of micro-cracked ECC with different volume of fly ash. Constr Build Mater 52:17–23
15. Yildirim G, Sahmaran M, Ahmed HU (2015) Influence of hydrated lime addition on the self-healing capability of high-volume fly ash incorporated cementitious composites. J Mater Civil Eng 27:1–11
16. ASTM C 215 (1997) Test method for fundamental transverse, longitudinal, and torsional frequencies of concrete specimens. ASTM Int., West Conshohocken
17. Peter M, Munteen A, Meier S, Bohm M (2008) Competition of several carbonation reactions in concrete: a parametric study. Cem Concr Res 38:1385–1393
18. Younsi A, Turcry P, Rozière E, Aït-Mokhtar A, Loukili A (2011) Performance-based design and carbonation of concrete with high fly ash content. Cem Concr Compos 33:993–1000
19. Papadakis VG, Vayenas CG, Fardis MN (1992) Hydration and carbonation of pozzolanic cements. ACI Mater J 89:119–130
20. Marie-Victoire E, Cailleux E, Texier A (2006) Carbonation and historical buildings made of concrete. J Phys Achiev IV 136:305–318
21. Castellote M, Fernandez L, Andrade C, Alonso C (2009) Chemical changes and phase analysis of OPC pastes carbonated at different CO_2 concentrations. Mater Struct 42:515–525
22. Dhir R, Limbachiya M, McCarthy M, Chaipanich A (2007) Evaluation of portland limestone cements for use in concrete construction. Mater Struct 40:459–473
23. Sahmaran M, Yildirim G, Noori R, Ozbay E, Lachemi M (2015) Repeatability and pervasiveness of self-healing in engineered cementitious composite. ACI Mater J 112:513–522
24. Song G, van Zijl GPAG (2004) Tailoring ECC for commercial applications. In: Proceedings of the 6th RILEM symposium on fiber reinforced concrete, BEFIB 2004, RILEM Pro039, RILEM Publications, Bagneux, pp 1391–1400
25. Barbhuiya SA, Gbagbo JK, Russell MI, Basheer PAM (2009) Properties of fly ash concrete modified with hydrated lime and silica fume. Constr Build Mater 23:3233–3239